D1703644

INSECT HYDROCARBONS

A unique and critical analysis of the wealth of research conducted on the biology, biochemistry and chemical ecology of the rapidly growing field of insect cuticular hydrocarbons. Authored by leading experts in their respective fields, the twenty chapters show the complexity that has been discovered of the nature and role of hydrocarbons in entomology. Covers, in great depth, aspects of chemistry (structures, qualitative and quantitative analysis), biochemistry (biosynthesis, molecular biology, genetics, evolution), physiology, taxonomy, and ecology. Clearly presents to the reader the array of data, ideas, insights and historical disagreements that have accumulated during the past half century. An emphasis is placed on the role of insect hydrocarbons in chemical communication is shown, especially among the social insects. Includes the first review on the chemical synthesis of insect hydrocarbons. The material presented is a major resource for current researchers and an unending source of ideas for new researchers.

GARY J. BLOMQUIST is chair of the Department of Biochemistry and Molecular Biology at the University of Nevada, Reno. He has published over 200 original research papers, reviews, chapters and books, including co-editing the books *Pheromone Biochemistry and Molecular Biology* (G. J. Blomquist and R. G. Vogt, 2003) and *Pheromone Biochemistry* (G. D. Prestwich and G. J. Blomquist, 1987). His work has been cited over 4500 times (ISI). Much of Blomquist's research career has involved the study of insect hydrocarbons, with an emphasis on their biosynthesis, endocrine regulation and chemical analysis. He published his first paper on insect hydrocarbons 40 years ago (1969) and has remained active in the field, collaborating with many of the early leaders including Larry Jackson, Dennis Nelson, Ralph Howard, Coby Schal and Anne-Geneviève Bagnères.

ANNE-GENEVIÈVE BAGNÈRES is Director of Research at the CNRS and team leader at the Institut de Recherche sur la Biologie de l'Insecte (IRBI) in Tours, France. She completed a PhD on the role of cuticular hydrocarbons in social insects at the University of Paris 6 in 1989, and received the Chancellerie of the Universities of Paris Prize for her Ph.D. work. She spent a year of postdoctoral studies in David Morgan's laboratory in 1990, and a sabbatical in 1996–97 in the laboratories of Gary Blomquist and Coby Schal. AGB is primarily interested in the chemical ecology of social insects, where she continues to be a leading contributor and proponent of the concept of chemical signature. While her primary research focuses on termites, she participates in several collaborative studies involving the chemistry of other insects. She is an active member of the International Society of Chemical Ecology (ISCE) and the bureau of the French section of the International Union for the Study of Social Insects. She has published nearly 100 original research papers, reviews and chapters.

INSECT HYDROCARBONS
BIOLOGY, BIOCHEMISTRY, AND
CHEMICAL ECOLOGY

GARY J. BLOMQUIST

University of Nevada

ANNE-GENEVIÈVE BAGNÈRES

Centre National de la Recherche Scientifique,
Université de Tours

CAMBRIDGE UNIVERSITY PRESS
Cambridge, New York, Melbourne, Madrid, Cape Town, Singapore,
São Paulo, Delhi, Dubai, Tokyo

Cambridge University Press
The Edinburgh Building, Cambridge CB2 8RU, UK

Published in the United States of America by Cambridge University Press, New York

www.cambridge.org
Information on this title: www.cambridge.org/9780521898140

First published 2010

Printed in the United Kingdom at the University Press, Cambridge

A catalogue record for this publication is available from the British Library

Library of Congress Cataloguing in Publication data
Insect hydrocarbons : biology, biochemistry, and chemical ecology / [edited by]
Gary J. Blomquist, Anne-Geneviève Bagnères.
p.   cm.
ISBN 978-0-521-89814-0 (hardback)
1. Insects–Physiology.   2. Insect biochemistry.   3. Cuticle.   4. Hydrocarbons.
I. Blomquist, Gary J.   II. Bagnères, Anne-Geneviève.   III. Title.
QL495.I4918 2010
595.7′01–dc22
2009038905

ISBN 978-0-521-89814-0 Hardback

Contents

Contributors

Bagnères, Anne-Geneviève, CNRS UMR 6035, I.R.B.I., University of Tours, Parc de Grandmont, 37200 Tours, France

Bartelt, Robert J., ARS, Crop Bioprotection Research, 1815 N University St Peoria, IL 61604–3902, USA

Blomquist, Gary J., Department of Biochemistry and Molecular Biology, University of Nevada, Reno, NV 89557–0014, USA

Buckner, James S., Red River Valley Agricultural Research Center, USDA-ARS, University Station, Fargo, ND 58105, USA

Chertemps, Thomas, University of Paris 6, UMR INRA 1272 Physiologie de l'Insecte Signalisation et Communication, 75252 Paris Cedex 05, France

Cobb, Matthew, Faculty of Life Sciences, University of Manchester, Oxford Road, Manchester, M13 9PT, UK

d'Ettorre, Patrizia, Centre for Social Evolution, Department of Biology, University of Copenhagen, Universitetsparken 15, 2100, Copenhagen, Denmark

Ferveur, Jean-François, UMR CNRS 5548, Développement-Communication Chimique, Université de Bourgogne, Faculté des Sciences, 6 bd. Gabriel, 21000 Dijon, France

Ginzel, Matthew D., Department of Entomology, Purdue University, 901 West State Street, West Lafayette, IN 47907–2089, USA

Gibbs, Allen G., Department of Biological Sciences, 4505 Maryland Parkway, University of Nevada, Las Vegas, Las Vegas, NV 89154, USA

Greene, Michael, Department of Integrative Biology, University of Colorado Denver, Denver, CO 80217, USA

Hefetz, Abraham, George S. Wise Faculty of Life Sciences, Department of Zoology, Tel Aviv University, Ramat Aviv 69978, Israel

Howard, Ralph W., 701 Pine Street, Wamego, Kansas 66547, USA

Liebig, Jürgen, School of Life Sciences, Arizona State University, Tempe, AZ 85287–4501, USA

Lorenzi, M., Cristina. Department of Animal and Human Biology, University of Turin, 10123 Torino, Italy

Millar, Jocelyn G., Department of Entomology, University of California, Riverside, CA 92521, USA

Ozaki, Mamiko, Department of Biology, Kobe University, Rokkodai, Nada-ku, Kobe, 657–8501, Japan

Rajpurohit, Subhash, Department of Biological Sciences, 4505 Maryland Parkway, University of Nevada, Las Vegas, NV 89154, USA

Trabalon, Marie, IPHC-DEPE, Physiologie du Comportement, UMR 7178 CNRS, Faculté des Sciences et des Techniques, B.P. 70239, 54 506 Vandoeuvre-Les-Nancy, France

van Zweden, Jelle S., Centre for Social Evolution, Department of Biology, University of Copenhagen, Universitetsparken 15, 2100, Copenhagen, Denmark

Wada-Katsumata, Ayako, Section of Insect Physiology, Kyoto University, Oiwake-cho, Kitashirakawa, Sakyo-ku, Kyoto 606–8502, Japan

Wicker-Thomas, Claude, CNRS UPR 9034. Laboratoire Evolution, Génomes et Spéciation, Avenue de la Terrasse, Bâtiment 13, Boite Postale 1, 91198 Gif sur Yvette, France

Foreword

As every young chemist once realized or was told, hydrocarbons are simple (even "boring") molecules of limited practical importance, and even less scientific interest. Certainly this was the situation in the early 1960s when the field of chemical ecology was established. Although it was known that insects and other arthropods seemed to have very high molecular weight hydrocarbons on their cuticle, nothing else was known of their chemistry or of their biological importance, other than that they were possibly involved in water retention. As in many cases in science, progress is sometimes stymied by lack of a particular tool or technique. The invention of gas chromatographs and their coupling with mass spectrometers in this same time period formed the critical impetus for the birth of a major field of science, and the realization that those "simple" hydrocarbon molecules are the end product of eons of evolution, that they are far from simple, and that they are critical elements in an amazing variety of biological roles not only in arthropods, but also in microorganisms, plants, and numerous animal phyla.

By 1980, enough progress had been made in the elucidation of the chemistry and ecological roles of arthropod hydrocarbons to warrant a comprehensive review in the *Annual Review of Entomology* (130 papers constituted the entire literature at that time). Progress continued unabated in the ensuing years, and in 2005 a second limited review of the progress since 1980 of our knowledge of the chemistry and biological roles of hydrocarbons was published in the *Annual Review of Entomology*. During this same interval, several other specialist reviews were also published. Despite the enormous amount of insect hydrocarbon research conducted worldwide during these years (literally thousands of publications and numerous symposia and presentations), no one had stopped long enough to write a monograph on the field. This book, therefore, meets a long-felt need by numerous scientists to stop and see where the field has been and where it is going.

The first generation of insect hydrocarbon scientists are now all either retired or near retirement, and subsequent younger generations have taken their place, using new technologies and building on the solid foundation they inherited to explore ever-expanding facets of these fascinating molecules. Clearly, this discipline has matured to a point where no single scientist is capable of covering all that has been discovered, nor is it possible to cover, even superficially, the entire field in a few book chapters. The authors of this monograph are all leading experts in their respective fields, and even a cursory examination of the Table of Contents will reveal the complexity that has been discovered of the nature and role of hydrocarbons in the science of life. The twenty chapters cover in great depth aspects of chemistry (structures, qualitative and quantitative analysis), biochemistry (biosynthesis, molecular biology, genetics, evolution), physiology, taxonomy, and ecology. They clearly present to the reader the enormous wealth of data, ideas, insights and historical disagreements that have accumulated during the past half century. The material so presented is a major resource for current researchers, an unending source of ideas for new researchers, and serves as an exemplar for surveys of similar rapidly developing areas of research in other biological disciplines.

I am privileged to have been one of the early researchers in this marvelous field of scientific endeavor and to have made some small contributions to its development over a period of roughly thirty years. The comradeship I have experienced on a worldwide basis with numerous very gifted scientists (including many of the authors in this book) has brought me many moments of great pleasure and a strong sense of accomplishment that I might never have known otherwise. The editors of this book, Gary Blomquist and Anne-Geneviève Bagnères, have traversed this journey with me almost from the first. I thank them for their many kindnesses over the years, and for bringing this monograph to fruition. Their enthusiasm and scholarship have never failed to be less than outstanding, nor have their gifts as friends and colleagues to not only me, but so many others.

Ralph W. Howard, USDA-ARS, Retired

Acknowledgments

We thank our mentors, collaborators, co-workers, post-docs and students for making work in this field both fun and rewarding. GJB especially thanks his Ph.D. adviser, Larry Jackson, for sparking a lifelong interest in insect hydrocarbons, and long-time collaborators Ralph Howard, Dennis Nelson and Coby Schal for making the journey more enjoyable and productive. AGB thanks Jean-Luc Clément for his special ability to have made science and working with people so enjoyable during their collaborative years, and for his visionary impact on deciphering the role of cuticular hydrocarbons in chemical communication. AGB also thanks Ralph Howard, David Morgan, and Coby Schal for their kindness and serving as catalysts for her work, and her team and Institute (IRBI) for accommodating her calendar. She thanks her colleagues of the CNRS GDR d'Ecologie Chimique for their support. We thank the authors of this work for preparing up-to-date and stimulating chapters in a timely manner. It was a great pleasure to interact and learn from them. We thank Cheri Blomquist and Andy Corsini for their help in editing a number of chapters. And finally AGB wishes to thank – and dedicate this book to – her family, Frédéric for his continued support, and to her children Lucas and Eva.

GJB acknowledges the financial support from the National Science Foundation, the USDA-NRI and the Nevada Agriculture Experiment Station for the work of his laboratory.

Part I

Chemistry, Biochemistry, and Physiology

Part I

Chemistry, Biochemistry, and Physiology

1

Introduction: history and overview of insect hydrocarbons

Gary J. Blomquist and Anne-Geneviève Bagnères

The long-chain hydrocarbons of insects play central roles in the waterproofing of the insect cuticle and function extensively in chemical communication where relatively non-volatile chemicals are required. The recognition of the critical roles that hydrocarbons serve as sex pheromones, kairomones, species and gender recognition cues, nestmate recognition, dominance and fertility cues, chemical mimicry, primer pheromones and task-specific cues has resulted in an explosion of new information in the past several decades, and, indeed, served as the impetus for this book.

A number of reviews and chapters on specific topics related to insect hydrocarbons have been published over the past few decades (Jackson and Blomquist, 1976a; Blomquist and Jackson, 1979; Howard and Blomquist, 1982, 2005; Blomquist and Dillwith, 1985; Blomquist *et al.*, 1987; Lockey, 1988, 1991; Howard, 1993; Nelson and Blomquist, 1995; Gibbs, 2002), and in this book we attempt to bring this information up-to-date and in one place. At the time the first insect hydrocarbons were chemically identified in the 1960s and early 1970s, no one could have predicted the amount of interest that they would generate. Indeed, a pioneer in this field advised one of the authors (GJB) to go into a field other than insect hydrocarbons as he began his independent research career in the early 1970s, as he saw no future in this area. This prophecy proved very wrong and illustrates how difficult it is to predict the future of any scientific field.

The ability of insects to withstand desiccation was recognized in the 1930s to be due to the epicuticular layer of the cuticle. Wigglesworth (1933) described a complex fatty or waxy substance in the upper layers of the cuticle which he called "cuticulin". The presence of hydrocarbons in this wax of insects was suggested by Chibnall *et al.* (1934) and Blount *et al.* (1937), and over the next few decades the importance of hydrocarbons in the cuticular wax of insects was established (Baker *et al.*, 1963 and references therein). The first relatively complete chemical analyses of the hydrocarbons from any insect, the American cockroach, *Periplaneta americana* (Baker *et al.*, 1963), occurred after the development of gas-liquid chromatography (GLC). The three major components of the hydrocarbons of this insect, *n*-pentacosane, 3-methylpentacosane and (Z,Z)-6,9-heptacosadiene, represent the three major classes of hydrocarbons on insects, *n*-alkanes, methyl-branched alkanes and alkenes. Baker and co-workers (1963) were able to identify *n*-pentacosane by its elution time on GLC to a standard and its inclusion in a 5-angstrom molecular sieve. 3-Methylpentacosane was identified by its failure to be absorbed in a 5-angstrom molecular sieve and mass spectral data. Silver nitrate column chromatography separated the C27 diene from other hydrocarbon components, and the double-bond positions and isomeric composition were determined by a variety of techniques including infrared analysis and oxidative cleavage of the double bonds. The composition of the hydrocarbons of *P. americana* is unique in their simplicity, with over 90% of the hydrocarbons comprised of only three components (Jackson, 1972), and it was a fortunate choice of insect species for Baker *et al.* (1963). In general, the hydrocarbon composition of insects is much more complex, and sometimes consists of well over a hundred components. In an earlier analysis of the cuticular wax of the Mormon cricket, *Anabrus simplex* (Baker *et al.*, 1960), the complexity of the hydrocarbon mixture (Jackson and Blomquist, 1976b) made it impossible for individual components to be characterized in the early 1960s (Baker *et al.*, 1960). The development and application of combined gas–liquid chromatography and mass spectrometry was key to the rapid and efficient analysis of insect hydrocarbons. In the late 1960s and during the next few decades, GC–MS analysis of insect hydrocarbons was established (Nelson and Sukkestad, 1970; Martin and MaConnell, 1970), and over the next several decades the hydrocarbons of hundreds of insect species were analyzed, first on packed columns and then much more efficiently on capillary columns. It was recognized that for many insect species, very complex mixtures of normal (straight-chain), methyl-branched and unsaturated components exist, with chain lengths ranging from 21 to 50 or so carbons. In retrospect, the occurrence of extremely complex mixtures of components might have suggested that hydrocarbons could play important roles in chemical communication, but only after the recognition of the number and variety of roles they do play have we come to more fully appreciate the importance of insect hydrocarbons in chemical communication. The variety of chain lengths and the number and positions of the methyl branches and double bonds provide the insect with the chemical equivalent of the visually variable colored plumage of birds. Cvačka *et al.* (2006) reported hydrocarbons of up to 70 carbons in chain length using MALDI–TOF mass spectrometry, indicating that the earlier chain-length limits of 50 or so carbons might be a limitation of GC–MS techniques and not the ability of the insect to make longer chain components. The structure and chemical analysis of insect hydrocarbons are covered in Chapter 2 (Blomquist, this book).

A series of studies in the 1960s demonstrated that labeled acetate was incorporated into insect cuticular hydrocarbons (Vroman *et al.*, 1965; Lamb and Monroe, 1968; Nelson, 1969) establishing the de novo biosynthesis of the majority of components. Because of the simple hydrocarbon composition and the ease of isolating 3-methylpentacosane and (Z,Z)-6,9-heptacosadiene from the *P. americana*, early *in vivo* biosynthetic studies concentrated on this insect (Conrad and Jackson, 1971; Blomquist *et al.*, 1975; Major and Blomquist, 1978; Dwyer *et al.*, 1981). Work with radiolabeled precursors established the elongation–decarboxylation pathway for hydrocarbon biosynthesis (Major and Blomquist, 1978) and the incorporation of propionate, as a methylmalonate unit, to form the methyl-branching unit (Blomquist *et al.*, 1975). Over the next several decades, studies evolved from *in vivo* studies to the use of microsomal preparations to gain an understanding of how the major components were made and regulated. In the final step of hydrocarbon formation, the elongated acyl-CoA is reduced to aldehyde and oxidatively decarbonylated to hydrocarbon (Reed *et al.*, 1994, 1995). The biosynthesis of insect hydrocarbons is presented in Chapter 3 (Blomquist, this book). Only recently have the powerful tools of molecular biology been applied to studies of insect hydrocarbon biosynthesis, and this is covered in Chapter 4 (Wicker-Thomas and Chertemps, this book).

In the mid-1960s, Locke (1965) presented an illustration of insect cuticular lipids in which the newly synthesized hydrocarbons exited the epidermal oenocyte cells where they were synthesized, were transported through pore canals and then formed an outer layer on the cuticle. He pictured the polar head groups of fatty acids interacting with the cuticle, and then the hydrocarbons layered on top of the acyl chains of fatty acids. This served as an excellent model in which to test a number of hypotheses, many of which are still unanswered. It is now clear that newly synthesized hydrocarbons are taken up first by lipophorin and transported via the hemolymph (Bagnères and Blomquist, Chapter 5, this book). How they get transferred to the surface of the insect is unknown, although a number of species are able to selectively transport shorter chain hydrocarbon pheromones and pheromone precursors to the pheromone gland on the abdomen, whereas longer chain cuticular hydrocarbons are transported to cover the entire cuticle (Schal *et al.*, 1998; Jurenka *et al.*, 2003). In many cases, hydrocarbons comprise the majority of the cuticular lipids, making the proposed role of fatty acids of lesser importance, and the arrangement of hydrocarbons on the surface of the insect is unknown, although suggestions have been made that the components of most importance in chemical communication may be on the outer surface (Ginzel *et al.*, 2003; Ginzel, Chapter 17, this book). A clear understanding of hydrocarbon transport to the cuticle and arrangement of hydrocarbons on the surface of insects is, at this time, unavailable.

The large surface-to-volume ratio of insects makes it important that excessive evaporation be prevented. The first recognized and perhaps still primary function of insect cuticular lipids, especially hydrocarbons, is to restrict water loss, a fact first recognized by Ramsay (1935). Early investigations centered on the measurement of transpiration of water and the role of cuticular lipids in preventing water loss, and these early studies were reviewed in Barton-Brown (1964), Beament (1961, 1964), Edney (1957), Richards (1951), and Locke (1965). It is now clear that the cuticle is permeable to water vapor, and that the cuticular lipid layer plays a major role in reducing transpiration. Cuticles from which the surface lipids are removed with organic solvents are relatively permeable to water. Vegetable oils, lecithin and a series of wetting agents and detergents show widely different effects on permeability of the cuticle to water (Beament, 1945). The waterproofing observed with intact insects is closely duplicated when extracted cuticular lipids are deposited on collodion membranes or intact wing membranes (Beament, 1945). The transpiration rate from an insect is found to increase rather abruptly at a temperature that closely corresponds to the transition point or change of phase point of the lipids on the cuticle of a particular species. This transition point is near the melting range of the extractable lipid (Wigglesworth, 1945; Beament, 1945). Rapid desiccation occurs as a result of scratching the outer surface

of the epicuticle with abrasive dusts. The abrasion needs to be only deep enough to penetrate the epicuticle. Adsorption of the lipids onto the dust may also play a role to a certain
extent, especially on insects where the lipids are soft (Wigglesworth, 1945). If the abraded
insect is kept in a moist atmosphere to prevent desiccation, the lipid layer is apparently
restored, along with the ability to resist desiccation. The lipid melting model achieved textbook status (Randall *et al.*, 1997; Chapman, 1998), although a number of researchers have
pointed out its limitation. In the last several decades, Gibbs and Rajpurohit (Chapter 6, this
book) have re-examined this phenomenon, and presents our current understanding of the
role of cuticular hydrocarbons and other lipids in restricting water loss from insects.

The cuticular hydrocarbons of many insect species are extremely complex and involve
mixtures of normal, mono-, di- and tri-unsaturated and mono-, di-, tri-, tetra- and
pentamethyl-branched components of chain lengths between 21 and 50 carbons. The
number of known components is very large and the number of possible isomers much
larger. Insects biosynthesize the vast majority of their hydrocarbon components, and thus
hydrocarbon composition may be considered a part of an insect's genotype and therefore
available for taxonomic use. The early studies by Jackson (1970), Jackson and Baker
(1970), Jackson and Blomquist (1976a), Lockey (1980) and Carlson and collaborators
(Carlson and Service, 1979, 1980; Carlson and Brenner, 1988; Carlson, 1988) recognized
the special role of cuticular hydrocarbons in chemical taxonomy. The importance of using
hydrocarbons in chemical taxonomy has continued to grow, with more recent efforts centered on Drosophilidae and termites. This work is covered in Chapter 7 (Bagnères and
Wicker-Thomas, this book).

In order to more fully understand and interpret mass spectra and retention indices of
methyl-branched and unsaturated hydrocarbons, standard known hydrocarbon compounds
were synthesized and analyzed (Pomonis *et al.*, 1978, 1980). The availability of standards
with one or more methyl branches and specific double bonds allowed for better understanding
of the roles individual components and hydrocarbon classes played in determining critical
transition temperatures and waterproofing characteristics of hydrocarbons (Gibbs and
Pomonis, 1995). With the increasing recognition that many hydrocarbon components play
important roles in chemical communication, the importance of synthesizing hydrocarbon
standards to determine the exact role of individual compounds became more important. The
first review of the chemical synthesis of long-chain hydrocarbons is presented in Chapter 8
(Millar, this book).

While the hydrocarbon fraction of insect cuticular lipids is certainly the most studied and
has been shown to play a key role in a wide range of chemical communication, other lipids
are often present on the surface of insects. The most common cuticular lipids in addition to
hydrocarbons include a variety of types of esters, free fatty acids, primary and secondary alcohols, ketones and sterols. Triacylglycerols and the more polar phospholipids are not common components of insect cuticular lipids. In some cases, hydrocarbons are hydroxylated and
metabolized to oxygenated components, and these products include some of the short range
and contact pheromones of the housefly (Blomquist, 2003) and the German cockroach (Schal
et al., 2003). The oxygenated cuticular lipids are discussed in Chapter 9 (Buckner, this book).

The recognition of the role of cuticular hydrocarbons in chemical communication has evolved over the last 38 years from a point where it was noteworthy when cuticular hydrocarbons functioned in chemical communication to where it is now noteworthy if the hydrocarbons don't serve a function in chemical communication. Carlson *et al.* (1971) first demonstrated the role of hydrocarbons in chemical communication when they showed that a component (Z-9-tricosene) of the hydrocarbons of the female housefly, *Musca domestica*, served as a short-range attractant for males. At about the same time, the *n*-alkanes docosane, tricosane, tetracosane and pentacosane (Jones *et al.*, 1973), and the branched chain hydrocarbon 13-methylhentriacontane (Jones *et al.*, 1971) were shown to be host-seeking stimulants (kairomones) for the egg parasite *Trichogramma evanescens* and the parasite *Microplitis croceipes*, respectively. The activity of the straight chain hydrocarbons, C_{22}, C_{23}, C_{24} and C_{25} might appear surprising at first glance, considering that these compounds are so widespread. On the other hand, a highly specific kairomone would hardly serve a parasite that has a broad latitude in the selection of insect hosts such as *T. evanescens*. Since the young of this parasite will develop in eggs of most Lepidoptera, a kairomone common to many species might be anticipated (Jones *et al.*, 1973).

It is becoming increasingly clear that a major function of cuticular hydrocarbons in arthropods is to serve as recognition signals between two or more individuals. One or more components of the complex mixture of hydrocarbons found on the cuticle of almost all arthropods is often the primary chemical cue that answers questions such as: "Are you a member of my species? Are you the same gender as me? For subsocial insects, are you a member of my family, cohort or group? For eusocial insects, are you a member of my colony? Are you a member of my nest? To which caste do you belong? Are you a queen or perhaps a brood? Are you a worker trying to convey to me the need to accomplish a certain task? Are you closely related kin? And, for many arthropods that exist as inquilines in the nest of social insects, can you recognize that I am alien?" (Howard and Blomquist, 2005).

Although a few early authors predicted that contact between cuticle and antenna was crucial for chemical recognition in insects, technical difficulties prevented proof of this concept until relatively recently. As a result of this delay, the study of volatile signals long dominated that of pheromone (and allomone) chemoreception. One of the first to implicate antennae in inter- and intra-specific recognition in ants was Adèle Fielde in 1904 and 1905. Shortly after, in 1907, W. Barrows (reference in Stocker and Rodrigues, 1999, and deBruyne, 2003) demonstrated that adult *Drosophila* detect odors using their antennae. Prokopy *et al.* (1982) appears to have been the first to use a combination of electrophysiological and behavioral techniques to identify the production source of contact pheromones in female *Rhagoletis pomonella* (from Stadler's 1984 review on contact chemoreception). Hodgson and Roeder (1956), Clément (1981, 1982), Rence and Loher (1977) and Le Moli *et al.* (1983) were among the first to implicate the antenna in the reception of chemical cues from insect cuticle. Further confirmation was obtained in experimental works by Venard and Jallon (1980), Antony and Jallon (1982), Howard *et al.* (1982a), Bonavita-Cougourdan *et al.* (1987), Getz and Smith (1983, 1987), Morel *et al.* (1988), and Bagnères *et al.* (1991) (see also later references in Chapter 10). The aforementioned authors almost consistently showed direct or indirect evidence that behavioral responses to total cuticular mixture extracts or individual compounds depended on antennal reception. Recent insect studies (Heifetz *et al.*, 1997; Brockmann *et al.*, 2003; Batista-Pereira *et al.*, 2004; Châline *et al.*, 2005; Ozaki *et al.*, 2005; Saïd *et al.*, 2005; Guerrieri and d'Ettorre, 2008) using one or two compounds or mixtures demonstrated direct olfactory sensitivity to long-chain hydrocarbons. Ozaki and Wada-Katsumata review the perception of cuticular hydrocarbons in Chapter 10.

Early studies on perception and olfaction of cuticular hydrocarbons were performed in social insects because of their sociobiological implications in kin and nestmate recognition processes. Adèle Fielde made her visionary predictions about the "power of recognition" (1904) and the "specific, progressive and incurred odor" (1905) based on observations in ants. Emerson (1929) and Dropkin (1946) suggested that species odor in social insects might be environmentally induced (in Howse, 1975). Wilson (1971) stated that conspecific recognition was acquired over time in insect colonies, whereas Wallis (1962, 1964, refs in Howse, 1975) found that environmental factors had little effect on colony odor that appeared soon after emergence. Crozier and Dix (1979), Hölldobler and Michener (1980), Shellman and Gamboa (1982), Carlin and Hölldobler (1983), Pfennig *et al.* (1983), Lacy and Sherman (1983), and Howse (1984), who were among the first to implicate contact pheromones as nestmate recognition cues, studied this aspect on various species of social insects.

Nestmate recognition is part of kin recognition in social insects. Genetic relatedness, as well as phenotype matching, play a key role in social insects (Hamilton, 1964; Lacy and Sherman, 1983). As stated by Howard (1993) and Passera and Aron (2005), the first proof that cuticular hydrocarbons play a role in colonial recognition was provided by Bonavita-Cougourdan *et al.* in 1987. Bonavita-Cougourdan and colleagues were also the first to suggest that methyl-branched hydrocarbons exhibited the greatest variation between colonies and to discuss preliminary data implicating the postpharyngeal gland in the colony odor. Nevertheless, proof of kin recognition using cuticular hydrocarbons, i.e., based on cues correlated with genetic relatedness, has been presented in only a few eusocial species, e.g., honeybees (Arnold *et al.*, 1996), wasps (Gamboa, 2004) and termites (Dronnet *et al.*, 2006). Most examples have involved nestmate recognition. There have been even fewer reports on kin recognition with contact cues in gregarious species such as cockroaches (Lihoreau *et al.*, 2007; Lihoreau and Rivault, 2009) and also solitary species (Lizé *et al.*, 2006), and most have involved incest avoidance and/or mate choice (Simmons, 1989; Lihoreau *et al.*, 2007). In this book, the role of cuticular hydrocarbons in nestmate recognition is presented in Chapter 11, along with a review of the nestmate recognition models proposed by van Zweden and d'Ettorre.

The use of cuticular hydrocarbon as task-specific cues was suggested by Howard *et al.* (1982a) and Bagnères *et al.* (1998) and illustrated by caste-specific mixtures in termites. Bonavita-Cougourdan *et al.* (1993) also reported evidence of a hydrocarbon cueing between functional castes of *Camponotus* ants. Several studies by Gordon, Greene and collaborators recently proved the involvement of cuticular compounds in task decisions in ants (Greene and Gordon, 2003, 2007). Green reviews this aspect of chemical communication of social insects in Chapter 12.

A particularly well-documented example of the cueing function of cuticular hydrocarbons involves fertility and dominance signals. Because age is often an important factor for task, fertility and dominance cues (Wilson, 1963; Robinson, 1992; Keller and Nonacs, 1993), some of the first studies were carried out using foundresses and young wasps as models (Bonavita-Cougourdan *et al.*, 1991; Brown *et al.*, 1991). There have also been studies on queen and brood odor in ants (Monnin and Peeters, 1997; Dahbi and Lenoir, 1998; Monnin *et al.*, 1998; Monnin, 2006; Liebig *et al.*, 1999, 2005), bees (Francis *et al.*, 1985) and bumble bees (Ayasse *et al.*, 1995). Recently cuticular hydrocarbons were even described as putative primer pheromones (Le Conte and Hefetz, 2008). Liebig reviews the literature describing social insects' use of cuticular hydrocarbons as dominance and fertility cues in Chapter 13.

Some insect (and non-insect) species use cuticular hydrocarbons for chemical deception/mimicry to gain some advantage from another species. In Chapter 14 Bagnères and Lorenzi review the wide range of multitrophic level mimicry relationships that have been described between plants and insects, social insects and nonsocial insects and non-insects, and various social insect interactions including other aspects of mimicry, for example artificial mixed social insect colonies and intraspecific mimicry. Howard *et al.* (1982b, 1990), Vander Meer and Wojcik (1982) and Vander Meer *et al.* (1989) were the first to report chemical mimicry with cuticular hydrocarbons. Subsequent reports included those of Espelie and collaborators on parabiotic associations in social insects (Espelie and Hermann, 1988), Franks *et al.* (1990) on chemical disguise phenomenon in a leptothoracine cuckoo ant, and Yamaoka (1990) on various chemical interactions. Scott (1986) described sexual mimicry in *Drosophila*. Since those different reports, two main theories have been proposed to account for the appearance of specific hydrocarbons used for chemical mimicry phenomena: de novo biosynthesis by the intruder or acquisition from the host by contact. These two processes are not mutually exclusive. Other possible integration processes could be involved for newborn insects with low amounts of hydrocarbon, parasite cuticular chemical insignificance that has often been observed, and odor change with age that was observed long ago by Fielde (1905).

The cuticular hydrocarbons of solitary insects have also been thoroughly studied. A wide range of functions has been documented in diptera behavior (aggregation dominance, courtship, mate discrimination) and reproduction (fecundity, sex-ratio). Ferveur and Cobb emphasize the evolutionary aspect of those various cuticular hydrocarbon functions in Chapter 15. A number of authors earlier studied dipteran cuticular hydrocarbons, including Rogoff *et al.* (1964), Butterworth (1969), Leonard *et al.* (1974) and Leonard and Ehrman (1976). Most authors focused on sexual selection using pheromones in *Drosophila* and, since the 1980s, the group of Jallon and associated people (Antony and Jallon, 1981, 1982; Jallon, 1984, reference found in Ferveur, 2005). The involvement of hydrocarbons in housefly sexual behavior was studied by Blomquist *et al.* (1987) and Carlson *et al.* (1971).

In Chapter 16 Trabalon and Bagnères present an overview of non-insect contact pheromones with emphasis on long-chain hydrocarbons and derivative compounds in Arachnids. Apart from studies on spider and scorpion venoms, the first semiochemical studies in these close insect relatives focused on permeability properties of cuticular lipids (Hadley, 1970, 1981; Hadley and Filshie, 1979; Hadley and Hall, 1980). Several early studies described contact pheromones on female spider webs and cuticle (Hegdekar and Dondale, 1969; Blanke, 1975; Dijkstra, 1976; Tietjen, 1979). Krafft (1982) showed that tactochemical stimuli obtained by body contact played an important role in spider recognition of conspecifics. One of the chapter authors (M. Trabalon) presented the first complete description of spider cuticular hydrocarbons. Since data on scorpion recognition mechanisms using cuticular contact pheromones has been rare, the other chapter author (A-C Bagnères) has provided personal results (obtained in collaboration with Phil Brownell) that have never been published elsewhere.

Chapter 18 by Millar describes the role of longer polyene hydrocarbons and their derivatives as sex pheromones in lepidopterans. Non-oxygenated polyenes were identified as moth pheromones before the oxygenated derivatives. Indeed, Roelofs *et al.* (1982) identified (Z,Z,Z)-1,3,6,9-nonadecatetraene as the sex pheromone of the winter moth *Operophtera brumata*. This finding was supported by trap tests showing that the synthetic compound was highly attractive. At that time, the finding that *O. brumata* males responded optimally to a conspecific female sex pheromone at temperatures between 7°C and 15°C was unique among pheromone studies. Conner *et al.* (1980) identified the sex pheromone produced by female *Utetheisa ornatrix* moths as (Z,Z,Z)-3,6,9-heneicosatriene. Shortly thereafter, further studies carried out by the same authors for the chemical determination of female sex attractants in other moth populations showed that a Z-C21 tetraene ((Z,Z,Z)-1,3,6,9-heneicosatetraene), along with a new C21 diene (Z,Z)-6,9-heneicosadiene, was a major component, and that both compounds were EAG-active (Jain *et al.*, 1983). In 2004, Ando *et al.* classified polyene hydrocarbons as type II pheromones (15% of the lepidopteran sex pheromones) generally found in the geometridae and arctiidae families.

Chapter 19 by Bartelt is devoted to the pheromonal role of short-chain hydrocarbons, especially short methyl/ethyl-branched and unsaturated components in beetles. The most abundant components in *Carpophilus hemipterus* (Coleoptera, Nitidulidae) have been identified as (2E,4E,6E,8E)-3,5,7-trimethyl-2,4,6,8-decatetraene and (2E,4E,6E,8E)-3,5,7-trimethyl-2,4,6,8-undecatetraene (Bartelt *et al.*, 1990). Later studies showed that male *C. hemipterus* emit nine all-*E* tetraene hydrocarbons and one all-*E* triene hydrocarbon in addition to the two previously reported pheromonally active tetraenes (Bartelt *et al.*, 1992). In their review of biologically active compounds in beetles, Francke and Dettner (2005) listed only a few dozen of those pheromonal compounds, most of which were identified by Bartelt.

While it is always difficult to predict where a given scientific field is headed, Chapter 20 summarizes some of the gaps in our understanding of insect cuticular hydrocarbons and points out areas where work is needed.

In addition to providing a thorough up-to-date compilation of current data, this book gives new perspective in several areas. Chapters 2–9 of this book significantly expand discussion of the chemistry, biochemistry and physiology of cuticular hydrocarbons. Chapter 8 provides the first review of the chemical synthesis of long-chain hydrocarbons. Chapters 11–13 provide new insight into the role of chemical communication in various social insect functions. Chapter 14 is specifically devoted to chemical mimicry. Chapters 15–19 focus on hydrocarbons as sex pheromones, including those of solitary insects such as Diptera, Coleoptera, and Lepidoptera as well as spiders and scorpions. In an effort to cover all aspects of cuticular hydrocarbons (and derivatives) functions and analyses, the authors of this book have included a wide range of data, not only from the large recent reviews but also from ongoing and unpublished studies.

References

Ando, T., Inomata, S. I. and Yamamoto, M. (2004). Lepidopteran sex pheromones. In *The Chemistry of Pheromones and other Semiochemicals I*. Topics in Current Chemistry, Vol. 229. Berlin: Springer, pp. 51–96.

Antony, C. and Jallon, J.-M. (1981). Evolution des hydrocarbures comportementalement actifs des *Drosophila melanogaster* au cours de la maturation sexuelle. *C. R. Acad. Sci. Paris Serie D*, **292**, 239–242.

Antony, C. and Jallon, J.-M. (1982). The chemical basis for sex recognition in *Drosophila melanogaster*. *J. Insect Physiol.*, **28**, 873–880.

Arnold, G., Quenet, B., Cornuet, J.-M., Masson, C., De Schepper, B., Estoup, A. and Gasqui, P. (1996). Kin recognition in honeybees. *Nature*, **379**, 498.

Ayasse M., Maelovits T., Tengö J., Taghizadeh T. and Francke W. (1995). Are there pheromonal dominance signals in the bumblebee *Bombus hypnorum* L. (Hymenoptera, Apidae)? *Apidologie*, **26**, 163–180.

Bagnères, A.-G., Killian, A., Clément, J.-L and Lange, C. (1991). Interspecific recognition among termites of the genus *Reticulitermes*: Evidence for a role for the hydrocarbons. *J. Chem. Ecol.*, **17**, 2397–2420.

Bagnères A.-G., Rivière G. and Clément J.-L. (1998). Artificial neural network modeling of caste odor discrimination based on cuticular hydrocarbons in termites. *Chemoecology*, **8**, 201–209.

Baker, G., Pepper, J. H., Johnson, L. H. and Hastings, E. (1960). Estimation of the composition of the cuticular wax of the Mormon cricket, *Anabrus simplex* Hald. *J. Insect Physiol.*, **5**, 47–60.

Baker, G. L., Vroman, H. E. and Padmore, J. (1963). Hydrocarbons of the American cockroach. *Biochem. Biophys. Res. Commun.*, **13**, 360–365.

Bartelt, R. J., Dowd, P. F., Plattner, R. D. and Weisleder, D. (1990). Aggregation pheromone of dried-fruit beetle, *Carpophilus hemipterus*. Wind-tunnel bioassay and identification of two novel tetraene hydrocarbons. *J. Chem. Ecol.*, **16**, 1015–1039.

Bartelt, R. J., Weisleder, D., Dowd, P. F. and Plattner, R. D. (1992). Male-specific tetraene and triene hydrocarbons of *Carpophilus hemipterus*: Structure and pheromonal activity. *J. Chem. Ecol.*, **18**, 379–402.

Barton-Browne, L. B. (1964). Water regulation in insects. *Annu. Rev. Entomol.*, **9**, 63–78.

Batista-Pereira, L. G., dos Santos, M. G., Corrêa, A. G., Fernandes, J. B., Arab, A., Costa-Leonardo, A.-M., Dietrich, C. R. R. C., Pereira, D. A. and Bueno O. C. (2004). Cuticular hydrocarbons of *Heterotermes tenuis* (Isoptera: Rhinotermitidae): Analyses and electrophysiological studies. *Z. Naturforsch.*, **59c**, 135–139.

Beament, J. W. L. (1945). The cuticular lipoids of insects. *J. Exp. Biol.*, **21**, 115–131.

Beament, J. W. L. (1961). The water relations of insect cuticle. *Biol. Rev.*, **36**, 281–320.

Beament, J.W.L. (1964). The active transport and passive movement of water in insects. *Adv. Insect Physiol.*, **2**, 67–129.

Blanke, R. (1975). Untersuchungen zum sexualverhalten von *Cyrtophora cicatrosa* (Stoliczka) (Araneae, Araneidae). *Zeitschrift Tierpsychol.*, **37**, 62–74.

Blomquist, G.J. (2003). Biosynthesis and ecdysteroid regulation of housefly sex pheromone production. In *Insect Pheromone Biochemistry and Molecular Biology – The Biosynthesis and Detection of Pheromones and Plant Volatiles*, ed. G.J. Blomquist and R.G. Vogt. London: Elsevier, pp. 231–252.

Blomquist, G.J. and Dillwith, J.W. (1985). Cuticular lipids. In *Comprehensive Insect Physiology, Biochemistry and Pharmacology*, ed. G.A. Kerkut and L.I. Gilbert, Vol. 3. *Integument, Respiration, and Circulation*. Oxford: Pergamon, pp. 117–154.

Blomquist, G.J., Major, M.A. and Lok, J.B. (1975). Biosynthesis of 3-methylpentacosane in the cockroach *Periplaneta americana*. *Biochem. Biophys. Res. Comm.*, **64**, 43–50.

Blomquist G.J. and Jackson, L.L. (1979). Chemistry and biochemistry of insect waxes. *Prog. Lipid Res.*, **17**, 319–345.

Blomquist, G.J., Nelson, D.R. and de Renobales, M. (1987). Chemistry, biochemistry, and physiology of insect cuticular lipids. *Arch. Insect Biochem. Physiol.*, **6**, 227–265.

Blount, B.K., Chibnall, A.C. and El Mangouri, H.A. (1937). The wax of the white pine chermes. *Biochem. J.*, **31**, 1375–1378.

Bonavita-Cougourdan, A., Clément, J.L. and Lange, C. (1987). Nestmate recognition: the role of cuticular hydrocarbons in the ant *Camponotus vagus* Scop. *J. Entomol. Sci.*, **22**, 1–10.

Bonavita-Cougourdan, A., Clément, J.-L. and Lange, C. (1993). Functional subcaste discrimination (foragers and brood-tenders) in the ant *Camponotus vagus* Scop: polymorphism of cuticular hydrocarbon patterns. *J. Chem. Ecol.*, **19**, 1461–1477.

Bonavita-Cougourdan, A., Theraulaz, G., Bagnères, A.-G., Roux, M., Pratte, M., Provost, E. and Clément, J.-L. (1991). Cuticular hydrocarbons, social organization and ovarian development in a polistine wasp: *Polistes dominulus* Christ. *Comp. Biochem. Physiol.*, **100B**, 667–680.

Brockmann, A., Groh, C. and Fröhlich, B. (2003). Wax perception in honeybees: contact is not necessary. *Naturwissenschaften*, **90**, 424–427.

Brown, W.V., Spradbery, J.P. and Lacey, M.J. (1991). Changes in the cuticular hydrocarbon composition during development of the social wasp, *Vespula germanica* (F.). (Hymenoptera, Vespidae). *Comp. Biochem. Physiol. B*, **99**, 553–562.

Butterworth, F.M. (1969). Lipids of *Drosophila*: a newly detected lipid in the male. *Science*, **163**, 1356–1357.

Carlin, N.F. and Hölldobler, B. (1983). Nestmate and kin recognition in interspecific mixed colonies of ants. *Science*, **222**, 1027–1029.

Carlson, D.A. (1988). Hydrocarbons for identification and phenetic comparisons: Cockroaches, honey bees and tsetse flies. *Florida Entomol.*, **71**, 333–345.

Carlson, D.A. and Brenner, R.J. (1988). Hydrocarbon based discrimination of three North American *Blattella* cockroach species using gas chromatography. *Ann. Entomol. Soc. Amer.*, **81**, 711–723.

Carlson, D.A., Mayer, M.S., Silhacek, D.L., James, J.D., Beroza, M. and Bierl, B.A. (1971). Sex attractant peheromone of the housefly: Isolation, identification and synthesis. *Science*, **174**, 76–77.

Carlson, D. A. and Service, M. W. (1979). Differentiation between species of the *Anopheles gambiae* Giles complex (Diptera: Culicdae) by analysis of cuticular hydrocarbons. *Ann. Trop. Med. Parasitol.*, **73**, 589–592.

Carlson, D. A. and Service, M. W. (1980). Identification of mosquitoes of *Anopheles gambiae* species complex A and B by analysis of cuticular hydrocarbons. *Science*, **207**, 1089–1091.

Châline, N., Sandoz J.-C., Martin S. J., Ratnieks F. L. W. and Jones, G. R. (2005). Learning and discrimination of individual cuticular hydrocarbons by honeybees (*Apis mellifera*). *Chem. Senses*, **30**, 327–335.

Chapman, R. F. (1998). *The Insects*, 4th edn. Cambridge: Cambridge University Press.

Chibnall, A. C., Piper, S. H., Pollard, A., Williams, E. F. and Sahai, P. N. (1934). The constitution of the priary alcohols, fatty acids and paraffins present in plant and insect waxes. *Biochem. J.*, **28**, 2189–2208.

Clément, J.-L. (1981). Comportement de reconnaissance individuelle dans le genre *Reticulitermes*. *C. R. Acad. Sci. Paris*, **292**, 931–933.

Clément, J.-L. (1982). Signaux responsables de l'agression interspécifique des termites du genre *Reticulitermes* (Isoptères). *C. R. Acad. Sci. Paris*, **294**, 635–638.

Conner, W. E., Eisner, T., Vander Meer, R. K., Guerrero, A., Ghiringelli, D. and Meinwald, J. (1980). Sex attractant of an arctiid moth (*Utetheisa ornatrix*): A pulsed chemical signal. *Behav. Ecol. Sociobiol.*, **7**, 55–63.

Conrad, C. W. and Jackson, L. L. (1971). Hydrocarbon biosynthesis in *Periplaneta americana*. *J. Insect Phys.*, **17**, 1907–1916.

Crozier, R. H. and Dix M. W. (1979). Analysis of two genetic models for the innate components of colony odor in social *Hymenoptera*. *Behav. Ecol. Sociobiol.*, **4**, 217–224.

Cvačka, J., Jiroš, P., Šobotník, J., Hanus, R. and Svatoš, A. (2006). Analysis of insect cuticular hydrocarbons using matrix-assisted laser desorption/ionization mass spectrometry. *J. Chem. Ecol.*, **32**, 409–434.

Dahbi A. and Lenoir A. (1998). Queen and colony odour in the multiple nest ant species, *Cataglyphis iberica* (Hymenoptera, Formicidae). *Insect. Soc.*, **45**, 301–313.

deBruyne, M. (2003). Physiology and genetics of odor perception in Drosophila. In *Insect Pheromone Biochemistry and Molecular Biology – The Biosynthesis and Detection of Pheromones and Plant Volatiles*, ed. G. J. Blomquist and R. G. Vogt. London: Elsevier, pp. 651–698.

Dijkstra, H. (1976). Searching behaviour and tactochemical orientation in males of the wolf spider *Pardosa amentata* (Cl.) (Araneae Lycosidae). *Proc. K. Ned. Akad. Wet.*, **79**, 235–244.

Dronnet S., Lohou, C., Christides, J.-P. and Bagnères, A.-G. (2006). Cuticular hydrocarbon composition reflects genetic relationship among colonies of the introduced termite *Reticulitermes santonensis* Feytaud. *J. Chem. Ecol.*, **32**, 1027–1042.

Dwyer, L. A., de Renobales, M., and Blomquist, G. J. (1981). Biosynthesis of (Z,Z)-6,9-heptacosadiene in the American cockroach. *Lipids*, **16**, 810–814.

Edney, E. B. (1957). *The Water Relations of Terrestrial Arthropods*. London: Cambridge University Press, p.6.

Espelie, K. E. and Hermann, H. R. (1988). Congruent cuticular hydrocarbons: biochemical convergence of a social wasp, an ant and a host plant. *Biochem. Systemat. Ecol.*, **16**, 505–508.

Ferveur, J.-F. (2005). Cuticular hydrocarbons: their evolution and roles in Drosophila pheromonal communication. *Behav. Genet.*, **35**, 279–295.

Fielde, A. (1904). Power of recognition among ants. *Biol. Bull.*, **7**, 227–250.

Fielde, A. (1905). The progressive odor of ants. *Biol. Bull.*, **10**, 1–16.

Francis, B. R., Blanton, W. E. and Nunamaker R. A. (1985). Extractable hydrocarbons of workers and drones of the genus *Apis*. *J. Apic. Res.*, **24**, 13–26.

Francke, W. and Dettner, K. (2005). Chemical signalling in beetles. In *The Chemistry of Pheromones and other Semiochemicals II*. Topics in Current Chemistry, Vol. 240. Berlin: Springer, pp. 85–166.

Franks, N., Blum, M. S., Smith, R.-K. and Allies, A. B. (1990). Behavior and chemical disguise of cuckoo ant *Leptothorax kutteri* in relation to its host *Leptothorax acervorum*. *J. Chem. Ecol.*, **16**, 1431–1444.

Gamboa, G. J. (2004). Kin recognition is eusocial wasps. *Ann. Zool. Fennici*, **41**, 789–808.

Getz, W. M. and Smith, K. B. (1987). Olfactory sensitivity and discrimination of mixtures in the honeybee *Apis mellifera*. *J. Comp. Physiol. A*, **160**, 239–245.

Getz, W. M. and Smith, K. B. (1983). Genetic kin recognition: honey bees discriminate between full and half sisters. *Nature*, **302**, 147–148.

Gibbs, A. G. (2002). Lipid melting and cuticular permeability: new insights into an old problem. *J. Insect Physiol.*, **48**, 391–400.

Gibbs, A. and Pomonis J. G. (1995). Physical properties of insect cuticular hydrocarbons: the effect of chain length, methyl-branching and unsaturation. *Comp. Biochem. Physiol.*, **112B**, 243–249.

Ginzel, M. D., Millar, J. G. and Hanks, L. M. (2003). (Z)-9-Pentacosene – contact sex pheromone of the locust borer, *Megacyllene robiniae*. *Chemoecology*, **13**, 135–141.

Greene M. J. and Gordon D. M. (2003). Cuticular hydrocarbons inform task decisions. *Nature*, **423**, 32.

Greene, M. J. and Gordon, D. M. (2007). Interaction rate informs harvester ant task decisions. *Behav. Ecol.*, **18**, 451–455.

Guerrieri, F. J. and d'Ettorre, P. (2008). The mandible opening response: quantifying aggression elicited by chemical cues in ants. *J. Experimental Biol.*, **211**, 1109–1113.

Hadley, N. F. (1970). Water relations of the desert scorpion, *Hadrurus arizonensis*. *J. Exp. Biol.*, **53**, 547–558.

Hadley N. F. (1981). Cuticular lipids of terrestrial plants and arthropods: a comparison of their structure, composition, and waterproofing function. *Biol. Rev.*, **56**, 23–47.

Hadley, N. F. and Filshie, B. K. (1979). Fine structure of the epicuticle of the desert scorpion, *Hadrurus arizonensis*, with reference to location of lipids. *Tissue Cell*, **11**, 263–275.

Hadley, N. F. and Hall R. L. (1980). Cuticular lipid biosynthesis in the scorpion, *Paruroctonus mesaensis*. *J. Exp. Zool.*, **212**, 373–379.

Hadley, N. F. and Jackson, L. L. (1977). Chemical composition of the epicuticular lipids of the scorpion, *Paruroctonus mesaensis*. *Insect Biochem.*, **7**, 85–89.

Hamilton, W. D. (1964). The genetical evolution of social behavior. *J. Theor. Biol.*, **7**, 1–16.

Hegdekar, B. M. and Dondale, C. D. (1969). A contact sex pheromone and some response parameters in lycosid spiders. *Can. J. Zool.*, **47**, 1–4.

Heifetz, Y., Boekhoff, I., Breer, H. and Appplebaum, S. W. (1997). Cuticular hydrocarbons control behavioural phase transition in *Schistocerca gregaria* nymphs and elicit biochemical responses in antennae. *Insect Biochem. Mol. Biol.*, **27**, 563–568.

Hodgson, E. S. and Roeder, K. D. (1956). Electrophysiological studies of arthropod chemoreception. 1. General properties of the labellar chemoreceptors of diptera. *J. Cell. Comp. Physiol.*, **48**, 51–75.

Hölldobler, B. and Michener C. D. (1980). Mechanisms of identification and discrimination in social Hymenoptera. In *Evolution of Social Behavior: Hypotheses and Empirical Tests*, ed. H. Markl. Weinheim: Verlag Chemie.

Howard, R. W. (1993). Cuticular hydrocarbons and chemical communication. In *Insect Lipids: Chemistry, Biochemistry and Biology*, ed. D. W. Stanley-Samuelson and D. R. Nelson. Lincoln: University of Nebraska Press, pp. 179–226.

Howard, R. W. and Blomquist, G. J. (1982). Chemical ecology and biochemistry of insect hydrocarbons. *Annu. Rev. Entomol.*, **27**, 149–172.

Howard, R. W. and Blomquist, G. J. (2005). Ecological, behavioral, and biochemical aspects of insect hydrocarbons. *Annu. Rev. Entomol.*, **50**, 371–393.

Howard, R. W., McDaniel, C. A. and Blomquist, G. J. (1982b). Chemical mimicry as an integrating mechanism for three termitophiles associated with *Reticulitermes virginicus. Psyche*, **89**, 157–168.

Howard, R. W., McDaniel, C. A., Nelson, D. R., Blomquist, G. J., Gelbaum, L. T. and Zalkow, L. H. (1982a). Cuticular hydrocarbons of *Reticulitermes virginicus* (Banks) and their role as potential species- and caste-recognition cues. *J. Chem. Ecol.*, **8**, 1227–1239.

Howard, R. W., Stanley-Samuelson, D. W. and Akre, R. D. (1990). Biosynthesis and chemical mimicry of cuticular hydrocarbons from the obligate predator, *Microdon albicomatus* Novak (Diptera: Syrphidae) and its ant prey, *Myrmica incompleta* Provancher (Hymenoptera: Formicidae). *J. Kansas Entomol. Soc.*, **63**, 437–443.

Howse, P. E. (1975). Chemical defense of ants, termites and other insects: some outstanding questions. In *Pheromones and Defensive Secretions in Social Insects*, ed. C. Noirot, P. E. Howse and G. LeMasne. Proceedings of the Symposium of the IUSSI, Dijon, pp. 23–40.

Howse, P. E. (1984). Alarm, defense and chemical ecology of social insects. In *Insect Communication*, ed. T. Lewis. Orlando: Academic Press, pp. 150–168.

Jackson, L. L. (1972). Cuticular lipids of insects IV. Hydrocarbons of the cockroaches *Periplaneta japonica* and *Periplaneta americana* compared to other cockroach hydrocarbons. *Comp. Biochem. Physiol.*, **41B**, 331–336.

Jackson, L. L. (1970). Cuticular lipids of insects – II. Hydrocarbons of the cockroaches *Periplaneta australasiae, Periplaneta brunea* and *Periplaneta fuliginosa. Lipids*, **5**, 38–41.

Jackson, L. L. and Baker, G. L. (1970). Cuticular lipids of insects. *Lipids,* **5**, 239–246.

Jackson, L. L and Blomquist, G. J. (1976a) Insect waxes. In *Chemistry and Biochemistry of Natural Waxes,* ed. P. E. Kolattukudy. Amsterdam: Elsevier, pp. 201–233.

Jackson, L. L. and Blomquist, G. J. (1976b). Cuticular lipids of insects: VIII – Alkanes of the mormon cricket *Anabrus simplex. Lipids*, **11**, 77–79.

Jain, S. C., Dussourd, D. E., Conner, W. E., Eisner, T., Guerrero, A. and Meinwald, J. (1983). Polyene pheromone components from an arctiid moth (*Utetheisa ornatrix*): Characterization and synthesis. *J. Org. Chem.*, **48**, 2266–2270.

Jones, R. L., Lewis, W. J., Beroza, M., Bierl, B. A. and Sparks, A. N. (1973). Host-seeking stimulants (kairomones) for the egg parasite *Trichogramma evanescents. Environ. Entomol.*, **2**, 593–596.

Jones, R. L., Lewis, W. J., Bowman, M. C., Beroza, M. and Bierl, B. A. (1971). Host-seeking stimulant for parasite of corn earworm: isolation, identification and synthesis. *Science*, **173**, 842–843.

Jurenka, R. A., Subchev, M., Abad, J.-L., Choi, M.-J. and Fabrias, G. (2003). Sex pheromone biosynthetic pathway for disparlure in the gypsy moth, *Lymantria dispar*. *Proc. Natl. Acad. Sci. USA*, **100**, 809–814.

Keller, L. and Nonacs, P. (1993). The role of queen pheromones in social insects – Queen control or queen Signal. *Animal Behav.*, **45**, 787–794.

Krafft, B. (1982). The significance and complexity of communication in spiders. In *Spider Communication: Mechanisms and Ecological Significance*, ed. P. N. Witt and J. S. Rovner. Princeton University Press., pp. 15–66.

Lacy, R. C. and Sherman, P. W. (1983). Kin recognition by phenotype matching. *Am. Nat.*, **121**, 489–512.

Lamb, N. J. and Monroe, R. E. (1968). Lipid synthesis from acetate-14C by the cereal leaf beetle, *Oulema malanopus*. *Ann. Ent. Soc. Am.*, **61**, 1164–1165.

Le Conte, Y. and Hefetz, A. (2008). Primer pheromones in social hymenoptera. *Annu. Rev. Entomol.*, **53**, 523–542.

Le Moli F., Mori, A. and Marpigiani, S. (1983). The effect of antennalectomy on attack behaviour of *Formica lugubris* Zett. (Hymenoptera: Formicidae). *Bollett. Zool.*, **50**, 201–206.

Leonard, J. E. and Ehrman, L. (1976). Recognition and sexual selection in *Drosophila*: classification, quantification, and identification. *Science*, **193**, 693–695.

Leonard, J. E., Ehrman, L., and Schorsch, M. (1974). Bioassay of a *Drosophila* pheromone influencing sexual selection. *Nature*, **250**, 261–262.

Liebig, J., Peeters, C. and Hölldobler, B. (1999). Worker policing limits the number of reproductives in a ponerine ant. *Proc. R. Soc. Lond. B*, **266**, 1865–1870.

Liebig, J., Monnin, T. and Turillazzi, S. (2005). Direct assessment of queen quality and lack of worker suppression in a paper wasp. *Proc. R. Soc. Lond. B*, **272**, 1339–1344.

Lihoreau, M. and Rivault, C. (2009). Kin recognition via cuticular hydrocarbons shapes cockroach social life. *Behav. Ecol.*, **20**, 46–53.

Lihoreau, M., Zimmer, C. and Rivault, C. (2007). Kin recognition and incest avoidance in a group-living insect. *Behav. Ecol.*, **18**, 880–887.

Lizé, A., Carval, D., Cortesero, A.-M., Fournet, S. and Poinsot, D. (2006). Kin discrimination and altruism in the larvae of a solitary insect. *Proc. R. Soc. Lond. B Biol. Sci.*, **273**, 2381–2386.

Locke, M. (1965). Permeability of insect cuticle to water and lipids. *Science*, **147**, 295–298.

Lockey, K. H. (1980). Insect cuticular hydrocarbons. *Comp. Biochem. Physiol.*, **B65**, 457–462.

Lockey, K. H. (1988). Lipids of the insect cuticle: origin, composition and function. *Comp. Biochem. Physiol.*, **89B**, 595–645.

Lockey, K. H. (1991). Insect hydrocarbons classes: implication for chemotaxonomy. *Insect Biochem.*, **21**, 91–97.

Major M. A. and Blomquist, G. J. (1978). Biosynthesis of hydrocarbons in insects: Decarboxylation of long chain acids to *n*-alkanes in *Periplaneta*. *Lipids*, **13**, 323–328.

Martin, M. M. and MacConnell, J. G. (1970). The alkanes of the ant, *Atta colombica*. *Tetrahedron*, **26**, 307–319.

Monnin, T. (2006). Chemical recognition of reproductive status in social insects. *Ann. Zool. Fennici*, **43**, 515–530.

Monnin, T. and Peeters, C. (1997). Cannibalism of subordinates' eggs in the monogynous queenless ant *Dinoponera quadriceps*. *Naturwissenschaften*, **84**, 499–502.

Monnin T., Malosse C., and Peeters C. (1998). Solid-phase microextraction and cuticular hydrocarbon differences related to reproductive activity in the queenless ant *Dinoponera quadriceps*. *J. Chem. Ecol.*, **24**, 473–490.

Morel, L., Vander Meer, R. K. and Lavine, B. K. (1988). Ontogeny of nestmate recognition cues in the red carpenter ant (*Camponotus floridanus*): Behavioral and chemical evidence for the role of age and social experience. *Behav. Ecol. Sociobiol.*, **22**, 175–183.

Nelson, D. R. (1969). Hydrocarbon synthesis in the American cockroach. *Nature*, **221**, 854–855.

Nelson, D. R. and Blomquist, G. J. (1995). Insect waxes. In *Waxes: Chemistry, Molecular Biology and Functions*, ed. R. J. Hamilton. Dundee: Oily Press, pp. 1–90.

Nelson, D. R. and Sukkestad, D. R. (1970). Normal and branched aliphatic hydrocarbons from the eggs of the tobacco hornworm. *Biochem.*, **9**, 4601–4611.

Ozaki, M., Wada-Katsumata, A., Fujikawa, K., Iwasaki, M., Yokohari, F., Satoji, Y., Nisimura, T. and Yamaoka, R. (2005). Ant nestmate and non-nestmate discrimination by a chemosensory sensillum. *Science*, **309**, 311–314.

Passera, L. and Aron, S. (2005). *Les fourmis: comportement, organisation sociale et évolution*. Ottawa: CNRC.

Pfennig, D. W., Gamboa, G. J., Reeve, H. K., Shellman-Reeve, J. S. and Ferguson I. D. (1983). The mechanism of nestmate discrimination in social wasps (*Polistes*, Hymenoptera: Vespidae). *Behav. Ecol. Sociobiol.*, **13**, 299–305.

Pomonis J. G., Fatland C. L., Nelson D. R. and Zaylskie, R. G. (1978). Insect hydrocarbons. Corroboration of structure by synthesis and mass spectrometry of mono- and dimethylallkanes. *J. Chem. Ecol.*, **4**, 27–39.

Pomonis, J. G., Nelson, D. R. and Fatland, C. L. (1980). Insect Hydrocarbons 2. Mass spectra of dimethylalkanes and the effect of the number of methylene units between methyl groups on fragmentation. *J. Chem Ecol.*, **6**, 965–972.

Prokopy, R. J., Averill, A. L., Bardinelli, C. M., Bowdan, E. S., Cooley, S. S., Crnjar, R. M., Dundulsis, E. A., Roitberg, C. A., Spatcher, P. J., Tumlinson, J. H. and Weeks, B. L. (1982). Site of production on of an oviposition-deterring pheromone component in *Ragoletis pomonella* flies. *J. Insect. Physiol.*, **28**, 1–7.

Ramsay, J. A. (1935). The evaporation of water from the cockroach. *J. Exp. Biol.*, **12**, 373–383.

Randall, D. J., Burggren, W. and French, K. (1997). *Eckert Animal Physiology: Mechanisms and Adaptations*, 4th edn. New York: Freeman.

Reed, J.R., Quilici, D. R., Blomquist, G. J. and Reitz, R. C. (1995). Proposed mechanism for the cytochrome-P450 catalyzed conversion of aldehydes to hydrocarbons in the housefly, *Musca domestica*. *Biochemistry*, **34**, 26221–26227.

Reed, J. R., Vanderwel, D., Choi, S., Pomonis, J. G., Reitz, R. C. and Blomquist, G. J. (1994). Unusual mechanism of hydrocarbon formation in the housefly: Cytochrome P450 converts aldehyde to the sex pheromone component (Z)-9-tricosene and CO_2. *Proc. Natl. Acad. Sci. USA*, **91**, 10000–10004.

Rence, B. and Loher, W. (1977). Contact chemoreceptive sex recognition in the male cricket, *Teleogryllus commodus*. *Physiological Entomol.*, **2**, 225 – 236.

Richards, A. G. (1951). *The Integument of Arthropods*. Minneapolis: University of Minnesota Press, p. 285.

Robinson, G. E. (1992). Regulation of division of labor in insect societies. *Annu. Rev. Entomol.*, **37**, 637–665.

Roelofs, W., Hill, A. S., Linn, C. E., Meinwald, J., Jain, S. C., Hebert, H. J. and Smith, R. F. (1982). Sex pheromone of the winter moth, a geometrid with unusually low temperature precopulatory responses. *Science*, **217**, 657–659.

Rogoff, W. M., Beltz, A. D., Johnson, J. O. and Plapp, F. W. (1964). A sex pheromone in the housefly, *Musca domestica* L. *J. Insect Physiol.*, **10**, 239–246.

Saïd, I., Gaertner, C., Renou, M. and Rivault, C. (2005). Perception of cuticular hydrocarbons by the olfactory organs in *Periplaneta americana* (L.)(Insecta, Disctyoptera). *J. Insect Physiol.*, **51**, 1384–1389.

Schal, C., Fan, Y. and Blomquist, G. J. (2003). Regulation of pheromone biosynthesis, transport, and emission in cockroaches. In *Insect Pheromone Biochemistry and Molecular Biology – The Biosynthesis and Detection of Pheromones and Plant Volatiles*, ed. G. J. Blomquist and R. G. Vogt. London: Elsevier, pp. 283–322.

Schal, C., Sevala, V. and Carde, R. T. (1998). Novel and highly specific transport of volatile sex pheromone by hemolymph lipophorin in moths. *Naturwissenschaften*, **85**, 339–342.

Scott, D. (1986). Sexual mimicry regulates the attractiveness of mated *Drosophila melanogaster* females. *Proc. Natl. Acad. Sci. USA*, **83**, 8429–8433.

Shellman, J. S. and Gamboa, G. J. (1982). Nestmate discrimination in social wasps. The role of exposure to nest and nestmates (*Polistes fuscatus*, Hymenoptera: Vespidae). *Behav. Ecol. Sociobiol.*, **11**, 51–53.

Simmons, L. W. (1989). Kin recognition and its influence on mating preferences of the field cricket *Gryllus bimaculatus* (DeGeer). *Anim. Behav.*, **38**, 68–77.

Stadler, E. (1984). Contact chemoreception. In *Chemical Ecology of Insects*, ed. W. J. Bell and R. T. Cardé. London: Chapman and Hall, pp. 3–35.

Stocker, R. F. and Rodrigues, V. (1999). Olfactory neurogenetics. In *Insect olfaction*, ed. B. S. Hansson. Berlin: Springer, pp. 283–314.

Tietjen, W. J. (1979). Tests for olfactory communication in four species of wolf spiders (Araneae, Lycosidae). *J. Arachnol.*, **6**, 197–206.

Vander Meer, R. K., Jouvenaz, D. P. and Wojcik, D. P. (1989). Chemical mimicry in a parasitoid (Hymenoptera: Eucharitidae) of fire ants (Hymenoptera: Formicidae). *J. Chem. Ecol.*, **15**, 2247–2261.

Vander Meer, R. K. and Wojcik, D. P. (1982). Chemical mimicry in the myrmecophilous beetle *Myrmecaphodius excavaticollis*. *Science*, **218**, 806–808.

Venard, R. and Jallon, J.-M. (1980). Evidence for an aphrodisiac pheromone of female *Drosophila*. *Experientia*, **36**, 211–213.

Vroman, H. E., Kaplanis, J. N. and Robbins, W. E. (1965). Effect of allatectomy on lipid biosynthesis and turnover in the female American cockroach, *Periplaneta americana* (L). *J. Insect Physiol.*, **11**, 897–903.

Wigglesworth, V. B. (1933). The physiology of the cuticle and of ecdysis in *Rhodnius prolixus* (Triatomidae, Hemiptera); with special reference to the function of the oenocytes and of the dermal glands. *Quart. J. Micr. Sci.*, **76**, 269–318.

Wigglesworth, V. B. (1945). Transpiration through the epicuticle of insects. *J. Exp. Biol.*, **21**, 97–114.

Wilson, E. O. (1963). The social biology of ants. *Annu. Rev. Entomol.*, **8**, 345–368.

Wilson E. O. (1971). *The Insect Societies*. Cambridge, MA: Harvard University Press.

Yamaoka R. (1990). Chemical approach to understanding interactions among organisms. *Physiol. Ecol. Jpn.*, **27**, 31–52.

2

Structure and analysis of insect hydrocarbons

Gary J. Blomquist

Hydrocarbons have evolved to play a plethora of roles in insects, primarily serving as a waterproofing cuticular layer and functioning extensively in chemical communication. This chapter will concentrate primarily on the structure and analysis of long-chain hydrocarbons, components with more than 20 carbons and often containing one or more methyl-branches or one or more double bonds. Until very recently, the chain length of cuticular hydrocarbons was thought to vary from a lower limit of about 21 carbons (shorter chain compounds are volatile) up to a maximum of about 50 carbons (Nelson and Blomquist, 1995). However, the apparent upper limit of the chain length of insect hydrocarbons appears to be a limitation based on analytical technique. Surprisingly, insect hydrocarbons with chain lengths above 60 carbans have been reported in a number of species using Matrix-assisted laser desorption/ionization–times of flight (MALDI–TOF) mass spectrometry (Cvačka *et al.*, 2006). A unique feature of insect hydrocarbons, compared to plant surface hydrocarbons, is the number and variety of positions in which methyl branches and double bonds occur. It seems reasonable to speculate that the variety of methyl-branch positions and number and positions of double bonds have evolved to increase the informational content of hydrocarbon mixtures, while still retaining their waterproofing capabilities. Much of this book will concentrate on the informational content of hydrocarbons in insects, and indeed, it is the tremendous increase in our understanding of the numerous roles in chemical communication that cuticular hydrocarbons play which has resulted in the rapid growth of this field. Cuticular hydrocarbons provide insects with the chemical equivalent of the visually variable colored plumage of birds.

Types of hydrocarbons

The hundreds of different cuticular hydrocarbon components reported on insects can be divided into three major classes, *n*-alkanes, methyl-branched components and unsaturated hydrocarbons. There are reports of methyl-branched alkenes (see below), but these are rare. The hydrocarbon components on the surface of insects are usually complex mixtures comprised of anywhere from a few to up to hundreds of different components in some species.

n-Alkanes

The *n*-alkanes usually range in chain length from 21 to 31 or 33 carbons. Hydrocarbons with fewer than 20 carbons commonly occur as pheromones, defensive compounds and intermediates to pheromones and defensive compounds, but their volatility makes them unsuited to function as cuticular components. *n*-Alkanes have been found on almost every insect species analyzed, and can range from less than one percent of the total hydrocarbons, as in tsetse flies (Nelson and Carlson, 1986; Nelson *et al.*, 1988) to almost all of the hydrocarbon fraction, as in the adult tenebrionid beetle, *Eurychora sp.* (Lockey, 1985). Depending upon the species, they can consist of essentially only one major component, such as *n*-pentacosane in the American cockroach, *Periplaneta americana* (Jackson, 1972) to a series of *n*-alkanes, such as the series from C_{23} to C_{33} in the housefly, *Musca domestica* (Nelson *et al.*, 1981), with trace amounts to C_{37} (Mpuru *et al.*, 2001). In all cases, the odd-numbered alkanes predominate, due to their formation from mostly two carbon units followed by a decarboxylation (Blomquist, Chapter 3, this book). Small amounts of even-numbered carbon chain *n*-alkanes often occur, and presumably arise from chain initiation with a propionyl-CoA rather than an acetyl-CoA. Occasionally, gas chromatographic analyses reveal similar amounts of even-numbered chain *n*-alkanes and odd-numbered chain components. This is a red flag that the samples must be checked for contamination.

Unsaturated hydrocarbons

Olefins are found in many insects (Blomquist and Dillwith, 1985; Nelson and Blomquist, 1995; Blomquist *et al.*, 1987; Howard and Blomquist, 2005), and are sometimes not completely characterized owing to the difficulties in determining the positions and stereochemistry of double bonds in the alkadienes and alkatrienes, especially when they are minor components. In many cases they are important in chemical communication. The first cuticular alkene shown to be a pheromone was (*Z*)-9-tricosene (Carlson *et al.*, 1971), which increased the number of mating strikes attempted by males on female houseflies, *Musca domestica,* or on treated dummies. The role of (*Z*)-9-tricosene as a pheromone is more pronounced in houseflies raised in a laboratory for many generations, and its importance in wild populations is open to question (Darbro *et al.*, 2005). A number of other Diptera use alkenes and alkadienes as pheromones, including (*Z,Z*)-7,11-heptacosadiene in *Drosophila melanogaster* (Blomquist *et al.*, 1987; Jallon and Wicker-Thomas, 2003). The ratio of different (*Z*)-9-alkenes can serve as a cue to recognize nestmates in *Formica exsecta* (Martin *et al.,* 2008). The positions of the double bonds can vary considerably, with the most common alkenes having the double bond in the 9-position. However, monoenes from different insects have been described with double bonds in almost every position. Likewise, the dienes and trienes have double bonds in a bewildering array of positions, including 6,9- *(Periplaneta americana)* (Jackson, 1972) and 7, 17-; 7, 19-; 7, 21-; 7, 23-; 9, 21-; 9, 23-; and 9, 25- as major components of the hydrocarbons of rice, maize and granary weevils (Baker *et al.*, 1984; Nelson *et al.*, 1984).

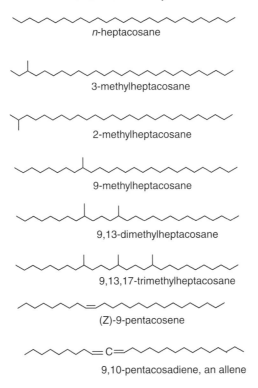

n-heptacosane

3-methylheptacosane

2-methylheptacosane

9-methylheptacosane

9,13-dimethylheptacosane

9,13,17-trimethylheptacosane

(Z)-9-pentacosene

9,10-pentacosadiene, an allene

Figure 2.1 Structures of major hydrocarbons present in insect cuticular lipids.

In most cases the double bonds are not conjugated. One exception is the conjugated diene (*Z,Z*)-7,9-pentacosadiene identified in the termite *Reticulitermes flavipes* (Howard *et al.*, 1978). In every case where determined, the double bonds are in the (*Z*) configuration (Blomquist *et al.*, 1987).

Fletcher *et al.* (2001, 2008) and McGrath *et al.* (2003) reported a novel group of allenic hydrocarbons from five Australian beetles, dienes in which the double bonds are on adjacent carbons, from 23 to 27 carbons. The major component is 9,10-pentacosadiene (Figure 2.1). Mass spectra of the methoxy derivatives support this structure. To my knowledge, these are the only reports of allenic hydrocarbons in insects. One of the components, 9,10-pentacosadiene, caused male *Antitrogus consanguineus* to grip microcentrifuge tubes coated with synthetic C_{25} allene in an apparent attempt to mate with the tube (Fletcher *et al.*, 2008), indicating that this component functions as a sex pheromone in this species.

While the vast majority of alkenes reported on the surface of insects are straight-chain molecules, there have been a few reports of methyl-branched alkenes. Warthen and Uebel (1980) found (*Z*)-2-methyl-24-hexatriacontene in the hydrocarbons of the house cricket, *Acheta domesticus*. Carlson and Schlein (1991) reported 19,23-dimethyltriacont-1-ene and other homologs in tsetse flies. Howard *et al.* (1990) found a homologous series of x-methyl-(Z)-4-enes and x,y-dimethyl-(Z)-4-enes in carpenter ants and an obligatory syrphid fly predatory inquiline of this ant. A series of mono- and dimethylalkenes were reported by Brown *et al.* (1990) from a primitive Australian ant. In this case, the mono- and dimethylalkenes were major components of the cuticular hydrocarbon. Bartlet has isolated a series of methyl- and ethyl-branched alkenes (see Chapter 19) that are not cuticular hydrocarbons but serve as aggregation pheromones in the dried-fruit beetles in the family Nitidulidae.

Methyl-branched alkanes

Commonly reported terminal alkanes have methyl groups on the 2-, 3-, 4- and 5- positions. The 2-methylalkanes can have either odd or even numbers of carbons in the alkyl chain, depending upon whether chain initiation occurred with the carbon skeleton of isoleucine or valine. The 3-methylalkanes, however, usually have odd-numbered carbon chains predominating. Less often, 4-methylalkanes are reported, but the mass spectra of 2-methylalkanes are very similar to those of 4-methylalkanes (discussed in more detail under analytical techniques), making the unambiguous determination of 2- and 4-methylalkanes difficult. Based on our understanding of the biosynthesis of methyl-branched alkanes, 4-methylalkanes would arise from the initiation of chain synthesis with a propionyl-CoA instead of an acetyl-CoA, followed by the insertion of a propionyl group (as a methylmalonyl-CoA unit), with the final step being the loss of a carbon dioxide molecule (Blomquist, Chapter 3; Dillwith *et al.*, 1982). Thus, 4-methylalkanes would have mostly even numbers of carbons in the alkyl chain.

Internally branched monomethylalkanes most often contain an odd number of carbons in the alkyl chain with the methyl-branch on an odd-numbered carbon, again due to their biosynthetic origin, in which a methylmalonyl-CoA unit replaces a malonyl-CoA unit during chain elongation (Blomquist, Chapter 3). Minor components can have an even number of carbons in the alkyl chain, and then have the methyl group on either odd- or even-numbered carbons. Internally branched monomethylalkanes are common and can be abundant components of insect cuticular lipids. They sometimes exist as a single isomer, whereas in other cases a single GC peak will contain a series of methylalkanes. The 2-methyl-, 3-methyl-, 4-methyl- and 5-methylalkanes will separate on a capillary GC column (2- and 4-methylalkanes have very similar retention times), whereas compounds with methyl-branches from position 7 to the middle of the chain co-elute as a single peak.

Dimethylalkanes have been identified in numerous insect species, and the major components usually have the methyl groups on odd-numbered carbons. A common motif is noted with the methyl groups separated by three methylene groups, but the number of methylene groups between methyl branches can be 5, 7, 9, 11, or 13 (Blomquist *et al.*, 1987; Lockey, 1985). Cuticular hydrocarbons with adjacent methyl groups have been reported, but no cases have been unambiguously confirmed (Blomquist *et al.*, 1987). Likewise, methyl-branched alkanes with one methylene group between the methyl branches are rare.

Another common type of dimethylalkane is the 2,X-, 3,X-, 4,X-, and 5-X dimethylalkane, where X is the second methyl-branch position and is separated from the first methyl-branch by an odd number of carbons. These dimethylalkanes usually occur as mixtures of isomers with 3, 5, 7, 9 or 11 methylenes between the methyl branches. In the Colorado potato beetle, *Leptinotarsa decemlineata*, forty percent of the egg hydrocarbons were comprised of 2,X-dimethylalkanes, with the most abundant being 2,10- and 2,6-dimethyloctacosane (Nelson *et al.*, 2003).

Tri- and tetramethylalkanes are present on some insects, with the first trimethylalkane reported by Martin and MacConnnell (1970) in the ant *Atta colombica*. Two types of pattern were present: 3, 7, 11-trimethylalkanes and 4, 8, 12-trimethylalkanes, all with carbon chain lengths of 34 to 39 carbons. Since then, tri- and tetramethylalkanes have been reported with methyl branches in a variety of positions, with either all of the methyl branches on odd-numbered carbons (most common) or all on even-numbered carbons. Abundant trimethylalkanes of *L. decemlineata* were 2,10,16 and 2,10,18-trimethyloctacosanes (Nelson *et al.*, 2003). Recently, multi-methylalkanes with a single methylene group between methyl branches have been reported as the major hydrocarbons of *Antitrogus parvulus*, the two major components being 4,6,8,10,16-pentamethyldocosane and 4,6,8,10,16,18-hexamethyldocosane. The structures were identified by comparison with synthetic standards (Fletcher *et al.*, 2008).

The confidence in our understanding of insect hydrocarbons has been shaken by the recent report of hydrocarbons on insect cuticles with up to 70 carbons (Cvačka *et al.*, 2006). This group used MALDI mass spectrometry to describe a series of hydrocarbons of much longer chain lengths than the hydrocarbon previously characterized by GC–MS, a technique which currently limits chain lengths to somewhere between 40 and 50 carbons. In order to more efficiently extract the very-long chain hydrocarbons, Cvačka *et al.* (2006) used chloroform rather than the usual hexane or pentane to extract cuticular lipids and then separated the hydrocarbons by TLC on silica. The MALDI–TOF identification of the hydrocarbons from the much studied American cockroach, *Periplaneta americana*, showed ions consistent with the major previously reported components *n*-pentacosane, 3-methylpentacosane and (Z,Z)-6,9-heptacosadiene (Jackson, 1972), along with other previously reported components with chain lengths up to 43 carbons. This demonstrates an agreement of the MALDI–TOF analyses with previous GC–MS analyses. *P. americana* did not have components longer than 43 carbons, but the agreement between GC–MS and MALDI–TOF data validates the MALDI–TOF approach. Many of the very-very-long chain hydrocarbons appear to be highly unsaturated or contain cyclic components. MALDI mass spectral analysis has the potential to be an alternative, or more likely, a complementary analytical technique for insect hydrocarbon analysis. More work is needed to elucidate the complete structures of the high-molecular-weight hydrocarbons and to more fully understand the usefulness of this technique in insect hydrocarbon analysis. The apparent presence of highly unsaturated or cyclic components and the current inability to handle mixtures and isomers underscore the need for further work on the structure and function of these high-molecular-weight compounds.

Analytical techniques

The extraction of cuticular hydrocarbons usually involves soaking or rinsing the insects in a non-polar solvent for specified time periods. Commonly used extraction procedures involve two or three consecutive 1–10 min extractions with hexane or pentane or three shorter extractions with chloroform. The larger the volume of solvent, the shorter the extraction time can be. An indication of the completeness of the extraction can be quickly obtained by analyzing each fraction by TLC. The presence of acylglycerols or phospholipids in the extract indicates that some internal lipid was extracted. Hydrocarbons can be easily separated from other cuticular components by silica gel TLC developed in hexane or another non-polar solvent, or by column chromatography eluting the hydrocarbons with hexane. A particularly useful and simple technique is to use Pasteur pipettes packed with several centimeters of activated silica gel (70–230 mesh) and then to elute the hydrocarbons with several volumes of hexane or pentane. Capillary GC–mass spectrometry using non-polar columns is often used to analyze cuticular hydrocarbons. GC with column temperatures of up to at least 300°C to 320°C and holding the temperature at the upper limit for a period of time is usually performed. The recent analysis of insect hydrocarbons with a MALDI–TOF instrument has indicated that hydrocarbons of up to 70 carbons may be present on some insects, and this indicates that improved techniques for GC–MS analysis are needed; to date, these techniques have been limited to components of below about 50 carbons. Unsaturated hydrocarbons can be easily separated from alkanes by either silver nitrate TLC or silver nitrate column chromatography with silica gel impregnated with 20 percent silver nitrate (Blomquist *et al.*, 1984). Likewise, *n*-alkanes can be readily removed from a mixture of straight chain and methyl-branched alkanes by the use of a 5-angstrom molecular sieve in an isooctane solvent. It is important to completely dry the molecular sieve prior to use, and we find that placing the molecular sieve in a GC oven for several days works well.

Alternatively, SPME can be used to obtain cuticular hydrocarbons, and there is evidence that hydrocarbons may be layered on the surface, as SPME appears to extract hydrocarbons with different percentage compositions than solvents (Ginzel *et al.*, 2003; Ginzel, Chapter 17, this book). More work is needed to determine if this is a general phenomenon. Also, the direct insertion of a piece of cuticle into a gas chromatograph using a pyroprobe has been shown to work with a single insect or a part of an insect (Brill and Bertsch, 1985). Similarly, pieces of insect cuticle or other selected parts of individual insects can be sealed in a glass capillary and introduced into the heated injector area of a gas chromatograph prior to crushing the glass capillary (Morgan and Wadhams, 1972; Attygalle *et al.*, 1985; Bagnères and Morgan, 1990, 1991). Morgan and co-workers have applied this technique to a wide variety of insects and to various parts of an insect and have had good success with it. However, neither of these techniques has been widely adopted, perhaps due to a lack of commercial sources for these injection devices.

Characterization of hydrocarbons

The characteristics of mass spectral fragmentation patterns for the interpretation of methylalkane mass spectra were first put forward by McCarthy *et al.* (1968). The effect of the position of the methyl branch on the gas chromatographic elution of long-chain monomethylalkanes was determined by Mold *et al.* (1966). Nelson and Sukkestad (1970) expanded on these approaches to methylalkane identification and used them to identify isomeric mixtures of long-chain mono-, di- and trimethylalkanes in the tobacco hornworm, *Manduca sexta*. Blomquist *et al.* (1987) suggested that three factors should be considered when proposing a structure. The diagnostic mass spectral even- and odd-mass ions allow a structure to be proposed. This structure must conform to values of the Kovats indices (KI) or the equivalent chain length (ECL) (Lockey, 1985; Carlson *et al.*, 1998; Katritzky and Chen, 2000; Zarei and Atabati, 2005). Finally, it is sometimes useful to consider the biosynthetic feasibility of the compound when proposing structures for cuticular hydrocarbons.

Mass spectral fragmentation patterns

The rationale used in the interpretation of the mass spectra of methylalkanes has been presented in several reports: 2- vs. 4-methylalkanes (Baker *et al.*, 1978; Scammells and Hickmott, 1976; McDaniel, 1990; Bonavita-Cougourdan *et al.*, 1991); 2,*X*- and 3,*X*-dimethylalkanes (Nelson *et al.*, 1980; Thompson *et al.*, 1981); and internally branched mono-, di- and trimethylalkanes (Blomquist *et al.*, 1987; Pomonis *et al.*, 1980). In the majority of reports, identification is based on GC and MS data, but the conclusions are not confirmed with standards or synthesis of the proposed structures. However, there are reports of chemical ionization (Howard *et al.*, 1980) and electron impact of synthetic methyl-branched hydrocarbons (Carlson *et al.*, 1978, 1984; Pomonis *et al.*, 1978, 1980) and these have been very useful in confirming mass spectral fragmentation patterns with chemical structures.

If a component is present as a single isomer in a GC peak, identification is usually straightforward. However, often a GC peak will contain isomers that do not completely separate, resulting in mass spectra which contain more than one component. A partial solution to this problem is to look at the mass spectra of a homologous series; one homolog will often contain a single isomer or the preponderance of a single isomer and this makes interpretation possible. Then, it is often possible to examine the spectra of other homologs in the series to determine the isomers present.

Often, the molecular ion of long-chain methylalkanes is not present in GC EI-MS, but is a prominent ion by GC CI-MS (Howard *et al.*, 1980). In GC EI-MS, the M−15 ion is the most intense ion in the high mass region of the mass spectrum, except in the case of the 2-, 3- and 4-methylalkanes, the 2,*X*-, 3,*X*-, 4,*X*-, and 5,*X*-dimethylalkanes and the 3,*X*,*Y*- and 4,*X*,*Y*-trimethylalkanes. For internally branched methylalkanes, M−43, M−29 and M−15 form a characteristic series of ions of increasing intensity, and because M+ is often not apparent, these ions indicate where the molecular ion should be. When possible, it is preferable to obtain the M+ ion by GC CI-MS, since chemical ionization usually gives a much stronger M+ ion than that given by EI-MS (Howard *et al.*, 1980). However, appropriate instrumentation is not always readily available.

For internally branched methylalkanes, α-cleavage at the branch point causes formation of two secondary carbonium ions, one an odd-numbered mass ion $[C_nH_{2n+1}]^+$ and the

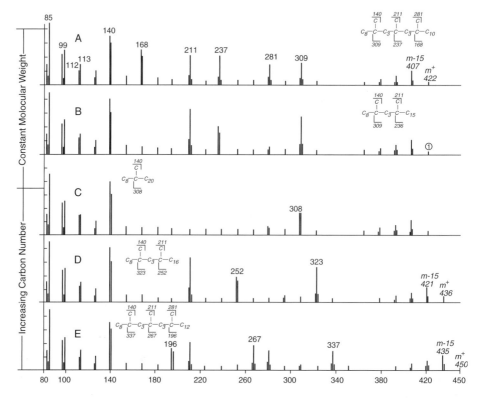

Figure 2.2 Idealized examples of electron impact mass spectra of mono-, di- and trimethylalkanes. (A) 9,13,17-trimethylheptacosane, (B) 9,13-dimethyloctacosane, (C) 9-methylnonacosane, (D) 9,13-dimethylnonacosane, (E) 9,13,17-trimethylnonacosane (from Blomquist *et al.*, 1987).

other an even-numbered ion $[C_nH_{2n}]^+$ one atomic mass unit (amu) less. For example, 9-methylnonacosane will undergo α-cleavage internal to the branch point to form a pair of ions at *m/z* 140:141. A second pair of ions at *m/z* 308:309 is caused by α-cleavage on the external side of the branch point. The preferred α-cleavage is that which results in the largest radical, i.e. the 20-carbon fragment (Figure 2.2C).

The ratio of the even- to odd-mass ion can aid in interpretation of spectra, and a series of idealized mass spectra of mono-, di- and trimethylalkanes is presented in Figure 2.2 (from Blomquist *et al.*, 1987). Depending upon the size of the molecule and the size of the secondary carbonium ion, the even-mass ion of the pair is frequently larger than the odd-mass ion (Blomquist *et al.*, 1987; Nelson and Sukkestad, 1970; Pomonis *et al.*, 1978). The formation of the primary (straight-chain) carbonium ion is much less favored than the secondary carbonium ion (which contains the methyl branch of the original molecule).

Hydrocarbons, whether straight-chain or methyl-branched, fragment to form clusters of ions every 14 amu, with the most intense ion in each cluster being of odd mass and corresponding to carbon–carbon bond cleavage. As the molecules get larger, or if other methyl groups are present, the number of bonds that are cleaved to give clusters of ions every 14 amu increases. This causes the relative intensities of the even-mass and odd-mass ions, in those ion pairs formed by cleavage at the methyl branch points, to shift in favor of the odd-mass ion. The amount of the shift increases as the size of the molecule increases and as the mass of the secondary carbonium ion decreases (Figure 2.2). Thus, the even-mass ion of a secondary carbonium ion is not always larger than the odd-mass ion, but is always more intense than that expected from carbon–carbon bond cleavage in straight-chain portions of the molecule.

A significant consideration in the interpretation of mass spectra is the fact that other methyl branches in the secondary carbonium ion suppress the formation of the even-mass ion. The 2,X-dimethylalkanes are an exception, where the 2-methyl group does not suppress the formation of the even-mass ion (Nelson *et al.*, 2002). The dimethylalkane 9,13-dimethyloctacosane (Figure 2.2B) preferentially cleaves at the branch points. However, external α-cleavage at the first branch point forms a secondary carbonium ion at *m/z* 309 containing both of the methyl branches. The presence of the second methyl branch suppresses the loss of hydrogen so that the even-mass ion at *m/z* 308 is not of significant intensity, in contrast to the case for an unbranched secondary carbonium ion, which has significant ions at both 308 and 309 formed by external α-cleavage of 9-methylnonacosane (Figure 2.2C).

Secondary carbonium ions containing two, three or more methyl branches all suppress the formation of the even-mass ions. Thus, in the spectra of single isomers of internally branched methylalkanes, the number of methyl branches can be determined simply by counting the number of diagnostically significant even- and odd-mass ion clusters. A monomethylalkane has two ion clusters, each with a significant even-mass ion, i.e. at 140:141 and 308:309 (Figure 2.2C). A dimethylalkane (Figure 2.2D) has four clusters – two clusters with a significant even-mass ion (at *m/z* 140:141 and 252:253) and two clusters with only a significant odd-mass ion (at *m/z* 211 and 323).

Methylalkanes with more than two methyl branches will form two odd-mass ions for each additional methyl branch, but will always have two ion clusters with a significant even-mass ion (formed by internal α-cleavage at the two external methyl branch points). Thus, 9,13,17-trimethylnonacosane will form six diagnostically significant ion clusters (Figure 2.2E): two ion clusters with an even-mass ion (at 140:141 and 196:197) and four ion clusters in which the ion of major intensity is of odd mass (at *m/z* 211, 267, 281 and 337).

The structures deduced from the mass spectra must account for the observed equivalent chain length (ECL) or Kovats indices (KI) (Carlson *et al.*, 1998; Katritzky and Chen, 2000; Zarei and Atabati, 2005). The values for monomethylalkanes (Mold *et al.*, 1966; Szafranek *et al.*, 1982) can be used to estimate the expected ECL or KI for a di-, tri- or tetramethylalkane structure proposed from a mass spectrum. For example, a dimethylalkane, such as 3,11-dimethylnonacosane, with 31 carbons, would have its elution time decreased by about 0.3 carbons for the 3-methyl group and about 0.7 carbons for the 11-methyl group. Thus, the predicted ECL is approximately 30 and this is the ECL observed.

Distinguishing between 2- and 4-methylalkanes is a continuing problem. The isomers elute from the gas chromatograph at nearly identical times and are usually not resolved by GC-MS, although partial separation has been reported.

Both 2- and 4-methylalkanes have an ion of low intensity at M−15, owing to loss of a methyl group. The major ion of significant intensity in the high-mass region of the mass spectra for both isomers is due to the M−43 ion: a primary carbonium ion formed by loss of isopropyl from the 2-methylalkane and a secondary carbonium ion formed by the loss of n-propyl from the 4-methylalkane. In addition, the 4-methylalkane has a primary carbonium ion of low intensity at the M−71:M−72 due to cleavage internal to the methyl branch point. The presence of this primary ion pair establishes the structure of a 4-methyl-alkane. However, in mass spectra of low intensity or in which other isomers are present, the ions at M−71: M−72 may not be readily apparent.

Scammells and Hickmott (1976) used the ratio of the intensity of the ion $[C_nH_{2n+1}]^+$ divided by the intensity of the re-arrangement ion $[C_nH_{2n}]^+$, formed by the loss of isopropyl, ethyl and propyl radicals, to distinguish between 2-, 3-, and 4-methylalkanes, respectively. They found that the ratio for a 2-methylalkane was greater than that for either a 3- or 4-methylalkane. This approach was used by Baker and co-workers (1979) to show that the ratio for the 3-methylalkane was always greater than that for the isomer in question, which, therefore, had to be a 4-methylalkane. However, in practice, and with components that do not completely separate on GC, it may be difficult to get exact ratios because of the presence of other components. Other ions have been examined in order to more correctly identify 2- and 4-methylalkanes (Nelson *et al.*, 1980), but in my opinion, no completely satisfactory method exists.

Positions of double bonds

It is often of interest to determine the positions of the double bonds in unsaturated hydro-carbons, particularly since these components are involved in chemical communication in some species. The two most common techniques used to determine double-bond positions in long-chain hydrocarbons make use of the methoxy (Blomquist *et al.*, 1980) and dimethyl-disulfide derivatives (Francis and Veland, 1981; Carlson *et al.*, 1989; Howard, 1993). To make the methoxy derivative, a sodium borohydride reduction of the methoxymercura-tion products yields a methoxy derivative with the methoxy group on either carbon of the original double bond. GC–MS of the methoxy derivative yields intense fragments from cleavage on both sides of the methoxy group which are characteristic of each isomer. This technique works well for monoenes and has been successfully applied to dienes (Syvertson *et al.*, 1995). The yields of methoxy products from this reaction are variable, and it is most successful when the products are separated by TLC or column chromatography prior to GC–MS analysis.

The second, even more convenient, method to determine double-bond positions in unsaturated hydrocarbons is the use of dimethyldisulfide derivatives (Francis and Veland, 1981; Carlson *et al.*, 1989; Howard, 1993). In this approach, alkenes are dissolved in a carbon disulfide and iodine solution and kept overnight. The reaction yields a derivative with methyl sulfide substituents on each of the carbons that comprised the double bond. If the double bonds are separated by four or more methylene groups, the reaction proceeds as if the compounds were monoenes, with dienes resulting in the addition of four –SCH$_3$ groups. The derivative fragments between the dimethyldisulfide groups to yield intense fragment ions which readily show the double-bond positions. There can be some difficulty in the gas chromatography of large molecules with multiple double bonds, as the derivatives increase the size of the molecule by two dimethyldisulfide groups (about the equivalent of four carbons) for each double bond present in the parent molecule. When the double bonds are closer to one another (1–4 methylene units), the reaction products are not simple thiomethylethers, but rather are sulfur heterocycles resulting from internal nucleophilic displacement (Howard, 1993). The resulting heterocyles, however, give very distinctive mass spectra which are readily interpretable in terms of the original double-bond positions in the alkene (Vicenti *et al.*, 1987; Howard, 1993).

Near-infrared spectroscopy (NIRS) in hydrocarbon analysis

A technique that has only relatively recently been applied to intact insect hydrocarbons is near-infrared spectroscopy (NIRS), and it has been used to identify insect colonies and nests (Newey *et al.*, 2008a), species and subspecies (Aldrich *et al.*, 2007), intercolony aggression (Newey *et al.,* 2008b), age (Perez-Mendoza *et al.*, 2002) and sex (Dowell *et al.*, 2005). This technique involves generating absorption spectra in the near-infrared, from about 4000 cm^{-1} to 12 500 cm^{-1} (800 to 2500 nm), and with hydrocarbons, gives pattern recognition based on C–H bonds (Foley *et al.*, 1998). The technique has the advantages of being very rapid as it can scan the hydrocarbons of intact insects without the need for extraction. It does not give the chemical specificity of GC–MS but it is very useful in examining differences in cuticular hydrocarbons among insects.

Distribution of hydrocarbons on the cuticle

Early depictions of the cuticular wax on insect cuticle envisioned an innermost layer of lipids with polar headgroups interacting with the cuticle, with crystals of hydrocarbons overlying this innermost layer (Locke, 1965). It was thought that the degree of organization of the outermost lipid layer decreased the more removed it was from the organizing effect of the innermost layer (Locke, 1965). Locke postulated that part of the epicuticle was penetrated by pore canals containing lipid filaments and proposed that transport of lipid from the site of synthesis across the endocuticle and epicuticle could be carried out in the hydrophobic center of the pore canal. He also proposed that the lipid molecules to which the cuticle was permeable (insecticides, juvenile hormone) could find their way into the insect through the pore canals.

While this model has certainly changed, features of it remain. A comparison of the timing of the uptake of apolar pesticides with the timing of cuticular hydrocarbons crossing the surface of the insect to the outside shows that they coincide, suggesting that the same route is used by both, and may explain why apolar pesticides cross insect cuticle so much better than vertebrate skin (Theisen *et al.*, 1991). The original model had newly synthesized hydrocarbon crossing the cuticle directly from the epidermal related cells where it was synthesized. From studies on a number of insects, it is now clear that the newly synthesized hydrocarbon is taken up by lipophorin and transported in the hemolymph, but the mechanism of how hydrocarbon and other cuticular lipid components are transported to the surface of the insect is not known. The discovery of very-long-chain hydrocarbons of up to 60 to 70 carbons by MALDI mass spectrometry further complicates a model for cuticular lipids. The recognition that in many insects the cuticular hydrocarbons serve in chemical communication suggests that the informational molecules might be closer to the surface than components not used in chemical communication. Indeed, Ginzel *et al.* (2003) provided evidence that this might be the case. A solvent extract of the hydrocarbons shows that (Z)-9-pentacosene, a contact pheromone of the locust borer, *Megacyllene robiniae*, comprises 16 percent of the hexane extractable hydrocarbons, whereas in SPME sampling, which presumably contains molecules closer to the surface, it is 38% of the total hydrocarbon. The effect of methyl branches and unsaturation in controlling water loss is dealt with by Gibbs and Rajpuhorit in Chapter 6 of this book.

Unanswered questions

Our understanding of insect cuticular hydrocarbons has made giant strides forward in the last several decades. The new discoveries have answered a number of long-standing questions and raised some very interesting new ones. Foremost among them are these: What are the structures, how widespread, how abundant and what is the role of the putative very-long-chain hydrocarbons with chain lengths up to 60 and 70 carbons that were recently reported (Cvačka *et al.*, 2006)? How are they made? How are hydrocarbons transported from the lipophorin in the hemolymph to the outside of the cuticle? Are hydrocarbons that are involved in chemical communication selectively deposited on the outer surface of the lipid layer? And, if so, how? Are most of the major classes of hydrocarbons identified? Will more novel structures, such as the allenes reported by Fletcher *et al.* (2001, 2008) be found? We expect answers to these and other questions as we gain a better understanding of insect cuticular hydrocarbons.

References

Aldrich, B. T., Maghirang, E. B., Dowell, F. E. and Kambhampati, S. (2007). Identification of termite species and subspecies of the genus *Zootermopsis* using near-infrared reflectance spectroscopy. *J. Insect Science*, **7**, 18–25.

Attygalle, A. B., Billen, J. P. J. and Morgan, E. D. (1985). The postpharyngeal glands of workers of *Solenopsis geminata* (Hym. Formicidae). *Act. Coll. Ins. Soc.* **II**, 79–86.

Bagnères, A.-G. and Morgan, E. D. (1990). A simple method for analysis of insect cuticular hydrocarbons. *J. Chem. Ecol.*, **16**, 3263–3276.

Bagnères, A.-G and Morgan, E. D. (1991). The postpharyngeal glands and the cuticle of Formicidae contain the same characteristic hydrocarbons. *Experientia*, **47**, 106–111.

Baker, J. E., Sukkestad, D. R., Nelson, D. R. and Fatland, C. L. (1979). Cuticular lipids of larvae and adults of the cigarette beetle, *Lasioderma serricorne*. *Insect Biochem.*, **9**, 603–611.

Baker, J. E., Sukkestad, D. R., Woo, S. M. and Nelson, D. R. (1978). Cuticular hydrocarbons of *Tribolium castaneum*: Effect of the food additive tricalcium phosphate. *Insect Biochem.*, **8**, 159–167.

Baker, J. E., Woo, S. M., Nelson, D. R., and Fatland, C. L. (1984). Olefins as major components of epicuticular lipids of three *Sitophilus* weevils. *Comp. Biochem. Physiol.*, **77B**, 877–884.

Blomquist, G. J., Adams, T. S. and Dillwith, J. W. (1984). Induction of female sex pheromone production in male houseflies by ovary implants or 20-hydroxyecdysone. *J. Insect Physiol.*, **30**, 295–302.

Blomquist, G. J. and Dillwith, J. W. (1985). Cuticular lipids. In *Comprehensive Insect Physiology, Biochemistry and Pharmacology*, ed. G. A. Kerkut and L. I. Gilbert, Vol. 3. *Integument, Respiration and Circulation*. Oxford: Pergamon, pp. 117–154.

Blomquist, G. J., Howard, R. W., McDaniel, C. A., Remaley, S., Dwyer, L. A. and Nelson, D. R. (1980). Application of methoxymercuration-demercuration followed by mass spectrometry as a convenient microanalytical technique for double-bond location in insect-derived alkenes. *J. Chem. Ecol.*, **6**, 257–269.

Blomquist, G. J., Nelson, D. R. and de Renobales, M. (1987). Chemistry, biochemistry, and physiology of insect cuticular lipids. *Arch. Insect Biochem. Physiol.*, **6**, 227–265.

Bonavita-Cougourdan, A., Theraulaz, G., Bagnères, A. G., Roux, M., Pratte, M., Provost, E. and Clèment, J.-L. (1991). Cuticular hydrocarbons, social organization and ovarian development in a polistine wasp: *Polistes dominulus* Christ. *Comp. Biochem. Physiol.*, **100B**, 667–680.

Brill, J. H. and Bertsch, W. (1985). A novel micro-technique for the analysis of the cuticular hydrocarbons of insects. *Insect Biochem.*, **15**, 49–53.

Brown, W. V., Jaisson, P., Taylor, R. W. and Lacey, M. J. (1990). Novel internally branched, internal alkenes as major components of the cuticular hydrocarbons of the primitive Australian ant *Nothomymecia macrops* Clark (Hymenoptera: Formicidae). *J. Chem. Ecol.*, **16**, 2623–2635.

Carlson, D. A., Bernier, U. R. and Sutton, B. D. (1998). Elution patterns from capillary GC for methyl-branched alkanes. *J. Chem. Ecol.*, **24**, 1445–1465.

Carlson, D. A., Langley, P. A. and Huyton, P. (1978). Sex pheromone of the tsetse fly: Isolaton, identification, and synthesis of contact aphrodisiacs. *Science*, **201**, 750–753.

Carlson, D. A., Mayer, M. S., Silhacek, D. L., James, J. D., Beroza, M. and Bierl, B. A. (1971). Sex attractant peheromone of the housefly: Isolation, identification and synthesis. *Science*, **174**, 76–77.

Carlson, D. A., Nelson, D. R., Langley, P. A., Coates, T. W., Davis, D. L. and Leegwater-Van Der Linden, M. E. (1984). Contact sex-pheromone in the tsetse fly *Glossina pallidipes* (Austen): Identification and synthesis. *J. Chem. Ecol.,* **10**, 429-450.

Carlson, D. A., Roan, C.-S. and Yost, R. A. (1989). Dimethyl disulfide derivatives of long chain alkenes, alkadienes, and alkatrienes for gas chromatography/mass spectrometry. *Anal. Chem.*, **61**, 1564–1571.

Carlson, D. A. and Schlein, Y. (1991). Unusual polymethyl alkenes in tsetse flies acting as abstinon in *Glossina moristans*. *J. Chem. Ecol.*, **17**, 267–284.

Cvačka, J., Jiroš, P., Šobotník, J, Hanus, R. and Svatoš, A. (2006). Analysis of insect cuticular hydrocarbons using matrix-assisted laser desorption/ionization mass spectrometry. *J. Chem. Ecol.*, **32**, 409–434.

Darbro, J. M., Millar, J. G., McElfresh, J. S. and mullens, B. A. (2005). Survey of muscalure [(Z)-9-tricosene] on house flies (Diptera: Muscidae) from field populations in California. *Environ. Entomol.*, **34**, 1418–1425.

Dillwith, J.W., Nelson, J.H., Pomonis, J. G, Nelson, D.R. and Blomquist, G.J. (1982). A ^{13}C NMR study of methyl-branched hydrocarbon biosynthesis in the housefly. *J. Biol. Chem.*, **257**, 11305–11314.

Dowell, F.E., Parker, A.G., Benedict, M.Q., Robinson, A.S., Broce, A.B. and Wirtz, R.A. (2005). Sex separation of tsetse fly pupae using near-infrared spectroscopy. *Bull. Entomol. Research*, **95**, 249–257.

Fletcher, M.T., Allsopp, P.G., McGrath, M.J., Chow, S., Gallagher, O.P., Hull, C., Cribb, B.W., Moore, C.J. and Kitching, W. (2008). Diverse cuticular hydrocarbons from Australian canebeetles (Coloeptera: Scarabaeidae). *Australian J. Entomol.*, **47**, 153–159.

Fletcher, M.T., McGrath, M.J., Konig, W.A., Moore, C.H., Cribb, B.W., Allsopp, P.G. and Kitching, W. (2001). A novel group of allenic hydrocarbons from five Australian (Melolonthine) beetles. *Chem. Commun.*, **2001**, 885–886.

Foley, W.J., McIlwee, A., Lawler, I., Aragones, L., Woolnough, A.P. and Berding N. (1998). Ecological applications of near infrared reflectance spectroscopy – a tool for rapid, cost-effective prediction of the composition of plant and animal tissues and aspects of animal performance. *Oecologia*, **116**, 293–305.

Francis, G.W. and Veland, K. (1981). Alkylthiolation for the determination of double-bond positions in linear alkenes. *J. Chromatogr.*, **219**, 379–384.

Ginzel, M.D., Millar, J.G. and Hanks, L.M. (2003). (Z)-9-Pentacosene – contact sex pheromone of the locust borer, *Megacyllene robiniae. Chemoecology*, **13**, 135–141.

Howard, R.W. (1993). Cuticular hydrocarbons and chemical communication. In *Insect Lipids: Chemistry, Biochemistry and Biology*, ed. D.W. Stanley-Samuelson and D.R. Nelson. Lincoln, NB: University of Nebraska Press, pp. 179–226.

Howard, R.W., Akre, R.D. and Garnett, W.B. (1990). Chemical mimicry in an obligate predator of carpenter ants (Hymenoptera: Formicidae). *Ann. Entomol. Soc. Am.*, **83**, 607–616.

Howard, R.W. and Blomquist, G.J. (2005). Ecological, behavioral, and biochemical aspects of insect hydrocarbons. *Annu. Rev. Entomol.*, **50**, 371–393.

Howard, R.W., McDaniel, C.A. and Blomquist, G.J. (1978). Cuticular hydrocarbons of the eastern subterranean termite, *Reticulitermes flavipes* (Kollar) (Isoptera: Rhinotermitidae). *J. Chem. Ecol.*, **4**, 233–245.

Howard, R.W., McDaniel, C.A., Nelson, D.R. and Blomquist, G.J. (1980). Chemical ionization mass spectrometry: application to insect-derived cuticular alkanes. *J. Chem. Ecol.*, **6**, 609–623.

Jackson, L.L. (1972). Cuticular lipids of insects IV. Hydrocarbons of the cockroaches *Periplaneta japonica* and *Periplaneta americana* compared to other cockroach hydrocarbons. *Comp. Biochem. Physiol.*, **41B**, 331–336.

Jallon, J.-M. and Wicker-Thomas, C. (2003). Genetic studies on pheromone production in *Drosophila*. In *Insect Pheromone Biochemistry and Molecular Biology. The biosynthesis and detection of pheromones and plant volatiles*, ed. G.J. Blomquist and R.G. Vogt. London: Elsevier, pp. 253–281.

Katritzky, A.R. and Chen, K. (2000). QSPR correlation and predictions of GC retention indexes for methyl-branched hydrocarbons produced by insects. *Anal. Chem.*, **72**, 101–109.

Locke, M. (1965). Permeability of insect cuticle to water and lipids. *Science*, **147**, 295–298.

Lockey, K.H. (1985). Cuticular hydrocarbons of adult *Eurychora sp.*

(Coleoptera: Tenebrionidae). *Comp. Biochem. Physiol.*, **81B**, 223–227.

Martin, M. M. and MacConnell, J. G. (1970). The alkanes of the ant, *Atta colombica. Tetrahedron*, **26**, 307–319.

Martin, S. J., Vitikainen, E., Helantera, H. and Drijfout, F. P. (2008). Chemical basis of nest-mate discrimination in the ant *Formica exsecta. Proc. Royal Soc. B*, **275**, 1271–1278.

McCarthy, E. D., Han, J. and Calvin, M. (1968). Hydrogen atom transfer in mass spectrometric fragmentation patterns of saturated aliphatic hydrocarbons. *Anal. Chem.*, **40**, 1475–1480.

McDaniel, C. A. (1990). Cuticular hydrocarbons of the Formosan termite *Coptotermes formosanus. Sociobiology*, **16**, 265–273.

McGrath, M. J., Fletcher, M. T., Konig, W. A., Moore, C. J., Cribb, B. W., Allsopp, P. G. and Kitching, W. (2003). A suite of novel allenes from Australian Melolonthine Scarab beetles. structure, synthesis and stereochemistry. *J. Organic Chem.*, **68**, 3739–3748.

Mold, J. D., Means, R. E., Stevens, R. K. and Ruth, J. M. (1966). The paraffin hydrocarbons of wool wax. Homologous series of methyl alkanes. *Biochem.*, **5**, 455–461.

Morgan, E. D. and Wadhams, L. J. (1972). Gas chromatography of volatile compounds in small samples of biological materials. *J. Chromatogr. Sci.*, **10**, 528–529.

Mpuru, S., Blomquist, G. J., Schal, C., Kuenzli, M., Dusticier, G. Roux, M. and Bagnères, A-G. (2001). Effect of age and sex on the production of internal and external hydrocarbons and pheromones in the housefly, *Musca domestica. Insect Biochem. Molec. Biol.*, **31**, 139–155.

Nelson, D. R., Adams, T. S. and Fatland, C. L. (2003). Hydrocarbons in the surface wax of eggs and adults of the Colorado potato beetle, *Leptinotarsa decemlineata. Comp. Biochem. Physiol.*, **134B**, 447–466.

Nelson, D. R. and Blomquist, G. J. (1995). Insect waxes. In *Waxes: Chemistry, Molecular Biology and Functions*, ed. R. J. Hamilton. Dundee, Scotland: The Oily Press, pp. 1–90.

Nelson, D. R. and Carlson, D. A. (1986). Cuticular hydrocarbons of the tsetse flies *Glossina morsitans, G. austeni* and *G. pallidipes. Insect Biochem.*, **16**, 403–416.

Nelson, D. A., Carlson, D. A. and Fatland, C. L. (1988). Cuticular hdrocarbons of the tsetse flies. II. *G. p. palpalis, G.p. gambiensis, G. fuscipes, G. tachinoides*, and *G. brevipalpis. J. Chem. Ecol.*, **14**, 963–987.

Nelson, D. R., Dillwith, J. W. and Blomquist, G. J. (1981). Cuticular hydrocarbons of the house fly, *Musca domestica. Insect Biochem.*, **11**, 187–197.

Nelson, D. R., Fatland, C. L. and Baker, J. E. (1984). Mass-spectral analysis of epicuticular normal-alkadienes in 3 *Sitopholus* weevils. *Insect Biochem.*, **14**, 435–444.

Nelson, D. R., Fatland, D. L., Howard, R. W., McDaniel, C. A. and Blomquist, G. J. (1980). Re-analysis of the cuticular methylalkanes of *Solenopsis invicata* and *S. richteri. Insect Biochem.*, **10**, 409–418.

Nelson, D. R., Olson, D. L., and Fatland, C. L. (2002). Cuticular hydrocarbons of the flea beetles, *Aphthona lacertosa* and *Aphthona nigriscutis*, biocontrol agents for leafy spurge (*Euphorbia esula*). *Comp. Biochem. Physiol.*, **133B**, 337–350.

Nelson, D. R. and Sukkestad, D. R. (1970). Normal and branched aliphatic hydrocarbons from the eggs of the tobacco hornworm. *Biochem.*, **9**, 4601–4611.

Newey, P. S., Robson, S. K. A. and Crozier, R. H. (2008a). Near-infrared spectroscopy identifies the colony and nest of origin of weaver ants, *Oecophylla smaragdina*.

Insect Soc., **55**, 171–175.

Newey, P. S., Robson, S. K. A. and Crozier, R. H. (2008b). Near-infrared spectroscopy as a tool in behavioural ecology: a case study of the weaver ant, *Oecophylla smaragdina*. *Anim. Behav.*, **76**, 1727–1733.

Perez-Mendoza, J., Dowell, F. E., Broce, A. B., Throne, J. F., Wirtz, R. A., Xie, F., Fabrick, J. A. and Baker, J. E. (2002). Chronological age-grading of house flies by using near-infrared spectroscopy. *J. Med. Entomol.*, **39**, 499–508.

Pomonis, J. G., Fatland, C. L., Nelson, D. R. and Zaylskie, R. G. (1978). Insect hydrocarbons. Corroboraton of structure by synthesis and mass spectrometry of mono- and dimethylalkanes. *J. Chem. Ecol.*, **4**, 27–39.

Pomonis, J. G., Nelson, D. R. and Fatland, C. L. (1980). Insect hydrocarbons 2. Mass spectra of dimethylalkanes and the effect of the number of methylene units between methyl groups on fragmentation. *J. Chem Ecol.*, **6**, 965–972.

Scammells, D. V. and Hickmott, B. (1976). Diagnostic trends in the mass spectra of some mono methyl alkanes. *Organic Mass Spectrom.*, **11**, 901–903.

Syvertson, T. C., Jackson, L. L., Blomquist, G. J. and Vinson, S. B. (1995). Alkadienes mediating courtship in the parasitoid *Cardiochiles nigriceps* (Hymenoptera: Braconidae). *J. Chem. Ecol.*, **12**, 1971–1989.

Szafranek, J., Kusmierz, J. and Czerwiec, W. (1982). Gas chromatographic–mass spectrometric investigations of high-boiling crude oil alkane fractions. *J. Chromatogr.*, **245**, 219–228.

Theisen, M. O., Miller, G. C., Cripps, C., de Renobales, M. and Blomquist, G. J. (1991). Correlation of carbaryl uptake with hydrocarbon transport to the cuticular surface during development in the cabbage looper, *Trichoplusia ni. Pesticide Biochem. Physiol.*, **40**, 111–116.

Thompson, M. J., Glancey, B. M., Robbins, W. E., Lofgren, C. S., Dutky, S. R., Kochansky, J., Vander Meer, R. K. and Glover, A. R. (1981). Major hydrocarbons of the post-pharyngeal glands of mated queens of the red imported fire ant *Solenopsis invicta. Lipids*, **16**, 485–495.

Vicenti, M., Guiglielmetti, G., Cassani, G. and Tonini, C. (1987). Determination of double bond position in diunsaturated compounds by mass spectrometry of dimethyl disulfide derivatives. *Anal. Chem.*, **59**, 694–699.

Warthen, J. D. Jr. and Uebel, E. C. (1980). Comparison of the unsaturated cuticular hydrocarbons of male and female house crickets, *Acheta domesticus* (L.) (Orthoptera: Gryllidae). *Insect Biochem.*, **10**, 435–439.

Zarei, K. and Atabati, M. (2005). Predictions of GC retention indexes for insect-produced methyl-substituted alkanes using an artificial neural network and simple structural descriptors. *J. Anal. Chem.*, **60**, 732–737.

3

Biosynthesis of cuticular hydrocarbons

Gary J. Blomquist

In vivo studies in the 1960s determined that labeled acetate was readily incorporated into insect cuticular lipids, especially hydrocarbons (Vroman *et al.*, 1965; Lamb and Monroe, 1968; Nelson, 1969), establishing the de novo synthesis of insect hydrocarbons. Later studies with specific radio-labeled precursors and careful analysis of metabolic products determined the biosynthetic pathways for the most common components. In vivo experiments with ^{13}C-labeled precursors extended and confirmed the conclusions based on radiochemical data. In vitro studies using microsomal preparations examined the elongation of fatty acyl-CoAs and the conversion of fatty acyl-CoAs to hydrocarbons. The mechanism of how long-chain fatty acyl-CoAs are converted to hydrocarbons has been controversial, and only recently have studies using the powerful techniques of molecular biology been applied to gaining a more complete understanding of the biosynthesis and regulation of insect hydrocarbons (Wicker-Thomas and Chertemps, Chapter 4, this book). The biosynthesis of hydrocarbons has been studied in relatively few insects, including the dipterans *Musca domestica* (Blomquist, 2003) and *Drosophila melanogaster* (Jallon and Wicker-Thomas, 2003), and considerable work has been done on cockroaches *Periplaneta americana* and *Blattella germanica*, the termite *Zootermopsis angusticollis* and several other insects (Nelson and Blomquist, 1995; Howard and Blomquist, 2005). Work has been done with the cabbage looper, *Trichoplusia ni* (Dwyer *et al.*, 1986; de Renobales *et al.*, 1988), southern armyworm *Spodoptera eridania* (Guo and Blomquist, 1991) and in cockroaches (Young and Schal, 1997) on the timing of hydrocarbon synthesis and its deposition on the insect cuticle. This chapter will concentrate on the biosynthesis of long-chain cuticular hydrocarbons. Millar (Chapter 18) describes the biosynthesis of the polyene hydrocarbons used by some lepidopterans as pheromones and a chapter by Bartelt (Chapter 19) covers the occurrence and biosynthesis of short-chain methyl- and ethyl-branched unsaturated hydrocarbons used by beetles as volatile pheromones.

Site of cuticular hydrocarbon biosynthesis

It is generally accepted that insects synthesize a majority of their cuticular hydrocarbons (Nelson and Blomquist, 1995), although studies have shown that dietary hydrocarbons are incorporated into cuticular lipids (Blomquist and Jackson, 1973a). However, for most species it appears that dietary lipid accounts for very small amounts of insect cuticular hydrocarbon. Some inquilines, which use cuticular hydrocarbons in chemical mimicry, synthesize hydrocarbons with a composition very similar to those of their host termites (Howard *et al.*, 1980; see also Chapter 14). A number of studies with widely diverse insect species have established that the major site of hydrocarbon biosynthesis occurs in the cells associated with the epidermal layer or peripheral fat body, specifically the oenocytes (see also Chapter 5).

The anatomical location of oenocytes varies among insect species and even across developmental stages. In some insects, such as *Tenebrio molitor* (Jackson and Locke, 1989), oenocytes are arranged in discrete clusters within the hemocoel and are readily accessible. In other insects, including the American cockroach, *P. americana,* the fruitfly, *Drosophila melanogaster*, and German cockroach, *B. germanica*, oenocytes are found within the abdominal integument. A variety of lines of evidence has localized the site of hydrocarbon synthesis to the oenocytes. In Drosophila, Ferveur *et al.* (1997) showed that the targeted expression of the *transformer* gene in oenocytes of male fruitflies resulted in the feminization of the hydrocarbon pheromone components. In the German cockroach, Gu *et al.* (1995) showed that only the abdominal sternites and tergites synthesize hydrocarbon, which is then loaded unto hemolymph lipophorin for transport to sites of deposition. More recently, the Schal laboratory (Fan *et al.*, 2003) enzymatically dissociated the oenocytes of the abdominal integument of the German cockroach and separated the oenocytes by Percoll gradient centrifugation. Using radiolabeled propionate, they showed that hydrocarbon synthesis was highest in the fractions most enriched in oenocytes. Because oenocytes in the German cockroach are localized to the abdominal integument, a system is necessary to transfer the hydrocarbons to other parts of the body, and this appears to be accomplished by a high-density lipoprotein (Chino, 1985; Schal *et al.*, 1998). This process is discussed in detail in Chapter 5 (Bagnères and Blomquist, this book). RNAi silencing of the putative cytochrome P450 that converts aldehydes to hydrocarbon in the final step in hydrocarbon biosynthesis in oenocytes of *Drosophila* almost completely shut off hydrocarbon biosynthesis (Wicker-Thomas *et al.*, unpublished data), providing further evidence for the important role of oenocytes in hydrocarbon production.

Biosynthetic pathways for hydrocarbons

It is now clear that insects synthesize hydrocarbons by elongating fatty acyl-CoAs to produce the very long-chain fatty acids that are then converted to hydrocarbon by loss of the carboxyl group (Nelson and Blomquist, 1995; Howard and Blomquist, 2005). Methyl-branched hydrocarbons (with the exception of 2-methylalkanes) arise from the incorporation of a propionyl-CoA group (as methylmalonyl-CoA) in place of an acetyl-CoA group at specific points during chain elongation (Nelson and Blomquist, 1995). 2-Methylalkanes arise from the elongation of the carbon skeleton of either valine (even number of carbons in the chain) or isoleucine (odd number of carbons in the chain) (Blailock *et al.*, 1976). Radiotracer and ^{13}C-NMR studies (Dwyer *et al.*, 1981; Chase *et al.*, 1990), and NMR plus mass spectrometry (Dillwith *et al.*, 1982) were useful in determining the precursors for methyl-branched lipids and the sequence of incorporation of the methyl branches during chain elongation.

Mechanism of hydrocarbon formation

Early studies in a termite (Chu and Blomquist, 1980a), a cockroach (Major and Blomquist, 1978) and the housefly (Tillman-Wall *et al.*, 1992) showed that tritium-labeled fatty acids were converted in vivo to hydrocarbons one carbon shorter. The mechanism of how this occurs has been controversial. Kolattukudy and co-workers have proposed a mechanism in which a fatty acyl-CoA is reduced to the aldehyde and, in the absence of cofactors, is decarbonylated to the hydrocarbon one carbon shorter and carbon monoxide. This has been demonstrated in plants, algae, vertebrates (Bognar *et al.*, 1984; Cheesbrough and Kolattukudy, 1984, 1988; Dennis and Kolattukudy, 1991) and the flesh fly *Sarcophaga crassipalpis* (Yoder *et al.*, 1992).

In housefly *M. domestica* microsomes, incubation of (Z)-15-[1–^{14}C]- and (Z)-15-[15,16–^{3}H$_2$]tetracosenoyl-CoA and the corresponding aldehydes in the presence of NADPH gave equal amounts of ^{14}CO$_2$ and [^{3}H]-(Z)-9-tricosene (Figure 3.1A) (Reed *et al.*, 1994). The formation of labeled carbon dioxide and not carbon monoxide was verified by both radio-GLC (Figure 3.1A) and trapping agents. A requirement for NADPH and O$_2$ (Figure 3.1B) and inhibition by CO and antibody to cytochrome P450 reductase strongly implicates a cytochrome P450 in the reaction. Chemical ionization mass spectrometry analysis of the products from [2,2–^{2}H$_2$, 2–^{13}C]tetracosanoyl-CoA demonstrated that the deuteriums on the 2,2 position of the acyl-CoA remained on the hydrocarbon product, suggesting a P450 mechanism in which the protons on the 2,2 positions of the acyl-CoA or aldehyde are retained on the hydrocarbon product (Reed *et al.*, 1995). Further work in a variety of insects (Mpuru *et al.*, 1996) showed that [9,10–^{3}H, 1–^{14}C]18:0 aldehyde was converted to a C17 hydrocarbon and CO$_2$ in the presence of O$_2$ and NADPH, indicating that the pathway for hydrocarbon formation in insects involved the reduction of the acyl-CoA to aldehyde followed by a cytochrome P450 mediated CO$_2$-producing oxidative decarbonylation (Figure 3.1C). The resolution of this controversy regarding the mechanism of hydrocarbon formation awaits the cloning, expression and characterization of the enzymes involved. Toward this end, Blomquist and Tittiger (unpublished data) discovered a cytochrome P450 that is integument enriched. The ortholog of this P450 was RNAi silenced in oenocytes of *Drosophila*, and the amount of hydrocarbon produced was decreased from about 1500 ng/fly in control insects to less than 100ng/fly with the P450 silenced (Wicker-Thomas, unpublished data). Studies are underway to express and characterize this cytochrome P450.

Chain length specificity

The regulation of the chain length specificity to produce the specific blend of hydrocarbons often used in chemical communication appears to reside in the microsomal fatty acyl-CoA

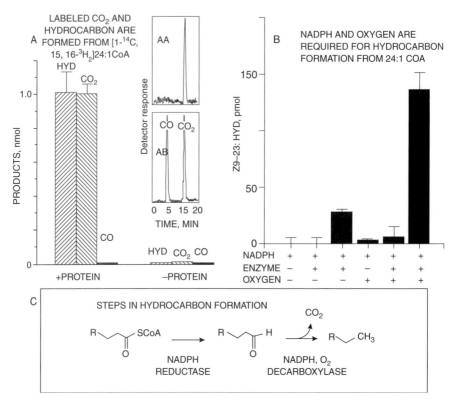

Figure 3.1 A. Formation of labeled hydrocarbon and carbon dioxide from [1–^{14}C, 15,16–^{3}H$_2$]-15-tetracosenoyl-CoA from housefly microsomes. B. Effect of NADPH, enzyme preparation and oxygen on hydrocarbon formation from [15,16–^{3}H]-15-tetracosenoyl-CoA. C. Proposed steps in hydrocarbon formation in insects.

elongase reaction and not in the reductive conversion of fatty acyl-CoAs to hydrocarbon. The American cockroach produces three major hydrocarbons: *n*-pentacosane, 3-methylpentacosane and (*Z,Z*)-9,12-heptacosadiene (Jackson, 1972). Studies with microsomes from integument tissue showed that stearyl-CoA was elongated up to a 26 carbon acyl-CoA that could serve as the precursor to *n*-pentacosane. In contrast, linoleoyl-CoA was readily elongated to 28 carbons to serve as the precursor to the 27:2 hydrocarbon (Vaz *et al.*, 1988). In the housefly, the picture is even clearer. Laboratory-reared female houseflies produce monoenes of 27 carbons and longer for the first two days after adult eclosion, and then switch to producing (*Z*)-9-tricosene under the influence of 20-hydroxyecdysone at three days post-eclosion (Dillwith *et al.*, 1983; Blomquist, 2003). Microsomes from one-day-old adult females readily elongated both 18:1-CoA and 24:1-CoA up to 28 carbons (Tillman-Wall *et al.*, 1992). Microsomes from day-4 females elongated 18:1 to 24:1 and did not effectively elongate 24:1-CoA (Figure 3.2 A). In contrast, males, which produce 27 and longer alkenes at all ages, readily elongated both 18:1-CoA and 24:1-CoA to fatty

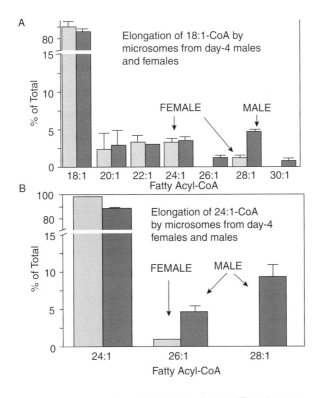

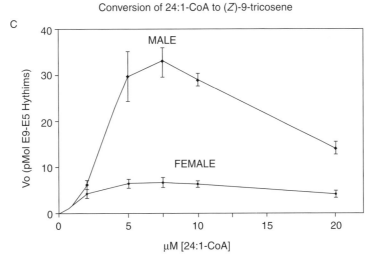

Figure 3.2 Elongation of 18:1-CoA (A) and 24:1-CoA (B) by male and female housefly microsomes in the presence of malonyl-CoA and NADPH. C. Effect of 24:1-CoA concentration on hydrocarbon production by male and female microsomal preparations.

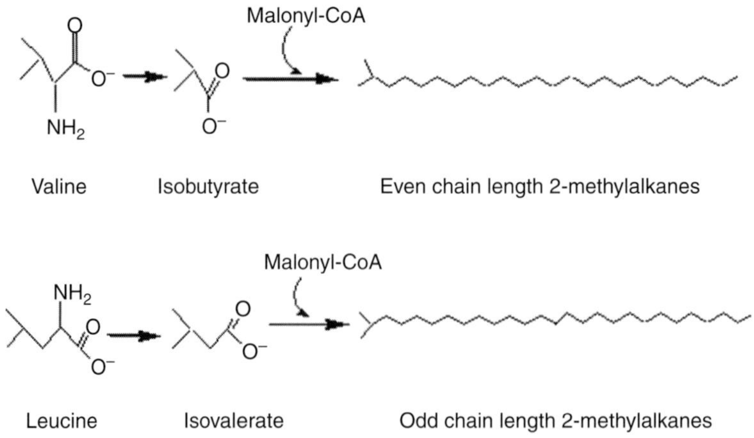

Valine   Isobutyrate   Even chain length 2-methylalkanes

Leucine   Isovalerate   Odd chain length 2-methylalkanes

Figure 3.3 Biosynthesis of 2-methylalkanes from leucine and valine carbon skeletons.

acids of 28 carbons (Figure 3.2B). Microsomes from males, which normally do not make very much C_{23} hydrocarbons, readily converted 24:1-CoA to C_{23} alkenes (Figure 3.2 C) (Tillman-Wall *et al.*, 1992). To further emphasize the importance of the acyl-CoA elongation reactions in controlling chain length, microsomes for day-1 and day-4 females and males of both ages all converted both 28:1-CoA, 24:1-CoA and the 28:1, 24:1 and 18:1 to hydrocarbons one carbon shorter (Reed *et al.*, 1996). Thus, the evidence strongly suggests that the reductive conversion of very-long-chain acyl-CoAs or aldehydes is not the step that regulates chain length specificity, but rather that the specificity resides in the acyl-CoA elongation system.

Methyl-branched hydrocarbon biosynthesis

The formation of 2-methylalkanes arises from the carbon skeletons of amino acids to initiate chain synthesis. The carbon skeleton of valine leads to even-numbered carbons in the backbone (Blailock *et al.*, 1976), and the carbon skeleton of leucine leads to odd-numbered carbons in the hydrocarbon backbone (Blailock *et al.*, 1976; Charlton and Roelofs, 1991) (Figure 3.3). The methyl-branches occurring in 3-methyl- and internally methyl-branched hydrocarbons arise from the substitution of a methylmalonyl-CoA unit for malonyl-CoA during chain elongation (Figure 3.4) (Blomquist *et al.*, 1975a; Blomquist and Kearney, 1976; Dillwith *et al.*, 1982; Chase *et al.*, 1990). The precursors for the methylmalonyl-CoA, especially in the insects which either lack or have low levels of vitamin B_{12}, are the amino acids valine, isoleucine and methionine (Dwyer *et al.*, 1981; Dillwith *et al.*, 1982; Halarnkar *et al.*, 1985; Chase *et al.*, 1990). The amino acids valine and isoleucine are also the precursors for the propionyl group that gives rise to ethyl-branched juvenile hormones (Brindle *et al.*, 1987, 1988) in Lepidoptera.

Termites contain high levels of vitamin B_{12} in contrast to most other insects (Wakayama *et al.*, 1984). In termites, succinate is converted to methylmalonyl-CoA and serves as the

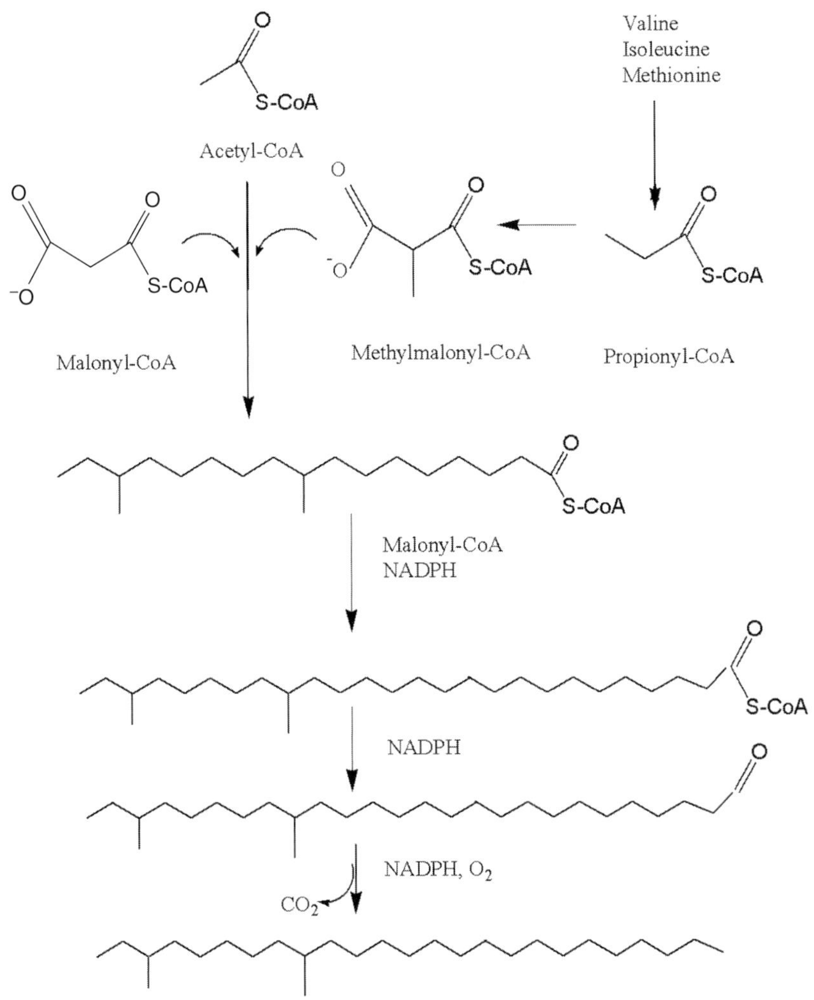

Figure 3.4 Biosynthetic pathway for internally methyl-branched hydrocarbons.

source of the methyl branches (Figure 3.4) (Blomquist *et al.*, 1980; Chu and Blomquist, 1980a, b; Halarnkar and Blomquist, 1989). The termite *Zootermopsis angusticollis* converts succinyl-CoA to methylmalonyl-CoA and then to propionate in the gut tract, probably via microorganisms (Guo *et al.*, 1991). The propionate is then apparently transported by the hemolymph to epidermal tissue, where it is reformed into methylmalonyl-CoA and used in methyl-branched hydrocarbon synthesis.

Carbon-13 NMR studies were performed with [1–^{13}C]propionate to determine whether the methyl groups of methylalkanes with methyl groups towards the end of the chain were incorporated early in chain synthesis or towards the end of the process. The data showed that the labeled carbon from [1–^{13}C]propionate was incorporated into carbon 4 and not carbon 2

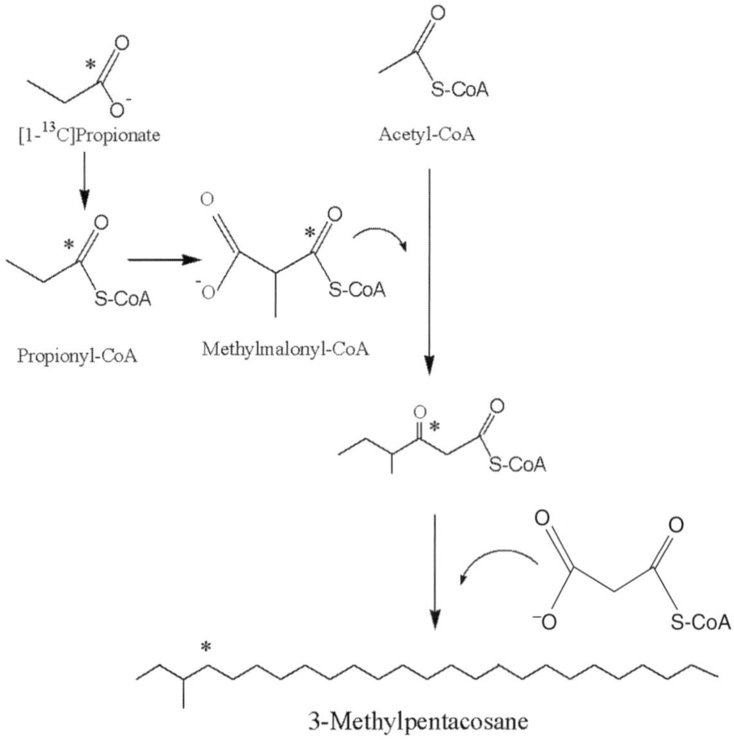

3-Methylpentacosane

Figure 3.5 Incorporation of the label from [1–^{13}C]propionate into position 4 of 3-methylpentacosane, indicating that the methyl-branching group is incorporated as the second unit in hydrocarbon formation rather than at the end of the chain synthesis.

of 3-methylalkanes (Figure 3.5), thus showing that it was incorporated as the second unit in 3-methylalkane synthesis in the housefly, the American cockroach and the German cockroach (Dillwith *et al.*, 1982; Dwyer *et al.*, 1981; Chase *et al.*, 1990). Similarly, [1,4–^{13}C$_2$]succinate labeled 5-methyl- and 5,17-dimethylalkanes of a termite were incorporated in a manner consistent with insertion early in chain synthesis for 5-methylalkanes (Guo *et al.*, 1991).

Methyl-branched fatty acids

If the methyl groups of 3- and 5- methyl, and by analogy the internal methyl branches of hydrocarbons, are inserted early in chain synthesis, then there should exist corresponding methyl-branched fatty acids in the integument tissue. A careful examination of the fatty acids from the integument of *B. germanica* and *M. domestica* showed that such fatty acids were present, although in very low amounts. In *B. germanica*, n-3, n-4, n-5, n-7, n-8 and n-9 monomethyl C_{15}, C_{16}, C_{17} and C_{18} and n-5,9 and n-3,11 dimethyl C_{16} and C_{18} fatty acids were present in low amounts (Juarez *et al.*, 1992). These fatty acids have a methyl branching pattern consistent with being the precursors to some of the major hydrocarbons of this insect (Jurenka *et al.*, 1989). Labeled propionate was incorporated into methyl-branched fatty acids in vivo (Chase *et al.*, 1990), and labeled methylmalonyl-CoA was incorporated into methyl-branched fatty acids by microsomal preparations (Juarez *et al.*, 1992). Likewise, methyl-branched fatty acids of 14 to 19 carbons, with methyl branches on n-2, n-3, n-4, n-5, n-6, n-7, n-8, n-9 and n-3,7 were characterized in *M. domestica* (Blomquist *et al.*, 1994). Evidence has accumulated that a microsomal fatty acid synthase is involved in synthesizing the methyl-branched fatty acid precursors to hydrocarbon in both *M. domestica* (Gu *et al.*, 1997) and *B. germanica* (Gu *et al.*, 1993; Juarez *et al.* 1992, 1996; Juarez and Fernandez, 2007).

Propionate metabolism in insects

Propionate is the source of methyl-branched hydrocarbons and ethyl-branched juvenile hormones in insects. During studies on the incorporation of ^{13}C-labeled propionate into hydrocarbons of the housefly (Dillwith *et al.*, 1982), it was observed that in addition to labeling the methyl-branch group, propionate labeled in the 2 and 3 positions labeled the straight-chain portion of hydrocarbons in the same manner as [2–^{13}C]acetate and [1–^{13}C] acetate, respectively. Subsequently, it was demonstrated that insects metabolize propionate by a different pathway to that observed in vertebrates. In insects, propionyl-CoA is dehydrogenated to acryl-CoA, hydrated to 3-hydroxypropionyl-CoA and then oxidized to acetyl-CoA, in which carbon 1 of propionate is lost as CO_2 (Halarnkar *et al.*, 1986; Halarnkar and Blomquist, 1989). All insects examined to date, including those with high levels of vitamin B_{12} and those with undetectable levels (Wakayama *et al.*, 1984) metabolize propionate to acetate, suggesting that this is a universal pathway in insects, as it is in plants (Lucas *et al.*, 2007).

Ecdysteroid regulation of hydrocarbon production in the housefly

The housefly has proven to be an exceptionally good model in which to study the regulation of the chain length of cuticular hydrocarbons. (Z)-9-Tricosene, present on the surfaces of females of many laboratory-reared colonies but less often present on wild flies (Darbro *et al.*, 2005), serves as a short-range sex attractant and stimulant (Carlson *et al.*, 1971). Newly emerged males and females of laboratory-reared strains have similar cuticular lipid profiles: the major hydrocarbon components are (Z)-9-alkenes of 27 carbons and longer. From vitellogenic stages 4 through 8, both the percentage of alkenes and the percentage of (Z)-9-heptacosene decrease, while the percentage of (Z)-9-tricosene increase. Newly emerged males and females have very low amounts of methyl-branched alkanes. Vitellogenic females begin producing methyl-branched alkanes, and this increase can be traced by a marked increase in the incorporation of labeled propionate into the surface hydrocarbons.

The endocrine regulation of sex pheromone production in a laboratory strain of the housefly has been summarized (Blomquist, 2003). Because housefly sex pheromone production was correlated with ovarian development, it was possible that both oogenesis and (Z)-9-tricosene synthesis were regulated by a common hormone. Such a common factor was found in cockroaches and beetles, where juvenile hormone (JH) induces both vitellogenesis and sex pheromone production.

Alternatively, a product of the developing ovary could initiate pheromone production. JH was ruled out because removal of the corpus allatum-corpus cardiacum within 6 h of eclosion did not inhibit the production of (Z)-9-tricosene. In contrast, flies overiectomized within 6 h of emergence do not produce detectable amounts of any of the C_{23} pheromone components (Dillwith et al., 1983). When ovariectomized flies received ovary implants from 24-h-old donors on day 4, C_{23} sex pheromone components were detected three days later. Housefly ovaries have been shown to produce ecdysteroids (Adams et al., 1997).

In many Diptera, ecdysteroids have been shown to play a role in regulating reproductive processes, including vitellogenin synthesis (Hagedom, 1985). Ovariectomized flies treated with 20-hydroxyecdysone produced the C_{23} sex pheromone components. Furthermore, males, which normally do not produce any of the C_{23} sex pheromone components, were induced to produce (Z)-9-tricosene and the C_{23} epoxide and ketone by implantation of ovaries or injection of 20-hydroxyecdysone (Blomquist et al., 1984). Thus, using the ability to manipulate the chain length of the hydrocarbon products, it was determined that the fatty acyl-CoA elongation reactions comprised the major factor that regulated hydrocarbon chain length (Blomquist, 2003).

Metabolism of cuticular hydrocarbons

The first demonstration of the metabolism of cuticular hydrocarbons was the hydroxylation of n-alkanes to secondary alcohols with the hydroxyl group located near the center of the molecule in the grasshopper *Melanoplus sanguinipes* (Blomquist and Jackson, 1973b). The secondary alcohols formed were then esterified to secondary alcohol wax esters (Blomquist et al., 1972). A chain-length specificity was evident, with the shorter-chain n-alkanes hydroxylated more readily that longer-chain components. A cytochrome P450 was implicated in the hydroxylation by the demonstration that the oxygen in the secondary alcohol was derived from O_2 (Blomquist et al., 1975b).

In addition to serving directly in chemical communication, the cuticular hydrocarbons of some insects serve as the precursors to contact pheromones. In the German cockroach, oxygenated derivatives of 3,11-dimethylnonacosane were shown to serve as the female-produced contact pheromone (Nishida and Fukami, 1983), and included 3,11-dimethyl-2-nonacosanone, 29-hydroxy-3,11-dimethyl-2-nonacosanone and 29-oxo-3,11-dimethyl-2-nonacosanone. The hydrocarbons of the German cockroach contain 3,11-dimethylnonacosane (Jurenka et al., 1989). In addition, 3,11-dimethylheptacosane was present on the German cockroach, and a re-examination of the pheromone components showed that 3,11-dimethyl-2-heptacosanone was present and functioned as a contact sex pheromone (Schal et al., 1990). The biosynthesis of the methyl-branched hydrocarbon and pheromone components showed that the carbon skeletons of the amino acids valine, isoleucine and methionine and succinate served as the precursor to the propionate that made up the methyl-branching unit, and that the branching unit of the 3,*X*-dimethylalkanes was inserted early in chain synthesis (Chase *et al.*, 1990). Tritium-labeled 3,11-dimethylnonacosane was hydroxylated to the 2-ol derivative, which was oxidized to the corresponding methyl ketone (Chase *et al.*, 1992). An age- and sex-specificity study showed that while both males and females of all ages converted labeled dimethyl-2-nonacosanol to the corresponding derivative, only females 5–9 days post-emergence converted the labeled 3,11-dimethylnonacosane to the corresponding methyl ketone, indicating that the hydroxylation reaction was sex-and age-specific. More recently, Eliyahu *et al.* (2008) showed that 27-oxo-3,11-dimethylheptacosan-2-one and 27-hydroxy-3,11-dimethylheptacosan-2-one are also part of the contact pheromone mixture of the German cockroach. Other species can also hydroxylate hydrocarbons, as was first demonstrated in the grasshopper *Melanoplus sanguinipes* (Blomquist and Jackson, 1973b).

M. domestica also hydroxylates and metabolizes alkenes and alkanes to oxygenated derivatives. The epoxide and ketone derivative of (*Z*)-9-tricosene, (9,10)-epoxytricosane and (*Z*)-14-tricosene-10-one are derived from the C_{23} alkene by a cytochrome P450 enzyme which epoxidizes at the 9,10 position and hydroxylates at carbon 10 from the other end of the molecule, and the secondary alcohol is then oxidized to the ketone (Ahmad *et al.*, 1987).

Developmental regulation of hydrocarbon synthesis and deposition

The timing of deposition of hydrocarbon on the cuticle was studied in the cabbage looper, *T. ni* (Dwyer *et al.*, 1986) and the southern armyworm, *Spodoptera eridania* (Guo and Blomquist, 1991). In both insects, the amount of internal hydrocarbon is low immediately after a molt and then increases dramatically during the instar. During the second half of the pupal state of *T. ni*, the amount of internal hydrocarbon is about three-fold that of the cuticular hydrocarbon (de Renobales *et al.*, 1988). About 50% of the internal hydrocarbon, both in larvae and pupae, appears on the cuticle of the newly molted insect, whereas the other 50% remains internal. Newly synthesized hydrocarbon (as measured by the incorporation of radiolabeled acetate) is also found among the internal lipids, although the amounts vary throughout development. In fifth instar larva, newly synthesized hydrocarbon appears primarily on the cuticle during the feeding period, but remains internal during the wandering period (Dwyer *et al.*, 1986). During the pupal stages, however, virtually no newly synthesized hydrocarbon is transported to the surface of either *T. ni* (de Renobales *et al.*, 1988) or *S. eridania* (Guo and Blomquist, 1991). The possibility that some of these "internal" lipids are associated with the innermost layers of the cuticle is likely, and indeed, "internal" hydrocarbons are probably present on the inner cuticle as well as in epidermal tissue, fat body, other organs, and especially lipophorin in the hemolymph (Chapter 5).

Long-chain methyl-branched hydrocarbons are the main components of the cuticular hydrocarbons in *T. ni* larvae (98%), pupae (98%) and adults (74%) and in *S. eridania* larvae (70%), pupae (75%) and adults (65%) (Guo and Blomquist, 1991), although large differences exist in their synthesis and transport to the cuticle throughout development. Immediately after a larval molt and during the feeding stages of the last two larval instars, hydrocarbons are actively synthesized and transported to the surface of the cuticle. However, during the wandering states, synthesis of hydrocarbon decreases dramatically, and the newly synthesized hydrocarbon remains inside the insect and does not appear on the surface of the cuticle. Two or three hours after the larval molt, the rate of hydrocarbon synthesis increases markedly (Dwyer *et al.*, 1986; Guo and Blomquist, 1991).

T. ni pupae synthesize methylalkanes from both labeled acetate and propionate between days 2 and 4 after the pupal mold, but labeled hydrocarbon is not recovered in extracts of surface hydrocarbons at any time during pupal development (de Renobales *et al.*, 1988). Thus, once the pupal case is formed, newly synthesized hydrocarbon is not added to the outside of the pupal cuticle. Similar findings were reported for *S. eridania* (Guo and Blomquist, 1991). The tissue associated with the pupal cuticle synthesizes almost all of the hydrocarbon, but most of this newly synthesized hydrocarbon is found in internal tissues, suggesting that after synthesis, hydrocarbon is transported from the integument to an unknown internal storage site (de Renobales *et al.*, 1988). Immediately before adult emergence, this hydrocarbon synthesized by the pharate adult within the pupal case begins to be deposited on the cuticle of the pharate adult, and deposition continues, although at a slower rate, during the first day of the adult stage. A similar study was done on the flesh fly, *Sarcophaga bullata*, and appears to show a similar time course of hydrocarbon formation from larvae to pupae to adult as was observed in the Lepidoptera, but interpretation is uncertain because analysis was done on the whole insect (Armold and Regnier, 1975).

These studies show that for *T. ni* and *S. eridania*, the internal hydrocarbon found in the insect immediately prior to a molt (larva to larva, larva to pupa and pupa to adult) is used as both cuticular and internal hydrocarbon immediately after the molt. The cast skin contains all the hydrocarbons that were found on the surface of the insect prior to the molt, indicating that the cuticular hydrocarbons were not re-adsorbed and re-utilized. Rather, the cuticular hydrocarbons of the newly ecdysed insect are from those synthesized and stored during the previous stage.

The storage and transport of hydrocarbons have only received attention in recent years (see Chapter 5). In a number of species, hemolymph hydrocarbons are carried by lipophorin (Chino, 1985; Schal *et al.*, 2003), and there has been a tacit assumption that most internal hydrocarbons are associated with the hemolymph. However, in *M. domestica,* Schal *et al.* (2001) showed that less than 20% of the internal hydrocarbons are in the hemolymph, and large amounts are found in the ovary. The recognition that newly synthesized hydrocarbons first appear in the hemolymph associated with lipophorin has required a paradigm shift in understanding hydrocarbon deposition on the surface of the insect. The old model, in which hydrocarbon is transported from the site of synthesis in epidermal related cells via pore canals, must be re-examined. Indeed, how hydrocarbon is transported from hemolymph lipophorin to the surface of the insect is not known. Work in the termite *Zootermopsis angusticollis* (Sevela *et al.*, 2000) suggests that specificity in the deposition of hydrocarbons can occur, and work is needed to explore the mechanism of how different hydrocarbons are shuttled to different places in the insect.

One of the most striking examples of specificity in the deposition of hydrocarbons occurs in several lepidopterans, in which shorter-chain pheromone or pheromone precursor hydrocarbons (Schal *et al.*, 1998; Jurenka, 2003; Jurenka *et al.*, 2003) are synthesized by oenocytes, and then co-transported by lipophorin along with the very-long-chain components destined to be cuticular hydrocarbons. The short-chain pheromone or pheromone precursor hydrocarbons are selectively transported to the pheromone gland, whereas the cuticular hydrocarbons become distributed over the cuticle (Jurenka *et al.*, 2003). How this remarkable specificity is achieved is unknown, but provides an interesting model in which to study hydrocarbon deposition.

While the site of synthesis and the metabolic pathways of hydrocarbon formation are known with some certainty for insects, our understanding of their regulation and the genes/ enzymes involved is in its infancy. The powerful tools of molecular biology are beginning to be applied to these systems (Chapter 4), and the next few decades should see giant leaps forward in our understanding of hydrocarbon formation.

References

Adams, T. S., Gerst, J. W. and Masler, E. P. (1997). Regulation of ovarian induced ecdysteroid production in the housefly, *Musca domestica. Arch. Insect Biochem. Physiol.*, **35**, 135–148.

Ahmad, S., Kirkland, K. E. and Blomquist, G. J. (1987). Evidence for a sex pheromone metabolizing cytochrome P-450 monooxygenase in the housefly. *Arch. Insect Biochem. Physiol.*, **6**, 121–140.

Armold, M. T. and Regnier, F. E. (1975). Stimulation of hydrocarbon biosynthesis by ecdysterone in the flesh fly *Sarcophaga bullata. J. Insect Physiol.*, **21**, 1581–1586.

Blailock, T. T., Blomquist, G. J. and Jackson, L. L. (1976). Biosynthesis of 2-methylalkanes in the crickets *Nemobius fasciatus* and *Gryllus pennsylvanicus. Biochem. Biophys. Res. Comm.*, **68**, 841–849.

Blomquist, G. J. (2003). Ecdysteroid regulation of pheromone production in the housefly, *Musca domestica*. In *Insect Pheromone Biochemistry and Molecular Biology*, ed. G. J. Blomquist and R. G. Vogt. London: Elsevier, pp. 231–252.

Blomquist, G. J., Adams, T. S. and Dillwith, J. W. (1984). Induction of female sex pheromone production in male houseflies by ovary implants or 20-hydroxyecdysone. *J. Insect Physiol.*, **30**, 295–302.

Blomquist, G. J., Chu, A. J., Nelson, J. H. and Pomonis, J. G. (1980). Incorporation of [2,3–^{13}C]succinate into methyl-branched alkanes in a termite. *Arch. Biochem. Biophys.*, **204**, 648–650.

Blomquist, G. J., Guo, L., Gu, P., Blomquist, C., Reitz, R. C. and Reed, J. R. (1994). Methyl-branched fatty acids and their biosynthesis in the housefly, *Musca domestica* L. (Diptera: Muscidae). *Insect Biochem. Mol. Biol.*, **24**, 803–810.

Blomquist, G. J. and Jackson, L. L. (1973a). Incorporation of labelled dietary *n*-alkanes into cuticular lipids of the grasshopper *Melanoplus sanguinipes. J. Insect Physiol.*, **19**, 1639–1647.

Blomquist, G. J. and Jackson, L. L. (1973b). Hydroxylation of n-alkanes to secondary alcohols and their esterification in the grasshopper *Melanoplus sanguinipes. Biochem. Biophys. Res. Comm.*, **53**, 703–708.

Blomquist, G. J. and Kearney, G. P. (1976). Biosynthesis of the internally branched monomethylalkanes in the cockroach *Periplaneta fulliginosa. Arch. Biochem. Biophys.*, **173**, 546–553.

Blomquist, G. J., Major, M. A. and Lok, J. B. (1975a). Biosynthesis of 3-methylpentacosane in the cockroach *Periplaneta americana. Biochem. Biophys. Res. Comm.*, **64**, 43–50.

Blomquist, G. J., McCain, D. C. and Jackson, L. L. (1975b). Incorporation of oxygen-18 into secondary alcohols of the grasshopper *Melanoplus sanguinipes. Lipids*, **10**, 303–306.

Blomquist, G. J., Soliday, C. L., Byers, B. A., Brakke, J. W. and Jackson, L. L. (1972). Cuticular lipids of insects: V. Cuticular wax esters of secondary alcohols from the grasshoppers *Melanoplus packardii* and *Melanoplus sanguinipes. Lipids,* **7**, 356–362.

Bognar, A. L., Paliyath, G., Rogers, L. and Kolattukudy, P. E. (1984). Biosynthesis of alkanes by particulate and solubilized enzyme preparations from pea leaves (*Pisum sativum*). *Arch. Biochem. Biophys.*, **235**, 8–17.

Brindle, P. A., Baker, F. C., Tsai, L. W., Reuter, C. C. and Schooley, D. A. (1987). Sources of propionate for the biogenesis of the ethyl-branched insect juvenile hormones: Role of isoleucine and valine. *Proc. Natl. Acad. Sci. USA*, **84**, 7906–7910.

Brindle, P. A., Schooley, D. A., Tsai, L. and Baker, F. C. (1988). Comaparative metabolism of branched-chain amino acids to precursors of juvenile hormone biogenesis in corpora allata of lepidopterous versus non-lepidopterous insects. *J. Biol. Chem.*, **263**, 10653–10657.

Carlson, D. A., Mayer, M. S., Silhacek, D. L., James, J. D., Beroza, M. and Bierl, B. A. (1971). Sex attractant pheromone of the housefly: Isolation, identification and synthesis. *Science*, **174**, 76–77.

Charlton, R. E. and Roelofs, W. L. (1991). Biosynthesis of a volatile, methyl-branched hydrocarbon sex-pheromone from leucine by arctiid moths (*Holomelina* spp.). *Arch. Insect Biochem. Physiol.*, **18**, 81–97.

Chase, J., Jurenka, R. A., Schal, C., Halarnkar, P. P. and Blomquist, G. J. (1990). Biosynthesis of methyl branched hydrocarbons in the German cockroach *Blattella germanica* (L.) (Orthoptera, Blattellidae). *Insect Biochem.*, **20**, 149–156.

Chase, J., Touhara, K., Prestwich, G. D., Schal, C. and Blomquist, G. J. (1992). Biosynthesis and endocrine control of the production of the German cockroach sex pheromone, 3,11-dimethylnonacosan-2-one. *Proc. Natl. Acad. Sci. USA*, **89**, 6050–6054.

Cheesbrough, T. M. and Kolattukudy, P. E. (1984). Alkane biosynthesis by decarbonylation of aldehydes catalyzed by a particulate preparation from *Pisum sativum. Proc. Natl. Acad. Sci. USA*, **81**, 6613–6617.

Cheesbrough, T. M. and Kolattukudy, P. E. (1988). Microsomal preparation from animal tissue catalyzes release of carbon monoxide from a fatty aldehyde to generate an alkane. *J. Biol. Chem.*, **263**, 2738–2742.

Chino, H. (1985). Lipid transport: biochemistry of hemolymph lipophorin. In *Comprehensive Insect Physiology, Biochemistry and Pharmacology*, Vol. 10, ed. G. A. Kerkut and L. I. Gilbert. Oxford: Pergamon, pp. 115–135.

Chu, A J. and Blomquist, G. J. (1980a). Decarboxylation of tetracosanoic acid to *n*-tricosane in the termite *Zootermopsis angusticollis. Comp. Biochem. Physiol.*, **66B**, 313–317.

Chu, A. J. and Blomquist, G. J. (1980b). Biosynthesis of hydrocarbons in insects: Succinate is a precursor of the methyl branched alkanes. *Arch. Biochem. Biophys.*, **201**, 304–312.

Darbro, J. M., Millar, J. G., McElfresh, J. S. and Mullens, B. A. (2005). Survey of muscalure [(Z)-9-tricosene] on house flies (Diptera: Muscidae) from field populations in California. *Environ. Entomol.*, **34**, 1418–1425.

Dennis, M. W. and Kolattukudy, P. E. (1991). Alkane biosynthesis by decarbonylation of aldehyde catalyzed by a microsomal preparation from *Botryococcus brauni*. *Arch. Biochem. Biophys.*, **287**, 268–275.

de Renobales, M., Nelson, D. R., Mackay, M. E., Zamboni, A. C. and Blomquist G. J. (1988). Dynamics of hydrocarbon biosynthesis and transport to the cuticle during pupal and early adult development in the cabbage looper *Trichoplusia ni* (Lepidoptera: Noctuidae). *Insect Biochem.*, **18**, 607–613.

Dillwith, J. W., Adams, T. S. and Blomquist, G. J. (1983). Correlation of housefly sex pheromone production with ovarian development. *J. Insect Physiol.*, **29**, 377–386.

Dillwith, J. W., Nelson, J. H., Pomonis, J. G., Nelson, D. R. and Blomquist, G. J. (1982). A ^{13}C-NMR study of methyl-branched hydrocarbon biosynthesis in the housefly. *J. Biol. Chem.*, **257**, 11305–11314.

Dwyer, L. A., Blomquist, G. J., Nelson, J. H. and Pomonis, J. G. (1981). A ^{13}C-NMR study of the biosynthesis of 3-methylpentacosane in the American cockroach. *Biochim. Biophys. Acta*, **663**, 536–544.

Dwyer, L. A., Zamboni, A. C. and Blomquist, G. J. (1986). Hydrocarbon accumulation and lipid biosynthesis during larval development in the cabbage looper, *Trichoplusia ni*. *Insect Biochem.*, **16**, 463–469.

Eliyahu, D., Nojima, S., Capracotta, S. S., Comins, D. L. and Schal, C. (2008). Identification of cuticular lipids eliciting interspecific courtship in the German cockroach, *Blattella germanica*. *Naturwissenschaften*, **95**, 403–412.

Fan, Y., Zurek, L., Dykstra, M. J. and Schal, C. (2003). Hydrocarbon synthesis by enzymatically dissociated oenocytes of the abdominal integument of the German cockroach, *Blattella germanica*. *Naturwissenschaften*, **90**, 121–126.

Ferveur, J.-F., Savarit, F., O'Kane, C. J., Sureau, G., Greenspan, R. J. and Jallon, J.-M. (1997). Genetic feminization of pheromones and its behavioral consequences in *Drosophila* males. *Science*, **276**, 1555–1558.

Gu, X., Quilici, D., Juarez, P., Blomquist, G. J. and Schal, S. (1995). Biosynthesis of hydrocarbons and contact sex pheromone and their transport by lipophorin in females of the German cockroach (*Blattella germanica*). *J. Insect Physiol.*, **41**, 257–267.

Gu, P., Welch, W. H. and Blomquist, G. J. (1993). Methyl-branched fatty acid biosynthesis in the German cockroach, *Blattella germanica*: kinetic studies comparing a microsomal and soluble fatty acid synthetase. *Insect Biochem. Mol. Biol.*, **23**, 263–271.

Gu, P., Welch, W. H., Guo, L., Schegg, K. M. and Blomquist, G. J. (1997). Characterization of a novel microsomal fatty acid synthetase (FAS) compared to a cytosolic FAS in the housefly, *Musca domestica*. *Comp. Biochem. Physiol.*, **118B**, 447–456.

Guo, L. and Blomquist, G. J. (1991). Identification, accumulation and biosynthesis of the cuticular hydrocarbons of the southern armyworm *Spodoptera eridania* (Lepidoptera: Noctuidae). *Arch. Insect Biochem. Physiol.*, **16**, 19–30.

Guo, L., Quilici, D. R., Chase, J. and Blomquist, G. J. (1991). Gut tract microorganisms supply the precursors for methyl-branched hydrocarbon biosynthesis in the termite, *Zootermopsis nevadensis*. *Insect Biochem.*, **21**, 327–333.

Hagedom, H. H. (1985). The role of ecdysteroids in reproduction. In *Comprehensive Insect Physiology, Biochemistry, and Pharmacology, Vol. 8,* ed. G.A. Kerkut and L.I. Gilbert. Oxford: Pergamon, pp. 205–262.

Halarnkar, P. P., Nelson, J. H., Heisler, C. R. and Blomquist, G. J. (1985). Metabolism of propionate to acetate in the cockroach *Periplaneta americana*. *Arch. Biochem. Biophys.*, **236**, 526–534.

Halarnkar, P. P. and Blomquist, G. J. (1989). Partial characterization of the pathway from propionate to acetate in the cabbage looper, *Trichoplusia ni*. *Insect Biochem.*, **19**, 7–13.

Halarnkar, P. P., Chambers, J. D. and Blomquist, G. J. (1986). Metabolism of propionate to acetate in nine insect species. *Comp. Biochem. Physiol.*, **84B**, 469–472.

Howard, R. W. and Blomquist, G. J. (2005). Ecological, behavioral and biochemical aspects of insect hydrocarbons. *Annu. Rev. Entomol.*, **50**, 371–393.

Howard, R. W., McDaniel, C. A. and Blomquist, G. J. (1980). Chemical mimicry as an integrating mechanism: Cuticular hydrocarbons of a termitophile and its host. *Science*, **210**, 431–433.

Jackson, L. L. (1972). Cuticular lipids of insects IV. Hydrocarbons of the cockroaches *Periplaneta japonica* and *Periplaneta americana* compared to other cockroach hydrocarbons. *Comp. Biochem. Physiol.*, **41B**, 331–336.

Jackson, A., and Locke, M. (1989). The formation of plasma membrane reticular systems in the oenocytes of an insect. *Tissue Cell*, **21**, 463–473.

Jallon, J.-M. and Wicker-Thomas, C. (2003). Genetic studies on pheromone production in *Drosophila*. In *Insect Pheromone Biochemistry and Molecular Biology: the Biosynthesis and Detection of Pheromone and Plant Volatiles*, ed. G. J. Blomquist and R. G. Vogt. London: Elsevier, pp. 253–281.

Juarez, M. P., Ayala, S. and Brenner, R. R. (1996). Methyl-branched fatty acid biosynthesis in *Triatoma infestans*. *Insect Biochem. Mol. Biol.*, **26**, 593–598.

Juarez, M. P., Chase, J. and Blomquist, G. J. (1992). A microsomal fatty acid synthetase from the integument of *Blattella germanica* synthesizes methyl-branched fatty acids, precursors to hydrocarbon and contact sex pheromone. *Arch. Biochem. Biophys.*, **293**, 333–341.

Juarez, M. P. and Fernandez, G. C. (2007). Cuticular hydrocarbons of triatomines. *Comp. Biochem. Physiol. A*, **147**, 711–730.

Jurenka, R. A. (2003). Biochemistry of female moth sex pheromones. In *Insect Pheromone Biochemistry and Molecular Biology: the Biosynthesis and Detection of Pheromones and Plant Volatiles*, ed. G. J. Blomquist and R. G. Vogt. London: Elsevier, pp. 53–80.

Jurenka, R. A., Schal, C., Burns, E., Chase, J. and Blomquist, G. J. (1989). Structural correlation between cuticular hydrocarbons and female contact sex pheromone of German cockroach *Blattella germanica* (L.). *J. Chem. Ecol.*, **15**, 939–949.

Jurenka, R. A., Subchev, M., Abad, J.-L., Choi, M.-Y. and Fabrias, G. (2003). Sex pheromone biosynthetic pathway for disparlure in the gypsy moth, *Lymantria dispar*. *Proc. Natl. Acad. Sci. USA*, **100**, 809–814.

Lamb, N. J. and Monroe, R. E. (1968). Lipid synthesis from acetate-1-C[14] by the cereal leaf beetle, *Oulema malanopus*. *Ann. Ent. Soc. Am.*, **61**, 1164–1166.

Lucas, K. A., Filley, J. R., Erb, J. M., Graybill, E. R. and Hawes, J. W. (2007). Peroxisomal metabolism of propionic acid and isobutyric acid in plants. *J. Biol. Chem.*, **34**, 24980–24989.

Major, M. A. and Blomquist, G. J. (1978). Biosynthesis of hydrocarbons in insects: Decarboxylation of long chain acids to *n*-alkanes in *Periplaneta*. *Lipids*, **13**, 323–328.

Mpuru, S., Reed, J.R., Reitz, R.C. and Blomquist, G.J. (1996). Mechanism of hydrocarbon biosynthesis from aldehyde in selected insect species: Requirement for O_2 and NADPH and carbonyl group released as CO_2. *Insect Biochem. Mol. Biol.*, **26**, 203–208.

Nelson, D.R. (1969). Hydrocarbon synthesis in the American cockroach. *Nature*, **221**, 854–855.

Nelson, D.R. and Blomquist, G.J. (1995). Insect waxes. In *Waxes: Chemistry, Molecular Biology and Functions*, ed. R.J. Hamilton. Dundee, Scotland: Oily Press, pp. 1–90.

Nishida, R. and Fukami, H. (1983). Female sex pheromone of the German cockroach, *Blattella germanica*. *Mem. Coll. Agric. Kyoto Univ.*, **122**, 1–24.

Reed, J.R., Hernandez, P., Blomquist, G.J., Feyereisen, R. and Reitz, R.C. (1996). Hydrocarbon biosynthesis in the housefly, *Musca domestica*: substrate specificity and cofactor requirement of P450hyd. *Insect Biochem. Mol. Biol.*, **26**, 267–276.

Reed, J.R., Quilici, D.R., Blomquist, G.J. and Reitz, R.C. (1995). Proposed mechanism for the cytochrome-P450 catalyzed conversion of aldehydes to hydrocarbons in the housefly, *Musca domestica*. *Biochemistry*, **34**, 26221–26227.

Reed, J.R., Vanderwel, D., Choi, S., Pomonis, J.G., Reitz, R.C. and Blomquist, G.J. (1994). Unusual mechanism of hydrocarbon formation in the housefly: Cytochrome P450 converts aldehyde to the sex pheromone component (Z)-9-tricosene and CO_2. *Proc. Natl. Acad. Sci. USA*, **91**, 10000–10004.

Schal, C., Fan, Y. and Blomquist, G.J. (2003). Regulation of pheromone biosynthesis, transport and emission in cockroaches. In *Insect Pheromone Biochemistry and Molecular Biology: the Biosynthesis and Detection of Pheromones and Plant Volatiles*, ed. G.J. Blomquist and R.G. Vogt. London: Elsevier Academic Press, pp. 283–322.

Schal, C., Burns, E.L., Jurenka, R.A. and Blomquist, G.J. (1990). A new component of the female sex pheromone of *Blattella germanica* (L.) (Dictyoptera: Blattellidae), and interaction with other pheromones components. *J. Chem. Ecol.*, **16**, 1997–2008.

Schal, C., Sevala, V., Capurro, M. de L., Snyder, T.E., Blomquist, G.J. and Bagnères A.-G. (2001). Tissue distribution and lipophorin transport of hydrocarbon and sex pheromones in the house fly, *Musca domestica*. *J. Insect Sci.*, **12**, 1–11.

Schal, C., Sevala, V. and Cardé, R.T. (1998). Novel and highly specific transport of volatile sex pheromone by hemolymph lipophorin in moths. *Naturwissenschaften*, **85**, 339–342.

Sevala, V., Bagnères, A.-G., Kuenzli, M., Blomquist, G.J. and Schal, C. (2000). Cuticular hydrocarbons of the termite *Zootermopsis nevadensis* (Hagen): caste differences and role of lipophorin in transport of hydrocarbons and hydrocarbon metabolites. *J. Chem. Ecol.*, **26**, 765–789.

Tillman-Wall, J.A., Vanderwel, D., Kuenzli, M.E., Reitz, R.C. and Blomquist, G.J. (1992). Regulation of sex pheromone biosynthesis in the housefly, *Musca domestica*: relative contribution of the elongation and reductive step. *Arch. Biochem. Biophys.*, **299**, 92–99.

Vaz, A.H., Jurenka, R.A., Blomquist, G.J. and Reitz, R.C. (1988). Tissue and chain length specificity of the fatty acyl-CoA elongation system in the American cockroach. *Arch. Biochem. Biophys.*, **267**, 551–557.

Vroman, H.E., Kaplanis, J.N. and Robbins, W.E. (1965). Effect of allatectomy on lipid biosynthesis and turnover in the female American cockroach, *Periplaneta americana* (L). *J. Insect Physiol.*, **11**, 897–903.

Wakayama, E.J., Dillwith, J.W., Howard, R.W. and Blomquist, G.J. (1984). Vitamin B12 levels in selected insects. *Insect Biochem.*, **14**, 175–179.

Yoder, J.A., Denlinger, D.L., Dennis, M.W. and Kolattukudy, P.E. (1992). Enhancement of diapausing flesh fly puparia with additional hydrocarbons and evidence for alkane biosynthesis by a decarbonylation mechanism. *Insect Biochem. Mol. Biol.*, **22**, 237–243.

4

Molecular biology and genetics of hydrocarbon production

Claude Wicker-Thomas and Thomas Chertemps

Hydrocarbons (HC) act as pheromones in a variety of orders including the Dictyoptera (Jurenka *et al.*, 1989; Schal *et al.*, 1994; Lihoreau and Rivault, 2009), Coleoptera (Ginzel *et al.*, 2003, 2006), Hymenoptera (Howard, 1993; Le Conte and Hefetz, 2008), Diptera (Carlson *et al.*, 1971; Antony and Jallon, 1982; Blomquist *et al.*, 1987) and several lepidopteran species (Roelofs and Cardé, 1971; Millar, 2000). In most insects, they are present on the cuticle and are synthesized in large cells called oenocytes located within or under the abdominal integument (Diehl, 1975; Ferveur *et al.*, 1997; Schal *et al.*, 1998; Fan *et al.*, 2003). In Lepidoptera, their synthesis occurs in tissues associated with the abdominal tegument (possibly oenocytes), and they are then released into a sex pheromone gland (Schal *et al.*, 1998).

Hydrocarbons have long straight or methyl-branched chains (C_{19} to C_{40}) that may be unsaturated or saturated and, depending on the chain length, are either volatile (in Lepidoptera: Roelofs and Cardé, 1971) or act over short distances or by contact (in other orders). Their biosynthesis has been studied by the use of radiolabeled precursors and the biosynthetic pathways have been established (Dillwith *et al.*, 1981, 1982; Blomquist *et al.*, 1987). Pathways involve the following: the synthesis of medium-chain fatty acids by a fatty acid synthetase (FAS), their desaturation by desaturase(s) followed by their elongation to very-long-chain fatty acids by elongase(s), and then decarboxylation to hydrocarbons. During the last ten years the use of molecular techniques has enabled a deeper understanding of fatty-acid-derived, short-chain (C_{10} to C_{18}) pheromone biosynthesis in Lepidoptera. However, such techniques have been applied less to the study of hydrocarbons, partly because their chain length has rendered functional studies more difficult, and partly because some biosynthesis enzymes, such as elongases used in the synthesis of very-long-chain fatty acids, are poorly conserved and difficult to characterize. That explains why they have only been characterized in *Drosophila*, whose genome has been entirely sequenced. In this chapter, the characteristics of biosynthesis enzymes are reviewed, with special reference to *Drosophila*. However, data available for other organisms are also reported to try to establish similarities between hydrocarbon biosynthesis in different species. The evolution of *Drosophila* hydrocarbons is discussed and their role in speciation is suggested.

Fatty acid synthetase

Biochemical studies have shown that, depending on the ionic strength, short-chain (C_{12} and C_{14}) or medium-chain (C_{14} to C_{18}) fatty acids are synthesized by the *Drosophila* FAS (de Renobales and Blomquist, 1984). In the German cockroach, *Blattella germanica*, and the housefly, *Musca domestica*, two different FAS have been characterized: a cytosolic and a microsomal FAS (Juarez *et al.*, 1992; Gu *et al.*, 1997). The microsomal FAS incorporates methylmalonyl more efficiently than the soluble FAS and has been suggested to be involved in forming the precursors to methyl-branched hydrocarbons, whereas the cytosolic FAS may be involved in forming the precursors to straight-chain hydrocarbons.

Desaturases

Fatty acyl-CoA desaturases are terminal oxidases of a membrane-bound enzyme complex that also includes cytochrome b5 and cytochrome b5 reductase (Bloomfield and Bloch, 1960). They remove substrate hydrogen atoms at a position determined by the specificity of the enzyme. They play essential roles in regulating membrane fluidity and are also involved in insect lipid and pheromone metabolism. They share the presence of three highly conserved histidine-rich sequences (H-boxes) that coordinate the diiron-oxo structure at the active sites (Shanklin and Cahoon, 1998) and four hydrophobic α helices that appear to anchor the protein into the lipid bilayer and situate the H-boxes in their correct position in the active site.

After the first three molecular characterizations of animal desaturases from rat liver (Thiede *et al.*, 1986), mouse adipose tissue (Ntambi *et al.*, 1988) and carp (Tiku *et al.*, 1996), a *Drosophila* desaturase was isolated in 1997 (Wicker-Thomas *et al.*, 1997). Since then, numerous studies have been made on lepidopteran desaturases involved in non-hydrocarbon short-chain pheromones (Knipple *et al.*, 1998; reviewed in Knipple and Roelofs, 2003). In *Drosophila*, there are seven fatty acyl-CoA desaturase genes, which are all located on chromosome III (Figure 4.1), but only three desaturases appear to be involved in hydrocarbon synthesis. In the cricket, a desaturase has been characterized but there is no evidence that this desaturase is involved in pheromone biosynthesis (Riddervold *et al.*, 2002). On the other hand, the desaturase isolated from the housefly is probably involved in both lipid and pheromone biosynthesis (Eigenheer *et al.*, 2002).

The isolation of the desaturase gene *ole1* from the yeast (Stukey *et al.*, 1989) and its replacement by the rat desaturase in an *ole1*-mutant yeast which resulted in functional complementation (Stukey *et al.*, 1990) have permitted the functional characterization of insect desaturases. While lepidopteran desaturases with unusual activities have been described (e.g. $\Delta 10$, $\Delta 11$, $\Delta 14$) (Knipple *et al.*, 1998; reviewed in Roelofs *et al.*, 2002), the three insect genes functionally characterized to date that are involved in hydrocarbon synthesis correspond to desaturases with a $\Delta 9$ specificity (Dallerac *et al.*, 2000; Eigenheer *et al.*, 2002). However, in some species, desaturases with other specificities that are involved in hydrocarbon synthesis seem likely to occur.

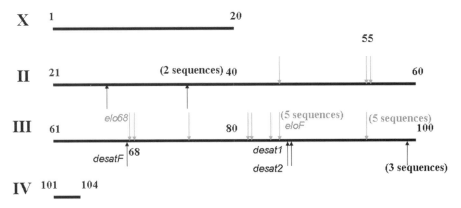

Figure 4.1 Cytogenetic localization of desaturase and elongase sequences in *D. melanogaster*. The four chromosomes are numbered with roman figures on the left. The arabic numbers on the chromosomes indicate the cytological positions. Arrows show the localizations of desaturases and elongases. Sixteen elongases and seven fatty acid desaturases are located on chromosome III. Three elongases, 1 sphingolipid delta-4 desaturase and two other putative desaturases are located on chromosome II.

Comparison of Musca *and* Drosophila *desaturases*

In *Musca*, only one fatty acyl-CoA desaturase with Δ9 specificity has been detected (Wang *et al.*, 1982). Sequences corresponding to a Δ9 desaturase have been isolated (Eigenheer *et al.*, 2002). This desaturase is similar to *D. melanogaster* Desat1 (Table 4.1). In yeast, it desaturates palmitic and stearic acids, producing approximately equal amounts of palmitoleic and oleic acids. As relatively high amounts of palmitoleic acid are present in the housefly (Wang *et al.*, 1982), the desaturation of palmitic acid might be used more for fat body metabolism, and that of stearic acid more for HC synthesis (Figure 4.2). At the gene level, *Musca* desaturase, like *desat1*, has several transcripts with identical coding regions, differing in their 5′ untranslated regions (UTRs), probably due to alternative splicing (Eigenheer *et al.*, 2002; Jallon and Wicker-Thomas, 2003). It contains three introns, located at sites identical to three of the four intron sites in *D. melanogaster desat1*. The second intron position is conserved across most vertebrate and insect desaturases, suggesting an ancestral character (Rosenfield *et al.*, 2001).

Drosophila *desaturases: genetic studies*

Role of Desat1

This desaturase transforms palmitic ($C_{16:0}$) and stearic ($C_{18:0}$) acids to palmitoleic (Δ9 $C_{16:1}$) and oleic (Δ9 $C_{18:1}$) acids, precursors of ω7 and ω9 fatty acids and 7- and 9-HC. Several *desat1* mutants have been described (Figure 4.3). The gene is expressed in both the fat body and oenocytes and plays a role both in general lipid metabolism and in hydrocarbon synthesis (Ueyama *et al.*, 2005; Marcillac *et al.*, 2005a; Figure 4.4). A piggyBac strain inserted in

Table 4.1 *Comparison of the desaturases from* D. melanogaster *and* M. domestica *involved in HC biosynthesis.*

	Drosophila melanogaster			*Musca*
Enzyme	Desat1	Desat2	DesatF	Desat
Enzyme specificity	Δ9	Δ9	N.S.	Δ9
Substrate specificity	C16:0, C18:0 (2,1)	C14:0	Monounsaturated fatty acids	C16:0, C18:0 (1,1)
Amino acid number	383	361	355	380
Identity to Desat1	100%	65%	53%	82%
Involved in HC biosynthesis	Z7-HC, Z7, Z11-HC	Z5, Z9-HC	Z5, Z9-HC Z7, Z11-HC	Z9-tricosene Z9-heptacosene
Involved in lipid metabolism	Yes	No	No	Yes
Gene: number of introns	4	3	0	3
Number of transcripts	5	1	1	2

desat2 has been described (Bonin and Mann, 2004); however, we found its site of insertion 280nt before the *desat1* orf. As the piggyBac vector contains a GFP encoding sequence fused to the transposon sequence, it is possible to visualize the expression of the gene where it is inserted. An expression pattern for *desat1* similar to that shown in Figure 4.4 (fat body and oenocytes) was found in this strain (data not shown).

Mutations in *desat1* affect HC, resulting in a very large decrease in 7-HC in males, and in 7- and 7,11-HC in females, with a parallel increase in saturated HC synthesis (Labeur *et al.*, 2002; Ueyama *et al.*, 2005; Marcillac *et al.*, 2005a,b). Lipid metabolism is impaired too, with both quantitatively and qualitatively altered fatty acid biosynthesis: the overall quantity of fatty acids was shown to be reduced by half and that of vaccenic acid, the common precursor to 7-HC in both sexes, reduced by a factor of six in a *desat1* mutant (Ueyama *et al.*, 2005).

Role of Desat2

In *D. melanogaster*, there is a female geographical polymorphism affecting the composition of cuticular hydrocarbons. Two distinct phenotypes occur: the "7,11-HC phenotype" in which females, more wide-spread, possess high levels of 7,11-heptacosadiene (7,11-HD) and the "5,9-HC phenotype", originating from sub-Saharan Africa, with only 1/10 the amount of 7,11-HD but high levels of its positional isomer 5,9-HD (Jallon and Péchiné, 1989; Ferveur *et al.*, 1996). Males from these different populations show the same main compounds unsaturated in position 7: 7-tricosene (7-T) and 7-pentacosene (7-P)

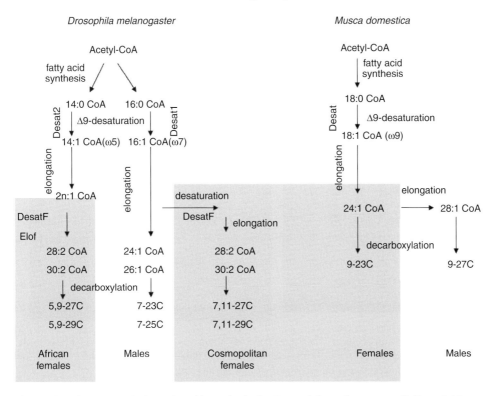

Figure 4.2 Pheromone hydrocarbon biosynthesis in *Drosophila melanogaster* (left) and *Musca domestica* (right). The steps within a grey background occur only in mature females.

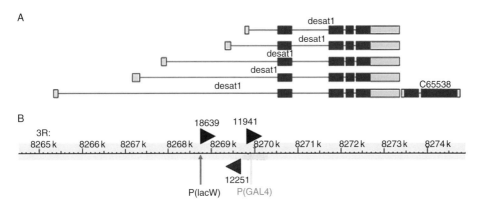

Figure 4.3 Representation of *desat1* gene. A. The five transcripts, corresponding to five different initiation sites of transcription. B. The P-strains reported in this study: P(lacW) (arrow) (Labeur *et al.*, 2002); three GS strains (arrows) (Ueyama *et al.*, 2005); one GAL strain: P(GawB)NP447 (arrow).

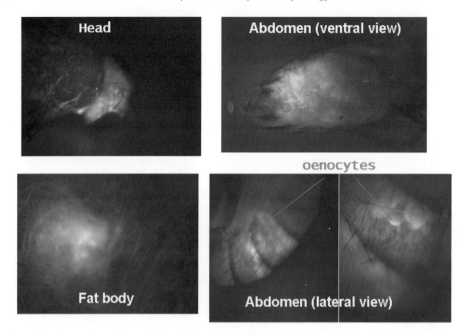

Figure 4.4 Expression of *desat1*. The P(GawB)NP447 strain was crossed to a P-UAS GFP strain and GFP expression was visualized in the progeny (four-day old flies). Head, ventral abdomen (fat body), fat body and oenocytes, visible on the lateral view of the abdomen, are labeled.

(Jallon, 1984). The HC polymorphism between populations of this species was mapped to the *desat* locus (Coyne *et al.*, 1999), which contains two Δ9-desaturase genes, *desat1* and *desat2*, located in tandem. We found that *desat2* is not expressed at all in Canton-S (7,11-HC strain), but is expressed in Tai females (5,9-HC strain) (Dallerac *et al.*, 2000). Desat2 desaturates myristic acid ($C_{14:0}$) to myristoleic acid (Δ9 $C_{14:1}$), leading to the synthesis of 5,9-HC in Tai females. A deficiency of 16 nucleotides present in 7,11-HC flies is responsible for the lack of *desat2* transcription in these females (Takahashi *et al.*, 2001).

The HC polymorphism in females seems to affect mating behavior, as females rich in 7,11-HD mate more rapidly than those rich in 5,9-HD (Ferveur *et al.*, 1996; Fang *et al.*, 2002) and a role for the inactivation of *desat2* in sexual isolation during incipient speciation has been suggested (Fang *et al.*, 2002).

In our initial experiments using reverse transcription-polymerase chain reaction (RT–PCR), no expression of *desat2* was detected in Tai males (Dallerac *et al.*, 2000). After modifying the RT–PCR conditions, we have, however, been able to detect faint expression of *desat2* in Tai males, but always much less than in females (data not shown). Using RT–PCR Greenberg *et al.* (2006) reported the expression of *desat2* in both males and females from two different 5,9-HC strains. In Tai, *desat2* appeared to be amplified less in males, while in Z30 strain, *desat2* amplification in males was similar to, or a little lower than, in females. We cannot explain the discrepancy between these studies. Although RT–PCR is not reliable for measuring mRNA quantities, *desat2* seems to be expressed in males from some 5,9-HC strains, and Coyne and Elwyn (2006) have suggested that *desat2* expression in males of different wild-type Z lines could vary from very low to very high. One other possibility is that genetic drift has occurred in the Tai strain; additionally, laboratory rearing conditions such as food might modify gene expression (Ferveur, 2005).

We have never found *desat2* expression in either sex of Canton-S, which is consistent with the data of Greenberg *et al.* (2006). Using microarrays and quantitative real-time RT–PCR, Michalak *et al.* (2007) found that *desat2* was down-regulated, but still expressed in three 7,11-HC strains (approximately half the amount expressed in 5,9-HC strains). That result raises a number of questions. The three 7,11-HC strains used in the latter study originated from different places (Japan, Ecuador, Crete). Is *desat2* expressed differently in different 7,11-HC populations, as is the case for 5,9-HC strains? What primers were used in this study? The two desaturase genes have very similar sequences and the open reading frames show overall co-hybridization with each other; most primers designed for *desat2* co-hybridize with *desat1* sequence and only primers complementary to the very beginning or end of the sequence are specific to *desat2*. It would be very useful to compare all the results and to investigate desat2 expression in other *D. melanogaster* strains to get more insight on the *desat2* gene and its expression.

Hydrocarbons from 5,9-HC and 7,11-HC strains differ only in females. In males, the same proportion of 5-HC (3–6%) occurs in both strains, suggesting that *desat2* has little or no influence on male HC biosynthesis.

The *desat2* gene has only one transcript and its regulation seems to be much simpler than *desat1*. It does not appear to have a role in fat-body metabolism; overall lipid composition is fairly similar in the Tai and Canton-S strains (Table 4.2). On the other hand, the loss of functional *desat2* in 7,11-HC strains might be involved in resistance to cold, leading to adaptation of these strains to more temperate regions (Greenberg *et al.*, 2003). However, Coyne and Elwyn (2006) could not reproduce these results. The resistance to cold of both males and females observed by Greenberg *et al.* (2003), if confirmed, might also be due to an indirect effect of *desat2* on other genes and not to a difference in HC.

Expression of *desat1* and *desat2* in the Tai strain

In African females rich in 5,9-HD, *desat1* and *desat2* genes are both expressed and the resulting enzymes coexist. The analysis of hydrocarbons shows a very high amount of 5,9-HD in Tai females (about 600 ng/fly) and a lower amount of 7,11-HD (about 100 ng/fly), compared to that of 7,11-HD in Canton-S females (about 400 ng/fly) (Jallon and Wicker-Thomas, 2003). These values might be the consequence of a greater representation of Desat2 enzyme in Tai females, or a better affinity for its substrate.

In order to test the first hypothesis, we tried to quantify the level of expression of both desaturase genes in Canton-S and Tai. The conventional methods for mRNA quantification such as northern blotting could not be used, as both genes show cross-hybridization over the entire open reading frame. Therefore, we chose to use an adaptation of PCR for accurate quantification of *desat* mRNA; this technique has the advantage of high sensitivity in the study of low-abundance mRNAs and has already been used for *desat1* quantification (Ueyama *et al.*, 2005). We developed the same technique for *desat2*.

Table 4.2 *Total fatty acid composition (percentages) and amount (μg/fly) in four-day-old Tai and Canton-S flies. Values ± SEM are means of 5 groups of 5 flies. In each line, values followed by the same letters (in brackets) are not significantly different at the 0.05 probability level.*

Fatty acids		Tai males	Canton-S males	Tai females	Canton-S females
12 :0		3.20±0.10 (b)	3.67±0.08 (a)	3.69±0.04 (a)	3.99±0.32 (a)
14 :0		24.00±0.36 (b)	20.88±0.28 (a)	24.45±0.34 (a)	23.97±1.13 (a)
14 :1	Δ5	0.35±0.03 (a)	0.42±0.04 (a)	0.38±0.02 (a)	0.47±0.02 (b)
14 :1	Δ7	0.20±0.04 (a)	0.24±0.06 (a,b)	0.39±0.03 (b)	0.34±0.04 (b)
14 :1	Δ9	1.06±0.08 (a)	0.43±0.07 (b)	1.14±0.04 (a)	0.89±0.19 (a)
15 :0		0.21±0.03 (b)	0.38±0.07 (a)	0.14±0.01 (b)	0.23±0.07 (a,b)
16 :0		25.84±0.31 (a)	27.97±0.15 (a)	24.63±0.09 (a)	23.59±1.14 (a)
16 :1	Δ7	0.31±0.02 (b)	0.55±0.09 (a)	0.25±0.01 (b)	0.47±0.03 (a)
16 :1	Δ9	22.35±0.14 (b)	18.73±0.38 (a)	22.08±0.14 (b)	20.84±0.57 (a)
17 :0		0.56±0.02 (b)	0.85±0.03 (a)	0.43±0.08 (b)	0.33±0.01 (a)
18 :0		2.14±0.09 (b)	3.31±0.07 (a)	1.36±0.05 (b)	3.61±1.82 (a)
18 :1	Δ9	14.06±0.11 (b)	13.29±0.15 (a)	16.0±0.17 (b)	14.73±0.60 (a)
18 :1	Δ11	0.29±0.01 (a)	0.28±0.02 (a)	0.16±0.01 (b)	0.20±0.01 (a)
20 :0		0.45±0.01 (b)	0.64±0.02 (a)	0.34±0.01 (b)	0.47±0.07 (a)
Total (μg/fly)		528±13 (b)	328±26 (a)	655±84 (c)	806±36 (b)

The quantitative analysis of *desat1* transcripts showed no significant difference, either between males and females or between Canton-S and Tai (Table 4.3). On the other hand, *desat2* RNA was only detected in Tai females, where it is present at a very low level (about 16% of *desat1* RNA level). This result seems surprising at first sight. However, unlike Desat2, Desat1 is also used for general metabolism. The very low abundance of *desat2* RNA favors the hypothesis of strict tissue specificity, although no expression study has yet been reported. Indeed, no *desat2* expressed sequence tag (EST) has been found in various databases, compared to more than 200 listed *desat1* ESTs (Jallon and Wicker-Thomas, 2003).

Table 4.3 *Transcript titers of* desat1 *and* desat2 *using competitive RT–PCR. The values represent the means of N determinations ± SEM of samples from separate experiments and are given in attomoles / μg RNA.*

Strain		N	*desat1*	*desat2*
Tai	Males	4	0.47 ± 0.05	0
	Females	6	0.42 ± 0.02	0.067 ± 0.008
Canton-S	Males	8	0.51 ± 0.07	0
	Females	4	0.45 ± 0.02	0

In 5,9-HC-producing tissues, competition between Desat1 and Desat2 for the substrate might occur. As *desat1* seems to be expressed at a similar level in 7,11- and 5,9-HC strains, the high production of 5,9-HC in the latter strains might also be due to the presence of a different FAS (producing more myristic than palmitic acid) or of different ionic strengths in the oenocytes (thus increasing the production of myristic acid). Alternatively, the expression of *desat1* might be lowered in oenocytes of 5,9-HC strains, without changing its expression in the fat body.

Role of DesatF

The production of diene pheromones in *D. melanogaster* females requires a second desaturation step. Characterization of the enzyme involved has taken time, although some genetic evidence has suggested that the cytological region 67E-69B is involved in this process (Coyne, 1996; Wicker-Thomas and Jallon, 2000; Gleason *et al.*, 2005). The main obstacle has been the inability to show substrate-specificity of the enzyme, and so RNA interference (RNAi) has been developed to allow the functional study of the gene (Chertemps *et al.*, 2006). The desaturase was found to be expressed only in *D. melanogaster* females, and was thus named *desatF*. As for *desat2*, no transcript has been reported in any of the databases from embryos, larvae, pupae, adult heads, ovaries, or testes. Only one full insert cDNA from whole flies has been described (BT023260). Again, this favors the strict tissue-specificity of the gene.

In some conditions, with varying genetic backgrounds of the *D. melanogaster* strains, faint *desatF* expression has been detected in males, with some dienes present in their HC (data not shown). However, expression remained low, while the ratio of female over male *desatF* expression reported in microarray studies was 6.9 (Parisi *et al.*, 2004) and 100 to 1000 (Table 4.4; Mackay *et al.*, 2005). Overexpression and knock-down of *desatF* resulted in more or less dienes in females, respectively. The courtship behavior of wild-type males toward desatF RNAi females decreased, resulting in fewer copulation attempts and increased copulation latency (Chertemps *et al.*, 2006). Interestingly, selection of flies for mating speed was followed by a divergence in RNA levels between the "fast" and "slow" replicate lines. Among the transcripts, *desatF* was significantly more abundant in "slow" males, compared with "fast" males, although there was much variation (Table 4.4; Mackay *et al.*, 2005).

Table 4.4 *Gene expression of desaturase (D) and elongase (E) genes differently expressed between* D. melanogaster *lines selected for fast and slow mating speeds (Mackay* et al., *2005). The cytological localization (cyt) of the genes is reported. Values are means (s.e.) of three replicates per fly.*

	Cyt.	Fast male	Slow male	Fast female	Slow female
D desatF	68	20.8 (10.5)	76.6 (148.0)	12758 (1037)	10199 (123)
M+C		18.4 (10.9)	19.0 (7.1)	9623 (693)	8403 (670)
D CG9743	99	1009 (38)	982 (37)	916 (52)	815 (29)
M+F+C		1050 (3.6)	873 (28)	1046 (40)	1020 (66)
E CG30008	46	13404 (832)	11494 (163)	7606 (581)	5538 (263)
M+F+C		13277 (580)	11835 (136)	6746 (350)	7137 (439)
E CG18609	55	2028 (91)	4089 (461)	1376 (135)	1758 (182)
M+F+C		2426 (324)	3345 (201)	1449 (84)	1950 (86)
E Elo68	68	1880 (104)	1918 (215)	12.7 (3.7)	6.3 (5.8)
M		1192 (150)	761 (295)	51.7 (1.5)	8.5 (10.7)
E Baldspot	73	2722 (19)	1910 (104)	2107 (109)	1586 (35)
C+M		2318 (147)	2206 (101)	1783 (58)	2094 (188)
E CG31522	82	1779 (89)	2092 (140)	1197 (75)	1366 (118)
F		2491 (156)	2211 (66)	1536 (181)	1831 (339)
E CG2781	84	2153 (65)	2739 (97)	899 (119)	1573 (88)
M+F+C		2436 (84)	2390 (456)	1194 (65)	1264 (171)
E CG16904	85	11468 (411)	10036 (201)	6293 (635)	6234 (47)
F		9561 (416)	10912 (595)	4725 (739)	6216 (674)
E CG5326	94	994 (72)	891 (46)	320 (28)	891 (46)
C+M		845 (78)	719 (69)	320 (12)	218 (49)
E CG33110	94	1518 (114)	1586 (72)	643 (34)	739 (44)
M+F+C		1249 (119)	1445 (22)	461 (139)	901 (34)

M and F denote male- and female-specific expression differences, and C denotes significant differences in transcript abundance pooled across sexes.

Elongases

In contrast to desaturases, whose highly conserved protein sequence has allowed extensive study of encoding genes and related enzyme function, the fatty acid elongases are still not well-characterized. Elongases mediate the first step of the elongation process: the condensation of malonyl-CoA with a long chain acyl-CoA (Blomquist, Chapter 3). Three subsequent steps yield the acyl-CoA elongated by two atom carbons. The condensation reaction is the rate-limiting and substrate-specific step (Cinti *et al.*, 1992; Moon *et al.*, 2001). Elongases are membrane-bound enzymes and contain one histidine-rich motif, which is also present in fatty acid desaturases.

Unlike elongases from *Caenorhabditis,* which cannot elongate fatty acids beyond 20C (Beaudoin *et al.*, 2000), the insect elongases that are involved in pheromone synthesis must

Table 4.5 *Number of transcripts reported for some elongase genes probably involved in pheromone biosynthesis. The cytological localization (cyt) of the genes is reported. Relative male vs. female expression is indicated (Parisi* et al., *2004; Mackay* et al., *2005).*

Genes	Cyt.	Whole adult	testis	head	embryo	Cell culture	M/F expression
CG30008	46	2	1	3		1	1.8
CG18609	55			1	1		1.5–2.8
CG17821	55	1	1				1.6
CG32072	68	13					50–200
CG16905	85	3					0.25
CG16904	85	4		4	2		2–3

have very high elongation capacities, which renders their study rather difficult. In the *D. melanogaster* genome, 19 elongase sequences have been found (Figure 4.1), sharing low amino acid identity (20–30%). Only two *Drosophila* elongases have yet been functionally characterized (none in other insects) and both are involved in the biosynthesis of pheromones. Given the complexity of the HC biosynthesis pathway, there must be other, as yet unknown, elongases involved.

Elongases involved in male 7-T/7-P ratio

A geographical polymorphism concerning the relative ratio of 7-T and 7-P has been described among *D. melanogaster* males (Rouault *et al.*, 2001, 2004) and *D. simulans* of both sexes (Jallon, 1984). This polymorphism – affecting the elongation of hydrocarbons – might be due to different elongase expression. An allele affecting this ratio in *D. simulans* has been reported and mapped at 2–65.3 (Ferveur, 1991). However, production also seems to be controlled by multiple loci from the four chromosomes (Scott and Richmond, 1988). In *D. melanogaster*, the 7-T to 7-P ratio appears to be under the strong control of chromosome II by two loci near positions 2–47.0 and 2–55.1 (Ferveur and Jallon, 1996). These locations correspond to cytological positions 49 B-C (locus in *D. simulans*) and 33F-34A and 41 A-B (loci in *D. melanogaster*). One elongase gene, CG 30008, is located nearby (in 46C), and two others, CG 17821 and CG 18609, farther away at 55F. We found the three genes expressed in both males and females, with higher expression in males (also reported in Parisi *et al.*, 2004 and Mackay *et al.*, 2005). Genes at 55F are in tandem and show 45% identity between the protein sequences; they might have occurred through duplication of the same gene.

Seven elongase transcripts have been reported at cytological position 46, and only two for each elongase at position 55 (Table 4.5) (flybase.bio.indiana.edu). One CG30008 transcript and one CG17821 transcript were found in the testis, but we verified that none of

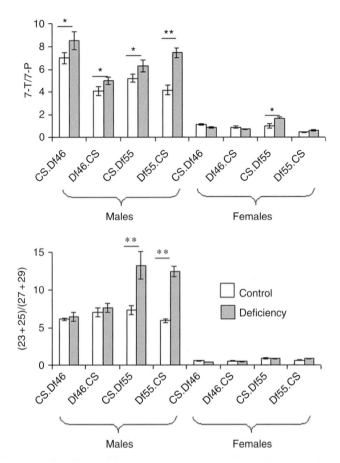

Figure 4.5 Effect of deficiency-mutants on HC elongation. 7-T / 7-P and (23 + 25) / (27 + 29) in either control or deficiency flies resulting from a cross between a wild type strain (CS: Canton-S) and a deficiency strain. Crosses are indicated on the X axis (Female X Male). Df46: Df(2R)01D09Y-MO92/SM6a (Df 45F5–6; 46C4–7); Df55: Df(2R) PC66/SM6a (Df 55D2-E1; 56B2). The values are means ± SEM of 10 individual measures. Means with * and ** were significantly different with the Mann–Whitney *U* test (p=0.05 and 0.01, respectively).

them had any effect on *cis*-vaccenyl acetate (cVA) synthesis. The two CG18609 transcripts were found in the adult head and the embryo; in the latter, CG18609 is specifically expressed in oenocytes (Gutierrez *et al.*, 2007). As no mutant was available for these genes, we used cytological deficiencies to address their role (Figure 4.5). The two loci at positions 46 and 55 might be involved in the elongation of short compounds in males (the 7-T to 7-P ratio was increased in mutants) and the locus at 55 might also be involved in the elongation of longer compounds in males (the (23+25) to (27+29) ratio was also increased in the Df55 mutant).

In *D. melanogaster* 7,11-HC strains, male 7-T has been shown to inhibit male–male courtship interactions (Scott and Jackson, 1988) and to stimulate conspecific females

(Scott, 1994; Sureau and Ferveur, 1999). If these elongase genes on chromosome II effect the elongation of fatty acids, resulting in higher production of 7-P in males, one could expect that lower expression of these genes would result in higher production of 7-T, and better mating fitness. Indeed, in males selected to mate quickly, CG18609 expression was reduced by 50%, but CG30008 expression was a little increased (Table 4.4). Both the effect of the Df55 mutant on HC elongation and the correlation between lower CG18609 expression in males and a quicker propensity for mating make this gene a candidate for the elongation of ω7-fatty acids from C_{24} to C_{26} (leading to 7-P). Steps have been taken to address this question.

Elongases involved in female long HC synthesis

We found that all the elongase genes showed higher expression in males, except CG16905, which was expressed about 20-fold more in females than in males. As *D. melanogaster* female HC are longer (27 and 29C), compared to male HC (23 and 25C), we investigated the role of CG16905 (which we named *eloF*), in this process.

EloF shows the highest homology with mouse Elovl1 (33% identity), which can elongate saturated and monosaturated fatty acids up to C_{24} and C_{26} (Tvrdik *et al.*, 2000). After heterologous expression in yeast, EloF was shown to elongate saturated and unsaturated fatty acids up to C_{30} (Chertemps *et al.*, 2007). In vivo, it is involved in female long-chain HC, particularly the synthesis of diene pheromones. It is expressed only in *D. melanogaster* and *D. sechellia* females, which have long-chain pheromones, and not in *D. simulans*, which has shorter chains (23 and 25C).

Is EloF the only elongase involved in long-chain HC synthesis? Experiments with females that were RNAi-knocked down for *eloF* show a large decrease in pheromone production, especially 7,11-ND (29C), which was replaced by 7,11-PD (25C). 7,11-HD levels were not much affected. This could stem from the driver used and/or the presence of another elongase involved in pheromone synthesis. EloF shows very low homology with other *Drosophila* elongases, with the exception of CG16904 (47% identity). Both genes are located at the same locus, lying in opposite directions. CG16904 is expressed twice as much in males as in females (Table 4.5), but the lower expression in females might be associated with a lower rate of mating (Table 4.4). This gene therefore seems to be another good candidate for long-chain HC synthesis.

In *Musca*, a similar elongation system occurs for HC, except that shorter monoenic HC (23 and 25C) are specific to mature females and longer HC (27 and 29C) are common to males and non-mature females (Dillwith *et al.*, 1983; Mpuru *et al.*, 2001). Unlike in *Drosophila*, where elongation is not hormonally regulated, *Musca* elongation is down-regulated by ecdysone (Adams *et al.*, 1984, 1995). It would be possible to isolate this *Musca* elongase using the relatively high homology that exists between elongases of different species that have similar specificities (mouse Elovl1 and *Drosophila* EloF). The characterization of elongases in different insect genera would really help the understanding of HC biosynthesis and regulation in insects.

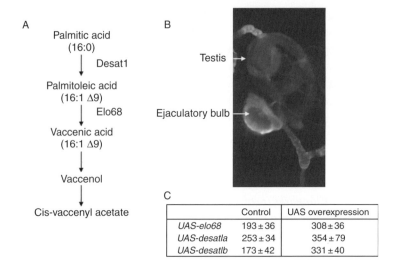

Figure 4.6 (A) Biosynthesis of cis-vaccenyl acetate. (B) Expression of *desat1* in male reproductive tract. The same strain and same protocol were used as in Figure 4.4. High *desat1* expression was seen in the ejaculatory bulb. (C) Effect of *desat1* or *elo68* overexpression in the ejaculatory bulb on cis-vaccenyl acetate production. The P(GawB)NP447 strain was crossed to a P-UAS /CyO strain. The progeny control desat1-GAL4/CyO was compared to desat1-GAL4/UAS. The values are means ± SEM of 10 individual measures (ng/fly).

cis-*vaccenyl acetate (cVA) biosynthesis*

cVA, also called (Z)-11-octadecenyl acetate, is not a hydrocarbon, since it carries an ester function. It is synthesized in the male reproductive tissue (ejaculatory bulb), transferred to the female during mating and inhibits its attractiveness during court-ship (Jallon *et al.*, 1981); it also functions as an aggregation pheromone (Bartelt *et al.*, 1985). Its synthesis is shown in Figure 4.6A.

We have characterized an elongase gene, *elo68* (CG32072), which shows male-biased expression, in the reproductive tract: testes and ejaculatory bulb. In yeast, Elo68 elon-gates both myristoleic (14:1) and palmitoleic (16:1) acids to fatty acids containing two additional carbons (16:1 Δ11 and 18:1 Δ11). Smaller but significant amounts of C18 Δ13 were also produced. Mutants were generated, which resulted in decreased production of cVA (Chertemps *et al.*, 2005). These results strongly suggest a role for *elo68* in cVA production.

To confirm this role, we wanted to study the effect of *elo68* overexpression on cVA syn-thesis. We first searched for Gal4 strains showing expression in the ejaculatory bulb. The desat1-Gal strain P(GawB)NP447, whose expression is shown in Figure 4.4, was found to be expressed in the fat body and oenocytes and also in the ejaculatory bulb (Figure 4.6B). The overexpression of *desat1* or *elo68* under the *desat1* promoter led to a 50% increase in cVA production (Figure 4.6C), leaving no doubt that *desat1* and *elo68* are indeed involved in cVA synthesis.

Are other genes involved in the HC synthesis pathway?

It is very difficult to answer this question. On the one hand, the genes that are reported in Table 4.4 are all good candidate genes, because they are expressed differently in flies with different mating speeds. On the other hand, these genes often show several sites of expression, and could affect mating behavior through action on the reception of the signal or on general metabolism, without affecting HC levels. For example, the elongase *baldspot* is more expressed in "fast" males compared to "slow" males. Baldspot is homologous to mammalian EloVL6 (Moon *et al.*, 2001; Jung *et al.*, 2007) and would thus convert palmitic to stearic acid. This gene is expressed in the brain (central brain, optic lobe), the third antennal segment and fat body (Vosshall, personal communication to Flybase, 1998) and also in testis (Jung *et al.*, 2007). A role in dorsal–ventral patterning during oogenesis (Rice and Duffy, 2001) and in male germline development (Jung *et al.*, 2007) has been reported. We did not find a role for *baldspot* in HC biosynthesis. Given its expression in the head and antenna, its role in courtship might involve pheromone perception.

An extensive study of elongases will take time, but is necessary to get a better insight into their roles. In this respect, the use of RNAi knockdown of genes specifically induced in biosynthesis tissues (oenocytes) could allow relatively fast determination of – at least – the genes involved.

Final steps involved in HC formation

Two steps seem to be required for the formation of a hydrocarbon from a fatty acid: the first step would reduce the fatty acid to an aldehyde, with the use of NADPH, the second step, which implies a cytochrome P450, would convert the aldehyde to the hydrocarbon (see Chapter 3). In the housefly a cytochrome P450 was isolated and shown to be integument enriched (Blomquist and Tittiger, unpublished data). The *Drosophila* ortholog of this gene has been inactivated in oenocytes, leading to a very drastic decrease in the amount of hydrocarbons (Wicker-Thomas *et al.*, unpublished data). Studies are in progress to characterize the gene and study its impact on physiology and courtship behavior.

Evolution of HC synthesis enzymes in *Drosophila*

Can a general pathway be drawn for HC synthesis in *Drosophila*? In mature *D. melanogaster*, only three desaturases seem to participate in pheromone biosynthesis, but as reported in the previous paragraphs, at least four elongases are involved (in cytological positions 55, 68, 85).

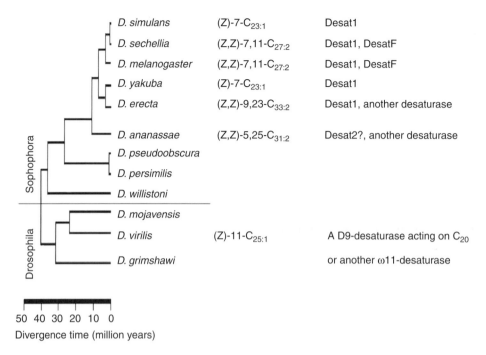

Figure 4.7 Phylogenetic tree of *Drosophila*, principal pheromones of the species and desaturases involved in their synthesis, (http://flybase.bio.indiana.edu/).

First desaturation

In the *melanogaster* subgroup, the same enzyme Desat1 seems to be involved in the first desaturation step of pheromone synthesis (Figure 4.7), even if the specificity concerning the desaturation is somewhat modified (for example, in *D. erecta*, stearic acid is used for the first desaturation, instead of palmitic acid as in other species). In other *Drosophila* species, other enzymes could be involved: Desat2 in *D. ananassae* and another yet unknown desaturase in *D. virilis*. The position of the double bond on carbon 11 in the latter species could indicate either that the desaturase acts on C_{20} saturated fatty acid, or that it has another unknown specificity, resembling the unusual specificities of some lepidopteran desaturases.

Second desaturation

The *desatF* gene is present in all genomes of the *melanogaster* subgroup, even when it is not expressed (unpublished data). It also seems to be present in the *Sophophora* subgenus but absent from the *Drosophila* one (Ritchie, unpublished data). In most species, pheromones have unusual desaturations: at position 23 in *D. erecta* (Péchiné *et al.*, 1988), 25 in *D. ananassae* (Doi *et al.*, 1997) and 27 in *D. pallidosa* (Nemoto *et al.*, 1994). In *D. virilis* other alkadienes, (Z,Z)-5,13- and (Z,Z)-7,15-pentacosadiene, and (Z,Z)-7,15-heptacosadiene, have additive pheromonal effects, combined with the major sex pheromone (Oguma *et al.*, 1992a, b). These compounds recall the singular HC found on immature *D. melanogaster* of both sexes, with unsaturations at positions 11 or 13 (first desaturation) and 21 or 23 (second desaturation) (Péchiné *et al.*, 1988). In species outside the *melanogaster* clade, the second desaturations might be caused by desaturases homologous to some *D. melanogaster* desaturases, which act only during the first hours after emergence in this species.

Evolution of HC synthesis enzymes in insects

Hydrocarbons are very diverse between species, but their synthesis depends on the same types of enzyme. Their regulation occurs through action on the desaturases (as in *Drosophila*) or the elongase(s) (in *Musca*). Therefore, knowledge of the genes involved is necessary to fully understand how this control is exerted. Desaturation enzymes are highly homologous within the animal kingdom, but elongases seem to have diverged rapidly even in the same species and are numerous (19 sequences in the *D. melanogaster* genome). However, these enzymes tend to share features which depend on their substrate specificity and it would be possible, using the homologies between elongases with similar specificities, to isolate elongases involved in the elongation of long-chain fatty acids, precursors of HC. The production of chains as long as 37C and even longer (Blomquist, Chapter 2) is unique in the animal kingdom and must be performed by special enzymes, either at the site of production or elsewhere, with a very powerful transport system to a pheromone gland or the cuticle. The study of elongases has just begun and will follow the same course over the next decade as the study of desaturases a decade ago. Other types of HC, with methyl-branched chains, occur in insects and play an important role in the recognition of species or strains (Hymenoptera). The genes involved in their synthesis are unknown and their characterization would allow us to understand how these compounds are regulated and permit development of a better strategy in pheromone-based control.

References

Adams, T., Dillwith, J. and Blomquist, G. (1984). The role of 20-hydroxyecdysone in housefly sex pheromone biosynthesis. *J. Insect Physiol.*, **30**, 287–294.

Adams, T. S., Nelson, D. R. and Blomquist, G. (1995). Effect of endocrine organs and hormones on (Z)-9-tricosene levels in the internal and external lipids of female house flies, *Musca domestica*. *J. Insect Physiol.*, **41**, 609–615.

Antony, C. and Jallon, J.-M. (1982). The chemical basis for sex recognition in *Drosophila melanogaster*. *J. Insect Physiol.*, **28**, 873–880.

Bartelt, R. J., Schaner, A. M. and Jackson, L. L. (1985). Cis-vaccenyl acetate as an aggregation pheromone in *Drosophila melanogaster*. *J. Chem. Ecol.*, **11**, 1744–1756.

Beaudoin, F., Michaelson, L. V., Hey, S. J., Lewis, M. J., Shewry, P. R., Sayanova, O. and Napier, J. A. (2000). Heterologous reconstitution in yeast of the polyunsaturated fatty acid biosynthetic pathway. *Proc. Natl. Acad. Sci. USA*, **97**, 6421–6426.

Blomquist, G. J., Dillwith, J. W. and Adams, T. S. (1987). Biosynthesis and endocrine regulation of sex pheromone production in Diptera. In *Pheromone Biochemistry*, ed. G. D. Prestwitch and G. Blomquist. London: Academic Press, pp. 217–250.

Bloomfield, D. K. and Bloch, K. (1960). The formation of delta 9-unsaturated fatty acids. *J. Biol. Chem.*, **235**, 337–345.

Bonin, C. P. and Mann, R. S. (2004). A *piggyBac* transposon gene trap for the analysis of gene expression and function in *Drosophila*. *Genetics*, **167**, 1801–1811.

Carlson, D. A., Mayer, M. S., Silhacek, D. L., James, J. D., Beroza, M. and Bierl, B. A. (1971). Sex attractant pheromone of the house fly: isolation, identification and synthesis. *Science*, **174**, 76–78.

Chertemps, T., Duportets, L., Labeur, C., Ueda, R., Takahashi, K., Saigo, K. and Wicker-Thomas, C. (2007). A female-biased expressed elongase involved in long-chain hydrocarbon biosynthesis and courtship behavior in *Drosophila melanogaster*. *Proc. Natl. Acad. Sci. USA.*, **104**, 4273–4278.

Chertemps, T., Duportets, L., Labeur, C., Ueyama, M. and Wicker-Thomas, C. (2006). A female-specific desaturase gene responsible for diene hydrocarbon biosynthesis and courtship behaviour in *Drosophila melanogaster*. *Insect Mol. Biol.*, **15**, 465–473.

Chertemps, T., Duportets, L., Labeur, C. and Wicker-Thomas, C. (2005). A new elongase selectively expressed in *Drosophila* male reproductive system. *Biochem. Biophys. Res. Commun.*, **333**, 1066–1072.

Cinti, D. L., Cook, L., Nagi, M. N. and Suneja, S. K. (1992). The fatty acid chain elongation system of mammalian endoplasmic reticulum. *Prog. Lipid Res.*, **31**, 1–51.

Coyne, J. A. (1996). Genetics of differences in pheromonal hydrocarbons between *Drosophila melanogaster* and *D. simulans*. *Genetics*, **143**, 353–364.

Coyne, J. A. and Elwyn, S. (2006). Does the *desaturase-2* locus in *Drosophila melanogaster* cause adaptation and sexual isolation? *Evolution, * **60**, 279–291.

Coyne, J. A., Wicker-Thomas, C. and Jallon, J.-M. (1999). A gene responsible for a cuticular hydrocarbon polymorphism in *Drosophila melanogaster*. *Genet. Res.*, **73**, 189–203.

Dallerac, R., Labeur, C., Jallon, J.-M., Knipple, D. C., Roelofs, W. L. and Wicker-Thomas, C. (2000). A delta 9 desaturase gene with a different substrate specificity is responsible for the cuticular diene hydrocarbon polymorphism in *Drosophila melanogaster*. *Proc. Natl. Acad. Sci. USA*, **97**, 9449–9454.

de Renobales, M., and Blomquist, G. J. (1984). Biosynthesis of medium chain fatty acids in *Drosophila melanogaster*. *Arch. Biochem. Biophys.*, **228**, 407–414.

Diehl, P. A. (1975). Synthesis and release of hydrocarbons by oenocytes of the desert locust, *Schistocerca gregaria*. *J. Insect Physiol.*, **21**, 1237–1246.

Dillwith, J. W., Adams, T. S and Blomquist, G. J. (1983). Correlation of housefly sex pheromone production with ovarian development. *J. Insect Physiol.*, **29**, 377–386.

Dillwith, J. W., Blomquist, G. J. and Nelson, D. R. (1981). Biosynthesis of the hydrocarbon components of the sex pheromone of the housefly, *Musca domestica* L. *Insect Biochem.*, **11**, 247–253.

Dillwith, J. W., Nelson, J. H., Pomonis, J. G., Nelson, D. R. and Blomquist, G. J. (1982). A 13C NMR study of methyl-branched hydrocarbon biosynthesis in the housefly. *J. Biol. Chem.*, **257**, 11305–11314.

Doi, M., Nemoto, T., Nakanishi, H., Kuwahara, Y. and Oguma, Y. (1997). Behavioral response of males to major sex pheromone component, (Z,Z)-5,25-hentriacontadiene, of *Drosophila ananassae* females. *J. Chem. Ecol.*, **23**, 2067–2078.

Eigenheer, A.L., Young, S., Blomquist, G.J., Borgeson, C.E., Tillman, J.A. and Tittiger, C. (2002). Isolation and molecular characterization of *Musca domestica* delta-9 desaturase sequences. *Insect Mol. Biol.*, **11**, 533–542.

Fan, Y., Zurek, L., Dykstra, M.J. and Schal, C. (2003). Hydrocarbon synthesis by enzymatically dissociated oenocytes of the abdominal integument of the German cockroach, *Blattella germanica*. *Naturwissenschaften*, **90**, 121–126.

Fang, S., Takahashi, A. and Wu, C. (2002). A mutation in the promoter of *desaturase 2* is correlated with sexual isolation between *Drosophila* behavioral races. *Genetics*, **162**, 781–784.

Ferveur, J.-F. (1991). Genetic control of pheromones in *Drosophila simulans*. I. *Ngbo*, a locus on the second chromosome. *Genetics*, **128**, 293–301.

Ferveur, J.-F. (2005). Cuticular hydrocarbons: their evolution and roles in *Drosophila* pheromonal communication. *Behav. Genet.*, **35**, 279–295.

Ferveur, J.-F. and Jallon, J.-M. (1996). Genetic control of male cuticular hydrocarbons in *Drosophila melanogaster*. *Genet. Res.*, **67**, 211–218.

Ferveur, J.-F., Cobb, M., Boukella, H. and Jallon, J.-M. (1996). World-wide variation in *Drosophila melanogaster* sex pheromone: behavioural effects, genetic bases and potential evolutionary consequences. *Genetica.*, **97**, 73–80.

Ferveur, J.-F., Savarit, F., O'Kane, C.J., Sureau, G., Greenspan, R.J. and Jallon, J.-M. (1997). Genetic feminization of pheromones and its behavioral consequences in *Drosophila* males. *Science*, **276**, 1555–1558.

Ginzel, M.D., Blomquist, G.J., Millar, J.G. and Hanks, L.M. (2003). Role of contact pheromones in mate recognition in *Xylotrechus colonus*. *J. Chem. Ecol.*, **29**, 533–545.

Ginzel, M.D., Moreira, J.A., Ray, A.M., Millar, J.G. and Hanks, L.M. (2006). (Z)-9-nonacosene-major component of the contact sex pheromone of the beetle *Megacyllene caryae*. *J. Chem. Ecol.*, **32**, 435–451.

Gleason, J.M., Jallon, J.-M., Rouault, J.D. and Ritchie, M.G. (2005). Quantitative trait loci for cuticular hydrocarbons associated with sexual isolation between *Drosophila simulans* and *D. sechellia*. *Genetics*, **171**, 1789–1798.

Greenberg, A.J., Moran, J.R., Coyne, J.A. and Wu, C.I. (2003). Ecological adaptation during incipient speciation revealed by precise gene replacement. *Science*, **302**, 1754–1757.

Greenberg, A.J., Moran, J.R., Fang, S. and Wu, C.I. (2006). Adaptive loss of an old duplicated gene during incipient speciation. *Mol. Biol. Evol.*, **23**, 401–410.

Gu, P., Welch, W.H., Guo, L., Schegg, K.M. and Blomquist, G.J. (1997). Characterization of a novel microsomal fatty acid synthetase (FAS) compared to a cytosolic FAS in the housefly, *Musca domestica*. *Comp. Biochem. Physiol. B*, **118**, 447–456.

Gutierrez, E., Wiggins, D., Fielding, B. and Gould, A.P. (2007). Specialized hepatocyte-like cells regulate *Drosophila* lipid metabolism. *Nature*, **445**, 275–280.

Howard, R.W. (1993). Cuticular hydrocarbons and chemical communication. In *Insect lipids: chemistry, biochemistry and biology*, ed. D.W. Stanley-Samuelson and D.R. Nelson. Lincoln: University of Nevada Press, pp. 176–226.

Jallon, J.-M. (1984). A few chemical words exchanged by *Drosophila* during courtship and mating. *Behav. Genet.*, **14**, 441–478.

Jallon, J.-M., Antony, C. and Benamar, O. (1981). Un anti-aphrodisiaque produit par les mâles de *Drosophila melanogaster* et transféré aux femelles lors de la copulation. *C. R. Acad. Sci. Paris,* **292**, 1147–1149.

Jallon, J.-M. and Péchiné, J.-M. (1989). Une autre race de chimique *Drosophila melanogaster* en Afrique. *C. R. Acad. Sci. Paris*, **309**, 1551–1556.

Jallon, J.-M. and Wicker-Thomas, C. (2003). Genetic studies on pheromone production in *Drosophila*. (Chapter 9). In *Insect Pheromone Biochemistry and Molecular Biology*. Part 1. Biochemistry and Molecular Biology of pheromone production, ed. G. J. Blomquist and R. G. Vogt. London: Elsevier Academic Press, pp. 253–281.

Juarez, P., Chase, J. and Blomquist, G. J. (1992). A microsomal fatty acid synthetase from the integument of *Blattella germanica* synthesizes methyl-branched fatty acids, precursors to hydrocarbon and contact sex pheromone. *Arch. Biochem. Biophys.*, **293**, 333–341.

Jung, A., Hollmann, M. and Schafer, M. A. (2007). The fatty acid elongase NOA is necessary for viability and has a somatic role in *Drosophila* sperm development. *J. Cell Sci.*, **120**, 2924–34.

Jurenka, R. A., Schal, C., Burns, E. L., Chase, J. and Blomquist, G. J. (1989). Structural correlation between cuticular hydrocarbons and female contact sex pheromone of German cockroach, *Blattella germanica* (L.). *J. Chem. Ecol.*, **15**, 939–949.

Knipple, D. C. and Roelofs, W. L. (2003). Molecular biological investigations of pheromone desaturases. (Chapter 4). In *Insect Pheromone Biochemistry and Molecular Biology*. Part 1. Biochemistry and Molecular Biology of pheromone production, ed. G. J. Blomquist and R. G. Vogt. London: Elsevier Academic Press, pp. 81–206.

Knipple, D. C., Rosenfield, C. L., Miller, S. J., Liu, W., Tang, J., Ma, P. W. and Roelofs, W. L. (1998). Cloning and functional expression of a cDNA encoding a pheromone gland-specific acyl-CoA Delta11-desaturase of the cabbage looper moth, *Trichoplusia ni*. *Proc. Natl. Acad. Sci. USA*, **95**, 15287–15292.

Labeur, C., Dallerac, R., and Wicker-Thomas, C. (2002). Involvement of *desat1* gene in the control of *Drosophila melanogaster* pheromone biosynthesis. *Genetica*, **114**, 269–274.

Le Conte, Y. and Hefetz, A. (2008). Primer pheromones in social hymenoptera. *Annu. Rev. Entomol.*, **53**, 523–542.

Lihoreau, M. and Rivault, C. (2009). Kin recognition via cuticular hydrocarbons shapes cockroach social life. *Behav. Ecol.*, **20**, 46–53.

Mackay, T. F., Heinsohn, S. L., Lyman, R. F., Moehring, A. J., Morgan, T. J. and Rollmann, S. M. (2005). Genetics and genomics of *Drosophila* mating behavior. *Proc. Natl. Acad. Sci. USA*, **102** Suppl 1, 6622–6629.

Marcillac, F., Bousquet, F., Alabouvette, J., Savarit, F. and Ferveur, J.-F. (2005b). A mutation with major effects on *Drosophila melanogaster* sex pheromones. *Genetics*, **171**, 1617–1628.

Marcillac, F., Grosjean, Y. and Ferveur, J.-F. (2005a). A single mutation alters production and discrimination of *Drosophila* sex pheromones. *Proc. Biol. Sci.*, **272**, 303–309.

Michalak, P., Malone, J. H., Lee, I. T., Hoshino, D. and Ma, D. (2007). Gene expression polymorphism in *Drosophila* populations. *Mol. Ecol.*, **16**, 1179–1189.

Millar, J. G. (2000). Polyene hydrocarbons and epoxides: a second major class of lepidopteran sex attractant pheromones. *Annu. Rev. Entomol.*, **45**, 575–604.

Moon, Y. A., Shah, N. A., Mohapatra, S., Warrington, J. A. and Horton, J. D. (2001). Identification of a mammalian long chain fatty acyl elongase regulated by sterol regulatory element-binding proteins. *J. Biol. Chem.*, **276**, 45358–45366.

Mpuru, S., Blomquist, G. J., Schal, C., Roux, M., Kuenzli, M., Dusticier, G., Clément, J.-L. and Bagnères, A.-G. (2001). Effect of age and sex on the production of internal and external hydrocarbons and pheromones in the housefly, *Musca domestica*. *Insect Biochem. Mol. Biol.*, **31**, 139–155.

Nemoto, T., Doi, M., Oshio, K., Mtsubayashi, H., Oguma, Y., Suzuki, T. and Kuwahara, Y. (1994). (*Z,Z*)-5,27-tritriacontadiene: major sex pheromone of *Drosophila pallidosa* (Diptera: Drosophilidae). *J. Chem. Ecol.*, **20**, 3029–3037.

Ntambi, J. M., S. A. Buhrow, K. H. Kaestner, R. J. Christy, E. Sibley, T. J. J. Kelly, and M. D. Lane. (1988). Differentiation-induced gene expression in 3T3-L1 preadipocytes. Characterization of a differentially expressed gene encoding stearoyl-CoA desaturase. *J. Biol. Chem.*, **263**,17291–17300.

Oguma, Y., Nemoto, T. and Kuwahara, Y. (1992a). A sex pheromone study of a fruit fly *Drosophila virilis* Sturtevant (Diptera: Drosophilidae): additive effect of cuticular alkadienes to major sex pheromone. *Appl. Entomol. Zool.*, **27**, 499–505.

Oguma, Y., Nemoto, T. and Kuwahara, Y. (1992b). (*Z*)-11-pentacosene is the major sex pheromone component in *Drosophila virilis* (Diptera). *Chemoecology*, **3**, 60–64.

Parisi, M., Nuttall, R., Edwards, P., Minor, J., Naiman, D., Lu, J., Doctolero, M., Vainer, M., Chan, C., Malley, J., Eastman, S. and Oliver, B. (2004). A survey of ovary-, testis-, and soma-biased gene expression in *Drosophila melanogaster* adults. *Genome Biol.*, **5**, R40.

Péchiné, J.-M., Antony, C. and Jallon, J.-M. (1988). Precise characterization of cuticular compounds in young *Drosophila* by mass spectrometry. *J. Chem. Ecol.*, **14**, 1071–1085.

Rice, A. H. and Duffy, J. B. (2001). Characterization of *baldspot*, a putative gene involved in dorsal-ventral patterning during *Drosophila* oogenesis. *A. Dros. Res. Conf.*, **42**, 667A.

Riddervold, M. H., Tittiger, C., Blomquist, G. J. and Borgeson, C. E. (2002). Biochemical and molecular characterizaton of house cricket (*Acheta domesticus*, Orthoptera: Gryllidae) Delta 9 desaturase. *Insect Biochem. Mol. Biol.*, **32**, 1731–1740.

Roelofs, W. L. and Cardé, R. T. (1971). Hydrocarbon sex pheromone in tiger moths (Arctiidae). *Science*, **171**, 684–686.

Roelofs, W. L., Liu, W., Hao, G., Jiao, H., Rooney, A. P. and Linn, C. E. Jr. (2002). Evolution of moth sex pheromones via ancestral genes. *Proc. Natl. Acad. Sci. USA*, **99**, 13621–13626.

Rosenfield, C. L., You, K. M., Marsella-Herrick, P., Roelofs, W. L. and Knipple, D. C. (2001). Structural and functional conservation and divergence among acyl-CoA desaturases of two noctuid species, the corn earworm, *Helicoverpa zea*, and the cabbage looper, *Trichoplusia ni. Insect Biochem. Mol. Biol.*, **31**, 949–964.

Rouault, J., Capy, P. and Jallon, J.-M. (2001). Variations of male cuticular hydrocarbons with geoclimatic variables: an adaptative mechanism in *Drosophila melanogaster*? *Genetica*, **110**, 117–130.

Rouault, J. D., Marican, C., Wicker-Thomas, C. and Jallon, J.-M. (2004). Relations between cuticular hydrocarbon (HC) polymorphism, resistance against desiccation and breeding temperature; a model for HC evolution in *D. melanogaster* and *D. simulans. Genetica*, **120**, 195–212.

Schal, C., Gu, X., Burns, E. L. and Blomquist, G. J. (1994). Patterns of biosynthesis and accumulation of hydrocarbons and contact sex pheromone in the female German cockroach, *Blattella germanica. Arch. Insect Biochem. Physiol.*, **25**, 375–391.

Schal, C., Sevala, V. and Cardé, R. T. (1998). Novel and highly specific transport of a volatile sex pheromone by hemolymph lipophorin in moths. *Naturwissenschaften*, **85**, 339–342.

Scott, D. (1994). Genetic variation for female mate discrimination in *Drosophila melanogaster*. *Evolution*, **48**, 112–121.

Scott, D. and Jackson, L. (1988). Interstrain comparison of male-predominant antiaphrodisiacs in *Drosophila melanogaster*. *J. Insect Physiol.*, **34**, 863–871.

Scott, D. and Richmond, R.C. (1988). A genetic analysis of male-predominant pheromones in *Drosophila melanogaster*. *Genetics*, **119**, 639–646.

Shanklin, J. and Cahoon, E.B. (1998). Desaturation and related modifications of fatty acids1. *Annu. Rev. Plant Physiol. Plant Mol. Biol.*, **49**, 611–641.

Stukey, J.E., McDonough, V.M. and Martin, C.E. (1989). Isolation and characterization of *OLE1*, a gene affecting fatty acid desaturation from *Saccharomyces cerevisiae*. *J. Biol. Chem.*, **264**, 16537–16544.

Stukey, J.E., McDonough, V.M. and Martin, C.E. (1990). The *OLE1* gene of *Saccharomyces cerevisiae* encodes the Δ9 fatty acid desaturase and can be functionally replaced by the rat stearoyl-CoA desaturase gene. *J. Biol. Chem.*, **265**, 20144–20149.

Sureau, G. and Ferveur, J.-F. (1999). Co-adaptation of pheromone production and behavioural responses in *Drosophila melanogaster* males. *Genet. Res.*, **74**, 129–137.

Takahashi, A., Tsaur, S.C., Coyne, J.A. and Wu, C.I. (2001). The nucleotide changes governing cuticular hydrocarbon variation and their evolution in *Drosophila melanogaster*. *Proc. Natl. Acad. Sci. USA* **98**, 3920–3925.

Thiede, M.A., Ozols, J. and Strittmatter, P. (1986). Construction and sequence of cDNA for rat liver stearyl coenzyme A desaturase. *J. Biol. Chem.*, **261**, 13230–13235.

Tiku, P.E., Gracey, A.Y., Macartney, A.I., Beynon, R.J., and Cossins, A.R. (1996). Cold-induced expression of delta 9-desaturase in carp by transcriptional and posttranslational mechanisms. *Science*, **271**, 815–818.

Tvrdik, P., Westerberg, R., Silve, S., Asadi, A., Jakobsson, A., Cannon, B., Loison, G. and Jacobsson, A. (2000). Role of a new mammalian gene family in the biosynthesis of very long chain fatty acids and sphingolipids. *J. Cell Biol.*, **149**, 707–718.

Ueyama, M., Chertemps, T., Labeur, C. and Wicker-Thomas, C. (2005). Mutations in the *desat1* gene reduces the production of courtship stimulatory pheromones through a marked effect on fatty acids in *Drosophila melanogaster*. *Insect Biochem. Mol. Biol.*, **35**, 911–920.

Wang, D.L., Dillwith, J.W., Ryan, R.O., Blomquist, G.J. and Reitz, R.C. (1982). Characterization of the acyl-CoA desaturase in the housefly, *Musca domestica* L. *Insect Biochem.*, **12**, 545–551.

Wicker-Thomas, C., Henriet, C. and Dallerac, R. (1997). Partial characterization of a fatty acid desaturase gene in *Drosophila melanogaster*. *Insect Biochem. Mol. Biol.*, **27**, 963–972.

Wicker-Thomas, C. and Jallon, J.-M. (2000). Role of *Enhancer of zeste* on the production of *Drosophila melanogaster* pheromonal hydrocarbons. *Naturwissenschaften*, **87**, 76–79.

5

Site of synthesis, mechanism of transport and selective deposition of hydrocarbons

Anne-Geneviève Bagnères and Gary J. Blomquist

Our understanding of the site of synthesis, transport and deposition of insect cuticular hydrocarbons has undergone a paradigm shift in the last few decades. In the model that prevailed until the early 1980s, newly synthesized hydrocarbons exited the epidermal cells where they were synthesized, were transported through pore canals and then formed an outer layer on the cuticle (Locke, 1965; Hadley, 1981). The picture presented polar head groups of fatty acids interacting with the cuticle, with the hydrocarbons layered on top of the acyl chains of fatty acids. This served as an excellent model in which to test a number of hypotheses, many of which are still unanswered. However, it is now clear that cuticular hydrocarbons are synthesized by oenocytes, either associated with epidermal cells or within the peripheral fat body, and that the newly synthesized hydrocarbons are taken up by lipophorin and transported via the hemolymph. The mechanism of how they get transferred from lipophorin to epidermal cells, cross the epidermal cells and then to the surface of the insect and to specific glands is unknown. There is great selectivity in the process, as a number of lepidopterans are able to selectively transport shorter-chain hydrocarbon pheromones and pheromone precursors to the pheromone gland on the abdomen, whereas longer-chain cuticular hydrocarbons are transported to cover the entire cuticle (Schal *et al.*, 1998a; Jurenka *et al.*, 2003). The transport process, even for long-chain cuticular components, can be somewhat selective; a quantitatively different composition of hydrocarbons can be present in the hemolymph compared to deposition sites on the surface of the insect or in the ovary or specific glands (Sevala *et al.*, 2000; Schal *et al.,* 2003), especially in social insects. In many cases, hydrocarbons comprise the majority of the cuticular lipids, making the proposed role of fatty acids on the surface of the insect of lesser importance. The arrangement of hydrocarbons on the surface of the insect is unknown. Suggestions have been made that the components of most importance in chemical communication may be on the outer surface (Ginzel *et al.*, 2003; Ginzel, Chapter 17, this book). This chapter will address the site of synthesis, transport and deposition of insect hydrocarbon, with special emphasis on the social insects. It must be emphasized that a clear understanding of the mechanism of hydrocarbon uptake to lipophorin and its deposition to epidermal cells and transit across the cuticle to the surface of the insect is, at this time, unavailable.

Site of synthesis

A number of studies with widely diverse species have established that the major site of cuticular hydrocarbon synthesis is within the cells associated with the epidermal layer or the peripheral fat body, particularly the oenocytes. In *Schistocerca gregaria*, Diehl (1973, 1975) separated the oenocyte-rich peripheral fat body from the central fat body tissue and observed the highest rate of hydrocarbon synthesis in the oenocyte-rich peripheral fat body. In *Tenebrio molitor*, Romer (1980) demonstrated that isolated oenocytes efficiently and specifically incorporated [^{14}C]acetate into hydrocarbon. Similar studies in *Periplaneta americana* (Nelson, 1969), *Sarcophaga bullata* (Armold and Regnier, 1975), and *Musca domestica* (Dillwith *et al.*, 1981) demonstrated that hydrocarbon synthesis occurs primarily in the epidermal tissue.

In the German cockroach, *Blattella germanica*, Fan *et al.* (2003) isolated an epidermal cell enriched sample that contained mostly epidermal cells and an oenocyte enriched sample that contained 60% oenocyte cells by using enzymic digestion of the basal lamina and the extracellular matrix. The intact cells in each sample were further separated into fractions by a Percoll gradient. The incorporation of [$1-^{14}$C]propionate into methyl-branched hydrocarbon was then used to assay each sample for hydrocarbon biosynthesis. The samples that contained the highest amount of oenocytes, large cells that are denser than other epidermal cells, synthesized the highest amount of hydrocarbon.

In *Drosophila melanogaster*, the hydrocarbons produced by adult insects are strikingly dimorphic, with the female flies producing diene hydrocarbons with 27 and 29 carbons that function as contact sex pheromones, components that are normally absent from males. In an elegant set of experiments, Ferveur *et al.* (1997), demonstrated that feminization of the cuticular hydrocarbon–pheromone mixture produced by males was induced by targeted expression of the *transformer* gene in adult oenocytes, providing strong evidence that the site of hydrocarbon synthesis is the oenocytes.

Further evidence that oenocytes are the cell type involved in hydrocarbon production was obtained by Wicker-Thomas *et al.* (unpublished data). The enzyme involved in the final step in hydrocarbon production, a cytochrome P450 that oxidatively decarbonylates aldehydes, was RNAi silenced specifically in oenocytes of both male and female *D. melanogaster*, resulting in marked inhibition of hydrocarbon production. The RNAi-silenced insects produced less than 10% of the hydrocarbon compared to control insects. Thus, the available evidence overwhelmingly supports oenocytes, whether associated with epidermal tissue or present in the peripheral fat body, as the site of cuticular hydrocarbon production. In some cases, even the shorter-chain hydrocarbon pheromones and pheromone precursors (see below) are produced in oenocytes.

Transport of hydrocarbons with lipophorin

The protein that we now refer to as lipophorin was originally referred to extensively as the insect diacylglycerol-carrying lipoprotein. It was the recognition in the 1970s that this protein also carried cholesterol (Chino and Gilbert, 1971) and in the early 1980s that it transported hydrocarbon (Chino *et al.*, 1981a) that led Chino *et al.* (1981b) to suggest the name lipophorin (Gr. lipos, fat; phoros, bearing) for this important lipid transport molecule in insects. The presence of relatively high amounts of hydrocarbon with the same composition as that found on the cuticle, along with the observation that radioactivity from injected [1–^{14}C]palmitic acid was recovered in hemolymph hydrocarbons in the American cockroach, *P. americana*, led Chino *et al.* (1981a) to suggest that the protein now called lipophorin was used to transport hydrocarbons from the site of synthesis to the site of deposition. Further work (Katase and Chino, 1982) demonstrated that the injection of labeled hydrocarbon associated with lipophorin into the hemolymph of *P. americana* resulted in labeled hydrocarbon appearing on the surface of the insect.

Schal *et al.* (1998a) reviewed the literature available at that time and concluded that the internal hydrocarbons in many insects were qualitatively similar to the cuticular hydrocarbons. They suggested that the internal hydrocarbons might represent pools of hydrocarbon in the oenocytes, epidermis, hemolymph, fat body and/or gonads. The oenocytes of insects can be associated with the epidermal tissue or the peripheral fat body tissue. For insects with oenocytes within the hemocoel, a lipid transport system would clearly be necessary for hydrocarbons to be transported to the surface of the cuticle. For insects with oenocytes that are within the basal membrane, such a system would not necessarily be required. Yet it appears that in both groups of insects, those with oenocytes within the hemocoel and those with oenocytes within the basal membrane, large amounts of hydrocarbon are found associated with lipophorin. It is now accepted that lipophorin transports hydrocarbons from the oenocytes for deposition on the surface of the insect. The mechanism of deposition, including uptake into and crossing the epidermal cells and integument, is unknown.

A remarkable specificity in the transport and deposition of hydrocarbons was demonstrated in the moths *Holomelina aurantiaca* (Schal *et al.*, 1998b), *Scoliopteryx libatrix* (Subchev and Jurenka, 2001) and *Lymantria dispar* (Jurenka *et al.*, 2003). In each insect, the hydrocarbon pheromone or, in the case of *L. dispar*, the hydrocarbon precursor to the epoxide pheromone, is biosynthesized in the oenocytes, transported by lipophorin to the pheromone gland on the abdomen, and then taken up and released from the pheromone gland, or, in the case of *L. dispar*, the alkene is converted to the epoxide in the pheromone gland and then released. It appears that the lipophorin of these insects has the remarkable specificity to deliver the shorter-chain hydrocarbon pheromone or pheromone precursor to the pheromone gland and the longer-chain cuticular hydrocarbons, which are also synthesized by the oenocytes, to other sites for deposition on the surface of the insect. More recently, Matsuoka *et al.* (2006) demonstrated that the female of a geometrid moth, *Ascotis solaria cretacea*, used lipophorin to transport a C_{19} triene hydrocarbon from the site of synthesis in the oenocytes to the pheromone gland where it is converted to the C_{19}, 3,4-epoxy-6,9-diene. Thus, it appears that in Lepidoptera that use C_{19} to C_{25} polyunsaturated hydrocarbons (Millar, Chapter 18, this book), or products derived from these hydrocarbons as pheromones, the parent hydrocarbon is biosynthesized in oenocyte cells from elongated products of dietary linoleic and/or linolenic acids, specifically transported to the pheromone gland as the hydrocarbon, and then, if released as other than the hydrocarbon, modified in the pheromone gland prior to release (for a review, see Chapter 18).

In laboratory-reared houseflies, *M. domestica,* the pheromone component (*Z*)-9-tricosene first appears on the surface of the female insect at about two days post-emergence (Dillwith *et al.*, 1983). (*Z*)-9-Tricosene appears in internal lipids prior to appearing on the surface of the insect, suggesting that it is transported in the hemolymph prior to deposition on the surface (Mpuru *et al.*, 2001). Mathematical analysis indicated that the time shift between internal production and external accumulation was about 24 hours. A C_{23} epoxide and ketone derived from the C_{23} alkene (Blomquist *et al.*, 1984) are not present in the internal lipids and only appear on the surface of the insect when (*Z*)-9-tricosene appears, suggesting that the epoxide and ketone are formed as the alkene reaches and is transported across the epidermal cells. The same phenomenon is also observed for those lepidopteran species that use epoxides derived from polyunsatured hydrocarbons as pheromone components (Millar, Chapter 18, this book). The observation that the unsaturated hydrocarbons in *M. domestica* are metabolized by a cytochrome P450 to form epoxides prior to deposition on the surface argues for a transport route from hemolymph through the epidermal cells, as it would be unlikely that the alkenes would encounter a cytochrome P450 if they somehow traveled between epidermal cells to reach and then cross the cuticle.

The most thoroughly studied system for the lipophorin transport of long-chain hydrocarbon is the German cockroach, *Blattella germanica* (Gu *et al.*, 1995; Schal *et al.*, 1998a; Sevala *et al.*, 1999; Young *et al.*, 1999; Fan *et al.*, 2002; reviewed in Schal *et al.*, 2003). The German cockroach lipophorin is similar to the lipophorin of many insect species and is a high-density lipoprotein with a molecular weight of 670 kDa, composed of two apolipophorin subunits of 212 kDa (ApoLp I) and 80 kDa (ApoLp II) (Sevala *et al.,* 1999). The lipid fraction represents a high percentage, 51%, of the total mass, and hydrocarbons constitute about 42% of the total lipids, which also contain diacylglycerol, cholesterol and phospholipid (Sevala *et al.,* 1999). In addition to their being transported in the hemolymph to eventually reach the surface of the insect, a large portion of the internal hydrocarbons is localized within oocytes, suggesting that an important function of hemolymph- and lipophorin-associated hydrocarbons is to provide the developing oocytes as well as the cuticular surface with hydrocarbons (Schal *et al.*, 2001).

One could ask why lipophorin would be necessary to transport hydrocarbons to the surface of the insect and why hydrocarbons are not simply directly transported to the surface in those species, including *B. germanica,* where oenocytes are associated with epidermal cells. An answer may be that certain parts of the insect do not synthesize hydrocarbons, and it is essential for lipophorin to deposit hydrocarbon in these places. For example, if the veins to the forewings of *B. germanica* are cut, much less hydrocarbon appears on the wings than in intact insects (Gu *et al.*, 1995). In *B. germanica,* most of the hydrocarbons produced are synthesized in the abdomen, with the sternites synthesizing considerably more hydrocarbon than the tergites. Thus, to transport hydrocarbon to the ovary and to parts of the insect that do not synthesize hydrocarbon, including the head, wings, legs and thorax, a highly efficient reusable shuttle system of lipophorin has evolved. In social insects, hydrocarbons play important roles in chemical communication, and the selective transport of hydrocarbons is especially important (see below).

A role for lipophorin in transporting hydrocarbon has been demonstrated in a number of insect species (Atella *et al.*, 2006; Pho *et al.*, 1996; and references therein), suggesting that this is a general role for lipophorin. In the transfer of diacylglycerol by lipophorin to the fat body, a lipid transfer particle is involved (Canavoso *et al.*, 2004 and references therein). Singh and Ryan (1991) demonstrated that a lipid transfer particle has the capacity to transfer hydrocarbon between lipophorin and a model acceptor lipid particle, human low-density lipoprotein. Two years later, Takeuchi and Chino (1993) showed that the lipid transfer particle from *P. americana* could catalyze the transfer of labeled hydrocarbon from labeled high-density lipoprotein bound with a transfer membrane to unlabeled high-density lipoprotein dissolved in saline. Likewise, Tsuchida *et al.* (1998) showed that a lipid transfer particle catalyzes the transfer of a different type of hydrocarbon, carotenoids, between lipophorins of *Bombyx mori*. To our knowledge, the role of the lipid transfer particle in lipophorin-mediated hydrocarbon uptake from oenocytes or deposition has not been demonstrated for any insect species and clearly needs further study.

Social insects, hydrocarbon transport and deposition

Social insects use hydrocarbons in chemical communication as a central feature in the organization of insect societies. Discrimination between nestmates and non-nestmates depends mainly on recognition of cuticular signatures during brief contacts between individuals (see other chapters). These signatures or odors are produced by various qualitative and quantitative variations of hydrocarbons that provide cues about species, gender, colony, task, fertility, etc., and can be used as primer pheromones (Howard and Blomquist, 2005; LeConte and Hefetz, 2008; present book). This hydrocarbon pattern, related to biological functions, has been designated by different names, including the following: signature, fingerprint, bar code, template, visa and label. An intriguing aspect of this pattern is "superpositioning", which allows insects to display specific, colonial and individual signals on the same channel, also called multipurpose signals (Denis *et al.*, 2006; d'Ettorre and Moore, 2008). Some odors are genetically fixed while others vary according to time, environment, and physiological state. Odor production, therefore, requires selective biosynthesis, transport, regulation, and transfer. While we do not have the complete picture for the transport of hydrocarbons in any species, the remainder of this chapter will attempt to put together our current understanding of hydrocarbon transport in social insects.

The development of the epicuticular signature is a dynamic process that can be studied using an endogenous or exogenous approach. The endogenous approach consists of examining the effect of factors such as hormonal levels, biosynthetic pathways, storage, and transport and the role of tissues, glands, and specific carriers. The exogenous approach focuses on the effect of food uptake, season, predators, etc. Studying signature formation and discrimination involves differentiating the impact of genetic factors from environmental factors. Arnold *et al.* (1996), Kaib *et al.* (2002), and Dronnet *et al.* (2006) discuss genetic factors: species, kin or specific colonial signature. Heinze *et al.* (1996), Liang and Silverman (2000), and Florane *et al.* (2004) explore environmental aspects: changes in social environment in artificial mixed colonies or food impact on hydrocarbon production. It should be pointed out, however, that food preference can be genetically influenced as shown for *Drosophila* larvae (Ryuda *et al.*, 2008). Ferveur (2005) reviewed the impacts of genetic vs. nongenetic factors on hydrocarbon production in a nonsocial insect, *Drosophila*.

The turnover of cuticular hydrocarbon patterns has been studied in several insects. Kent *et al.* (2007) estimated cyclic turnover rates in *D. melanogaster* and reported that they were correlated with male sexual behavior. Similar findings on a cyclic turnover have been reported in various social insects such as leptothoracine ants (Provost *et al.*, 1993) and *Camponotus* ants (Meskali *et al.*, 1995). In their report on ants, Vander Meer and Morel (1998) reviewed the dynamic (vs. static) process of hydrocarbon cues/templates. We determined in subterranean termite workers a cyclic turnover of the total cuticular hydrocarbons in one month, whatever the experimental conditions (Bagnères *et al.*, in prep.). Steiner *et al.* (2007) reported rapid modification of the pheromonal blend in the parasitic wasp *L. distinguendus* and were able to differentiate between the bioactive and bioinactive hydrocarbon blends, allowing sex pheromonal differentiation in adults. It would be interesting to carry out similar studies in a social insect model to determine if differential transport of the two blends also occurs. Recent studies have focused on the involvement of certain classes of hydrocarbons, alkenes or methyl-branched alkanes vs. *n*-alkanes in chemical signatures (see Chapter 11 for review on this aspect from a nestmate recognition standpoint).

Another interesting mechanism, suggested in several studies, is that selective transport pathways exist for the export of various constituents to the exterior or to special tissues. Schal *et al.* (1998a) and others (see above) documented pathways for the specific transport of volatile sex pheromone in moths. Similar pathways have been observed in various ant species for colonial signature (Lucas *et al.*, 2004), egg marking (Lommelen *et al.*, 2008), and nest area marking by feces (Grasso *et al.*, 2005). In a study on the regulation of reproduction in the ant *Camponotus floridanus*, Endler *et al.* (2004) implicated pheromone-binding proteins in the transport of hydrocarbons that identify eggs from the surface of the antennae to the receptors on sensory neurons. Another likely example of a regulatory process based on selective transport was observed in mixed experimental colonies comprised of two termite species (Vauchot *et al.*, 1996, 1998).

Using a theoretical approach, Millor *et al.* (2006) showed that a weak level of interactions between ants is enough to lead to aggregation around the same food source and indirectly accelerates the appearance of a common blend. Errard *et al.* (2006) demonstrated that the signal learning process must be reinforced at a very young stage, cannot take place at any time between heterospecific or heterocolonial groups, and is dependent on the degree of chemical similarity between groups. Differential mixture adsorption has been observed in artificial mixed heterospecific termite colonies (Vauchot *et al.*, 1998) and can occur naturally between young and mature insects, as demonstrated in *Polistes* wasps (Lorenzi *et al.*, 2004). These findings show that template/cue acquisition and modification do not conform to a simple rule, but rather are, complex processes related to the physicochemical nature of a specific signature and the behavioral regulation of social interactions.

Role of endocrine and physiological factors in regulation of biosynthesis and transport in social insects

Endocrine factors play a key role in the regulation processes of social interactions in insects. The correlation between pheromone production and endocrine factors was described many years ago in solitary insects such as dipterans (Dillwith *et al.*, 1983; Wicker and Jallon, 1995; Mpuru *et al.*, 2001; Chapters 3 and 4). In social insects, nestmate recognition cues, task-specific cues, and dominance and fertility cues provide good backgrounds for studying the endocrine factors affecting hydrocarbon signatures (see related chapters). For instance, the central neuromediator octopamine has been shown to decrease ant trophallaxis, resulting indirectly in a decrease in hydrocarbon transfer but not hydrocarbon synthesis (Boulay *et al.*, 2000). Octopamine levels in the ant brain could be an important mediator, playing a role in social cohesion and so, indirectly, in hydrocarbon exchange. In a recent study in *Drosophila* by Wicker-Thomas and Hamann (2008), a Ddc mutant was used and found to implicate another amine, i.e., dopamine, in a complex process regulating female hydrocarbons, male courtship behavior, and locomotion. Since the ability to obtain mutants has not yet been achieved in social insects, most studies on hormonal or pheromonal regulation of HCs have been based on correlation.

A number of studies in social insects including the ant *Camponotus floridanus* (Endler *et al.*, 2004, 2006, 2007), the queenless ant *Streblognathus peetersi* (Cuvillier-Hot *et al.*, 2004, 2005), the bulldog ant *Myrmecia gulosa* (Dietemann *et al.*, 2003, 2005), and Polistine wasps (Dapporto *et al.*, 2007a, b) have demonstrated, among other things, that cuticular hydrocarbons mediate discrimination of reproductives and egg marking for policing by workers (see Monnin, 2006, for review). These studies have also shown a strong correlation between ovarian development and hydrocarbon patterns that change according to ant status (queen vs. workers, worker egg vs. queen egg, dominant worker vs. non-dominant), thus providing a "status badge" which allows a clear distinction between a queen (when present) and a fertile worker. Variations in hydrocarbon signatures as a function of hormonal and dominance status have also been well documented in *Polistes* wasps (Bonavita-Cougourdan *et al.*, 1991; Sledge *et al.*, 2004). Thus, the synthesis of hydrocarbons depends on physiological traits and must, therefore, change rapidly to allow this important recognition process, as shown in Hora *et al.* (2008), where cuticular chemistry as well as visual appearance changes drastically for the mated queen of an ectatommine ant.

Several authors have investigated the role of juvenile hormone (JH) in hydrocarbon signatures. Sledge *et al.* (2004) performed one such study based on measuring the size of the corpora allata as an indicator of JH levels. Since caste determination depends strictly on the colony social structure, it can be assumed that topical JH application mimics natural caste development and thus modifies the cuticular caste-specific odor. Based on this assumption,

Bortolotti *et al.* (2001) studied the role of JH in caste determination and colony processes in the bumblebee *Bombus terrestris*. We (AGB and colleagues) are undertaking a similar study to assess the effect of JH on worker/soldier induction following change of the cuticular signature, in *Reticulitermes* termites (Bagnères *et al.*, personal communication). In another study, Henderson (1998) suggested that the corpora allata might produce the main primer pheromone in lower termites, based on the gland's critical involvement in caste differentiation. Mohamed and Prestwich (1988) showed an indirect link between pheromone and the effect of methoprene, a JH analog. Lengyel *et al.* (2007) studied the effect of JH III on cuticular hydrocarbon blends and the division of labor in *Myrmicia eumenoides* ant colonies. Their data indicated that JH III titer was correlated with at least two key processes: the acceleration of cuticular hydrocarbon profile changes and the long-term modulation of task shifting. It is interesting to note that task-specific hydrocarbon profiles in this ant appear to be controlled by two independent processes, depending upon a differential mode of action (short-term or long-term) of JH.

A recent study differentiated the role of JH and ecdysteroids in the ant *Streblognathus peetersi*. Ecdysone was correlated with dominant ants and reproduction while JH was more commonly associated with low-ranking ants and sterility (Brent *et al.*, 2006).

Site of synthesis

As stated by Blomquist *et al.* (1998) in their chapter, "the line of demarcation between glandular or cuticular release of semiochemical signals is not always clear". This statement echoes an earlier one by Blum (1985), who reported that insect exocrine glands consisting of modified epidermal cells located throughout the body could perform de novo biosynthesis and secretion of behavioral chemicals. Later, Blum (1987) put forth a unified chemosociality concept proposing that epicuticular lipids carried numerous exocrine compounds and that the cuticle could be compared to a thin layer phase. Nevertheless, it is known that in various non-social insects epicuticular hydrocarbons are synthesized by modified cells often associated with the epidermis, the oenocytes (see above), and that these oenocytes can be located in several sites within insects.

Soroker and Hefetz (2000) proposed a model for hydrocarbon circulation in the ant *Cataglyphis niger* in which hydrocarbons are synthesized by the fat body, probably the abdominal fat body (free oenocyte cells?), and released to the hemolymph with the crop serving as a transport site and the postpharyngeal gland as a releaser site (see role of PPG reviewed below). They also stated that the dissimilarity in hydrocarbon composition between the PPG and the cuticle might be due to different transport mechanisms and/or additional biosynthetic pathways. A recent study on fat body composition in some genera of leafcutter ants showed that trophocytes and oenocytes were the most common cells (Roma *et al.*, 2006), suggesting that the oenocytes are the site of hydrocarbon synthesis.

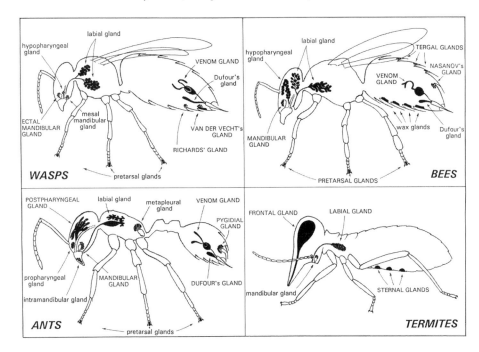

Figure 5.1 Commonly exocrine glands found in social insects. Glands with capital lettering indicate a putative pheromonal function (author J. Billen) (with Westview Press permission).

Role of various glands in hydrocarbon production and storage

Many social insects are particularly rich in exocrine glands (Figure 5.1). The variety and role of these glands have been thoroughly described in Hymenoptera and Isoptera. Ants have a rich exocrine gland network which is described in several reports: Billen (2006), Billen and Morgan, (1998), Gobin *et al.* (2001; 2003) and also Morgan (2008) reviewing their chemicals. Quennedey (1975; 1978) and Clément and Bagnères (1998) found that no large glandular structures existed in termites (except in soldiers and alates). Gobin *et al.* (2003) described by electron microscopy subepithelial glands in various ant species and suggested that they were probable hydrocarbon producing cells and not simply a reservoir for hydrocarbons. However, no biochemical data was presented. Subsequently, Cabrera *et al.* (2004) analyzed the metapleural gland of *Solenopsis* and showed that it has a species specific chemical composition, as present in other glands. The significant amount of hydrocarbons in the gland reservoir is similar to that founds in the postpharyngeal gland (see below). Eltz *et al.* (2007) published an interesting paper describing enfleurage, lipid recycling, and the origin of perfume collection in orchid bees. The bees use a lipophilic carrier (here for straight chain lipids such as alkanes) produced by the bee's labial gland to capture and collect orchid perfume in their hind leg pockets. This advanced perfume collection could evolve from the scent marking of ancestral bees.

The content of the Dufour gland has been studied in a number of Hymenoptera: in bumblebees for the close similarity between gland and cuticular hydrocarbons, in *Polistes* wasps for possible involvement in nestmate recognition or ant repellent effect (Singer *et al.*, 1998; Dani *et al.*, 1996a, 1996b, 2003), and in honeybees as a source of caste-specific secretions (Katzav-Gozansky *et al.*, 1997). In his review on non-*Apis* bee exocrine glands, Hefetz (1998) described the Dufour gland as the best-documented glandular structure with respect to chemistry and function in Halictine bees. The Dufour secretions are rich in long-chain *n*-alkanes that appear to provide a waterproof lining. In *Proxylocopa olivieri,* the Dufour gland secretions contain a series of hydrocarbons ranging from C_{23} to C_{29} that are clearly involved in brood cell lining. Studies of Dufour gland secretions in various large carpenter bees of the genus *Xylocopa* have revealed a chemospecific pattern formed exclusively by alkanes and alkenes. One of the postulated functions of Dufour secretions is as a marker for precise location of the bee nest, in particular the nest entrance. Indeed, the use of hydrocarbon to mark the nest entrance has been shown in several eusocial insects. The biosynthetic capabilities of the Dufour gland in bees should have changed through evolution, so that exudates contain not only alkanes but also lactones and numerous other compounds (Hefetz, 1998). The Dufour gland has also been studied in ants. In a study of the contents of various glands in the jumping ant *Harpegnatos salvator*, Do Nascimento *et al.* (1993) calculated that *(Z)*-9-tricosene in the Dufour gland accounted for half of the complex mixture of linear hydrocarbons observed. In addition to pointing out that the Dufour gland present in all ant species was possibly the least understood of all endocrine glands, Morgan (2008) also showed that shorter-chain hydrocarbons in the gland were mostly saturated alkanes, with an increasing proportion of alkenes in the longer-chain hydrocarbons. This process could have a specific purpose, since short-chain alkanes are solids at room temperature while alkenes are more liquid, resulting in mixtures with lower melting points than either component. The role of these hydrocarbons may be to enhance the persistence of more volatile compounds in the environment by acting as a "keeper" to avoid evaporation rather than as a pheromone (Morgan, 2008). No data are available on the biosynthesis of those compounds by the Dufour's gland or elsewhere.

Role of the postpharyngeal gland in ants

The postpharyngeal gland is the most studied glandular structure with respect to hydrocarbon content in social insects. It is the largest gland in the ant's cephalic capsule, with a direct connection to the pharynx and with an external opening to the mouth cavity. Herzner *et al.* (2007a) first described a postpharyngeal gland in a non-ant species, the solitary wasp *Philanthus triangulum* (the European beewolf), and compared its function to that of the gland in ants (Strohm *et al.*, 2007). In a second report (Herzner *et al.*, 2007b), the same authors showed that female beewolves wrap food with large amounts of this gland secretion (see Chapter 14). The postpharyngeal gland appears to be specific to Formicidae (exception noted above) with regard to its large volume, glove-shaped structure (Figure 5.2) and

Figure 5.2 Photo of postpharyngeal glands of *Camponotus fellah* (right) and *Aphaenogaster senilis* (left). With permission of Alain Lenoir (IRBI,Tours).

important role in chemical communication. Since there are several hundred articles on the ant postpharyngeal gland, it is impossible to cover them all in this chapter.

The first authors to describe the lipid content and anatomy of the postpharyngeal gland were Peregrine *et al.* (1972, 1973), Peregrine and Mudd (1974), and Delage-Darchen (1976). This gland was initially described as a digestive organ, even though earlier results based on analysis of head extracts from *Myrmica* queens by Brian and Blum (1969) had suggested a role in caste differentiation. Work carried out on the postpharyngeal gland in the fire ant *Solenopsis invicta* by Vinson *et al.* (1980), Thompson *et al.* (1981), and Vander Meer *et al.* (1982) demonstrated the presence of hydrocarbons with a caste-specific pattern. Bagnères and Morgan (1991) showed a qualitative similarity between cuticular and glandular blends in several ant species that set off a decade of studies to determine the function of the postpharyngeal gland in numerous ants. The same authors also noted quantitative similarity in some species and divergence in others. These studies opened a debate over the gland function as a passive or active reservoir or possibly as a hydrocarbon-producing organ.

The role of the postpharyngeal gland was discussed in recent reviews by Lenoir *et al.* (1999), Blomquist and Howard (2003), and Howard and Blomquist (2005). Schoeters and Billen (1997) noted that ant age influences the color and shape of the gland as in diponerine ants. Findings in *Monomorium* queens have also demonstrated that gland thickness varies with age, and release of secretion differs according to whether the queens are virgin or mated, with mated queens accumulating larger amounts of lipid secretions and virgins releasing secretions earlier at seven days (Eelen *et al.,* 2006). Ultrastructural study of the gland indicates involvement in metabolism and de novo synthesis of lipids, but no biochemical or molecular data are available regarding de novo synthesis of hydrocarbons. Eelen *et al.* (2006) have offered different views on the possible biological functions of the postpharyngeal gland: feeding of larvae, providing nutritious oils for egg production in mated queens, and signaling species, nestmate, caste identity and reproductive capacity of queens. Numerous authors have suggested that the gland plays an indirect role in the ant's hierarchical-physiological-worker policing active involvement of hydrocarbons (see beginning of the social insect's part of the present chapter). Lenoir *et al.* (1999) reviewed the role of the postpharyngeal gland in mediating individual and colony identity, i.e., the most frequently studied aspect.

Since postpharyngeal gland content is species-specific, it can be used as a chemotaxonomic tool to assist in phylogeny resolution as reported in the *Cataglyphis* species complex (Dahbi *et al.,* 1996) and described in Chapter 7 of this book. Following a previous report (Cabrera *et al.,* 2004) showing that the hydrocarbon content of the postpharyngeal and metapleural glands in two *Solenopsis* species (*S. invicta* and *S. geminata*) was similar, Attygalle *et al.* (2006) compared the postpharyngeal gland composition of *Solenopsis geminata* workers and soldiers and concluded that gland content was also caste-specific. Vargo and Hulsey (2000) performed bioassays to identify the sources of queen pheromones in *S. invicta* and showed that the postpharyngeal gland was also the source of the queen recognition pheromone. Kaib *et al.* (2000) studied variations of postpharyngeal and cuticular hydrocarbon compositions as a function of task (brood-tenders, foragers, scouts) in *Myrmicaria eumenoides* ant colonies. Even though variations were more extensive on the cuticle than in the gland, the authors reported that they were strongly correlated and concluded that there was a dynamic relationship between the postpharyngeal gland and cuticular hydrocarbons. This finding added new evidence to an ongoing discussion opened by Meskali *et al.* (1995), who investigated regulation and homogeneity of the chemical signature between *Camponotus vagus* members. By topical application in a solvent (*n*-pentane) of *n*-tetracosane, an endogenous compound existing in trace quantity, and of an exogenous hydrocarbon, (Z)-9-tricosene, on marked ants, Meskali *et al.* (1995) tracked new content on the cuticle and in the postpharyngeal gland of the marked ants and in some unmarked nestmates. Quantities of the endogenous and exogenous compounds varied on the basis of their natural presence in the ant, with the endogenous hydrocarbon being metabolized much later than the exogenous one. The data suggested that the turnover of the different compounds was highly dynamic, since the washed cuticle recovered its original odor relatively quickly (Meskali *et al.*, 1995). In early studies on heterospecific (artificial or natural) colonies, the process leading to underlying congruency between the cuticle and the gland hydrocarbons was discussed from the standpoint of acquisition by passive transfer or de novo biosynthesis (Hefetz *et al.*, 1992; Vienne *et al.*, 1995; Bonavita-Cougourdan *et al.*, 2004; Chapter 14 of this book). The current data indicate that acquisition depends mainly on transfer, but that de novo biosynthesis has not been completely ruled out yet. Most studies have focused on hydrocarbon circulation in natural ant colonies to gain insight into nestmate recognition (Soroker and Hefetz, 2000; Lucas *et al.*, 2004; Chapter 11 of this book). A number of these studies have been devoted to the ponerine ant *Pachycondila apicalis* because trophallaxis is not as well established in this primitive species as it is in higher-order subfamilies, and thus, it should be easier to differentiate hydrocarbon circulation by allogrooming vs. trophallaxis and to explain the role of the postpharyngeal gland in colony odor formation. The data showed that transfer to the gland in this species occurs during allogrooming (Soroker *et al.*, 1998). The data also indicate that the postpharyngeal gland secrets hydrocarbons to the cuticular hair brush on the front legs (front basitarsal brush) from where they are either distributed over the body surface or cleared via the alimentary canal (Hefetz *et al.*, 2001). Hydrocarbon transfer to the postpharyngeal gland was lower in this species than others; higher transfer to the cuticle is attributed to the important role of the basitarsal brushes (Soroker *et al.*, 2003). Contribution to the uniform odor of large colonies, i.e. the "gestalt odor" (Crozier and Dix, 1979) was confirmed in the ant *Cataglyphis cursor* (Soroker *et al.*, 1994, 1995a, 1995b). A study comparing gestalt odor dynamics in two ant species was carried out on *Camponotus fellah* that performs trophallaxis and on *Aphaenogaster senilis*, which does not. The goal was to gain understanding of the respective importance of trophallaxis and allogrooming in formation of the gestalt odor in two differently evolved species (Lenoir *et al.*, 1999, 2001). Taken together, the current study indicates that the postpharyngeal gland is used for the active storage/release (and not de novo synthesis) of colonial odor hydrocarbons, and also conclusively demonstrates its adaptative use as a gestalt organ.

The postpharyngeal gland has also been implicated in egg marking functions. Studies in the ant *Gnaptogenys striatula* (Lommelen *et al.*, 2008) showed that the same hydrocarbons were present on the egg surface and in the ovaries. Similar findings have been reported in cockroaches (Schal *et al.*, 1998b; Fan *et al.*, 2002), mosquitoes (Atella *et al.*, 2006) and houseflies (Schal *et al.*, 2001). In the case of ant egg marking, it has been proposed that this marking process may help ants to identify the eggs from dominant egg layers. In consequence, this mechanism could also be implicated in egg cannibalism, something that has been observed by various authors including Dietemann, Cuvillier-Hot, and Endler (for further reading see in Chapters 12 and 13 of this book and review by Monnin, 2006). Egg cannibalism is an important policing tactic used to control worker reproduction (Monnin and Peeters, 1997; Monnin, 2006).

Analysis of hydrocarbons from anal fluids (feces) involved in nest marking by the ant *Messor capitatus* (Grasso *et al.*, 2005) suggests that territory marking using abdominal secretions issued from the rectal sac may be another labeling function by the way of abdominal secretions. The same authors reported that the hydrocarbons in the feces were similar to those hydrocarbons on the cuticle of workers.

The transport of hydrocarbons by social insects can be involved in creating the "hydrocarbon signature". Evidence was first obtained in the termite *Zootermopsis nevadensis* (Sevala *et al.*, 2000). Comparison of cuticular lipids with internal and hemolymph hydrocarbons in different castes showed that, as in other species, the content was qualitatively similar. However, quantitative differences were observed between hemolymph and cuticular hydrocarbon profiles. Sevala *et al.* (2000) showed that hemolymph hydrocarbons were associated with a dimeric high-density lipoprotein (HDLp) lipophorin, similar to those described from other insects (see above). This lipoprotein consisted of two subunits: apolipophorin I and apolipophorin II. They had molecular weights of 220 kDa and 82 KDa respectively in *Z. nevadensis*, and 210 and 85 kDa respectively in *R. flavipes*, similar to those of other insect species. In the same study, evidence was presented of internalization of exogenous/supplementary hydrocarbons and transportation of polar hydrocarbon metabolites by the same lipoprotein (Sevala *et al.*, 2000). Demonstration of cuticle-hemolymph transport helps to explain the formation of caste-specific profiles. Termite lipophorin and its involvement in hydrocarbon transport have been further characterized using the anti-lipophorin antiserum generated with lipophorin isolated from the hemolymph of the subterranean termite *R. flavipes* (Fan *et al.*, 2004). Immunoprecipitation tests confirmed that this antiserum specifically recognized lipophorin from various termite species and cockroaches. Immunodiffusion tests showed that cross-reactivity with hemolymph from termites and cockroaches was closely correlated with the evolutionary proximity of *R. flavipes* to cockroaches. An ELISA developed using this antiserum showed that lipophorin levels in *R. flavipes* varied significantly as a function of caste and developmental stage. A methyl-branched hydrocarbon precursor (propionate, see above references) was injected into *Z. nevadensis* to monitor de novo hydrocarbon biosynthesis and transport. The transfer process was hindered by injection of the lipophorin-antiserum in that it suppressed externalization of the newly synthesized hydrocarbons (Fan *et al.*, 2004). These data support the hypothesis that newly synthesized hydrocarbons are carried by lipophorin prior to deposition on the surface of the insect.

In a paper on involvement of fat body in hydrocarbon synthesis, Soroker and Hefetz (2000) showed that hydrocarbons were released internally into the hemolymph and transported to the cuticle and the postpharyngeal gland (where they could be applied to the cuticle by self-grooming or cleared via the alimentary canal). Lucas *et al.* (2004) presented a comparative study designed to gain a better understanding of the role of hemolymph lipophorin, cuticle, postpharyngeal gland (PPG) and self-grooming in hydrocarbon circulation and formation of the colonial signature in the ant *Pachycondyla villosa* (see paragraph on PPG). Their data showed a difference in the hydrocarbon composition of the ant lipophorin (dimeric 820 kDa Lp, and two 245 and 80 kDa apoproteins) and the postpharyngeal gland, thus suggesting selectivity in this transfer between normal and methyl-branched alkanes. The hydrocarbons in both cuticle and postpharyngeal gland differed only in quantity between ant colonies and appeared to be essential for chemical communication, and in particular in nestmate recognition. This finding on selective transfer could explain the differing biological roles of some of the hydrocarbons in a mixture. Discrimination of individual cuticular hydrocarbons from different classes was tested using proboscis extension response on the honeybee (Châline *et al.*, 2005). The tested hydrocarbons were classified into those learned and well discriminated (mostly alkenes) and those which were not discriminated well (alkanes and some alkenes). Alkenes may constitute important compounds used in the honeybee as cues in the social recognition process. Chan *et al.* (2006) reported the results of a quantitative comparison of caste differences in the honeybee hemolymph protein composition. Although among the first insects in which lipophorins were identified (Robbs *et al.*, 1985), the honeybee is not the most suitable model, since cuticular

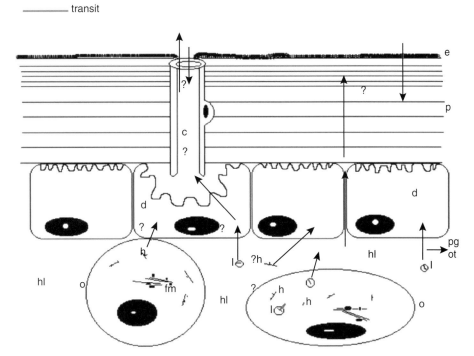

Figure 5.3 Schematic drawing showing transport of hydrocarbons (and other lipids) from site of synthesis (oenocytes) to cuticle surface (epicuticle) and various tissues and glands. Arrows represent hypothetical transport of hydrocarbons (and/or precursors) [legends: e: epicuticule; p: procuticule; h: hydrocarbons (and/or precursors); d: epidermal cell; c: canal issuing from an epidermal cell; o: oenocytes; l: lipophorins; fm: microsome fraction (reticulum endoplasmic of oenocytes, site of hydrocarbon biosynthesis); hl: hemolymph; pg: pheromone glands; ot: other tissues (ovaries)] (updated from Bagnères, 1996).

hydrocarbons may not be the major compounds involved in different biological functions (Breed, 1998). However, the honeybee is important from an ecological and agronomic standpoint and appears to be the only social insect in which a low-density lipoprotein receptor (LDLR) family has been identified as the putative ortholog for the lipophorin receptor (Amlpr) in the genome (Guidugli-Lazzarini *et al.*, 2008).

To illustrate current concepts of hydrocarbon transportation in insects, we present a scheme (Figure 5.3) showing the hypothetical transport pathways of hydrocarbons and other lipids to various tissues (Bagnères (1996) and Jurenka (2004)).

An interesting study carried out by Guntur *et al.* (2004) on antennal proteins of male *Solenopsis invicta* fire ants is worth mentioning in closing. Their findings showed that the isolated protein corresponded to the sequence of an apolipophorin III, i.e., an exchange-able lipid-binding protein expressed in the antenna as well as in the head and abdomen of the ant. Although this protein could act as a pheromone-binding protein, the authors were unable to identify the source: hemolymph circulation, fat body cells, or other. Data also showed that the protein was also caste differing in expression, with lower amounts in workers. Further study will be required to determine if such caste differences are related to multiple functions. More work will also be necessary to clarify whether this lipophorin involved in phospholipid binding could also be involved in olfaction in fire ants, and may be used as a hydrocarbon binding protein. Also worth noting, the work of Böröczky *et al.* (2008) on antenna differences in cuticular lipid composition (not only hydrocarbons) shows a clear male and female cuticular profile difference in two non-social insect species (Lepidoptera).

Conclusion

A number of questions about the machinery of chemical communication remain largely unanswered. A recent study on iridoid-producing chrysomeline leaf beetles (Kunert *et al.*, 2008) provided insight into the question of sequestration vs. de novo biosynthesis of semiochemicals. Findings showed that beetle larvae are able to both sequester precursors from food and synthesize the compounds de novo. This dual strategy could enable the beetle larvae to shift from one to another plant host. The fundamental question of whether insect hydrocarbons are biosynthesized by the insect or whether some can be adsorbed, sequestered and released must be resolved in social insects. The issue of sequestration vs. biosynthesis has important implications from the evolutionary standpoint, e.g. chemical mimicry (see Chapter 14) and glandular content as in the PPG (current chapter). Other unanswered questions regarding the chemical machinery underlying chemical communication in insects include hydrocarbon selection (see above), pathways of internalization and externalization through the cuticle, and mechanisms of transfer of hydrocarbons between lipophorin and tissues.

References

Armold, M. T. and Regnier, F. E. (1975). Stimulation of hydrocarbon biosynthesis by ecdysterone in the flesh fly *Sarcophaga bullata*. *J. Insect Physiol.*, **21**, 1581–1586.

Arnold, G., Quenet, B., Cornuet, J.-M., Masson, C., de Schepper, C., Estoup, A. and Gasquet, P. (1996). Kin recognition in honeybees. *Nature*, **379**, 498.

Atella, G. C., Silva-Neto, A. C., Golodne, D. M., Arefin, S. and Shahabuddin, M. (2006). *Anopheles gambiae* lipophorin: Characterization and role in lipid transport to developing oocyte. *Insect Biochem. Mol. Ecol.*, **36**, 375–386.

Attygalle, A., Jham, G. and Morgan, E. D. (2006). Contents of the hypertrophied postpharyngeal gland of workers and soldiers of of the fire ant *Solenopsis geminata*. (Hymenoptera: Myrmicinae). *Sociobiology*, **47**, 471–482.

Bagnères, A.-G. (1996). *Composition, Variation et Dynamique de la Signature Chimique chez les Insectes*. Habilitation à diriger des Recherches, unpublished HDR thesis, Université St. Charles, Aix-Marseille I.

Bagnères, A.-G and Morgan, E. D. (1991). The postpharyngeal glands and the cuticle of Formicidae contain the same characteristic hydrocarbons. *Experientia*, **47**, 106–111.

Billen, J. (2006). Signal variety and communication in social insects. *Proc. Neth. Entomol. Soc. Meet.*, **17**, 9–25.

Billen, J. and Morgan, E. D. (1998). Pheromone communication in social insects: sources and secretion. In *Pheromone Communication in Social Insects. Ants, Wasps, Bees and Termites*, ed. R. K. Vander Meer, M. D. Breed, K. E. Espelie and M. L. Winston. New York: Westview Press, pp. 3–33.

Blomquist, G. J., Dillwith, J. W. and Pomonis, J. G. (1984). Sex pheromone of the housefly: metabolism of (Z)-9-tricosene to (Z)-9,10-epoxytricosane and (Z)-14-tricosene-10-one. *Insect Biochem.*, **14**, 279–284.

Blomquist, G. J. and Howard, R. W. (2003). Pheromone biosynthesis in social insects. In *Insect Pheromone Biochemistry and Molecular Biology*, ed. G. J. Blomquist and R. Vogt. Amsterdam: Elsevier, pp. 323–340.

Blomquist, G. J., Tilman, J. A., Mpuru, S. and Seybold, S. J. (1998). The cuticle and cuticular hydrocarbons of insects: Structure, function and biochemistry. In *Pheromone Communication in Social Insects. Ants, Wasps, Bees and Termites*, ed. R. K. Vander Meer, M. D. Breed, K. E. Espelie and M. L. Winston. New York: Westview Press, pp. 34–54.

Blum, M. S. (1985). Exocrine systems. In *Fundamentals of Insect Physiology*, ed. M. S. Blum. New York: Wiley, pp. 535–579.

Blum, M. S. (1987). Specificity of pheromonal signals: A search for its recognitive bases in terms of a unified chemisociality. In *Chemistry and Biology of Social Insects*, ed. J. Eder and H. Rembold, Munich: Peperny, pp. 401–405.

Bonavita-Cougourdan, A., Provost, E., Rivière, G., Bagnères, A.-G. and Dusticier, G. (2004). Regulation of cuticular and postpharyngeal hydrocarbons in the slave-making ant *Polyergus rufescens*: effect of *Formica rufibarbis* slaves. *J. Insect Physiol.*, **50**, 285–293.

Bonavita-Cougourdan, A., Theraulaz, G., Bagnères, A.-G., Roux, M., Pratte, M., Provost, E. and Clément J.-L. (1991). Cuticular hydrocarbons, social organization and ovarian development in a *Polistes* wasp: *Polistes dominulus* Christ. *Comp. Biochem. Physiol. B*, **100**, 667–680.

Böröczky, K., Park, K. C., Minard, R. D., Jones, T. H., Baker, T. C. and Tumlinson, J. H. (2008). Differences in cuticular lipid composition of the antennae of *Helicoverpa zea*, *Heliothis virescens*, and *Manduca sexta*. *J. Insect Physiol.*, **54**, 1385–1391.

Bortolotti, L., Duchateau, M. J. and Sbrenna, G. (2001). Effect of juvenile hormone on caste determination and colony processes in the bumblebee *Bombus terrestris*. *Entomol. Experim. Applic.*, **101**, 143–158.

Boulay, R., Soroker, V., Godzinska, E. J. A., Hefetz, H. and Lenoir. A. (2000). Octopamine reverses the isolation-induced increase in trophallaxis in the carpenter ant *Camponotus fellah*. *J. Exp. Biol.*, **203**, 513–520.

Breed, M. D. (1998). Chemical cues in kin recognition: criteria for identification, experimental approaches, and the honey bee as an example. In *Pheromone Communication in Social Insects. Ants, Wasps, Bees and Termites*, ed. R. K. Vander Meer, M. D. Breed, K. E. Espelie and M. L. Winston. New York: Westview Press, pp. 57–78.

Brent, C., Peeters, C., Dietmann, V., Crewe, R. and Vargo, E. D. (2006). Hormonal correlates of reproductive status in the queenless ponerine ant, *Streblognathus peetersi*. *J. Comp. Physiol.*, **192A**, 315–320.

Brian, M. V. and Blum, M. S. (1969). The influence of *Myrmica* queen head extracts on larval growth. *J. Insect Physiol.*, **15**, 2213–2223.

Cabrera, A., Williams, D., Hernández, J. V., Caetano, F. H. and Jaffe, K. (2004). Metapleural- and postpharyngeal-gland secretions from workers of the ants *Solenopsis invicta* and *S. geminata*. *Chem. Biodivers.*, **1**, 303–311.

Canavoso, L. E., Yun, H. K., Jouni, Z. E. and Wells, M. A. (2004). Lipid transfer particle mediates the delivery of diacylglycerol from lipophorin to fat body in larval *Manduca sexta*. *J. Lipid Res.*, **45**, 456–465.

Châline, N., Sandoz J.-C., Martin S. J., Ratnieks F. L. W. and Jones G. R. (2005). Learning and discrimination of individual cuticular hydrocarbons by honeybees (*Apis mellifera*). *Chem. Senses*, **30**, 327–335.

Chan, Q. W. T., Howes, C. G. and Foster L. J. (2006). Quantitative comparison of caste differences in honeybee hemolymph. *Mol. Cell. Proteomics*, **5**, 2252–2262.

Chino, H., Downer, R. G. H., Wyatt, G. R. and Gilbert, L. I. (1981b). Lipophorins, a major class of lipoproteins of insect haemolymph. *Insect Biochem.*, **11**, 491.

Chino, H. and Gilbert, L. I. (1971). The uptake and transport of cholesterol by haemolymph lipoproteins. *Insect Biochem.*, **1**, 337–347.

Chino, H., Katase, H., Downer, R. G. H. and Takahashi, K. (1981a). Diacylglycerol-carrying lipoprotein of hemolymph of the American cockroach: purification, characterization, and function. *J. Lipid Res.*, **22**, 7–15.

Clément J.-L. and Bagnères, A.-G. (1998). Nestmate recognion in termites. In *Pheromone Communication in Social Insects. Ants, Wasps, Bees and Termites*, ed. R. K. Vander Meer, M. D. Breed, K. E. Espelie and M. L. Winston. New York: Westview Press, pp. 126–155.

Crozier, R. H. and Dix, M. W. (1979). Analysis of two genetic models for the innate components of colony odour in social hymenoptera. *Behav. Ecol. Sociobiol.*, **4**, 217–224.

Cuvillier-Hot, V., Lenoir, A., Crewe, R. Malosse, C. and Peeters, C. (2004). Fertility signalling and reproductive skew in queenless ants. *Anim. Behav.*, **68**, 1209–1219.

Cuvillier-Hot, V., Renault, V. and Peeters, C. (2005). Rapid modification in the olfactory signal of ants following a change in reproductive status. *Naturwissenschaften*, **92**, 73–77.

Dahbi, A., Lenoir, A., Tinaud, A., Taghizadeh, T., Francke, W. and Hefetz, A. (1996). Chemistry of the postpharyngeal gland secretion and its implication for the phylogeny of the Iberian *Cataglyphis* species (Hymenoptera: Formicidae). *Chemoecology*, **7**, 163–171.

Dani, F., Fratini, S. and Turillazzi, S. (1996a). Behavioural evidence for the involvement of Dufour's gland secretion in nestmate recognition in the social wasp *Polistes dominulus* (Hymenoptera: Vespidae). *Behav. Ecol. Sociobiol.*, **38**, 311–319.

Dani, F., Jones, G. R., Morgan, E. D. and Turillazzi, S. (2003). Reevaluation of the chemical secretion of the sternal glands of *Polistes* social wasps (Hymenoptera Vespidae). *Ethol. Ecol. Evol.*, **15**, 73–82.

Dani, F., Morgan, E. D. and Turillazzi, S. (1996b). Dufour gland secretion of *Polistes* wasps: chemical composition and possible involvement in nestmate recognition (Hymenoptera: Vespidae). *J. Insect Physiol.*, **42**, 541–548.

Dapporto, L., Dani, F. R. and Turillazzi, S. (2007a). Social dominance molds cuticular and egg chemical blends in a paper wasp. *Current Biology*, **17**, 502–503.

Dapporto, L., Santini, A., Dani, F. R. and Turillazzi, S. (2007b). Workers of a *Polistes* wasp detect the presence of their queen by chemical cues. *Chem. Senses*, **32**, 795–802.

Delage-Darchen, B. (1976). Les glandes post-pharyngiennes des fourmis. Connaissances actuelles sur leur structure, leur fonctionnement, leur rôle. *Ann. Biol.*, **15**, 63–76.

Denis, D., Blatrix, R. and Fresneau, D. (2006). How an ant manages to display individual and colonial signals by using the same channel. *J. Chem. Ecol.*, **32**, 1647–1661.

D'Ettorre, P. and Moore, A. J. (2008). Chemical communication and the coordination of social interactions in insects. In *Sociobiology of Communication*, ed. P. d'Ettore and D. P. Hughes. Oxford: Oxford University Press, pp. 81–117.

Diehl, P. A. (1973). Paraffin synthesis in the oenocytes of the desert locust. *Nature*, **243**, 468–470.

Diehl, P. A. (1975). Synthesis and release of hydrocarbons by the oenocytes of the desert locust *Schistocercaq gregaria*. *J. Insect Physiol.*, **21**, 1237–1246.

Dietemann, V., Liebig, J., Hölldobler, B. and Peeters, C. (2005). Changes in the cuticular hydrocarbons of incipient reproductives correlate with triggering of worker policing in the bulldog ant *Myrmecia gulosa*. *Behav. Ecol. Sociobiol.*, **58**, 486–496.

Dietemann, V., Peeters, C., Liebig, J., Thivet, V. and Hölldobler, B. (2003). Cuticular hydrocarbons mediate recognition of queens and reproductive workers in the ant *Myrmecia gulosa*. *Proc. Natl. Acad. Sci. USA*, **100**, 10341–10346.

Dillwith, J. W., Adams, T. S. and Blomquist, G. J. (1983). Correlation of housefly sex pheromone production with ovarian development. *J. Insect Physiol.*, **29**, 377–386.

Dillwith, J. W., Blomquist, G. J. and Nelson, D. R. (1981). Biosynthesis of the hydrocarbon components of the sex pheromone of the housefly *Musca domestica* L. *Insect Biochem.*, **11**, 247–253.

Do Nascimento, R., Billen, J. R., and Morgan, E. D. (1993). The exocrine secretions of the jumping ant *Harpegnathos saltator*. *Comp. Biochem. Physiol.*, **104B**, 505–508.

Dronnet, S., Lohou, C., Christides J.-P. and Bagnères, A.-G. (2006). Cuticular hydrocarbon composition reflects genetic relationship among colonies of the introduced termite *Reticulitermes santonensis* Feytaud. *J. Chem. Ecol.*, **32**, 1027–1042.

Eelen, D., Børgesen, L. and Billen, J. (2006). Functional morphology of the postpharyngeal gland of queens and workers of the ant *Monomorium pharaonis* (L.). *Acta Zoologica*, **87**, 101–111.

Eltz, T., Zimmermann, Y., Haftmann, J., Twele, R., Francke, W., Quezada-Euan, J. J. and Lunau, K. (2007). Enfleurage, lipid recycling and the origin of perfume collection in orchid bees. *Proc. R. Soc. B*, **274**, 2843–2848.

Endler, A., Hölldobler, B. and Liebig, J. (2007). Lack of physical policing and fertility cues in egg-laying workers of the ant *Camponotus floridanus*. *Anim. Behav.*, **74**, 1171–1180.

Endler, A., Liebig, J. and Hölldobler, B. (2006). Queen fertility, egg-marking and colony size in the ant *Camponotus floridanus*. *Behav. Ecol. Sociobiol.*, **59**, 490–499.

Endler, A., Liebig, J., Schmitt, T., Parker, J. E., Jones, G. R., Schreier, P. and Hölldobler, B. (2004). Surface hydrocarbons of queen eggs regulate worker reproduction in a social insect. *Proc. Natl. Acad. Sci. USA*, **101**, 2945–2950.

Errard, C., Hefetz, A. and Jaisson, P. (2006). Social discrimination tuning in ants: template formation and chemical similarity. *Behav. Ecol. Sociobiol.*, **59**, 353–363.

Fan, Y., Chase, J., Sevala, V. L. and Schal, C. (2002). Lipophorin-facilitated hydrocarbon uptake by oocytes in the German cockroach *Blattella germanica*. *J. Exp. Biol.*, **205**, 781–790.

Fan, Y., Schal, C., Vargo E. L. and Bagnères, A.-G. (2004). Characterization of termite lipophorin and its involvement in hydrocarbon transport. *J. Insect Physiol.*, **50**, 609–620.

Fan, Y., Zurek, L., Dykstra, M. J. and Schal, C. (2003). Hydrocarbon synthesis by enzymatically dissociated oenocytes of the abdominal integument of the German cockroach *Blattella germanica*. *Naturwissenschaften*, **90**, 121–126.

Ferveur, J.-F. (2005). Cuticular hydrocarbons: Their evolution and roles in *Drosophila* pheromonal communication. *Behav. Genet.*, **35**, 279–295.

Ferveur, J.-F., Savarit, F., O'Kane, C. J., Sureau, G., Greenspan, R. J. and Jallon, J-M. (1997). Genetic feminization of pheromones and its behavioral consequences in *Drosophila* males. *Science*, **276**, 1555–1558.

Florane, C. B., Bland, J. M., Husseneder, C. and Raina, A. K. (2004). Diet-mediated inter-colonial aggression in the Formosan termite *Coptotermes formosanus*. *J. Chem. Ecol.*, **30**, 2559–2574.

Ginzel, M. D., Millar, J. G. and Hanks, L. M. (2003). (Z)-9-Pentacosene – contact sex pheromone of the locust borer, *Megacyllene robiniae*. *Chemoecology*, **13**, 135–141.

Gobin, B., Ito, F. and Billen, J. (2003). The subepithelial gland in ants: a novel exocrine gland closely associated with the cuticle surface. *Acta Zoologica*, **84**, 285–291.

Gobin, B. Rüppell, O., Hartmann, A., Jungnickel, H., Morgan, E. D. and Billen, J. (2001). A new type of exocrine gland and its function in mass recruitment in the ant *Cylindromyrmex whymperi* (Formicidae, Cerapachynae). *Naturwissenschaften*, **88**, 323–326.

Grasso, D. A., Sledge, M. F., Le Moli, F. Mori, A. and Turillazzi, S. (2005). Nest-area marking with faeces: a chemical signature that allows colony-level recognition in seed harvesting ants (Hymenoptera, Formicidae). *Insect Soc.*, **52**, 36–44.

Gu, X., Quilici, D., Juarez, P., Blomquist, G. J. and Schal, C. (1995). Biosynthesis of hydrocarbon and contact sex pheromone and their transport by lipophorin in females of the German cockroach (*Blattella germanica*). *J. Insect Physiol.*, **41**, 257–267.

Guidugli-Lazzarini, K. R., do Nascimento, A. M., Tanaka, E. D., Piulachs, M. D., Hartfelder, K., Bitondi, M. G. and Simões, Z. L. (2008). Expression analysis of putative vitellogenin and lipophorin receptors in honey bee (*Apis mellifera* L.) queens and workers. *J. Insect Physiol.*, **54**, 1138–1147.

Guntur, K. V. P., Velasquez, D., Chadwell, L., Carroll, C., Weintraub S., Cassill, J. A. and Renthal, R. (2004). Apolipophorin-III-like protein expressed in the antenna of the red imported fire ant, *Solenopsis invicta* Buren (Hymenoptera: Formicidae). *Arch. Insect Biochem. Physiol.*, **57**, 101–110.

Hadley, N. F. (1981). Cuticular lipids of terrestrial plants and arthropods: a comparison of their structure, composition, and waterproofing function. *Biol. Rev.*, **56**, 23–47.

Hefetz, A. (1998). Exocrine glands and their products in non-*Apis* bees: chemical, functional and evolutionary perspective. In *Pheromone Communication in Social Insects. Ants, Wasps, Bees and Termites*, ed. R. K. Vander Meer, M. D. Breed, K. E. Espelie and M. L. Winston. New York: Westview Press, pp. 236–256.

Hefetz, A., Errard, C. and Cojocaru, M. (1992). Heterospecific substances in the postpharyngeal gland secretion of ants reared in mixed groups. *Naturwissenschaften*, **79**, 417–420.

Hefetz, A., Soroker, V., Dahbi, A., Malherbe, M.-C. and Fresneau, D. (2001). The front basitarsal brush in *Pachycondyla apicalis* and its role in hydrocarbon circulation. *Chemoecology*, **11**, 17–24.

Heinze, J., Foitzik, S., Hippert, A. and Hölldobler, B. (1996). Apparent dear-enemy phenomenon and environment-based recognition cues in the ant *Leptothorax nylanderi*. *Ethology*, **102**, 510–522.

Henderson, G. (1998). Primer pheromones and possible soldier caste influence on the evolution of sociality in lower termites. In *Pheromone Communication in Social Insects. Ants, Wasps, Bees and Termites*, ed. R. K. Vander Meer, M. D. Breed, K. E. Espelie and M. L. Winston. New York: Westview Press, pp. 314–330.

Herzner, G., Goettler, W., Kroiss, J., Purea, A., Webb, A. G., Jakob, P. M., Rössler, W. and Strohm, E. (2007a). Males of a solitary wasp possess a postpharyngeal gland. *Arthropod Struct. Develop.*, **36**, 123–133.

Herzner, G., Schmitt, T., Peschke, K, Hilpert, A. and Strohm, E. (2007b). Food wrapping with the postpharyngeal gland secretion by females of the European beewolf *Philanthus triangulum. J. Chem. Ecol.*, **33**, 849–859.

Hora, R. R, Ionescu-Hirsh, A., Simon, T., Delabie, J., Robert, J., Fresneau, D. and Hefetz, A. (2008). Postmating changes in cuticular chemistry and visual appearance in *Ectatomma tuberculatum* queens (Formicidae: Ectatomminae). *Naturwissenshaften*, **95**, 55–60.

Howard, R. W. and Blomquist, G. J. (2005). Ecological, behavioral, and biochemical aspects of insect hydrocarbons. *Annual Rev. Entomol.*, **50**, 371–393.

Jurenka, R. (2004). Insect Pheromone Biosynthesis. In *Topics in Current Chemistry, I*, **239**, 97–132.

Jurenka, R. A., Subchev, M., Abad, J.-L., Choi, M.-J., and Fabrias, G. (2003). Sex pheromone biosynthetic pathway for disparlure in the gypsy moth, *Lymantria dispar. Proc. Natl. Acad. Sci. USA*, **100**, 809–814.

Kaib, M., Eisermann, B., Schoeters, E., Billen, J., Franke, S. and Francke, W. (2000). Task-related variation of postpharyngeal and cuticular hydrocarbon compositions in the ant *Myrmicaria eumenoides. J. Comp. Physiol. A*, **186**, 939–948.

Kaib, M., Franke, S., Francke, W. and Brandl, R. (2002). Cuticular hydrocarbons in a termite: phenotypes and a neighbour-stranger effect. *Physiol. Entomol.*, **27**, 189–198.

Katase, H. and Chino, H. (1982). Transport of hydrocarbons by the lipophorin of insect hemolymph. *Biochim. Biophys. Acta,* **710**, 341–348.

Katzav-Gozansky, T., Soroker, V., Hefetz, A., Cojocaru, M. D., Erdmann, H. and Francke W. (1997). Plasticity of caste-specific Dufour's gland secretion in the honey bee (*Apis mellifera* L.). *Naturwissenschaften*, **84**, 238–241.

Kent, C., Azanchi, R., Smith, B., Chu, A. and Levine, J. (2007). A modelbased analysis of chemical and temporal patterns of cuticular hydrocarbons in male *Drosophila melanogaster. PLoS One*, **2**, e962.

Kunert, M., Søe, A., Bartram, S., Discher, S., Tolzin-Banasch, K., Nie, L., David, A., Pasteels, J. and Boland. W. (2008). *De novo* biosynthesis versus sequestration: A network of transport systems supports in iridoid producing leaf beetle larvae both modes of defense. *Insect. Biochem. Mol. Biol.*, **38**, 895–904.

LeConte, Y. and Hefetz, A. (2008). Primer pheromones in social hymenoptera. *Annu. Rev. Entomol.*, **53**, 523–542.

Lengyel, F., Westerlund, S. A. and Kaib, M. (2007). Juvenile hormone III influences task-specific cuticular hydrocarbon profile changes in the ant *Myrmicaria eumenoides. J. Chem. Ecol.*, **33**, 167–181.

Lenoir, A., Fresneau, D., Errard, C. and Hefetz, A. (1999). Individuality and colonial identity in ants: the emergence of the social representation concept. In *Information processing in social insects*, ed. C. Detrain, J.-L. Deneubourg and J.-M. Pasteels. Basel, Switzerland: Birkhäuser Verlag.

Lenoir, A., Hefetz, A., Simon, T. and Soroker, V. (2001). Comparative dynamics of gestalt odour formation in two ant species *Camponotus fellah* and *Aphaenogaster senilis* (Hymenoptera: Formicidae). *Physiol. Entomol.*, **26**, 275–283.

Liang, D. and Silverman, J. (2000). "You are what you eat": diet modifies cuticular hydrocarbons and nestmate regognition in the Argentine ant, *Linepithema humile*. *Naturwissenschaften*, **87**, 412–416.

Locke, M. (1965). Permeability of insect cuticle to water and lipids. *Science*, **147**, 295–298.

Lommelen, E., Johnson, C. A., Drijfhout, F. P., Billen, J. and Gobin, B. (2008). Egg marking in the facultative queenless ant *Gnamptogenys striatula*: The source and mechanism. *J. Insect Physiol.*, **54**, 727–736.

Lorenzi, M. C., Sledge, M. F., Laiolo, P., Strurlini, E. and Turillazzi, S. (2004). Cuticular hydrocarbon dynamics in young adult *Polistes dominulus* (Hymenoptera: Vespidae) and the role of linear hydrocarbons in nestmate recogniton systems. *J. Insect Physiol.*, **50**, 935–941.

Lucas, C., Pho, D. B., Fresneau, D. and Jallon, J.-M. (2004). Hydrocarbon circulation and colonial signature in *Pachycondyla villosa*. *J. Insect Physiol.*, **50**, 595–607.

Matsuoka, K., Tabunoki, H., Kawai, T., Ishikawa, S., Yamamoto, M., Sato, R. and Ando, T. (2006). Transport of a hydrophobic biosynthetic precursor by lipophorin in the hemolymph of a geometrid female moth which secretes an epoxyalkenyl sex pheromone. *Insect Biochem. Mol. Biol.*, **36**, 576–583.

Meskali, M., Bonavita-Cougourdan, A., Provost, E., Bagnères, A.-G., Dusticier, G and Clément, J.-L. (1995). Mechanism underlying cuticular hydrocarbon homogeneity in the ant *Camponotus vagus* (Scop) (Hymenoptera: Formicidae): role of postpharyngeal glands. *J. Chem. Ecol.*, **21**, 1127–1148.

Millor, J., Halloy, J., Amé, J.-M. and Deneubourg, J.-L. (2006). Individual discrimination capability and collective choice in social insects. In *Ant Colony Optimization and Swarm Intelligence, Lecture Notes in Computer Science*. Berlin: Springer, pp. 167–178.

Mohamed, M. and Prestwich, G. D. (1988). Hemolymph juvenile hormone binding proteins of *Reticulitermes flavipes*: induction by methoprene and suppression by pheromones. In *Endocrinological frontiers in physiological insect ecology*, ed. F. Sehnal and D.L. Denlinger. Wroclaw: Wroclaw Technical University Press, pp. 975–981.

Monnin, T. (2006). Chemical recognition of reproductive status in social insects. *Ann. Zool. Fennici*, **43**, 515–530.

Monnin, T. and Peeters, C. (1997). Cannibalism of subordinates'eggs in the monogynous queen-less ant *Dinoponera quadriceps*. *Naturwissenschaften*, **84**, 499–502.

Morgan, E. D. (2008). Chemical sorcery for sociality: Exocrine secretions of ants (Hymenoptera: Formicidae). *Myrmecol. News*, **11**, 79–90.

Mpuru, S., Blomquist, G. J., Schal, C., Roux, M., Kuenzli, M. Dusticier, G., Clément, J.-L. and Bagnères, A.-G. (2001). Effect of age and sex on the production of internal and external hydrocarbons and pheromones in the housefly, *Musca domestica. Insect Biochem. Mol. Biol.*, **31**, 139–155.

Nelson, D. R. (1969). Hydrocarbon synthesis in the American cockroach. *Nature (London)*, **221**, 854–855.

Peregrine, D. J. and Mudd, A. (1974). The effects of diet on the composition of the postpharyngeal glands of *Acromyrmex octospinosus* (Reich). *Insect. Soc.*, **21**, 417–424.

Peregrine, D. J., Mudd, A. and Cherrett, J. M. (1973). Anatomy and preliminary chemical analysis of the postpharyngeal glands of the leaf-cutting ant, *Acromyrmex octospinosus* (Reich.) (Hym. Formicidae). *Insect. Soc.*, **20**, 355–363.

Peregrine, D. J., Percy, H. C. and Cherrett, J. M. (1972). Intake and possible transfer of lipid by the postpharyngeal glands of *Atta cephalotes* L. *Ent. Exp. Appl.*, **15**, 248–258.

Pho, D. B., Pennanec'h, M. and Jallon, J.-M. (1996). Purification of adult *Drosophila melanogaster* lipophorin and its role in hydrocarbon transport. *Arch. Insect Biochem. Physiol.*, **31**, 289–303.

Provost, E., Rivière, G., Roux, M., Morgan, E. D. and Bagnères A.-G. (1993). Change in the chemical signature of the ant *Leptothorax lichtensteini* Bondroit with time. *Insect Biochem. Mol. Biol.*, **23**, 945–957.

Quennedey, A. (1975). Morphology of exocrine glands producing pheromones and defensive substances in subsocial and social insects. In *Pheromones and defensive secretions in social insects*. Dijon: Proceedings of the IUSSI, pp. 1–21.

Quennedey, A. (1978). *Les glandes exocrines des termites. Ultrastructure comparée des glandes sternales et frontales*. Thèse d'Etat université de Dijon.

Robbs, S. L., Ryan, R. O., Schmidt, J. O., Keim, P. S. and Law, J. H. (1985). Lipophorin of the larval honeybee, *Apis mellifera. J. Lipid Res.*, **26**, 241–247.

Roma, G. C., Camargo-Mathias, M. I. and Bueno, O. C. (2006). Fat body in some genera of leaf-cutting ants (Hymenoptera: Formicidae). Proteins, lipids and polysaccharides detection. *Micron*, **37**, 234–242.

Romer, F. (1980). Histochemical and biochemical investigations concerning the function of larval oenocytes of *Tenebrio molitor* L. (Coleoptera Insecta). *Histochemistry,* **69**, 69–84.

Ryuda, M., Tsuzuki, S., Tanimura, T., Tojo, S. and Hayakama, Y. (2008). A gene involved in the food preferences of larval *Drosophila melanogaster. J. Insect Physiol.*, **54**, 1440–1445.

Schal, C., Fan, Y. and Blomquist, G. J. (2003). Regulation of pheromone biosynthesis, transport, and emission in cockroaches. In *Insect Pheromone Biosynthesis and Molecular Biology: The Biosynthesis and Detection of Pheromones and Plant Volatiles*, ed. G. J. Blomquist and R. G. Vogt. Amsterdam: Elsevier, pp. 283–322.

Schal, C., Sevala, V. and Cardé, R. T. (1998b). Novel and higly specific transport of volatile sex pheromone by hemolymph lipophorin in moths. *Naturwissenschaften*, **85**, 339–342.

Schal, C., Sevala, V. L., Young, H. and Bachmann, J. A. S. (1998a). Synthesis and transport of hydrocarbons: Cuticle and ovary as target tissues. *Am. Zoologist*, **38**, 382–393.

Schal, C., Sevala, V., Capurro, M. L., Snyder, T. E., Blomquist, G. J. and Bagnères, A.-G. (2001). Tissue distribution and lipophorin transport of hydrocarbons and sex pheromones in the house fly, *Musca domestica. J. Insect Sci.*, **1**,12. Available online: insectscience.org/1.12

Schoeters, E. and Billen, J. (1997). The postpharyngeal gland in *Dinoponera* ants (Hymenoptera: Formicidae): Unusual morphology and changes during the secretory process. *Int. J. lnsect Morphol. Embryol.*, **25**, 443–447.

Sevala, V., Bagnères, A.-G, Kuenzli, M., Blomquist, G. J. and Schal, C. (2000). Cuticular hydrocarbons of the termite *Zootermopsis nevadensis* (Hagen): caste differences and role of lipophorin in transport of hydrocarbons and hydrocarbon metabolites. *J. Chem. Ecol.*, **26**, 765–790.

Sevala, V., Shu, S., Ramaswamy, S. B. and Schal, C. (1999). Lipophorin of female *Blattella germanica* (L): characterization and relation to hemolymph titers of juvenile hormone and hydrocarbons. *J. Insect Physiol., **45**, 431–441.

Singer, T. L., Espelie, K. E. and Gamboa, G. J. (1998). Nest and nestmate discrimination in independent-founding paper wasps. In *Pheromone Communication in Social Insects. Ants, Wasps, Bees and Termites*, ed. R. K. Vander Meer, M. D. Breed, K. E. Espelie and M. L. Winston. New York: Westview Press, pp. 104–125.

Singh, T. K. A. and Ryan, R. O. (1991). Lipid transfer particle-catalyzed transfer of lipoprotein-associated diacylglycerol and long chain aliphatic hydrocarbons. *Arch. Biochem. Biophys., **286**, 376–382.

Sledge, M. F., Trinca, I., Massolo, A., Boscaro, F. and Turillazzi, S. (2004). Variation in cuticular hydrocarbon signatures, hormonal correlates and establishment of reproductive dominance in a polistine wasp. *J. Insect Physiol.*, **50**, 73–83.

Soroker, V., Fresneau, D. and Hefetz, A. (1998). Formation of colony odor in ponerine ant *Pachychondyla apicalis. J. Chem. Ecol.*, **24**, 1077–1090.

Soroker, V. and Hefetz, A. (2000). Hydrocarbon site of synthesis and circulation in the desert ant *Cataglyphis niger. J. Insect Physiol.*, **46**, 1097–1102.

Soroker, V., Hefetz, A., Cojocaru, M., Billen, J., Franke, S. and Franke, W. (1995a). Structural and chemical ontogeny of the postpharyngeal gland of the desert ant *Cataglyphis niger. Physiol. Entomol.*, **20**, 323–329.

Soroker, V., Lucas, C., Simon, T., Fresneau, D., Durand, J.-L. and Hefetz, A. (2003). Hydrocarbon distribution and colony odour homogenisation in *Pachycondyla apicalis. Insect. Soc.*, **50**, 212–217.

Soroker, V., Vienne, C. and Hefetz, A. (1995b). Hydrocarbon dynamics within and between nestmates in *Cataglyphis niger* (Hymenoptera, Formicidae). *J. Chem. Ecol.*, **21**, 365–378.

Soroker, V., Vienne, C., Hefetz, A. and Nowbahari, E. (1994). The postpharyngeal gland as a "gestalt" organ for nestmate recognition in the ant *Cataglyphis niger. Naturwissenschaften*, **81**, 510–513.

Steiner, S., Mumm, R. and Ruther, J. (2007). Courtship pheromones in parasitic wasps: Comparison of bioactive and inactive hydrocarbon profiles by multivariate statistical methods. *J. Chem. Ecol.*, **33**, 825–838.

Strohm, E., Herzner, G. and Goettler, W. (2007). A 'social' gland in a solitary wasp? The postpharyngeal gland of female European beewolves (Hymenoptera, Crabronidae). *Arthropod Struct. Develop.*, **36**, 113–122.

Subchev, M. and Jurenka, R. A. (2001). Identification of the pheromone in the hemolymph and cuticular hydrocarbons from the moth *Scoliopteryx libatrix* L. (Lepidoptera: Noctuidae). *Arch. Insect Biochem. Physiol.*, **47**, 35–43.

Takeuchi, N. and Chino, H. (1993). Lipid transfer particle in the hemolymph of the American cockroach: evidence for its capacity to transfer hydrocarbons between lipophorin particles. *J. Lipid Res.*, **34**, 543–551.

Thompson, M. J., Glancey, B. M., Robbins, W. E., Lofgren, C. S., Dutky, S. R., Kochansky, J., Vander Meer, R. K. and Glover A. R. (1981). Major hydrocarbons of the postpharyngeal glands of mated queens of the red imported fire ant *Solenopsis invicta. Lipids*, **16**, 485–495.

Tsuchida, K., Arai, M., Tanaka, Y., Ishihara, R., Ryan, R. O. and Maekawa, H. (1998). Lipid transfer particle catalyzes transfer of carotenoids between lipophorins of *Bombyx mori. Insect Biochem. Mol. Biol.*, **28**, 927–934.

Vander Meer, R. K., Glancey, B. M. and Lofgren, C. S. (1982). Biochemical changes in the crop, oesophagus and postpharyngeal gland of colony-founding red imported fire ant queens (*Solenopsis invicta*). *Insect Biochem.*, **12**, 123–127.

Vander Meer, R. K. and Morel, L. (1998). Nestmate recognition in ants. In *Pheromone Communication in Social Insects. Ants, Wasps, Bees and Termites*, ed. R. K. Vander Meer, M. D. Breed, K. E. Espelie and M. L. Winston. New York: Westview Press, pp. 79–103.

Vargo, E. L. and Hulsey, C. D. (2000). Multiple glandular origins of queen pheromones in the fire ant *Solenopsis invicta*. *J. Insect Physiol.*, **46**, 1151–1159.

Vauchot, B., Provost, E., Bagnères, A.-G. and Clément, J.-L. (1996). Regulation of the chemical signatures of two termite species, *Reticulitermes santonensis* and *R. (l.) grassei*, living in mixed colonies. *J. Insect Physiol.*, **42**, 309–321.

Vauchot, B., Provost, E., Bagnères, A.-G., Rivière, G., Roux, M. and Clément, J.-L. (1998). Differential adsorption of allospecific hydrocarbons by the cuticles of two termite species, *Reticulitermes santonensis* and *R. lucifugus grassei*, living in a mixed colony. *J. Insect Physiol.*, **44**, 59–66.

Vienne, C., Soroker, V. and Hefetz, A. (1995). Congruency of hydrocarbon patterns in heterospecific groups of ants: tranfer and/or biosynthesis? *Insect Soc.*, **42**, 267–277.

Vinson, S. B., Phillips, Jr, S. A. and William, H. J. (1980). The function of the post-pharyngeal glands of the red imported fire ant, *Solenopsis invicta* Buren. *J. Insect Physiol.*, **26**, 645–650.

Wicker, C. and Jallon, J.-M. (1995). Hormonal control of sex pheromone biosynthesis in *Drosophila melanogaster*. *J. Insect Physiol.*, **41**, 65–70.

Wicker-Thomas, C. and Hamann, M. (2008). Interaction of dopamine, female pheromones, locomotion and sex behavior in *Drosophila melanogaster*. *J. Insect Physiol.*, **54**, 1423–1431.

Young, H. P., Bachmann, J. A. S., Sevala, V. and Schal, C. (1999). Site of synthesis, tissue distribution, and lipophorin transport of hydrocarbons in *Blattella germanica* (L.) nymphs. *J. Insect Physiol.*, **45**, 305–315.

6

Cuticular lipids and water balance

Allen G. Gibbs and Subhash Rajpurohit

Epicuticular lipids serve many roles in insects and other terrestrial arthropods (see other chapters in this volume), but the first to be recognized was as a barrier to transpiration through the surface of the animal. Surface-area to volume ratios increase as size decreases, so that smaller animals become increasingly susceptible to dehydration. Kühnelt (1928; cited by Wigglesworth, 1933) noted the presence of hydrophobic substances on the insect cuticle and proposed that these reduce water-loss. Their importance in water conservation is made apparent by the fact that even a brief treatment with organic solvents to remove surface lipids can result in water-loss rates increasing 10–100 fold (Hadley, 1994). On the other hand, water can be lost via other routes. These losses can be substantial (e.g. blood-sucking insects excrete huge amounts of fluid after a meal). Thus, under certain conditions, cuticular waterproofing may be a minor part of the overall water budget.

We address several issues in this chapter. First, is cuticular water-loss significant? The importance of spiracular control and discontinuous gas exchange in insect water balance has been challenged in recent years, leading to a need to reassess insect water budgets in general. Second, if cuticular transpiration is important for the overall water budget of insects, how does variation in surface lipid composition affect transpiration through the cuticle? We will discuss the relationships between lipid composition, lipid physical properties, and cuticular permeability. Third, what is the biophysical explanation for the critical temperature phenomenon, a rapid increase in water-loss at high temperatures? We consider both experimental results and a few theoretical issues that address this question. Finally, do other cuticular structures besides lipids significantly affect insect water budgets? Our goal is both to describe what we do know and to indicate gaps in our knowledge of cuticular function.

Is cuticular permeability important? Cuticular and respiratory water-loss

Consideration of the waterproofing function of cuticular lipids first requires an assessment of cuticular transpiration relative to the overall water budget. The fact that organismal water-loss rates increase greatly when surface lipids are removed does not necessarily mean that increased cuticular permeability is responsible. Insects can lose water by transpiration through the cuticle, by evaporation from the tracheal system through open spiracles, and by various excretions (oral secretions, feces, eggs, and even "sweat"; Hadley, 1994; Toolson and Hadley, 1987). Respiratory water-loss can be particularly important. Many insects exhibit cyclic breathing patterns, in which spiracles are held closed for some period, then open to allow release of accumulated carbon dioxide. The textbook explanation for this phenomenon is that it is an important water conservation mechanism. However, a number of recent studies have challenged this idea, alternative hypotheses have been put forward, and insect respiration has become an active and sometimes contentious field (see Chown, 2002; Chown *et al.*, 2006; Quinlan and Gibbs, 2006 for more discussion).

In the context of this chapter, respiratory behavior is important only to the extent that it affects our understanding of cuticular water-loss. Organic solvents tend to kill insects, causing spiracular control to be lost. It has often been assumed that spiracles close upon death, and therefore water-loss from dead animals reflects cuticular transpiration, but water-loss rates can increase after death (e.g., they approximately double in *Drosophila melanogaster*; A. G. Gibbs, unpublished observations), suggesting that spiracles may open in some species. Thus, it is conceivable that solvent-extracted animals lose significant amounts of water via the spiracles after death.

With flow-through respirometry, water-loss associated with spiracular opening can be detected easily in large enough insects. Water-loss rates can more than double when spiracles open (e.g., Lighton *et al.*, 1993; Figure 6.1). In *Manduca* pupae, opening of a single "master" spiracle is sufficient for whole-organism gas exchange (Slama, 1988). With few exceptions (e.g., Duncan and Byrne, 2002; Byrne and Duncan, 2003), we do not know how many or which spiracles open to allow gas exchange (and respiratory water-loss). Based on the *Manduca* example, however, theoretically a solvent-extracted insect with 10 spiracles could lose water 10 times faster simply because of loss of spiracular control.

Because of issues regarding respiratory control, water-loss measurements in live animals are preferable to those using dead animals, particularly when experimental techniques allow cuticular, respiratory and other components of water-loss to be distinguished from each other. This is relatively easy when insects exhibit discontinuous gas exchange (Hadley and Quinlan, 1993); when they do not, alternative methods have been developed (Gibbs and Johnson, 2004; Lighton *et al.*, 2004). Such studies have generally shown that, while respiratory water-loss constitutes a significant fraction of the water budget (especially in active insects), cuticular water-loss generally accounts for >80% of overall water-loss (Hadley, 1994; Chown *et al.*, 2006; Quinlan and Gibbs, 2006). Figure 6.2 illustrates the potential importance of cuticular transpiration for overall water balance. In this example, differences in cuticular water-loss accounted for 97% of inter-individual variation in total water-loss (Johnson and Gibbs, 2004).

We have, to some extent, set up a straw man in this section, but it is important to recognize that most studies have not distinguished cuticular transpiration from other components of the overall water budget. This is probably not a serious problem for work with inactive insects; it may be in other cases. The permeability of the cuticle to water is clearly an important aspect of insect water balance, but rigorous analysis requires a good quantitative understanding of cuticular and other routes for water-loss. Below, we first discuss the role

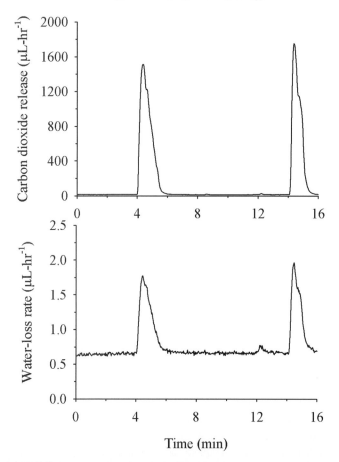

Figure 6.1 Discontinuous gas exchange in a grasshopper, *Trimerotropis pallidipennis*. Note that total water-loss more than doubles when the spiracles open to allow gas exchange. In this case, cuticular transpiration is easily estimated by measuring water-loss when CO_2 release is negligible.

of surface lipids in waterproofing the cuticle, then other aspects of cuticle architecture that may also affect water-loss.

Correlations between cuticular lipids and water-loss rates

The composition of cuticular lipids varies at all levels of organization in insects, from among species to within individuals. The amount of cuticular lipid can also vary substantially. For example, wax blooms of desert tenebrionid beetles are associated with reduced water-loss (Hadley, 1994). High densities of wax may also serve to reduce heat load by reflecting solar radiation (Hadley, 1994) or to deter predators (Eigenbrode and Espelie, 1995); thus, it cannot be assumed that water conservation is the primary function of wax

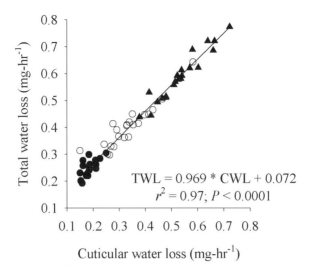

Figure 6.2 Correlation between cuticular water-loss and total water-loss in female ants, *Pogonomyrmex barbatus*. Differences in cuticular transpiration account for 97% of the variation in total water-loss. Different symbols indicate differences in mating status. From Johnson and Gibbs (2004); reproduced with permission.

blooms. It should be noted that blooms often do not form a homogeneous lipid layer across the cuticle (Nelson *et al.*, 2000), which may result in varying permeability from one region to another. The effects of spatial heterogeneity on cuticular transpiration have received very little attention (e.g., Hadley and Quinlan, 1987; Toolson and Hadley, 1987; Machin and Lampert, 1989).

A number of studies have compared cuticular lipids and water-loss rates among related species (e.g. Hadley, 1978; Hadley and Schultz, 1987). Much of this work was done before the importance of accounting for phylogenetic relationships in inter-specific comparisons was widely recognized (Felsenstein, 1985; Garland and Adolph, 1994; see also Chapter 7, this book). Intraspecific studies reduce phylogenetic concerns (although one still has to consider genetic relationships between populations) and should provide a better understanding of how changes in lipid composition affect cuticular transpiration.

Intraspecific variation in water-loss rates has often been associated with differences in cuticular lipids (e.g., Toolson and Hadley, 1979; Toolson, 1984; Johnson and Gibbs, 2004). At the population level, grasshoppers (*Melanopus sanguinipes*) from lower latitudes lose water slowly and have surface lipids with higher melting points (Rourke, 2000), and long-term maintenance of *Drosophila* populations in culture affects both water-loss and HC composition (Toolson and Kuper-Simbrón, 1989). Summer-collected scorpions and tenebrionid beetles have lower water-loss rates than fall/winter-collected animals, as well as a different complement of cuticular hydrocarbons (Hadley, 1977; Toolson and Hadley, 1979; see also Chapter 19). Thermal acclimation to higher temperatures in the laboratory causes similar changes (Hadley, 1977; Toolson, 1982), suggesting that seasonal temperature changes in nature are responsible for both lipid variation and differences in water-loss rates. It should be noted, however, that climatic changes are not the only factors affecting surface waxes. The hydrocarbon composition of *Drosophila mojavensis*, for example is affected by host plant (Stennett and Etges, 1997), and, in desert harvester ants, HC composition and water-loss rate are correlated with queen mating status (Johnson and Gibbs, 2004).

The phase transition model for cuticular permeability

What makes a good cuticular waterproofing barrier? Is it simply enough to have a hydrophobic layer, or do lipids differ in their ability to reduce transpiration? Ramsay (1935) proposed the first mechanistic hypothesis, beyond the mere fact of hydrophobicity, for the function of cuticular waxes. An "incautious use of a tap" (Maddrell, 1990) splattered one of Ramsay's cockroaches (*Periplaneta americana*) with water. Ramsay noticed that beads of water on the surface of the cockroach evaporated much more slowly than drops on the lab bench. After further investigation, he concluded that lipids from the cuticle formed a hydrophobic coating around the droplets that reduced evaporation. He also noted that the properties of this coating were highly dependent upon temperature, with evaporative water-loss increasing very rapidly above 30°C. After microscopic observations, Ramsay (1935) proposed that the lipids undergo a phase change at 30°C from a solid, relatively impermeable state to a melted, permeable condition.

Ramsay (1935) had the good fortune to be working with an animal with an unusual complement of cuticular lipids. The surface lipids of cockroaches include high levels of relatively polar compounds (by the standards of other insects), including alcohols, ketones, etc. (Schal *et al.*, 1990; Buckner, 1993; see also Buckner's Chapter 9). These are able to form hydrogen bonds with water molecules and set up monolayers on aqueous surfaces, with the hydrophobic ends oriented away from the water. In contrast, many insects appear to contain only hydrocarbons that do not form hydrogen bonds with water and therefore cannot form oriented monolayers. Despite these considerations, the model proposed by Ramsay (1935) has held up well. The general picture that has developed is one in which cuticular lipids typically occur in a solid state at moderate temperatures and provide an excellent barrier to cuticular transpiration. As temperature increases (sometimes within the physiological range), the lipids melt and become more permeable, and water-loss increases rapidly (Gibbs, 2002). The point at which water-loss begins to increase is termed the critical temperature (T_C).

Experimental tests of the phase transition model

To rigorously test the phase transition model, one must demonstrate that surface lipids actually do melt at the critical temperature, and that melted lipids are in fact more permeable to water than solid ones. A variety of biophysical techniques have been applied to this problem (Table 6.1). These have included direct biophysical measurements of lipid properties, and associated measurements of water-loss. In general, these studies have supported

Table 6.1 *Biophysical methods used to study cuticular lipids.*

Biophysical technique	Support for T_m model?	Reference
Capillary melting point	Yes	Beament (1945)
Electron diffraction	Mixed	Holdgate and Seal (1956)
Surface film compressibility	Mixed	Lockey (1976)
Electron paramagnetic spectroscopy	Yes	Toolson *et al.* (1979)
Differential scanning calorimetry	Mixed	Machin and Lampert (1990)
Infrared spectroscopy	Yes	Rourke and Gibbs (1999)

the phase transition model, at least to the extent that some change in lipid physical properties occurs close to an observed change in water-loss (Gibbs, 2002).

One limitation of several studies listed in Table 6.1 is that the surface lipids must be extracted from the animal for analysis. It is possible that extraction will affect lipid properties by removing lipid–cuticle interactions, but direct comparison of lipid melting *in situ* and in lipid extracts suggests that this is not a major problem (Gibbs and Crowe, 1991; A. G. Gibbs, unpublished observations). A more important concern is regional variation in lipid composition and physical properties. For example, lipids on the forewings of the grasshopper *Melanoplus sanguinipes* melt at ~65°C, whereas those on the body and hind wings melt at ~45°C (Gibbs and Crowe, 1991). This type of variation makes attempts to correlate bulk lipid properties with cuticular permeability problematic. In addition, transitions in rates of water-loss have been observed in animals whose surface lipids have been removed, suggesting that other components of the cuticle can affect transpiration (Yoder *et al.*, 2005a).

Conversely, studies that investigate cuticular lipids *in situ* face difficulties if what they measure includes a signal from the underlying cuticle. For example, are changes in electron diffraction patterns (Holdgate and Seal, 1956) caused by lipid melting or thermal disruption of the underlying cuticle? Differential scanning calorimetry (DSC, Machin and Lampert, 1990) definitely generates a mixed signal of heat absorption by lipids and cuticle. An interesting recent application of calorimetry has been to study the physical properties of beeswax (Buchwald *et al.*, 2006, 2008). Although beeswax does not have a waterproofing function, it has an important role in maintaining the structural integrity of honeycombs. In these studies, DSC provided detailed information on biophysical changes related to temperature. The ability to examine small samples (milligrams) has improved tremendously since Machin and Lampert's (1990) work, and further DSC studies of cuticular lipids are warranted.

Each of the studies mentioned so far is fundamentally correlative, with lipid-phase behavior and water-loss rates being measured with different animals. Lipid composition and physical properties can vary substantially within species, so a close correlation between these parameters is not necessarily expected. One possible explanation is a "file drawer" problem of the type alluded to by Lighton (1998); results that conflict with the dominant paradigm may be dismissed as resulting from experimental errors. Rourke and Gibbs (1999) instead used inter-individual variation in surface lipids of *M. sanguinipes* as an experimental variable to test the phase behavior model. Lipid melting points (T_m) varied by ~15°C among animals, and were highly correlated with T_C. In a model membrane system, T_m and T_C were again strongly correlated (Rourke and Gibbs, 1999).

Lipid composition and physical properties: size doesn't matter (much)

The majority of publications on cuticular lipids involve analyses of lipid composition. Which compounds are present, and what is their function? Correlations between lipid composition and water-loss have provided indirect tests of the phase transition hypothesis, under the assumption that changes in lipid composition predictably affect lipid properties. In this section, we summarize available information on how specific structural changes affect the physical properties of pure surface lipids, as well as how different lipids interact with each other.

We know the most about cuticular hydrocarbons, because they are abundant and because it is relatively easy to isolate and identify them. They are also the most hydrophobic lipid components, and so should provide the best barrier to water-loss. *n*-Alkanes isolated from insect cuticles typically have chain lengths of 20–40 carbons. These can be modified by insertion of *cis* double bonds, or addition of one or more methyl groups. Relatively polar surface lipids include alcohols, aldehydes, ketones and wax esters (see Chapter 9). Given this diversity, is it possible to predict lipid phase behavior (and, by extension, waterproofing characteristics) from composition alone? If so, a large body of literature would become instantly interpretable in the context of water balance. Unfortunately, this is not the case.

Hadley (1977, 1978) was the first to try to relate variation in lipid composition to cuticular permeability via the link of lipid phase behavior. He used the relative mobility of hydrocarbons by gas chromatography (i.e. equivalent chain length) to predict the relative physical properties of HCs *in situ*. The basic assumption was that melting points (the physiologically relevant property) and boiling points (approximately measured by gas chromatography (GC) retention time) would be highly correlated, and many subsequent authors have made the same assumption. As an example of this approach, a (Z)-9-tricosene molecule has a GC retention time like that of an *n*-alkane of chain length 22.3, and a methyltricosane molecule has an equivalent chain length of ~23.7. Under this assumption, the melting point of tricosene would be ~45°C, between those of *n*-docosane (44°C) and *n*-tricosane (46.5°C) (Maroncelli *et al.*, 1982; Gibbs and Pomonis, 1995), and methyltricosanes would melt at ~48°C. Instead, (Z)-9-tricosene melts at ~0°C, and methylalkanes melt 10–30°C below *n*-alkanes with the same chain length, depending on the location of the methyl branch (Gibbs and Pomonis, 1995).

The previous examples illustrate an important point regarding the effects of HC chain modification on melting points: not all HC structural changes are equivalent. In particular, chain length is the *least* important factor affecting T_m. *n*-Alkanes have high melting points because they pack well together in a crystalline state (Maroncelli *et al.*, 1982). Lengthening

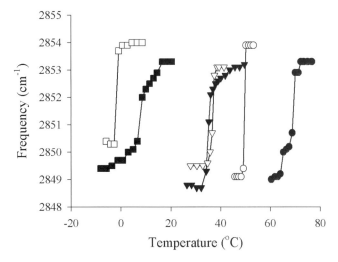

Figure 6.3 Effects of hydrocarbon chain modifications on melting points of similar-sized cuticular lipids. When lipids melt, the absorption frequency of C-H symmetric stretching vibrations increases from ~2849 cm^{-1} to ~2854 cm^{-1}. From right to left, compounds are (chemical change relative to *n*-alkane, molecular mass in daltons): filled circles, *n*-dotriacontane (no change, 450); open circles, palmitic acid myristyl ester (wax ester, 452); filled triangles, 13-methylhentriacontane (methyl-branched alkane, 450); open triangles, (Z)-13-tritriacontene (double bond, 462); filled squares, 9,13-dimethylhentriacontane (2 methyl branches, 464); open squares, oleic acid oleyl ester (2 double bonds and an ester link, 532). Data from Gibbs and Pomonis (1995) and Patel *et al.* (2001).

of an alkane chain increases T_m by ~2°C per CH$_2$, unit. Insertion of a *cis* double bond disrupts crystalline structure, so that melting can occur nearly 50°C lower (see above; Gibbs and Pomonis, 1995). Other chain modifications (methyl branching, insertion of an ester linkage) also disrupt lipid packing and lower T_m.

Figure 6.3 illustrates how structural differences greatly outweigh the effects of molecular size on melting point. The lowest molecular weight compound shown, *n*-dotriacontane (MW = 450), has the highest melting point (~69°C), and the highest molecular weight compound, oleyl acid oleic ester (MW = 532), melts below 0°C. In *M. sanguinipes*, increased melting point is correlated with decreased methyl-branching, despite the fact that methylalkanes in this species have longer chains (Gibbs *et al.*, 1991; Gibbs and Mousseau, 1994). More extensive disruptions of HC chain structure have even greater effects. The compounds considered in Figure 6.3 are essentially linear; *M. sanguinipes* synthesizes "T-shaped" compounds (wax esters of secondary alcohols; Blomquist *et al.*, 1972). Despite having molecular weights averaging >600, they melt at 5–10°C (Patel *et al.*, 2001). Populations with greater proportions of wax esters also have lower T_m values (A.G. Gibbs, unpublished observations). In summary, molecular weight is generally well correlated with GC retention time, but not with T_m.

Of course, insects contain mixtures of many different HCs and other hydrophobic compounds, and it is their interactions that determine the overall physical properties of the

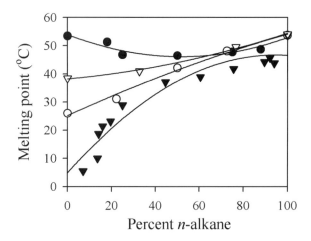

Figure 6.4 Melting points of binary lipid mixtures, each containing an *n*-alkane component, as indicated by FTIR spectroscopy. Filled circles, *n*-pentacosane and myristic acid stearyl ester. These compounds melt individually at ~53.5°C, but mixtures have lower melting points. Open circles, *n*-pentacosane and 13-methylpentacosane. The methyl branch of the latter compound reduces its melting point, but mixtures melt at intermediate temperatures equivalent to the weighted average of the components' T_m values. Filled triangles, *n*-tricosane and (Z)-9-tricosene. Melting points of mixtures appear higher than predicted, but this may reflect phase separation between the component lipids (Small, 1986). Open triangles, *n*-pentacosane and dodecyl acid myristoyl ester. Data from Gibbs (1995) and Patel *et al*. (2001).

surface lipids. Figure 6.4 depicts T_m values for binary mixtures of *n*-alkanes with other types of cuticular lipid. Mixtures of two different alkanes (either two straight chains or a straight-chain and a branched compound) have melting points that approximate those of the weighted average of the components, with melting occurring over a relatively wide range of temperatures (Bonsor and Bloor, 1977; Gibbs, 1995). 50:50 Mixtures of wax esters with *n*-alkanes melt up to 10°C below the expected temperature, consistent with disruption of lipid packing by the ester link (Patel *et al.*, 2001).

Mixtures of alkenes with alkanes present a more complicated picture. Based on Fourier transform infrared [spectroscopy] (FTIR) alone, mixtures appear to melt at higher than predicted temperatures (Gibbs, 1995), but other techniques reveal a much more complicated picture (Small, 1986). Alkenes found on insects generally melt at lower temperatures than the alkanes present (Gibbs and Pomonis, 1995). Saturated and unsaturated HCs crystallize separately below the alkene's melting point, then the alkene melts at its respective T_m (Small, 1986). As temperature increases further, the alkane crystals begin melting into the fluid phase, leaving a state of phase separation until the alkane is completely melted. The implications of this state for cuticular transpiration, chemical communication, defense, etc. are not understood. Given the fact that alkenes and alkanes are found on so many insects, and the communication function frequently exhibited by unsaturated HCs (Howard, 1993), further investigation of their interactions is certainly warranted.

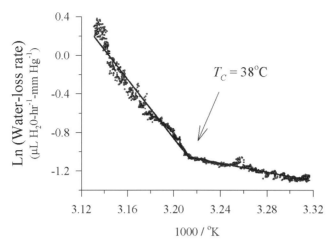

Figure 6.5 Arrhenius plot of water-loss from a live grasshopper, *Melanoplus sanguinipes*. Cuticular lipids extracted from this individual melted at ~42°C. Data from Rourke and Gibbs (1999).

Theoretical consideration of the transition phenomenon

The physical–chemical theory behind cuticular permeability has been considered on several occasions (Toolson, 1978, 1980; Gelman *et al.*, 1988; Noble-Nesbitt, 1991; Gelman and Machin, 1994; Yoder *et al.*, 2005b). One topic of continued discussion is how T_C is determined. Critical temperature has often been estimated by visual inspection of plots of water-loss rate vs. temperature. This approach is somewhat subjective, and the apparent T_C may depend upon the range of experimental temperatures. Because water-loss rates increase in a somewhat exponential fashion with temperature, larger experimental temperature ranges compress the *Y*-axis at low temperatures, leading to the impression of higher T_C values. Plotting water-loss rates on a logarithmic scale reduces this problem to some extent, but estimating T_C remains subjective. To allow a more objective measure, T_C has been estimated as the break-point on biphasic Arrhenius plots (Holdgate and Seal, 1956; Machin and Lampert, 1989; Rourke and Gibbs, 1999). In many cases, apparent critical temperatures disappear when Arrhenius plots are used (Yoder *et al.*, 2005b; Yoder and Tank, 2006).

Arrhenius plots are familiar to us from basic chemistry courses. The theory was derived in the late-nineteenth century for simple chemical reactions. It has since been applied to more complicated situations, from enzymatic reactions to entire ecosystems. Figure 6.5 shows an Arrhenius plot of Ln (water-loss rate) vs. 1000/(absolute temperature) for a grasshopper, *Melanoplus sanguinipes*. At ~38°C, water-loss rates began to increase more rapidly with inverse temperature (the melting point for surface lipids extracted from this animal was ~42°C). Not all insects exhibit such a transition, at least not within the range of experimental temperatures, even when such transitions appear evident on linear–linear plots (Gibbs, 1998; Yoder *et al.*, 2005b; Yoder and Tank, 2006).

The general formula for Arrhenius relationships is

$$\text{Rate} = A * \exp(-\Delta G^{\ddagger}/RT) = A * \exp(-(\Delta H^{\ddagger} - T\Delta S^{\ddagger})/RT) \tag{6.1}$$

or

$$\text{Ln(Rate)} = \text{Ln}(A) - \Delta G^{\ddagger}/RT = \text{Ln}(A) - \Delta H^{\ddagger}/RT + \Delta S^{\ddagger}/R \tag{6.2}$$

where R is the gas constant, $\Delta G^{\ddagger}$ is the *apparent* free energy of activation (i.e. the difference in Gibbs free energy between the transition and ground states of the reaction) and A is a reaction-specific factor that takes into account steric factors (i.e. the likelihood that the reactants will be in the correct orientation to react successfully) and other variables, including properties of the medium in which the reaction occurs. $\Delta H^{\ddagger}$ and $\Delta S^{\ddagger}$ are, respectively, the apparent enthalpy and entropy of activation. An important point is that A, $\Delta G^{\ddagger}$, $\Delta H^{\ddagger}$ and $\Delta S^{\ddagger}$ may all depend upon temperature, even in very simple chemical reactions (Berry *et al.*, 1980).

For a simple reaction, the slope of an Arrhenius plot is proportional to $\Delta G^{\ddagger}$, with a steeper slope implying a higher energetic barrier to overcome. Yoder *et al.* (2005b) have recently challenged this common interpretation when applied to cuticular water-loss. The problem is the upward inflection at high temperatures, with the higher slope being indicative of a higher energetic barrier to water-loss. Yoder *et al.* (2005b) argue that, as cuticular waxes melt at the T_C and become more permeable, the energetic barrier to diffusion cannot possibly become higher. Their point is well taken, and we discuss below some factors that may allow a better theoretical understanding of cuticular permeability. Our discussion is largely based on theoretical reviews by Silvius and McElhaney (1981) and Klein (1982), developed in the context of membrane transport processes. These are, as theoretical works go, relatively accessible to experimental physiologists, and we recommend them to mathematically oriented readers.

Transpiration through the cuticle involves more than just the single step of diffusion through the epicuticular lipid layer. Molecules of water must leave the tissues adjacent to the cuticle, diffuse through the cuticle itself, enter the lipid layer, diffuse across the lipids, and enter the gas phase outside the animal. Each step is likely to be affected by temperature to a different extent. Lipid composition and physical properties can also differ from one region of the cuticle to the next, so that the biophysical details of cuticular transpiration may not be homogeneous across the entire animal. Thus, transpiration at the organismal level involves multiple steps, and parallel routes for water flux.

Consider first a simple two-step transport reaction. It can be demonstrated theoretically (Klein, 1982) that the step with the higher value of $\Delta G^{\ddagger}$ will be the slow step at low temperatures, and the low $\Delta G^{\ddagger}$ step will be rate-limiting at higher temperatures, leading to a downwardly concave Arrhenius plot (note that Arrhenius plots for cuticular transpiration are concave upward). A similarly shaped curve will be obtained if the apparent activation energy decreases at high temperature, for example if a cell membrane undergoes a melting phase transition that decreases the energy barrier to activity of a membrane enzyme, but does not change the rate-limiting step. Thus, Klein (1982) provides the mathematical underpinnings for the argument by Yoder *et al.* (2005b) that activation energy cannot increase with temperature. This is a relatively familiar example in which changes in lipid phase behavior can result in non-linear Arrhenius behavior, but several other scenarios can result in similar behavior, or even upward concavity. I will discuss the latter here, as these are potentially relevant to cuticular transpiration.

Partitioning of transport between different lipid phases. Klein (1982) considers the case of a transport reaction, in which the transporter preferentially exists in either the fluid or gel region of a membrane. An important conclusion is that, even if the apparent activation energy is the same in both phases, non-linear Arrhenius behavior can be observed. In fact, the Arrhenius curve is triphasic when a broad enough range of temperatures is considered. In addition, depending on the partitioning of the transporter, neither of the two break-points need correspond to the actual melting point of the membrane. Effectively, the *A* term in Equation (6.1) changes with phase state. A potential mechanistic explanation might be that reduced membrane viscosity allows the transporter to turn over more rapidly when it enters the fluid-phase membrane, although the activation energy does not change. How might this relate to cuticular transpiration? Water moves through membranes via transient defects in lipid packing (Carruthers and Melchior, 1983; Deamer and Bramhall, 1986). These defects increase as lipids melt, and indeed what FTIR spectroscopy actually measures is lipid defects (*trans–cis* isomerizations of hydrocarbon chains). The more defects available, the more water molecules will partition into the lipid layer, and the better the chance that molecules will diffuse through the lipid layer. Melting of cuticular lipids therefore may effectively increase the solubility of water in the lipid layer, thereby increasing the overall flux. In analogy to the models derived by Klein (1982), more defects ("transporters") are available in fluid lipids, and so water flux will increase more rapidly than if one considers the permeability of just solid or melted lipids alone.

Multiple routes for transpiration. The insect cuticle is a complex structure. Different types of cuticular lipids may be deposited on different regions of the body, resulting in different physical properties (Gibbs and Crowe, 1991; Young *et al.*, 2000; see also discussion in Chapter 20). At even finer scales, surface lipids may form non-homogeneous mixtures of melted and solid regions (Small, 1986; Gibbs, 2002). Additionally, the amount of lipid deposited may vary, and underlying cuticular layers may have differing permeabilities (Hadley and Quinlan, 1987). Each route may differ in its temperature sensitivity, and this may cause non-linear Arrhenius behavior. In particular, at low temperatures, cuticular lipids may provide such a strong barrier to water-loss that other water-loss routes dominate. If transpiration through lipid barriers has a higher apparent activation energy ($\Delta G^{\ddagger}$) than flux through other routes, trans-lipid fluxes will increase more rapidly with temperature, and an upward inflection of an Arrhenius plot will be observed. This can happen whether or not the lipids actually melt.

Although originally derived for membrane transport phenomena, the mechanisms outlined above can also provide potential explanations for non-linear, upwardly-concave Arrhenius plots for passive diffusion (Silvius and McElhaney, 1981; Klein, 1982). Other biophysical explanations can also explain these findings (e.g. a negative heat capacity of activation). The take-home message is that Arrhenius plots alone provide little information regarding underlying physical mechanisms of reactions or processes. Sharp break-points on Arrhenius plots may not occur, even if there is a sharp lipid-phase transition (Silvius and McElhaney, 1981). Break-points can occur without a change in rate-limiting step, and changes in rate-limiting steps can occur without break-points (Klein, 1982). Finally, $\Delta G^{\ddagger}$ is affected by temperature, even in simple reactions (Berry *et al.*, 1980). The temperature dependence of $\Delta G^{\ddagger}$ is usually small, but differs depending on the reaction mechanism. Movement of water through the insect cuticle is a complex process, and there is clearly much to be learned from further experimental and theoretical work.

Are cuticular lipids the entire story? Melanization and water-loss

Not all studies find a relationship between water-loss and surface lipid amounts, composition or melting point, including numerous studies using *Drosophila*. A comparative study found that the longest chain lengths and highest melting points occurred in species from cool boreal forests, species which lose water the fastest (Gibbs *et al.*, 2003). Within species, Indian populations of *D. immigrans* differ in water-loss rate, but not in surface lipid amounts (Parkash *et al.*, 2008c). In a laboratory selection experiment, populations selected for desiccation resistance lost water ~50% less rapidly than unselected controls, but the two groups exhibited minor differences in lipid composition and T_m (Gibbs *et al.*, 1997). Thermal acclimation of the desert fly *D. mojavensis* results in substantial changes in HC composition, but relatively little change in water-loss rates (Gibbs *et al.*, 1998). It must be noted that not all studies result in negative findings (e.g. Toolson, 1982; Toolson and Kuper-Simbrón, 1989).

Why do so many *Drosophila* studies report no relationship between surface waxes and water-loss? There are several possibilities. One is that *Drosophila* have relatively high respiratory water-loss rates (Lehmann *et al.*, 2000; Lehmann, 2001), so cuticular water-loss is relatively unimportant. Inter-specific correlations between water-loss and metabolic rate are consistent with this idea (Gibbs *et al.*, 2003). Cuticular water-loss accounts for >80% of total water-loss in most insects studied to date (Quinlan and Gibbs, 2006). *Drosophila* can regulate spiracular tone to some extent (Lehmann, 2001; Gibbs *et al.*, 2003), but whether they normally do so is unknown. *Drosophila* may also be constrained in the degree to which they can evolve surface lipids that restrict water-loss better. Cuticular hydrocarbons in *Drosophila* are used in chemical communication as species- and sex-recognition pheromones (Howard, 1993). Changes that conserve water but reduce a fly's mating success will not spread in a population unless water stress is an important factor affecting survival. Even in deserts, *Drosophila* usually have access to rotting plant material to replenish lost water (Breitmeyer and Markow, 1998).

Finally, it must be noted that the cuticle contains other components besides lipids. The cuticle is a complex structure, and its components vary greatly between species and populations. With regard to water balance, melanin may be a particularly important component. Significant structural integrity of the cuticle is provided by aromatic cross-links inserted between adjoining polypeptide chains, causing progressive hardening, dehydration, and close packing (Hopkins and Kramer, 1992). Hardening and darkening of the cuticle share the same biochemical processes (Fraenkel and Rudall, 1940; Pryor, 1940). Dark cuticle achieves its color largely because of the deposition of melanin granules (polymers of dopa and other tyrosine derivatives; True, 2003). Melanin is also hydrophobic and therefore may reduce the permeability of the cuticle. In this light, it is interesting to note that light-colored *yellow Drosophila* mutants lose water faster than darker wildtype flies, while dark *ebony* mutants are more desiccation resistant (Kalmus, 1941; Da Cunha, 1949).

A series of recent studies have examined clines in melanization and water balance in Indian populations of *Drosophila* (reviewed by Rajpurohit *et al.*, 2008a). Temperature is negatively correlated with latitude and altitude on the Indian subcontinent (Parkash *et al.*, 2008d; Rajpurohit *et al.*, 2008b). Populations of *Drosophila melanogaster* and *D. immigrans* exhibit increased melanization in cooler regions (Parkash *et al.*, 2008c,d), consistent with a need to maintain a high body temperature to allow flight and other activities. The importance of melanization for temperature regulation in drosophilids and other small insects needs to be established experimentally, however. Willmer and Unwin (1981) reported that insects the size of *Drosophila* do not attain temperatures even 1°C above ambient, whereas large (>100 mg), dark insects with poorly reflective bodies can achieve body temperatures >10°C above ambient.

An alternative explanation for clines in melanization is suggested by the fact that humidity also decreases with latitude and altitude in India (Parkash *et al.*, 2008b, d). Darker populations of several *Drosophila* species have relatively low water-loss rates compared to lighter populations, despite a lack of differences in surface lipid quantities (Parkash *et al.*, 2008a, c; Rajpurohit *et al.*, 2008a). In contrast, populations of a light-bodied drosophilid, *Zaprionus indianus*, exhibit the same latitudinal cline in water-loss, without pigmentation variation, but with an increase in HC amounts in drier areas (Figure 6.6; Parkash *et al.*, 2008a). These results suggest that there can be alternative "strategies" for reducing cuticular transpiration and coping with arid environments.

Additional evidence supporting a role for melanization in water conservation comes from a study of *D. polymorpha* populations in Brazil (Brisson *et al.*, 2005). Dark flies were most abundant in open, dry environments, despite the fact that these were warmer than more humid forest habitats, where light flies predominated. In this case, melanization patterns were opposite to those expected for thermal regulation, but consistent with a role in water conservation. Thus, in combination with studies by Parkash and colleagues, there is increasing evidence that melanization significantly affects water balance, at least in *Drosophila*. The fact that transitions in water-loss can be observed in solvent-extracted animals (Yoder and Tank, 2006) also supports a role for non-lipid layers in determining cuticular permeability. Lipids are probably the primary barrier to cuticular transpiration, but other cuticular barriers may be important and should not be ignored.

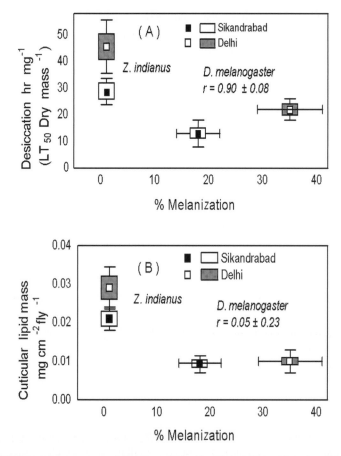

Figure 6.6 Within- and between-population variation in desiccation resistance (A) and cuticular lipid quantity (B) as a function of body melanization in northern and southern Indian populations of *Zaprionus indianus* and *Drosophila melanogaster*. In both species, populations from drier (northern) habitats are more desiccation resistant. The desiccation-resistant population of *D. melanogaster* is melanic, whereas desiccation resistance in *Z. indianus* is correlated with higher amounts of surface lipids. From Parkash *et al.* (2008a); reproduced with permission.

Finally, even when HC composition and cuticular transpiration are correlated, causation cannot be assumed. For example, higher cuticular water-loss rates in the desert ant, *Pogonomyrmex barbatus*, are correlated with a decrease in abundance of an *n*-alkane and an increase in a methylalkane (Figure 6.2; Johnson and Gibbs, 2004). This is exactly what one would expect if lipid melting points affect cuticular permeability, but this increase is also accompanied by a change in mating status. Mated, de-alate queens that have founded colonies lose water most rapidly, but they also have undergone the physical stress of repeated mating in large aggregations, followed by soil abrasion during colony founding

(Johnson, 2000). Both HC composition and cuticular damage may contribute to higher cuticular water-loss in this species (Johnson and Gibbs, 2004).

Summary

The function of epicuticular lipids in reducing insect water-loss has been recognized for almost a century. Despite this early recognition, our understanding of the process by which water moves through the cuticle is limited. Transpiration rates are strongly affected by temperature, probably via temperature's effects on lipid phase behavior. Solid-phase lipids provide a tighter barrier to transpiration than melted ones, but we know relatively little about how different compounds interact to determine cuticular permeability. Many biophysical techniques have achieved orders of magnitude improvements in sensitivity and precision since they were last applied to insects, and some (e.g. NMR, atomic force microscopy) have never been applied. More speculatively, researchers in nanotechnology have developed the ability to create and study objects and materials smaller than the smallest insects; some of these tools should be applicable to studying cuticular structure and function. Finally, several recent studies have highlighted the potential importance of other cuticular properties, specifically melanization, in conserving water. Although surface lipids are clearly important in determining cuticular transpiration, the cuticle is a complex structure, and lipids are only part of the picture.

Acknowledgments

We thank numerous students and collaborators for their assistance in collecting and interpreting data. Research in the Gibbs lab has been supported by the National Science Foundation, most recently awards IOB-0514402 and IOB-0723930.

References

Beament, J. W. L. (1945). The cuticular lipoids of insects. *J. Experim. Biol.*, **21**, 115–131.

Berry, R. S., Rice, S. A. and Ross, J. (1980). *Physical Chemistry*. New York: Wiley and Sons.

Blomquist, G. J., Soliday, C. L., Byers, B. A., Brakke, J. W. and Jackson, L. L. (1972). Cuticular lipids of insects: V. Cuticular wax esters of secondary alcohols from the grasshoppers *Melanoplus packardii* and *Melanoplus sanguinipes*. *Lipids*, **7**, 356–362.

Bonsor, D. H. and Bloor, D. (1977). Phase transitions of *n*-alkane systems. 2. Melting and solid state transition of binary mixtures. *J. Mater. Sci.*, **12**, 1559–1564.

Breitmeyer, C. M. and Markow, T. A. (1998). Resource availability and population size in cactophilic *Drosophila*. *Function. Ecol.*, **12**, 14–21.

Brisson, J. A., Toni, D. C. D., Duncan, I. and Templeton, A. R. (2005). Abdominal pigmentation variation in *Drosophila polymorpha*: geographical variation in the trait and underlying phylogeography. *Evolution*, **59**, 1046–1059.

Buchwald, R., Breed, M. D., Greenberg, A. R. and Otis, G. (2006). Interspecific variation in beeswax as a biological construction material. *J. Experim. Biol.*, **209**, 3984–3989.

Buchwald, R., Breed, M. D. and Greenberg, A. R. (2008). The thermal properties of beeswaxes: unexpected findings. *J. Experim. Biol.*, **211**, 121–127.

Buckner, J. S. (1993). Cuticular polar lipids of insects. In *Insect Lipids: Chemistry, Biochemistry and Biology*, ed. D. W. Stanley-Samuelson and D. R. Nelson. Lincoln: University of Nebraska Press, pp. 227–270.

Byrne, M. J. and Duncan, F. D. (2003). The role of the subelytral spiracles in respiration in the flightless dung beetle *Circellium bacchus*. *J. Experim. Biol.*, **206**, 1309–1318.

Carruthers, A. and Melchior, D. L. (1983). Studies of the relationship between bilayer water permeability and bilayer physical state. *Biochemistry*, **22**, 5797–5807.

Chown, S. L. (2002). Respiratory water loss in insects. *Comp. Biochem. Physiol.* A, **133**, 791–804.

Chown, S. L., Gibbs, A. G., Hetz, S. K., Klok, C. L., Lighton, J. R. B. and Marais, E. (2006). Discontinuous gas exchange in insects: a clarification of hypotheses and approaches. *Physiol. Biochem. Zool.*, **79**, 333–343.

Da Cunha, A. B. (1949). Genetic analysis of the polymorphism of color patterns in *Drosophila polymorpha*. *Evolution*, **3**, 239–252.

Deamer, D. W. and Bramhall, J. (1986). Permeability of lipid bilayers to water and ionic solutes. *Chem. Phys. Lipids*, **40**, 167–188.

Duncan, F. D. and Byrne, M. J. (2002). Respiratory airflow in a wingless dung beetle. *J. Experim. Biol.*, **205**, 2489–2497.

Eigenbrode, S. D. and Espelie, K. E. (1995). Effects of plant epicuticular lipids on insect herbivores. *Annu. Rev. Entomol.*, **40**, 171–194.

Felsenstein, J. (1985). Phylogenies and the comparative method. *Amer. Natur.*, **125**, 1–15.

Fraenkel, G. and Rudall, K. M. (1940). A study of the physical and chemical properties of the insect cuticle. *Proc. Royal Soc.* B, **129**, 1–35.

Garland, T. and Adolph, S. C. (1994). Why not to do 2-species comparative studies – limitations on inferring adaptation. *Physiol. Zool.*, **67**, 797–828.

Gelman, N. and Machin, J. (1994). Diffusion through the water barrier of arthropod cuticles: A statistical mechanical approach to the analysis of temperature effects. *J. Theoret. Biol.*, **167**, 361–372.

Gelman, N., Machin, J. and Kestler, P. (1988). The nature of driving forces for passive water transport through arthropod cuticle. *J. Thermal Biol.*, **13**, 157–162.

Gibbs, A. (1995). Physical properties of insect cuticular hydrocarbons: model mixtures and interactions. *Comp. Biochem. Physiol.* B, **112**, 667–672.

Gibbs, A. G. (1998). Waterproofing properties of cuticular lipids. *Amer. Zool.*, **38**, 471–482.

Gibbs, A. G. (2002). Lipid melting and cuticular permeability: new insights into an old problem. *J. Insect Physiol.*, **48**, 391–400.

Gibbs, A. and Crowe, J. H. (1991). Intra-individual variation in cuticular lipids studied using Fourier transform infrared spectroscopy. *J. Insect Physiol.*, **37**, 743–748.

Gibbs, A. G. and Johnson, R. A. (2004). The role of discontinuous gas exchange in insects: the chthonic hypothesis does not hold water. *J. Experim. Biol.*, **207**, 3477–3482.

Gibbs, A. and Mousseau, T. A. (1994). Thermal acclimation and genetic variation in cuticular lipids of the lesser migratory grasshopper (*Melanoplus sanguinipes*): effects of lipid composition on biophysical properties. *Physiol. Zool.*, **67**, 1523–1543.

Gibbs, A. and Pomonis, J. G. (1995). Physical properties of insect cuticular hydrocarbons: the effects of chain length, methyl-branching and unsaturation. *Comp. Biochem. Physiol.* B, **112**, 243–249.

Gibbs, A., Mousseau, T. A. and Crowe, J. H. (1991). Genetic and acclimatory variation in biophysical properties of insect cuticle lipids. *Proc. Natl. Acad. Sci. USA*, **88**, 7257–7260.

Gibbs, A. G., Chippindale, A. K. and Rose, M. R. (1997). Physiological mechanisms of evolved desiccation resistance in *Drosophila melanogaster*. *J. Experim. Biol.*, **200**, 1821–1832.

Gibbs, A. G., Louie, A. K. and Ayala, J. A. (1998). Effects of temperature on cuticular lipids and water balance in a desert *Drosophila*: Is thermal acclimation beneficial? *J. Experim. Biol.*, **201**, 71–80.

Gibbs, A. G., Fukuzato, F. and Matzkin, L. M. (2003). Evolution of water conservation mechanisms in desert *Drosophila*. *J. Experim. Biol.*, **206**, 1183–1192.

Hadley, N. F. (1977). Epicuticular lipids of the desert tenebrionid beetle, *Eleodes armatus*: seasonal and acclimatory effects on chemical composition. *Insect Biochem.*, **7**, 277–283.

Hadley, N. F. (1978). Cuticular permeability of desert tenebrionid beetles: correlations with epicuticular lipid composition. *Insect Biochem.*, **8**, 17–22.

Hadley, N. F. (1994). *Water Relations of Terrestrial Arthropods*. San Diego: Academic Press.

Hadley, N. F. and Quinlan, M. C. (1987). Permeability of arthrodial membrane to water: A first measurement using in vivo techniques. *Experimentia*, **43**, 164–166.

Hadley, N. F. and Quinlan, M. C. (1993). Discontinuous carbon dioxide release in the eastern lubber grasshopper *Romalea guttata* and its effect on respiratory transpiration. *J. Experim. Biol.*, **177**, 169–180.

Hadley, N. F. and Schultz, T. D. (1987). Water loss in three species of tiger beetles (*Cicindela*): correlations with epicuticular hydrocarbons. *J. Insect Physiol.*, **33**, 677–682.

Holdgate, M. W. and Seal, M. (1956). The epicuticular wax layers of the pupa *Tenebrio molitor* L. *J. Experim. Biol.*, **33**, 82–108.

Hopkins, T. L. and Kramer, K. J. (1992). Insect cuticle sclerotization. *Annu. Rev. Entomol.*, **37**, 273–302.

Howard, R. W. (1993). Cuticular hydrocarbons and chemical communication. In *Insect Lipids: Chemistry, Biochemistry and Biology*, ed. D. W. Stanley-Samuelson and D. R. Nelson. Lincoln: University of Nebraska Press, pp. 179–226.

Johnson, R. A. (2000). Water loss in desert ants: caste variation and the effect of cuticle abrasion. *Physiol. Entomol.*, **25**, 48–53.

Johnson, R. A. and Gibbs, A. G. (2004). Effects of mating stage on water balance, metabolism and cuticular hydrocarbons of the desert harvester ant, *Pogonomyrmex barbatus*. *J. Insect Physiol.*, **50**, 943–953.

Kalmus, H. (1941). The resistance to desiccation of *Drosophila* mutants affecting body colour. *Proc. Royal Soc.* B, **130**, 185–201.

Klein, R. A. (1982). Thermodynamics and membrane processes. *Quart. Rev. Biophys.*, **15**, 667–757.

Kühnelt, W. (1928). Uber den bau des Insektenskelettes. *Zool. Jahrbr. Abteil. Anat. Ontog. Tiere*, **50**, 219–278.

Lehmann, F.-O. (2001). Matching spiracle opening to metabolic need during flight in *Drosophila*. *Science*, **294**, 1926–1929.

Lehmann, F.-O., Dickinson, M. H. and Staunton, J. (2000). The scaling of carbon dioxide release and respiratory water loss in flying fruit flies (*Drosophila* spp.). *J. Experim. Biol.*, **203**, 1613–1624.

Lighton, J. R. B. (1998). Notes from the underground: towards ultimate hypotheses of cyclic, discontinuous gas-exchange in tracheate arthropods. *Amer. Zool.*, **38**, 483–491.

Lighton, J. R. B., Garrigan, D., Duncan, F. D. and Johnson, R. A. (1993). Spiracular control of respiratory water loss in female alates of the harvester ant, *Pogonomyrmex rugosus*. *J. Experim. Biol.*, **179**, 233–244.

Lighton, J. R. B., Schilman, P. E. and Holway, D. A. (2004). The hyperoxic switch: assessing respiratory water loss rates in tracheate arthropods with continuous gas exchange. *J. Experim. Biol.*, **207**, 4463–4471.

Lockey, K. H. (1976). Cuticular hydrocarbons of locusta, Schistocerca, and Periplaneta, and their role in waterproofing. *Insect Biochem.*, **6**, 457–472.

Machin, J. and Lampert, G. J. (1989). Energetics of water diffusion through the cuticular water barrier of *Periplaneta*: the effect of temperature, revisited. *J. Insect Physiol.*, **35**, 437–445.

Machin, J. and Lampert, G. J. (1990). Evidence for water-organized structure in the cuticular water barrier of the American cockroach, *Periplaneta americana*. *J. Comp. Physiol. B*, **159**, 739–744.

Maddrell, S. H. P. (1990). James Arthur Ramsay. *Biograph. Mem. Fell. Royal Soc.*, **36**, 420–433.

Maroncelli, M., Qi, S. P., Strauss, H. L. and Snyder, R. G. (1982). Nonplanar conformers and the phase behavior of solid *n*-alkanes. *J. Amer. Chem. Soc.*, **104**, 6237–6247.

Nelson, D. R., Freeman, T. P. and Buckner, J. S. (2000). Waxes and lipids associated with the external waxy structures of nymphs and pupae of the giant whitefly, *Aleurodicus dugesii*. *Comp. Biochem. Physiol. Part B*, **125**, 265–278.

Noble-Nesbitt, J. (1991). Cuticular permeability and its control. In *The Physiology of the Insect Epidermis*, ed. K. Binnington and A. Retnakaran. East Melbourne, Victoria, Australia: CSIRO Publications, pp. 252–283.

Parkash, R., Kalra, B. and Sharma, V. (2008a). Changes in cuticular lipids, water loss and desiccation resistance in a tropical drosophilid – analysis of within population variation. *Fly*, **2**, 187–197.

Parkash, R., Rajpurohit, S. and Ramniwas, S. (2008b). Changes in body melanization and desiccation resistance in highland vs. lowland populations of *Drosophila melanogaster*. *J. Insect Physiol.*, **54**, 1050–1056.

Parkash, R., Ramniwas, S., Rajpurohit, S. and Sharma, V. (2008c). Variations in body melanization impact desiccation resistance in *Drosophila immigrans* from Western Himalayas. *J. Zool.*, **276**, 219–227.

Parkash, R., Sharma, V. and Kalra, B. (2008d). Climatic adaptations of body melanization in *Drosophila melanogaster* from Western Himmalayas. *Fly*, **2**, 111–117.

Patel, S., Nelson, D. R. and Gibbs, A. G. (2001). Chemical and physical analyses of wax ester properties. *J. Insect Sci.*, **1**, 4. Epub 24 April 2001.

Pryor, M. G. M. (1940). On the hardening of the cuticle of insects. *Proc. Royal Soc. B*, **128**, 393–407.

Quinlan, M. C. and Gibbs, A. G. (2006). Discontinuous gas exchange in terrestrial insects. *Resp. Physiol. Neurobiol.*, **154**, 18–29.

Rajpurohit, S., Parkash, R. and Ramniwas, S. (2008a). Body melanization and its adaptive role in thermoregulation and tolerance against desiccating conditions in drosophilids. *Entomol. Res.*, **38**, 49–60.

Rajpurohit, S., Parkash, R., Ramniwas, S. and Singh, S. (2008b). Variations in body melanisation, ovaviole number and fecundity in highland and lowland populations of *Drosophila melanogaster* from the Indian subcontinent. *Insect Sci.*, **15**, 553–561.

Ramsay, J. A. (1935). The evaporation of water from the cockroach. *J. Experim. Biol.*, **12**, 373–383.

Rourke, B. C. (2000). Geographic and altitudinal variation in respiratory and cuticular water loss from the lesser migratory grasshopper, *Melanoplus sanguinipes*. *J. Experim. Biol.*, **203**, 2699–2712.

Rourke, B. C. and Gibbs, A. G. (1999). Effects of lipid phase transitions on cuticular permeability: model-membrane and in situ studies. *J. Experim. Biol.*, **202**, 3255–3262.

Schal, C., Burns, E. L., Jurenka, R. A. and Blomquist, G. J. (1990). A new component of the sex pheromone of *Blattella germanica* (Dictyoptera: Blattellidae), and interaction with other pheromone components. *J. Chem. Ecol.*, **16**, 1997–2008.

Silvius, J. R. and McElhaney, R. N. (1981). Non-linear Arrhenius plots and the analysis of reaction and motional rates in biological membranes. *J. Theor. Biol.*, **88**, 135–152.

Slama, K. (1988). A new look at insect respiration. *Biol. Bull.*, **175**, 289–300.

Small, D. M. (1986). *Physical Chemistry of Lipids*. Handbook of Lipid Research, vol. 4. New York: Plenum Press.

Stennett, M. D. and Etges, W. J. (1997). Premating isolation is determined by larval rearing substrates in cactophilic *Drosophila mojavensis*. III. Epicuticular hydrocarbon variation is determined by use of different host plants in *Drosophila mojavensis* and *Drosophila arizonae*. *J. Chem. Ecol.*, **23**, 2803–2824.

Toolson, E. C. (1978). Diffusion of water through the arthropod cuticle – thermodynamic consideration of the transition phenomenon. *J. Thermal Biol.*, **3**, 69–73.

Toolson, E. C. (1980). Thermodynamic and kinetic aspects of water flux through the arthropod cuticle. *J. Thermal Biol.*, **5**, 1–6.

Toolson, E. C. (1982). Effects of rearing temperature on cuticle permeability and epicuticular lipid composition in *Drosophila pseudoobscura*. *J. Experim. Zool.*, **222**, 249–253.

Toolson, E. C. (1984). Interindividual variation in epicuticular hydrocarbon composition and water loss rates of the cicada *Tibicen dealbatus* (Homoptera: Cicadidae). *Physiol. Zool.*, **57**, 550–556.

Toolson, E. C. and Hadley, N. F. (1979). Seasonal effects on cuticular permeability and epicuticular lipid composition in *Centruroides sculpturatus* Ewing 1928 (Scorpiones: Buthidae). *J. Comp. Physiol.* B, **129**, 319–325.

Toolson, E. C. and Hadley, N. F. (1987). Energy-dependent facilitation of transcuticular water flux contributes to evaporative cooling in the Sonoran Desert cicada, *Diceroprocta apache* (Homoptera: Cicadidae). *J. Experim. Biol.*, **131**, 439–444.

Toolson, E. C. and Kuper-Simbrón, R. (1989). Laboratory evolution of epicuticular hydrocarbon composition and cuticular permeability in *Drosophila pseudoobscura*: effects of sexual dimorphism and thermal-acclimation ability. *Evolution*, **43**, 468–473.

Toolson, E. C., White, T. R. and Glaunsinger, W. S. (1979). Electron paramagnetic resonance spectroscopy of spin-labelled cuticle of *Centruroides sculpturatus* (Scorpiones: Buthidae): correlation with thermal effects on cuticular permeability. *J. Insect Physiol.*, **25**, 271–275.

True, J. R. (2003). Insect melanism: the molecules matter. *Trends Ecol. Evol.*, **18**, 640–647.

Wigglesworth, V. B. (1933). The physiology of the cuticle and of ecdysis in *Rhodnius prolixus* (Triatomidae, Hemiptera); with special reference to the oenocytes and function of the dermal glands. *Quart. J. Microscop. Sci.*, **76**, 270–318.

Willmer, P. G. and Unwin, D. M. (1981). Field analyses of insect heat budgets: reflectance, size and heating rates. *Oecologia*, **50**, 250–255.

Yoder, J. A. and Tank, J. L. (2006). Similarity in critical transition temperature for ticks adapted for different environments: studies on the water balance of unfed ixodid larvae. *Int. J. Acarol.*, **32**, 323–329.

Yoder, J. A., Benoit, J. B., Ark, J. T. and Rellinger, E. J. (2005a). Temperature-induced alteration of cuticular lipids is not required for transition temperature phenomenon in ticks. *Int. J. Acarol.*, **31**, 175–181.

Yoder, J. A., Benoit, J. B., Rellinger, E. J. and Ark, J. T. (2005b). Critical transition temperature and activation energy with implications for arthropod cuticular permeability. *J. Insect Physiol.*, **51**, 1063–1065.

Young, H. P., Larabee, J. K., Gibbs, A. G. and Schal, C. (2000). Relationship between tissue-specific hydrocarbon profiles and their melting temperatures in the cockroach *Blatella germanica* (L.). *J. Chem. Ecol.*, **26**, 1245–1263.

7

Chemical taxonomy with hydrocarbons

Anne-Geneviève Bagnères and Claude Wicker-Thomas

Comparative chemistry is an old discipline originally designed to "use chemistry to name and classify plants". The first reports appeared in 1960, and there has been a steady stream ever since (see Reynolds, 2007 for review). Although plants have been the most studied chemotaxonomy, many researchers have applied the same approach to invertebrates, especially insects.

One of the most widely studied chemotaxonomic characters in insects has been cuticular hydrocarbons (CHCs). The earliest descriptive papers on CHCs were by Jackson (1970), Jackson and Bayer (1970), Jackson and Blomquist (1976), Lockey (1976), and Lange *et al.* (1989). Carlson and collaborators (Carlson and Service 1979, 1980; Carlson *et al.,* 1978; Carlson, 1988) were the first to recognize the special role of CHCs in chemical taxonomy. Since then, numerous reviews have devoted sections to CHCs (Lockey, 1980, 1988; Howard and Blomquist, 1982, 2005; Blomquist and Dillwith, 1985; Blomquist *et al.,* 1987; Kaib *et al.,* 1991; Singer, 1998; Howard, 1993). Since the introduction of molecular techniques, the number of insect chemotaxonomy papers about hydrocarbons has dropped, but there are still many submissions and publications. Explanations for this continued interest in insect CHCs include not only their importance as chemical cues in recognition processes, but also the fact that insects are excellent bioindicators of nature conservation, and inversely, pests to control.

Although systematic and phylogeny analyses have become multidisciplinary, involving a broad range of markers (Clément *et al.,* 2001; Lucas *et al.,* 2002; Austin *et al.,* 2007; Everaerts *et al.,* 2008), there have also been new techniques for chemical identification, perception, as well as the development of mathematical tools for phenetic comparisons of mixtures (Bagnères and Morgan, 1990; Bagnères *et al.,* 1998; Steiner *et al.,* 2002; Lavine *et al.,* 2003; Akino, 2006; Cvačka *et al.,* 2006; Ozaki *et al.,* 2005). These tools have enabled discovery and description of new species and assisted classification and identification of sibling or cryptic species. Speciation mechanisms, evolutionary questions and description of invasive species can be well implemented by data from the different variations of chemical signatures used as complementary phenotypes (Dronnet *et al.,* 2006; Torres *et al.,* 2007).

The purpose of this chapter is to provide an overview of all aspects of chemotaxonomy with hydrocarbons from fundamental to applied research, in solitary as well as eusocial insects.

Table 7.1 *Hydrocarbons useful for discriminating close* Drosophila *species.*

Subgenus/ Group	Species	Charateristic hydrocarbons	References
Sophophora/ melanogaster	*D. simulans* (males) (populations)	7-T ; 7-P	Luyten (1982); Jallon (1984)
	D. melanogaster (males) (populations)	7-T ; 7-P	Jallon (1984); Ferveur and Jallon (1996)
	D. melanogaster African/Cosmopolitan	5,9-HD/ 7,11-HD	Jallon and Péchiné (1989)
	D. melanogaster/ D. simulans	7,11-HD/ 7-T	Jallon and David (1987)
	D. sechellia (males)/ other species (males)	5-T / 7-T	Cobb and Jallon (1990); Jallon and Wicker-Thomas (2003)
	D. santomea (males)/ *D. yakuba* (males)	n-heneicosane	Mas and Jallon (2005)
	D.elegans (brown/ black morph)	7- and 9-pentacosenes	Ishii *et al.* (2002)
Drosophila/ repleta	*D. mojavensis* cluster (3 species)	Alkadienes (C_{33}, C_{35}, C_{37}) Methylalkene (C_{33})	Etges and Jackson (2001)
Drosophila/ Hawaiian Drosophila	*D. paulistorum* (6 species) adiastola subgroup (3 species)	11-docosenylacetate 2-methyltriacontane cuticular HC (C_{21}–C_{30}) (male and female) anal droplet HC (C_{19}–C_{30}) (male)	Kim *et al.* (2004) Tompkins *et al.* (1993)

Chemotaxonomy and sex pheromones

A great deal of information on CHCs as chemotaxonomic characters and sex pheromones is available for Drosophilidae. Several species and populations within species have been identified on the basis of hydrocarbon patterns. Studies have shown that CHCs differing between close species or populations often act as pheromones and may participate in prezygotic isolation. This section presents examples in which sex-pheromone polymorphism has been used as a basis for quick determination of strain/species (see Table 7.1). In section four of this chapter we deal with the possible role of pheromones in speciation.

Pheromone polymorphism in the Sophophora subgroup (Table 7.1)

D. melanogaster *and* D. simulans *polymorphism*

The *D. melanogaster* subgroup consists of nine species. Eight of them are remarkably conservative in their morphology, except for their terminalia. Although not actually needed for species differentiation, hydrocarbon determination is without ambiguity and easy to develop. Two other chapters have been devoted to hydrocarbons in *Drosophila* (Chapter 4; Chapter 15; see also Figure 4.1 of Chapter 4) and describe the main characteristics of *Drosophila* pheromones. Seven species in the *D. melanogaster* subgroup are restricted to Africa. The other two, *D. melanogaster* and *D. simulans*, are cosmopolitan (Lachaise *et al.,* 2000) and display geographic polymorphism in their hydrocarbon patterns.

D. melanogaster males show geographical variation in the amounts of 7-tricosene (7-T; 23C) and 7-pentacosene (7-P; 25C) (Jallon, 1984; Ferveur and Jallon, 1996). Similar findings have been observed in *D. simulans* populations (Luyten, 1982; Jallon, 1984). For *D. melanogaster*, the level of 7-P increases and that of 7-T decreases in populations close to the equator. Thus it is possible to define chemical races based on HC polymorphism. Rouault *et al.* (2001, 2004) analyzed 85 *D. melanogaster* and 29 *D. simulans* populations and confirmed the same correlation found in *D. melanogaster* (Figure 7.1). In all strains from Europe, Africa and America, latitude was well correlated with balanced 7-T vs. 7-P and tricosane vs. pentacosane ratios. This correlation was not found in Asian populations. For *D. simulans*, there was no correlation with latitude and all three populations displaying high 7-P were restricted to West Equatorial Africa.

Geographical polymorphism has also been observed in females (Jallon and Péchiné, 1989; Ferveur *et al.,* 1996). African and Caribbean females have high 5,9-heptacosadiene (5,9-HD; 27C), whereas cosmopolitan females have high 7,11-heptacosadiene levels (7,11-HD; 27C) (Figure 7.2). Chapter 4 presents a detailed description of this polymorphism, and later in this chapter we discuss its implications in phylogeny and speciation.

D. santomea *and* D. yakuba *polymorphism*

D. santomea is a recently discovered species (Lachaise *et al.,* 2000) in São Tomé, an island in the Gulf of Guinea. It lives with other species, including a putative sister species, *D. yakuba*. These two sister species, i.e. *D. santomea* and *D. yakuba*, show strong reproductive isolation (Coyne *et al.,* 2002). *D. santomea* is the only species of the subgroup that displays a distinctive morphologic trait, i.e. absence of the black abdominal banding. However, study shows that CHCs and, especially, *n*-heneicosane – that can be up to seven-fold higher in *D. santomea* than *D. yakuba* males – are more important than this visual cue in species recognition and reproductive isolation (Mas and Jallon, 2005).

Pheromone polymorphism in D. elegans

Two morphs have been identified in this species. The black morph is found in the Ryukyu islands and Taiwan, while the brown morph occurs in southern China, the Philippines and Indonesia (Lemeunier *et al.,* 1986). Partial reproductive isolation, based mainly on

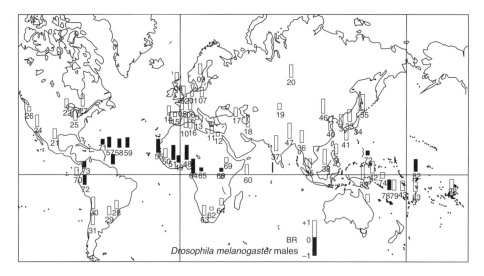

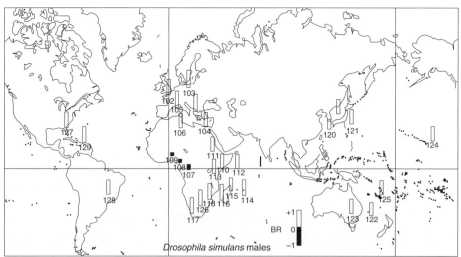

Figure 7.1 (7-T – 7-P)/(7-T + 7-P) in males from 85 *D. melanogaster* populations and 29 *D simulans* populations (in Rovault *et al.*, 2001).

recognition of CHCs, has developed between these morphs (Hirai and Kimura, 1997). Black morph females have more 7-pentacosene and 9-pentacosene, and males discriminate females based on this difference (Ishii *et al.,* 2002).

Pheromone polymorphism in the Drosophila *subgroup (Table 1)*

D. mojavensis *cluster*

The *D. mojavensis* cluster belongs to the *D. repleta* group and includes three closely related species with CHCs from C28 to 40 (Etges and Jackson, 2001). Some CHCs are

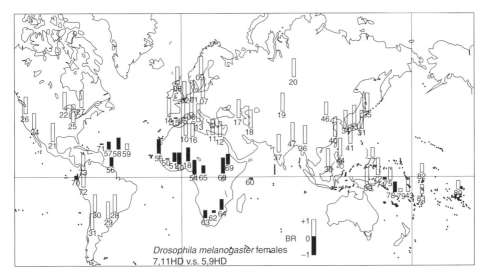

Figure 7.2 7,11-HD / 5,9-HD in females from 85 *D. melanogaster* populations (in Rouault *et al.*, 2004).

present in only one species (e.g. 31-methyldotriacont-6-ene in *D. navojoa*) or in two species (e.g. 8,24-tritriacontadiene in *D. mojavensis* and *D. arizonae*). Canonical discrimination function analysis has been used to separate all three species into distinct, non-overlapping groups, suggesting that CHCs are species-specific. These data also indicate that reproductive isolation between the three species is strong, and that epicuticular CHCs may have played a barrier role prior to species divergence (Etges and Jackson, 2001).

D. paulistorum

D. paulistorum is a species complex composed of six semispecies, displaying significantly different quantities and ratios of CHCs, especially 2-methyl triacontane and 11-docosenylacetate. These differences probably contribute to reproductive isolation (Kim *et al.*, 2004).

Pheromone polymorphism in the adiastola *subgroup*

The *adiastola* subgroup comprises several picture-wing fly species that produce acoustic and pheromonal mating signals (Tompkins *et al.*, 1993). During courtship, males curl their abdomens up over their heads and emit anal droplets that may provide females with information about courting males. Droplet-associated CHCs, as well as male and female CHCs, differ between three closely related species belonging to this subgroup and could act as courtship-stimulating pheromones (Tompkins *et al.*, 1993).

Chemotaxonomy and insects of economic importance

Insects at the interfaces between agricultural and urban pests, e.g. termites, and between medical and urban pests, e.g. flies, wasps, and cockroaches, have been extensively studied. Since morphological identification of many close species is difficult, not only at various larval stages but also at the adult stage, detection of even small differences in cuticular compounds can be useful. A full presentation of all evidence to support this claim is beyond the scope of this chapter and would be repetitive. Instead, this section presents a few typical examples in which CHC analysis has been used to study insects with economic impact in various sectors.

Food processing and agronomy
Orthoptera, Mantodea (Table 7.2)

Analysis of CHCs has been used in taxonomic studies of grasshoppers. Results in *Schistocerca shoshone* revealed the same CHCs in populations from different localities, but in variable relative abundance. These differences appeared to be genetically determined, since sex, maturity, and food type had little effect on CHC patterns (Chapman *et al.*, 1995). In a chemotaxonomic study of thirteen species belonging to the *Locusta* or *Schistocerca* genus (Lockey and Oraha, 1990), and five species belonging to the *Schistocerca americana* complex (Grunshaw *et al.*, 1990), CHC analysis confirmed species status previously demonstrated by hybridization and established the possible origins and evolution of the *S. americana* complex. Similarly, comparison of CHCs provided evidence to support separation of the *Gryllotalpa* complex into two species (Broza *et al.*, 2000). Conversely study of hydrocarbon profiles demonstrated an absence of a taxonomic relationship between two species or genera: *Halemus* that contains few methylalkanes and a high proportion of squalene (17%) and *Schistocerca* (Chapman *et al.*, 2000). CHC composition has also allowed discrimination of five mantid species (Jones *et al.*, 1997).

Homoptera (Table 7.2)

Identification of aphid species is often difficult due to the low number of adults as compared to larvae. Another problem is that lipid and hydrocarbon composition appear highly dependent on the host plant (Neal *et al.*, 1994). Analysis of CHCs has been used alone or in combination with molecular techniques to differentiate close Aphidae species (Lazzari *et al.*, 1991; Raboudi *et al.*, 2005). For example, analysis of the lipid and hydrocarbon compositions of pupal exuviae has been used to differentiate greenhouse and sweetpotato whiteflies (Neal *et al.*, 1994).

Coleoptera (Table 7.3)

Phylogeny based on CHC phenotype has been used to corroborate putative taxonomic identities of Coleoptera species characterized by the presence of various CHCs (alkanes, alkenes, and mono-, di-, tri-, tetra-methylalkanes). Some families exhibit particular

Table 7.2 Main hydrocarbons and hydrocarbons used in taxonomy for Orthoptera, Hemiptera and Homoptera.

Order/Family	Genera	Hydrocarbons	Species	Discrimination based on	References
Orthoptera/ Acrididae	Halemus Schistocerca	n-alkanes, mono-, di- tri-methylalkanes (C_{28} to C_{53})	Halemus sp., Schistocerca spp.	methylalkanes squalenes	Chapman et al. (1995)
	Locusta Schistocerca		13 species	alkanes dimethylalkanes	Lockey and Oraha (1990)
	Schistocerca		5 species (S. americana complex)	alkanes, di-, tri-methylalkanes	Grunshaw et al. (1990)
Orthoptera/ Gryllotalpidae	Gryllotalpa	n-alkanes, methylalkanes	G. tali, G. marismortui	alkanes (C_{27} and C_{28})	Broza et al. (2000)
Mantodea/ Mantidae	Bruneria, Mantis Stagnomantis, Tenodera	n-alkanes (C_{29}, C_{30}, C_{31}, C_{32}, C_{33})	B. borealis, M. religiosa, S. carolina, T. sinensis, T. angustipennis	n-alkanes (C_{31}, C_{33}) aldehydes (C_{30}) alcohols (C_{30})	Jones et al. (1997)
Homoptera/ Aphididae	Rhopadosiphon Shizaphis	n-alkanes, mono-, di-methylalkanes (C_{25} to C_{37})	R. maidis, R. padi, R. insertum, S. graminum	general hydrocarbon pattern	Lazzari et al. (1991)
	Aphis, Myzus, Macrosiphum	n-alkanes, compounds (C_{25} to C_{46}) not determined	A. gossypii, A. craccivora, M. persicae, M. euphorbiae	general hydrocarbon pattern	Raboudi et al. (2005)
Homoptera/ Aleyrodidae	Trialeurodes Bennisia	n-alkanes (C_{25}–C_{33}) (pupal exuviae)	T. vaporariorum, B. tabaci	Hydrocarbons, alcohols, aldehydes, acetates	Neal et al. (1994)

Table 7.3 *Main hydrocarbons and hydrocarbons used in taxonomy for Coleoptera.*

Family	Genera	Hydrocarbons	Species	Discrimination based on	References
Tenebrionidae	*Onymacris*	*n*-alkanes, mono-, di-, tri- and tetra-methylalkanes (C_{19} to C_{42})	*O. bicolor* *O. boschimana*	tetramethylalkanes, 3,5-dimethylalkanes	Lockey (1991)
Tenebrionidae	*Epiphysa* *Onymacris*	*n*-alkanes, methylalkanes (C_{20} to C_{35})	*E. spp.* *O. unguicularis*	alkanes, 3-methylalkanes	Lockey (1992)
Scolytidae	*Conophtorus*	*n*-alkanes, alkenes, alkadienes, mono-, di-, tri-methylalkanes (C_{20} to C_{43})	8 species	heptacosene, two dimethylhexacosanes	Page *et al.* (1990a)
Scolytidae	*Dendroctonus*	*n*-alkanes, alkenes, monomethylalkanes, dimethylalkanes (C_{22} to C_{35})	*D. ponderosae* *D. jeffreyi* *D. brevicomis* *D. frontalis*	3,7-dimethylalkanes 3,X-dimethylnonacosanes *n*-alkanes (C_{31} to C_{35})	Page *et al.* (1990b) Page *et al.* (1990b)
Scolytidae	*Ips, Orthotominus*	*n*-alkanes, alkenes, alkadienes, mono-, di-, tri- and tetra-methylalkanes (C_{21} to C_{37})	*I. latidens* *I. pini* *O. caelatus*	Specific alkanes and dimethylalkanes, 3,7,11- and 3,7,15-trimethyl heptacosane	Page *et al.* (1997)
Chrysomelidae	*Diabrotica*	*n*-alkanes, alkenes, mono-, di-methylalkanes, (C_{19} to C_{41})	*D. longicomis* *D. barberi*	General hydrocarbon pattern	Golden *et al.* (1992)

features attributed to their habitat. For instance, the overall hydrocarbon patterns of aquatic Dytiscidae are similar to those of terrestrial Coleoptera but with fewer methylalkanes and more alkenes (Alarie *et al.*, 1998). Based on analysis of 37 Coleoptera species in eight different families, a good correlation was observed between systematic characters and CHC patterns (Jacob and Hanssen, 1986). This correlation makes CHC composition a particularly useful tool for discrimination of congeneric Tenebrionidae species. However, it is less valuable for more heterogeneous families like Cucujoidea (Jacob and Hanssen, 1986). Hydrocarbon profiles have been most extensively studied in Tenebrionidae and Scolytidae. Results show that dimethylalkanes are important diagnostic characters in sibling species of both families (Table 7.3).

Other data in Coleoptera show that in two *Onymacris* species (Tenebrionidae), i.e. *O. boschimana* and *O. bicolor*, the presence and absence of 3,5-dimethylalkanes (C_{29} and C_{31}) respectively and of tetramethylalkanes (C_{39}, C_{40} and C_{42}), respectively is sufficient to achieve discrimination (Lockey, 1991). In *Epiphysa sp.* and *O. unguicularis*, the complete hydrocarbon pattern is necessary for differentiation (Lockey, 1992). In Scolytidae hydrocarbon phenotypes have been used to determine the relationship of eight *Conophthorus* species that were difficult to separate on the basis of morphological characteristics alone. *C. cembroides* and *C. edulis* can be easily distinguished, based on the presence of heptacosene in large quantity in *C. edulis* and of two dimethylhexacosanes in *C. cembroides* (Page *et al.*, 1990a). Likewise, the phylogeny of seven *Ips* species belonging to the *grandicollis* group and to three outgroup species (*Ips* and *Orthotominus* genera) and of four *Dendroctonus* species has been constructed based on CHC composition (Page *et al.*, 1990b, 1997). Relatively little information is available for other insect families. In Chrysomelidae, two sibling species have been discriminated by CHC pattern (Golden *et al.* 1992). In this group, hydrocarbon analysis seems to be a powerful tool for discriminating close species, with the exception of some heterogeneous families, as stated above.

Other taxa

Hymenoptera. Since some hymenoptera are parasitoids of other insects, it is important to identify these species. *Rhopalinus* is an ectoparasitoid of bark beetles. Its hydrocarbons, including alkanes and mono- and di-methylalkanes from C_{25} to C_{37}, have evolved with the host and populations originating from different regions of the United States (e.g. Georgia and California), and can show geographic differences, in particular in the proportion of dimethylalkanes. This finding suggests that the parasitoids from both regions have diverged, possibly due to different climatic conditions, and may now constitute different species (Espelie *et al.*, 1990). In *Muscidifurax*, a parasitoid of Diptera, hydrocarbon analysis proved reliable for identification, not only of the three North American species, but also of populations within species (Bernier *et al.*, 1998; Geden *et al.*, 1998). This capability could make CHC analysis a useful tool for detecting problems of species purity on farms using parasitoids for fly control.

Diptera. Some dipteran species are farm pests. In the event of fruit fly infestation, the species must be identified quickly. If adult flies are absent, accurate identification of larvae can be difficult or impossible on the basis of morphological characters. One study has shown that the Caribbean fruit fly, *Anastrepha suspense* and the Mediterranean fruit fly, *Ceratitis capitata*, have different CHC patterns at the larval stage and thus can be easily distinguished from each other (Sutton and Steck, 1994). This finding suggests that hydrocarbon analysis could be extended to identification of other species with high economic impact.

Urban pests

Blattaria/Blattodea. Cockroaches are the most common household pests. Although they can be disease vectors, they are generally commensal insects, living gregariously in human dwellings. The first two chemotaxonomic papers published the same year (Baker *et al.*, 1963; Gilby and Cox, 1963) described cuticular hydrocarbons in the American cockroach (*Periplaneta americana*), a major urban pest in the USA. A few years later, Jackson in 1970 provided the first succinct comparison of three species of *Periplaneta* (i.e., *P. australasiae, P. brunnea* and *P. fuliginosa)* and showed that *cis*-9-tricosene was a major hydrocarbon only in males. In 1972, the same author compared *P. japonica* and *P. americana* to the other species (Jackson, 1972). Since these early publications, there have been surprisingly few chemotaxonomic papers focusing on cockroaches.

Only recently, the Blattidae family was the focus of a study by Saïd *et al.* (2005), who compared the hydrocarbon profiles of four species of the *Periplaneta* genus. Findings showed that *P. brunnea, P. fuliginosa* and *P. australasiae* have more cuticular components in common with one another (from C_{21} to C_{41} carbon atoms) than with *P. americana,* which displays components with higher molecular weight (from C_{24} to C_{43} carbon atoms). Data also showed that unconjugated diene 6,9-$C_{27:2}$ was present in *P. americana* but absent in the other three species. In their discussion, the authors suggested that these compounds may be involved in aggregation and interspecific recognition.

Tartivita and Jackson (1970) compared the CHCs of *Blatta orientalis*, another household pest in the Blattidae family, with those of *Leucophaea maderae* in the Blaberidae family. Findings showed that the major hydrocarbons in the surface lipids of both cockroaches were *n*-heptacosane, 11-methylheptacosane, 13-methylheptacosane and 3-methylheptacosane, and indicated that these components were qualitatively identical and quantitatively similar.

Everaerts *et al.* (1997) reported species-specific profiles based on comparison of CHCs from *L. maderae* with those from *Nauphoeta cinerea*, not an urban pest but widely used by herpetologists. In the same paper the authors mentioned unpublished results on a good species-specific correlation of CHCs in 30 cockroach species.

Carlson and Brenner (1988) carried out several studies in a third household cockroach family, Blattellidae, from the same Blaberoidea superfamily as Blaberidae. One study was aimed at identifying the various life stages of three North American species of Blattella:

B. asahinai, *B. germanica*, and *B. vaga*. The same authors performed chemotaxonomic identification of different *Blattella* cockroaches in Taiwan and various Pacific Basin locations (Brenner *et al.*, 1993).

A large amount of data on cockroach hydrocarbons has been included in descriptive papers designed for evaluation of behavior, physiology, or biochemistry of household pest species, rather than for taxonomic purposes. Schal and colleagues at North Carolina State University investigated *B. germanica* with the aim of developing extermination techniques for this major urban pest, which is an allergenic vector infesting kitchens and bathrooms. Liang *et al.* (2001) investigated the brownbanded cockroach *Supella longipalpa*, another blattellidae occurring inside dwellings. Findings showed similarity to the German cockroach, but most hydrocarbons were longer than in Blattella, with 35 and 37 carbons with one to three methyl branches. Youngsteadt *et al.* (2005) studied CHC synthesis and maternal provisioning of embryos in the viviparous blaberinae cockroach *Diploptera punctata*. It should also be mentioned that numerous oxygenated cuticular compounds, e.g. sex pheromones, have been shown to have hydrocarbon precursors in their biosynthesis process.

Isoptera. Many termite species are structural pests. Isopteran are currently considered to be an order close to cockroaches, indeed recently the Eggleton team even proposed that they should be classified as social cockroaches (Inward *et al.*, 2007). Unlike those of the Blattaria, the CHCs of termites have been studied and reported extensively, and have been involved in many precursor works on taxonomy and sociobiology. They have proven to be as effective for termite classification (Clément *et al.*, 1985; Bagnères *et al.*, 1988, 1990a, b; Kaib *et al.*, 1991; Haverty *et al.*, 1999a) as trail following pheromones (Sillam-Dussès *et al.*, 2007) or soldier frontal gland secretions (Chuah, 2005). Hydrocarbons are also important cues in termite species and caste recognition (Howard *et al.*, 1982; Blum, 1987, Bagnères *et al.*, 1991b, 1998). Recently, invasion of urban areas by various termites has generated renewed interest in CHCs as taxonomic markers. Most species identified have been subterranean species of three Rhinotermitidae genera, i.e., *Heterotermes*, *Coptotermes* and *Reticulitermes*. All of these structural pest genera are found in urban settings, usually in relatively temperate climates all over the world. It is important to note that morphological features are generally of little taxonomic value for these genera and can lead to misidentification. Since these subterranean termites have been the subject of numerous chemotaxonomic papers, a large body of data is available. For presentation of these data, this section is organized according to geographic area rather than genus.

Asia. Takematsu and Yamaoka (1999) reported a study of CHCs in *Reticulitermes* from Japan and neighboring countries. They identified three distinct phenotypes among five subspecies of *R. speratus*: (1) three subspecies from the Japanese mainland, (2) *R. s. okinawanus* and (3) *R. s. yaeyamanus*, the last two being considered as real species by the same author (from Takematsu, 1999, in Park *et al.*, 2006). In addition, the hydrocarbon components of *R. flaviceps flaviceps* have been shown to differ from those of *R. f. amamianus*, and those of *R. miyatakei* differed from those of three undetermined species. Based on hydrocarbon analysis, a total of nine chemical phenotypes have been identified in the region. Recently, Park *et al. (*2006) published a phylogeographic work on the species status of *R. speratus*. Using the COII and COIII mitochondrial genes, all species in the region were classified into two major clades comprising the Korean/southern Japanese population and the northern Japanese population. All were of Chinese ancestral origin from the early Pleistocene. The two distinct lineages are congruent with two of the three distinct cuticular phenotypes named *R. s. speratus* and *R. s. kyuhuensis*. The third lineage, *R. s. leptolabralis*, could not be confirmed because of overlapping geographic distribution and possible sampling errors. The same study also showed that *R. okinawanus* and *R. yaeyamanus* are different from the *R. speratus* clade and from *R. miyatakei*.

Australia. The Australian termite fauna is known mainly for the nonsubterranean primitive genus *Mastotermes* (mainly *Mastotermes darwiniensis*). Most reports have focused on their impact as agro/horticulture pests, but CSIRO groups have carried out also a few chemical taxonomy studies on the *Heterotermes* and *Coptotermes* subterranean genera. Study of *Heterotermes* taxonomy with analysis of CHCs in workers/soldiers and alate morphology in southeastern Australia (Watson *et al.*, 1989) showed that *Heterotermes ferox* could be a complex of species. The same study also revealed two other species named *H. brevicatena* and *H. longicatena*. The preliminary findings of two other papers on Australian *Coptotermes* distinguished *C. lacteus* from *C. frenchi* and identified three species – *C. michaelseni*, *C. brunneus* and *C. dreghorni* – with hydrocarbon patterns differing from those of *C. lacteus* and *C. frenchi* (Brown *et al.*, 1990, 1994). A phylogenetic study (Lo *et al.*, 2006) showed that *C. lacteus*, *C. frenchi* and *C. michaelseni* form closely related monophyletic groups, but indicated that *C. acinaciformis* forms a distinct complex of species. *C. acinaciformis* is recognized to be the most important subterranean pest in Australia.

Europe. European *Reticulitermes* was one of the first groups of species to undergo chemosystematic study with CHCs (Clément *et al.*, 1985, 1988). Initial chemical data distinguishing three of the European taxons, i.e., *R. lucifugus grassei, R. lucifugus banyulensis,* and *R. santonensis* were subsequently confirmed by Bagnères *et al.* (1990a, 1991b). In a more recent biosystematic review on European species, the same authors (Clément *et al.*, 2001) used various taxonomic markers, including cuticular compounds, to show that the first two species were true species named *R. grassei* and *R. banyulensis* (Figure 7.3). The latter finding was confirmed by phylogenetic study (Kutnik *et al.*, 2004). In the same 2001 paper, Clément *et al.* identified a novel phenotype that was later elevated to species status as *R. urbis* (Bagnères *et al.*, 2003; Uva *et al.*, 2004a). *R. urbis* appears to be closely related to the balkanic species *R. balkanensis* (Leniaud *et al.*, 2009a,b). Recently the status of another subspecies included in the Italian *lucifugus* complex was confirmed on the basis of chemical and molecular findings, i.e., *R. l. corsicus* is related to the continental *R. l. lucifugus* (Uva *et al.*, 2004b; Lefebvre *et al.*, 2008) (see also Figure 7.3). All these species are considered to be structural pests in Europe. The *R. santonensis* chemotype was shown to be close to the main American termite pest *R. flavipes* (Bagnères *et al.*, 1990b). This chemotaxonomic finding was confirmed by Clément *et al.* in 2001, and has been discussed by other authors (Nelson *et al.*, 2001). The Bagnères group is currently working on the possibility that the two taxa are synonymous, and considers *R. santonensis* to be an invasive variant of *R. flavipes*. The different CHC profiles are shown in Figure 7.3. Study of invasive insect species has revived interest in developing criteria to describe invasion processes and in understanding the ecological consequences of invasion.

Russian researchers (Klochkov *et al.*, 2005) recently published a descriptive paper on the specificity of CHCs in two termite genera: the drywood termites *Kalotermes* and various *Reticulitermes* populations in Eastern Europe. Despite mistakenly referring to *R. lucifugus* as a *Reticulitermes* species, this paper provided the first information about *Reticulitermes* termites from Azerbaijan, Armenia, Turkmenistan and Ukraine, and also showed that these populations are sister clades of *R. balkanensis* and *R. urbis*. This finding is in agreement with GC and GC–MS data, showing that the chemical profiles of *R. urbis* and *R. balkanensis* are identical (Clément *et al.*, 2001), but totally different from the GC profile of the Italian species *R. lucifugus*. *R. urbis* is now considered to be a new invasive species in Europe (Leniaud *et al.*, 2009a, b).

America. Another invasive termite species that has been extensively studied is *Coptotermes formosanus,* now a major structural pest in the USA. In the first description of the *Coptotermes* cuticular compounds (McDaniel, 1990), the author distinguished 4-methylC$_{25}$ from 2-methylC$_{25}$. These two compounds are generally confused due to the great similarity between their respective mass spectra. GC–MS analysis shows only a tiny difference in the 70/71 ion ratio, the *m/z* 70 ion being more abundant in the 4-methyl branched compound. Another distinguishing feature for the two compounds was a small difference in retention time, since the 2-methylbranched (the one present in *C. formosanus*) elutes between 4-methyl and 3-methyl branched pentacosane (this was determined by authentic sample comparison). Most *C. formosanus* compounds have 27 carbons in the parent chain, and analysis of quantitative differences allows castes and populations to be separated according to different concentration profiles (Haverty *et al.*, 1990a). Another interesting finding concerning quantitative differences is the existence of two concentration profiles in Louisiana, suggesting the existence of two separate points of introduction. In a comparative study, using different *Coptotermes* species collected outside of the USA, Haverty *et al.* (1991) showed no hydrocarbon profiles to be species-specific. *C. curvignathus* from Thailand was the most similar to the Formosan profile. *C. acinaciformis* from Australia showed a unique C$_{27:1}$ peak. *C. lacteus* from Australia also had long-chain hydrocarbons. *C. testaceus* from Trinidad presented a different pattern but only one sample was collected. The 13,15-diMeC29 component was unique to *C. formosanus* and was considered as diagnostic. In a subsequent study of *C. formosanus,* Haverty *et al.* (1996c) showed that quantitative variations could be used to discriminate between different colonies and seasonal samples.

Reticulitermes has been most extensively studied using chemotaxonomic analysis of CHCs, particularly in the USA. One reason for this interest is the great economic impact of the genus and difficulty in describing the different species, which are similar morphologically, especially with their worker caste. The first description of CHCs in a *Reticulitermes* species involved *R. flavipes*, another major pest species in the USA (Howard *et al.*, 1978). It was shown that *R. flavipes* displayed a specific conjugated diene, i.e., (Z,Z)7,9-C25:2, that was

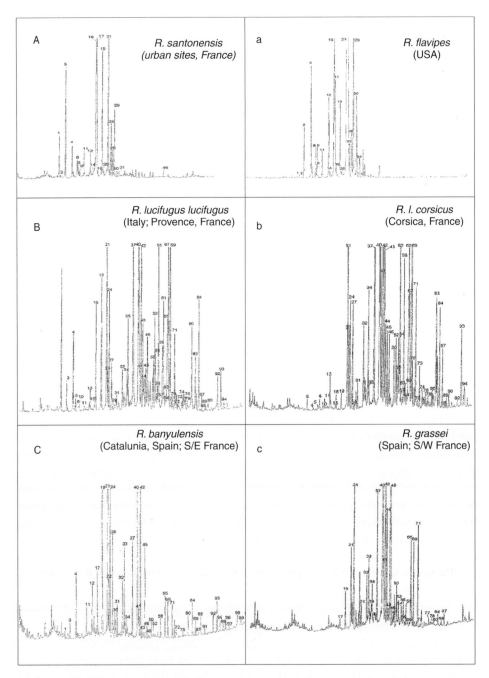

Figure 7.3 Gas chromatograms of different *Reticulitermes* species with various evolutive relationships (A and a: synonymous species; B and b: subspecies; C and c: closely related species) (see Clément *et al.*, 2001 for peak determination).

also present in *R. virginicus,* along with the non-conjugated diene 6,9-$C_{25:2}$ (Howard *et al.,* 1982). The same paper also provided the first evidence that CHCs were species cues. In the paper by Clément *et al.* (1986), it was preliminarily described as a new species in the southern USA; more recently Austin *et al.* (2007) provided a full description and classified it as a valid species: *R. malletei.* In this species numerous trienes were described for the first time in *Reticulitermes* (in the PhD thesis of Bagnères, 1989). Bagnères *et al.* (1990b) described different phenotypes of *R. flavipes* only in the state of Georgia (three based on cuticular hydrocarbons, six based on defensive compounds of soldiers). Since these early works, most papers describing new taxa based on cuticular phenotypes have come from Haverty and colleagues, who published numerous papers between 1996 and 2007 (see Haverty references). Haverty *et al.* (1996a) confirmed the chemical polymorphism of *R. flavipes* and speculated that various undescribed taxa were present in Georgia. This hypothesis was confirmed by a study correlating mitochondrial haplotypes with cuticular hydrocarbon phenotypes (Jenkins *et al.,* 2000). In California, where only two species (i.e., *R. hesperus* and *R. tibialis)* had been described, Haverty *et al.* used cuticular hydrocarbon profiles to suggest that there are at least two undescribed taxa and that *R. hesperus* might have two variants (Delphia *et al.,* 2003; Haverty and Nelson, 1997; Haverty *et al.,* 1999a, b, 2003). Another paper describing termites in the Californian region (Copren *et al.,* 2005) corroborated taxonomic designations based on phylogenetic analyses of mt DNA sequences and CHC phenotypes, classifying six existing clades: *R. tibialis* (clade 1), *R. hesperus* (clade 6), and four new clades (clades 2 to 5). By showing five additional CHC phenotypes (i.e., AZ-A,-B,-C,-D and NM-A), a similar study in Arizona and neighboring states, where only *R. tibialis* had been previously described, indicated that there were additional taxa. Several compounds allowed differentiation, e.g. $C_{29:1}$ for AZ-C and 9,13-diMeC25 for AZ-D and AZ-A. AZ-D likely represents the *R. hesperus* complex, and *R. tibialis* may support a complex species chemotype hypothesis (Haverty and Nelson, 2007). In other papers, Haverty compared cuticular hydrocarbon mixtures of *Reticulitermes* samples from various regions of the USA: the southwest (California, Nevada, and Arizona), south central region (New Mexico) and southeast (Georgia). In Haverty *et al.* (1999a), the authors suggested that there were 4 additional undescribed taxa in Georgia, one in New Mexico, 3 or 4 in Arizona and one in Nevada. Comparison of soldier defense secretion mixtures with cuticular hydrocarbon phenotypes suggested that there were numerous undescribed species of *Reticulitermes* in North America (Nelson *et al.,* 2001). Based on PAUP analysis, to infer degrees of relatedness between chemical phenotypes, it was estimated that around 25 phenotypes from 141 colonies could be classified into three lineages: lineage I, with a preponderance of internally branched monomethylalkanes and 11,15-dimethylalkanes; lineage II, with a preponderance of 5-methylalkanes and 5,17-dimethylalkanes; and lineage III, characterized by a predominance of olefins (alkenes and alkadienes) (Page *et al.,* 2002). Based on these findings, it appears that the *Reticulitermes* genus in the USA is complex and that only about one-third of US species have been described up to now. In addition to showing the need for further study in this region, these findings underline the utility of CHCs in identifying species (Haverty and Nelson, 2007; Nelson *et al.,* 2008).

Regarding the synonymy of European *R. santonensis* and American *R. flavipes*, several comments are noteworthy. First, research in the USA has not identified the same quantitive cuticular hydrocarbon chenotype (even if very close as shown in Figure 7.3), or same soldier defensive compounds (Bagnères *et al.*, 1990b) as have *R. santonensis* in France, Chile and Germany (Clément *et al.*, 2001). Nevertheless, *R. flavipes* seems to be one of the most invasive species of termites, not only on non-native continents, but also in the Americas (presence in Toronto, Canada, and in Santiago and Valparaiso, Chile). Indeed, *R. flavipes* could be one of the taxa present in California (Bagnères and colleagues, personal communication). It is likely that transport of materials by humans is a major factor in this species complexity. Further study linking DNA with hydrocarbons will be needed to map the species in the USA.

Insects of medical importance

Nematocera (Diptera) suborder

Culicidae. Mosquitoes are vectors of malaria. Analysis of hydrocarbons has been essential for discrimination between closely related species and/or populations. It allows accurate and quick identification of the *Aedes* species (Pappas *et al.*, 1994). Five of seven North American populations and nine of fourteen North American and Asian populations of *Aedes albopictus* were distinguished based on discriminant analysis of hydrocarbons (Kruger *et al.*, 1991; Kruger and Pappas, 1993). Studies on *Aedes* have shown important geographic variations in hydrocarbons.

Anopheles species. The most extensively studied species of mosquitoes is *Anopheles*. The *An. arabiensis* prevails in arid savannas, whereas *An. gambiae* has colonized all ecological areas (arid, humid savannas and irrigated or flooded zones), and is undergoing a speciation process. *An. gambiae* and *An. arabiensis* often occur sympatrically and cannot be distinguished morphologically. These two species can, however, be readily distinguished based on CHCs that are qualitatively the same (linear, mono- and di-methylalkanes from C_{23} to C_{45}) but with variations in relative proportions (Carlson and Service, 1979, 1980). *An. gambiae* can be divided into three different morphs showing chromosomal inversion polymorphism. These chromosome arrangements (designated B, M and S) have led to complete or partial reproductive isolation. The M and S forms live in sympatry with *An. arabiensis*. All three form swarms, and mating takes place predominantly during swarming (Charlwood and Jones, 1980). Mixed swarms exist between the two species (Marchand, 1984) and between M and S forms (Diabate *et al.*, 2006), but hybridiziation is rare.

The same 48 CHC compounds have been found in populations from different geographic areas and in taxa from the same geographic origin (Caputo *et al.*, 2007). However, relative proportions vary between both geographic area and species or populations within the same geographic area. The most variable compounds are *n*-hentriacontane, MeC_{29}, MeC_{30} and MeC_{31}. Similar variations have been observed between the two morphs of *An. gambiae* and between *An. gambiae* and *An. arabiensis* species. Based on hydrocarbon pattern,

Table 7.4 *Main hydrocarbons used in taxonomy for Nematocera (Diptera).*

Family	Genera	Main hydrocarbons	Species / Population	Discrimination based on	References
Culicidae	Aedes	n-alkanes, mono-, di-, methylalkanes (C_{16} to C_{45})	A. albopictus (14 populations)	general hydrocarbon pattern	Kruger and Pappas (1993), Kruger et al. (1991)
			A. hendersoni, A. triseriatus	general hydrocarbon pattern	Pappas et al. (1994)
	Anopheles	n-alkanes, mono-, di-, methylalkanes (C_{23} to C_{47})	A. gambiae complex	Ratios of 3 pairs of peaks	Carlson and Service (1979, 1980)
			A. gambiae, A. arabiensis	general hydrocarbon pattern	Anyanwu et al. (1994, 2001), Caputo et al. (2007)
			A. quadrimaculus complex sp. A. maculipennis complex sp.		Carlson et al. (1997) Phillips et al. (1990a)
Simulidae	Simulium	n-alkanes, methylalkanes (C_{22} to C_{35})	S. damnosum complex (4 species) (adults)	general hydrocarbon pattern	Phillips et al. (1985)
			S. damnosum complex (3 species) (larvae)	general hydrocarbon pattern	Mafuyai et al. (1994)
Psychodidae	Phlebotomus Sergentomiya Psychodopygus	Terpenes	Ph. and Se. (9 species)	general cuticular compound pattern	Mahamat and Hassanali (1998)
					(Hamilton et al. 1996)
			Ph. ariani (populations)		Kamhawi et al. (1987)
			Ps. (2 species)		Phillips et al. (1990b)

genotype can be assigned with nearly 80% probability. Interestingly, Me-C_{29} is a candidate pheromone in *Ae. aegypti* (Horne and Priestman, 2002), suggesting the possibility that these compounds, considered as population markers, may play a role in the recognition of mating partners. In *An. gambiae*, malathion-susceptible or -resistant populations also show a high degree of segregation (95%) in hydrocarbons (Anyanwu *et al.*, 2000). Even more interestingly, analysis of cuticular and internal lipids and hydrocarbons of larvae has shown important differences that could be useful for distinguishing species at an early stage (Hamilton and Service, 1983). Likewise, two populations of *An. gambiae* and one of *An. arabiensis* can be separated at larval stages by their hydrocarbon profile (Anyanwu *et al.*, 1994, 2001).

Several species and populations from other *Anopheles* complexes have been discriminated based on CHC patterns. Examples include all five species of the *An. quadrimaculus* complex (Carlson *et al.*, 1997), some species of the *An. maculipennis* complex (Phillips *et al.*, 1990a), malaria-vector and non-vector forms of the *An. maculates* complex (Kittayapong *et al.*, 1990, 1993), and *An. stephensi* strains susceptible or resistant to DDT and malathion (Anyanwu *et al.*, 1993, 1997). CHCs have been used in combination with isoenzyme analysis to successfully differentiate populations of *An. darlingi* (Rosa-Freitas *et al.*, 1992). All these findings demonstrate that hydrocarbon analysis is a powerful tool for distinguishing mosquito species and populations. This is particularly important for disease vectors, since it can facilitate interpretation of epidemiological data and assist implementation of control measures.

Simuliidae. Successful use of CHC analysis to identify Simuliidae, vectors of onchocerciasis, has been reported (Phillips *et al.*, 1985; Mafuyai *et al.*, 1994).

Psychodidae. Psychodidae (sandflies) are major vectors of leishmaniasis. Males and females produce terpene sex pheromones (Hamilton *et al.*, 1996). *Psychodopygus wellcomei* and *P. complexus* are sympatric species that cannot be distinguished by morphometrics or isoenzyme profiles. Discriminant function analysis of their cuticular compounds allows easy identification of both species (Ryan *et al.*, 1986). Successful separation has also been achieved for nine species belonging to the *Phlebotomus* and *Sergentomiya* genera (Mahamat and Hassanali, 1998), for populations of *Phlebotomus ariasi* (Kamhawi *et al.*, 1987) and for the *Psychodopygus carrerai* and *P. yucumensis* species (Phillips *et al.*, 1990b). However, it has also been shown that hydrocarbon profiles are not exactly the same in laboratory specimens as they are in field specimens, due to differences in environmental conditions (temperature, relative humidity), dietary factors, or genetic changes following selection during laboratory colonization (Gebre-Michael *et al.*, 1994).

Diptera – Brachycera Suborder (Table 7.5)

Tabanidae and Glossinidae. The *Tabanus* and *Glossina* species can be differentiated based on hydrocarbon composition (Hoppe *et al.*, 1990; Sutton and Carlson, 1997a, b). Interestingly, di- and tri-methylalkanes are good species markers in females (Sutton and Carlson, 1997b) and have also been shown to act as sex pheromones (Carlson *et al.*, 1978). In males, alkanes and di-or tri-methylalkanes differ between species (Carlson and Schlein,

Table 7.5 *Main hydrocarbons used in taxonomy for Brachycera (Diptera).*

Family	Genera	Main hydrocarbons	Species / Population	Discrimination based on	References
Tabanidae	*Tabanus*	*n*-alkanes, methyl-alkanes (C_{21} to C_{35})	11 species *T. nigrovittatus* complex	general hydrocarbon pattern	Hoppe *et al.* (1990) Sutton and Carlson (1997a)
Glossinidae	*Glossina*	*n*-alkanes, mono-, di- and tri-methylalkanes (26C to 39C) (females)	4 species	general hydrocarbon pattern + di- and tri-methylalkanes (C_{31} to C_{39})	Sutton and Carlson (1997b)
		n-alkanes di-, tri-methylalkenes (males)	*G. morsitans* *G. pallipises* (subspecies)	19,23-dimethyl- tritriacont-1-ene 4,8,12-trimethyl- hexacos-25-ene	Carlson and Schlein (1991)
Muscidae	*Hydrotaea*	*n*-alkanes, *n*-alkenes, *n*-alkadienes, mono-, di- methylalkanes	*H. aenescens* (populations)	general hydrocarbon pattern	Carlson *et al.* (2001)
	Haematobia	*n*-alkanes, *n*-alkenes (C_{21} to C_{29})	*H. exigua* *H. irritans*	5-, 7-, 9-, 11- tricosene	Urech *et al.* (2005)
Calliphoridae	*Aldrichina* *Chrysomya, Lucilia* *Achoetandrus* *Boettcherisca* *Parasarcophaga*	*n*-alkanes, mono-, di-, methylalkanes	*Al. grahami,* *C. megacephala,* *L. sericata,* *Ac. rufifacies, B. peregrina,* *P. crassipalpis*	alkanes (C_{23}, C_{28}), mono-, di-, methylalkanes	Ye *et al.* (2007)
Tephritidae	*Anastrepha Ceratitis*	*n*-alkanes, monomethylalkanes	*A. suspense* *C. capitata*	2-monomethylalkanes (C_{29}, C_{31}) (larvae)	Sutton and Steck (1994, 2005)

1991). It has been speculated that male-specific methylalkanes, which have an antiaphrodisiac effect on males, are transmitted to females during mating and probably mask female sex pheromone (Langley *et al.,* 1987).

Muscidae. Analysis of hydrocarbons has also been a determinant for distinguishing between closely related species or populations in other Diptera. The relationship between the buffalo fly, *Haematobia exigua,* and the horn fly, *H. irritans* (Muscidae), was long controversial, with some considering the buffalo fly to be a subspecies of *H. irritans* and others as a separate species. Study of other characters along with CHC analysis leaves no doubt that they are separate species. *Haematobia exigua* is rich in 11- and 5-tricosene, while *H. irritans* is rich in 9- and 5-tricosene (Urech *et al.,* 2005). Hydrocarbon analysis has also been used to assist biological control by discriminating strains of the dump fly *Hydrotaea aenescens* (Carlson *et al.,* 2001).

Calliphoridae. One study in Calliphoridae indicated a good correlation between phylogeny and female hydrocarbon composition (Roux *et al.,* 2006). Another interesting study used hydrocarbon analysis of pupal exuviae to discriminate between six necrophageous flies (Ye *et al.,* 2007) (Table 7.5). CHC composition included over 50 compounds (*n*-alkanes, monomethylalkanes, and dimethylalkanes), but eight were sufficient for clear-cut discrimination of these species.

Hemiptera

Triatomines. CHC determination has been applied to the taxonomy of Triatomines, vectors of Chagas disease. A particularly interesting finding is that *T. infestans* contains squalene, while *T. mazzotti* does not (Juarez and Blomquist, 1993). Various *Triatoma* and *Rhodnius* species also display different hydrocarbon patterns (Juarez and Fernandez, 2007). *Triatoma dimidiata* populations from different geographic areas can be differentiated by hydrocarbon patterns. Populations living in the same geographical vicinity show more similarities in hydrocarbon profiles than populations originating from distant areas (Fernandez *et al.,* 2005a, b).

Hymenoptera

Formicidae. Inadvertently introduced ants can become major pests. Fire ants are extremely aggressive insects that inject powerful venom, causing occasional pain and allergic reaction in humans. They are dangerous imported pests that have been able to spread to large areas of Australia with severe consequences, not only for the environment and outdoor lifestyle, but also for the agriculture and tourism industries. Two South American species of fire ant, *Solenopsis richteri* and *S. invicta* (RIFA), were imported into the southeastern United States at the beginning of the last century. For many years *Solenopsis richteri,* which is black, was thought to either be a subspecies or color variation of *Solenopsis invicta,* which is red. *Solenopsis richteri* and *Soleopsis invicta* are now recognized as separate species with two different hydrocarbon patterns (Nelson *et al.,* 1980). *S. invicta* is more poisonous than *S. richteri,* which appears to be less invasive than *Solenopsis invicta,* certainly because it is less tolerant to cold, and less expansive and dominant in behavior. One study showed that natural hybridization is possible, and that hybrid ants show an intermediate pattern of cuticular compounds from the parent ones (Vander Meer *et al.*, 1985; Vander Meer and Lofgren, 1989).

Apidae. The aggressive behavior of the Africanized bee *Apis mellifera scutelata,* that constitutes a danger for man and domestic animals (Lavine and Carlson, 1987), raised the need to develop a method to rapidly distinguish *scutelata* from the European honeybee *A. mellifera mellifera.* Lavine *et al.* (1988, 1989, 1990) and Lavine and Vora (2005) showed that analysis of hydrocarbon patterns allowed discrimination of the two species and their hybrids.

Non-insects

Ticks. Chemical taxonomy has been used to classify tick species. Findings show a broad range of morphological variation that is further complicated by hybridization between species. All species present hydrocarbon profiles comprising a mixture of *n*-alkanes, methyl-alkanes and a few alkenes (Estrada-Peña *et al.*, 1992a). Although they can vary with age and physiological state, species-specific hydrocarbon patterns have been identified in the four *Rhipicephalus* species (Estrada- Peña *et al.*, 1992a, b) and in two closely related *Argas* species (Estrada-Peña and Dusbabek, 1993).

Chemotaxonomy, biodiversity and bioindicators

Lepidoptera

Like other insects, lepidopterans bear cuticular hydrocarbons to protect them against desiccation. These hydrocarbons can be used as taxonomic markers (Table 7.6). In the *Danaus* genus, *D. erippus* and *D. plexippus* show reproductive isolation. Some taxonomists view them as separate species, but others consider *D. erippus* as a subspecies of *D. plexippus*. Hybridization experiments showing prezygotic and postzygotic isolation, as well as differences in CHC chromatograms, strongly support the hypothesis that the two species are separate (Hay-Roe *et al.*, 2007).

Coleoptera

Several studies using CHC analysis have been carried out in Coleoptera. One study involved three subspecies of the tiger beetle *Neocicindela perhispida*, from different ecological beach areas, which were recognized on the basis of dorsal coloration. Successful differentiation was achieved on the basis of the cuticular wax composition correlated with water loss rates (Hadley and Savill, 1989). In another study (Hadley and Schultz, 1987), water loss was correlated with CHC variations in three species (eight populations) of *Cicidela* (*C. oregona maricopa, C. o. navajoensis, C. tranquebarica admiscens, C. t. kirbyi, C. obsoleta santaclarae*) living in different counties of Arizona and Nevada. For further detail, see the paragraph on Chrysomelidae chemical differentiation in the next section.

Table 7.6 *Hydrocarbons used in chemotaxonomy for potential insect biodiversity and bioindicator purposes.*

Orders	Species	Identification	References
Lepidoptera	*Danaus erippus & D. plexippus*	CHCs	Hay-Roe *et al.*, 2007
Coleoptera	*Neocicindela perhispida Cincidela spp.*	CHCs	Hadley and Savill, 1989 Hadley and Schultz, 1987
Dictyoptera	*Cryptocercus punctatus*	CHCs	Everaerts *et al.*, 2008
Social Hymenoptera			
Vespidae	*Polistes spp.*	venoms	Bruschini *et al.*, 2007
Formicidae	*Pachycondyla villosa* complex	CHCs	Lucas *et al.*, 2002
	Tetramorium spp.	CHCs	Steiner *et al.*, 2002
	Cataglyphis (bicolor) spp.	PPG /Dufour gl.	Oldham *et al.*, 1999; Gökçen *et al.*, 2002; Keegans *et al.*, 1992; Dahbi *et al.*, 2008
	Leptothorax spp.	CHCs	Provost *et al.*, 2009
	Myrmica spp.	CHCs	Elmes *et al.*, 2002
	Fungus-growing ants (*Atta colombica, Acromyrmex echinatior, Sericomyrmex*)	CHCs	Richard *et al.*, 2007
Apidae	*Apis mellifera* complex	CHCs / venoms	McDaniel *et al.*, 1984
	Melipona spp.	Wax	Pianaro *et al.*, 2007
	Bombus spp.	CHCs	Hadley *et al.*, 1981; Dronnet *et al.*, 2005
	Bombus spp.	CHCs / Dufour gl.	Oldham *et al.*, 1994
Isoptera			
Families	Termitidae, Kalotermitidae, Rhinotermitidae	CHCs	Haverty *et al.*, 1990, 1996b, 1997, 2000, 2005; Thorne *et al.*, 1994
Species	*Zootermopsis spp.*	CHCs	Blomquist *et al.*, 1979; Haverty *et al.*, 1988; Thorne *et al.*, 1993
	Drepanotermes spp.	CHCs	Brown *et al.*, 1996a,b
	Macrotermes spp.	CHCs	Bagine *et al.*, 1994

Dictyoptera

The *Cryptocercus punctatus* wood-feeding cockroach complex in the eastern United States was recently revised based on cuticular hydrocarbon profiles, chromosome count and DNA sequencing (16S and COII mitochondrial genes and ITS nuclear gene). Findings showed that there were five distinct hydrocarbon phenotypes that were not totally congruent with chromosome count. This finding is almost certainly related to evolution. Some congruency between molecular clades and hydrocarbon phenotypes could be seen on the first axis of a PCA hydrocarbon analysis plot (82% of total variance), with the different mitotypes on the combined COII-16S tree and unrooted ITS tree (Everaerts *et al.*, 2008).

Social Hymenopterans

Hydrocarbon analysis has not been extensively used for chemotaxonomy in social hymenopterans. Most studies have focused on the physiological functions of CHCs or on chemical variations in their numerous glands.

*Vespidae.*Although the role of CHCs as recognition cues has been intensively studied in a dozen *Polistes* wasp species (Bonavita-Cougourdan *et al.*, 1991; Lorenzi *et al.*, 1996, 1997; Bagnères *et al.*, 1996a; Dani, 2006), as well as in the Vespula wasp (Butts *et al.*, 1993), chemotaxonomy based on venom volatile chemical profiles was reported recently (Bruschini *et al.*, 2007). There is currently no review paper comparing CHCs from the different social wasps studied.

*Formicidae.*Only a few Formicidae taxa have been investigated, one being the *Pachycondyla villosa* species complex, which has undergone multidisciplinary investigation including analysis of variations in CHC chain length and patterns according to taxa. Findings support the existence of three valid species (Lucas *et al.*, 2002). Steiner *et al.* (2002) reported the chemotaxonomic implications of data obtained by analysis of CHC samples from 63 nests of five different species of *Tetramorium* ants from Central Europe. Self-organizing maps (SOM) were used to classify GC–MS results into different phenotypes and demonstrated the need to revise the classification of two species, *T. caespitum* and *C. impurum*. Studies have been carried out in the *Cataglyphis* (*bicolor*) species group, to assess species identification by analysis of volatile Dufour gland secretion from workers of different colonies, vs. analysis of hydrocarbons from the post-pharyngeal gland (PPG). Findings showed hydrocarbons to be less reliable in some cases (Oldham *et al.*, 1999) and better or complementary in other cases (Gökçen *et al.*, 2002; Keegans *et al.*, 1992·). The utility of Dufour gland secretions as chemotaxonomic markers has been the focus of numerous papers, in particular those from the E. D. Morgan group at Keele University in the United Kingdom, but postpharyngeal gland content, because of its similarity in composition to cuticular lipids, has also been used to separate species (Bagnères and Morgan, 1991; Bagnères *et al.*, 1991; Dahbi *et al.*, 1996; see also Chapter 5 for review of the role of PPG). Recently, Dahbi *et al.* performed a study on ants from one of the more complicated groups of the genus, the *Cataglyphis bicolor* group, in Morocco and Burkina Faso (Dahbi *et al.*, 2008). They analyzed the contents of the postpharyngeal and Dufour glands, and identified several rare methyl alkenes in the PPG of one of the species. Based on their findings, the authors suggested that, unlike the Iberian *Cataglyphis,* the Dufour gland content would be a better phylogenetic marker than postpharyngeal gland content. They also concluded that the bicolor group has evolved into different species in the north and south of the Sahara (Dahbi *et al.*, 2008). One thesis (Provost, 1994) devoted a chapter to the identification of Leptotoracine ant species by CHCs. One of us (Anne Geneviève Bagneres) performed this chemical identification recently, the results of which have been published in a review on different CHC roles by Provost *et al.* (2008). Elmes *et al.* (2002) analyzed the cuticular hydrocarbons of five common *Myrmica* species to gain insight into host specificity by *Maculinea* butterflies. Findings showed that the cuticular blends of *Myrmica rubra, M. ruginodis, M. sabuleti, M. scabrinodis,* and *M. schrenki* mixtures were highly distinctive, and that the butterfly is able to follow host specificity within the full range of European *Maculinea* (see also Chapter 14 on chemical mimicry). Richard *et al.* (2007) compared the chemical profiles of three different fungus-growing ants, *Atta colombica, Acromyrmex echinatior,* and *Sericomyrmex amabilis,* showing well-differentiated specific chemical signatures that allowed discrimination at the species level.

Apidae. Since an early paper by McDaniel *et al.* (1984) devoted to identification and preliminary evaluation of the CHCs, sting apparatus and sting shaft description of *Apis mellifera*, few chemotaxonomic studies have been performed in Apidae. As for other insects, however, the impact of imported and invasive species has led to a renewal of interest in Apidae chemosystematics (see previous section). Honeybee lipids have also been studied within the framework of research on honeybee parasites such as *Varroa destructor*. These studies showed that European honeybee lipids can be distinguished from those of the Asian bee *A. cerana* resistant to *Varroa* (Bagnères *et al.*, 2006b). Pianaro *et al.* (2007) compared wax constituents produced by colony workers of *Melipona scutellaris* and *M. rufiventris,* two Meliponinae bees, and showed numerous unsaturated compounds in both species, but with more long-chain dienes in *M. rufiventris* and more long-chain esters in *M. scutellaris*, thus giving clear specificity between those closed stingless honeybee species.

Analysis of cuticular lipids has also been carried out in several Bombinae species, i.e., *Bombus appositus* and *B. occidentalis* (Hadley *et al.*, 1981) and *B. terrestris terestris, B. t. audax, B. pascuorum floris, B. pratorum,* and *B. lapidaries* (Oldham *et al.*, 1994). The study by Oldham *et al.* showed a close correspondence between hydrocarbons from the Dufour gland and CHCs. Each bumble bee species showed a characteristic composition that did not vary in function over time or with place. Another interesting hydrocarbon analysis study, aimed at improving understanding of the inquilinism phenomenon, was performed on *B. sylvestris*, an obligate social inquiline, and *B. pratorum* its host (Dronnet *et al.*, 2005). As observed in other social parasite models, chemotaxonomy was of limited value, since some parasites are able to mimic the host so well that their own signature is either suppressed or absent (see Chapter 14 for a review).

Isoptera

Analysis of hydrocarbons in Isoptera has been performed for a number of reasons. Some ecological taxa, such as the arboreal nesting *Nasutitermes* genus, have been studied with the aim of standardizing sampling methodology for taxonomic and ecological studies in tropical ecosystems (Haverty *et al.*, 1996b). Taxonomy and biogeographical studies using various features, including hydrocarbons, have been carried out in different species on different continents. Another reason for studying Isoptera has been to assist management of tropical termites in regions where lack of taxonomic data on species considered as pests in agrosystems or forestry has been a major impediment. The combined data from these studies indicate that species are numerous and many are poorly or not at all described (http://www. chem.unep.ch/pops/termites/termite_ch4.htm). Findings have also shown that termites can have positive impact on ecosystems. Extensive surveys of tropical termites using hydrocarbons for taxonomic diagnosis have been performed in the British Virgin Islands (Haverty *et al.,* 1990b, 1997) and in Central America and the Caribbean (Thorne *et al.,* 1994). Total mixtures were characterized for each taxa, i.e. three species of Kalotermitidae, one species of Rhinotermitidae and five species of Termitidae. Only one of the Kalotermitidae, i.e., *Incisitermes* spp., could not be well characterized. One of the Termitidae, i.e., *Nasutitermes costalis*, displayed a rarely observed termite profile characterized by an enormous peak (13,17-dimethyl C_{31}) forming more than 40% of the total mixture. Another study from the Hawaiian islands (Haverty *et al.,* 2000) used CHCs to differentiate seven species belonging to three families and indicated that some had been imported. For identification of endemic vs. introduced taxa, CHC analysis showed correlation of habitat requirements (subterranean, drywood and dampwood) and environmental conditions (moist vs. dry area) with the presence or absence of olefins (unsaturated alkanes), variations in the quantity of methylbranches, chain-length of alkanes, etc. Recent work using the same species from the British Virgin Islands and Hawaiian Islands used hydrocarbons in feces to identify species in wooden structures without the presence of termites (Haverty *et al.,* 2005).

Hydrocarbon analysis of various dampwood termite species endemic to forests on the west coast of the USA confirmed phenotypes including *Z. angusticolli* and *Z. laticeps,* but clearly subdivided *Zootermopsis nevadensis* into two chemotypes (Blomquist *et al.,* 1979; Haverty *et al.,* 1988; Thorne *et al.,* 1993). The first attempts to identify these chemotypes using allozyme markers failed (Korman *et al.,* 1991), but more recent studies successfully discriminated two subspecies, i.e., *Z. nevadensis nevadensis* and *Z. n. nuttingi* (Aldrich and Kambhampati, 2007; Aldrich *et al.,* 2007), thus confirming the four cuticular phenotypes as species-specific. In Australia, Brown *et al.* (1996a, b) carried out chemotaxonomic surveys using cuticular hydrocarbons as taxonomic characters in *Drepanotermes* termites. In Africa, several species including *Macrotermes* (Isoptera, Macrotermitidae) have been identified using CHCs (Bagine *et al.,* 1994).

Chemotaxonomy, phylogeny and speciation

It is tempting to speculate that differences in hydrocarbon phenotypes are correlated with speciation. These variations have been shown to be important for mating or species recognition in a number of species (Drosophilidae and Culicidae in Diptera, Isoptera, etc.). However, it cannot be ruled out that these differences act on the quantity of a specific hydrocarbon and are due to environmental factors affecting hydrocarbon production. In some cases, hydrocarbon production seems to be inherited, suggesting that external factors may not be the only parameters involved. Hybridism and, on the contrary, sexual isolation, can be used to illustrate that chemotaxonomy has an evolutionary basis. This section presents different examples in various orders to illustrate this issue, i.e., Diptera, Isoptera, Orthoptera, and Coleoptera.

Diptera. A possible explanation for the 7-T/7-P gradient observed in *Drosophila mela-nogaster* populations is a decrease in temperature when moving away from the equator. However, the fact that hydrocarbon profiles do not change when the same populations are reared in the laboratory indicates that the 7-T/7-P ratio is now genetically determined. It should be noted that an evolutionary event, such as this, has never been induced in the laboratory, probably because evolution takes long periods of time to proceed. According to Wu and Ting (2004), "in the genic view of speciation, speciation genes are those that contribute to reproductive isolation, often in the form of hybrid invariability, sterility or behavioral aberration. This definition can include genes that cause isolation owing to physiological, behavioral or even ecological factors."

As described in the first section of this chapter, the geographic polymorphism observed in female *D. melanogaster* (5,9-HD in African females and 7,11-HD in cosmopolitan females) has led to two ecologically/behaviorally different races considered by some authors as incipient species (Wu *et al.,* 1995; Hollocher *et al.,* 1997). The 5,9-HD phenotype has been linked to a desaturase gene, i.e., *desat2*, in African populations. The 7,11-HD phenotype is due to the inactivation of the *desat2* gene caused by a small deletion in the promoter (Dallerac *et al.,* 2000; Takahashi *et al.,* 2001). Based on these findings, *desat2* must be considered as a speciation gene.

According to the definition of Wu and Ting (2004), *desatF* can also be considered as a speciation gene. It is expressed in *D. melanogaster* females but not in *D. simulans* and is responsible for the synthesis of 7,11-HD and 5,9-HD (Chertemps *et al.,* 2006). Since these dienes have a repellent effect (Coyne *et al.,* 1994; Coyne and Oyama, 1995; Savarit *et al.,* 1999), on *D. simulans* males, they rarely court and even more rarely mate with *D. mela-nogaster* females. Similar findings have been observed with the rare hybrid females obtained by crossing the two species. *D. simulans* males rarely court or mate with hybrids except when they lack a functional *desatF*, in which case courting and mating are the same as for *D. simulans* females (Legendre *et al.,* 2008). Although *desatF* cannot be the sole cause of reproductive isolation between *D. melanogaster* and *D. simulans*, it certainly contributes to prezygotic isolation. Further studies are needed to characterize other genes involved in pheromone biosynthesis and speciation in Drosophila.

Isoptera. In Europe, two species, i.e. *Reticulitermes grassei* and *R. banyulensis,* are derived from an ancestral reliquary taxon in southern Spain (Kutnik *et al.,* 2004). *R. grassei* is the main Iberic species and closest to the ancestral form. *R. banyulensis* is a sister clade differing from the main species with regard to several parameters, including higher aggressivity, more colonies showing a simple family structure, low polymorphism of populations, and different geographical distribution (eastern location). As shown in Figure 7.3, these two species also exhibit different hydrocarbon profiles. In *R. grassei,* C_{27} is a major peak and C_{25} and C_{29} are minor peaks. In *R. banyulensis,* C_{25} and C_{27} are equivalent, with inhibition of C_{29} chains and amplification of C_{25} chains. Interestingly, the ITS haplotypes follow the chemical differentiation (Kutnik *et al.,* 2004). Study of mitochondrial genes, which are more ancestral and variable in *Reticulitermes* termites, gives a better indication of the paleogeographical history of the two species. A multimarker analysis (Lefebvre *et al.,* in prep.) is under way to gain further insight into this complex evolution rarely described in social insects. Nevertheless it is likely that CHCs, by their role in species recognition, intervene in reproductive isolation of these two species, as has been demonstrated for the *Danaus* butterflies (see above).

Orthoptera. An interesting recent example to illustrate the utility of CHC analysis for chemotaxonomy, phylogeny and speciation involves the Hawaiian swordtailed crickets *Laupala.* Study of cuticular hydrocarbon profiles (Mullen *et al.,* 2007, 2008) demonstrated rapid recent development of seven species. Long-distance isolation appeared to involve acoustic signals, while short-distance isolation required contact pheromone differences. Biogeographic differences in the hydrocarbon mixtures of lineages from two islands (Big Island and Maui) seemed to be strongly correlated with a founder effect inducing a loss of biochemical variation (due mainly to a decrease in di- and tri-unsaturated hydrocarbons). The presence of two major clades (one with two species and the other with three) on the Big Island seemed to result more from a post-colonization process and reinforcement of pheromonal sexual selection, as in *Chrysochus* (see next paragraph).

Coleoptera. A recent work on chrysomelidae (Peterson *et al.,* 2007) evaluated the evolution of sexual isolation between two leaf beetles, *Chrysochus cobaltinus* and *C. auratus,* in a hybrid zone in Washington state (USA). By painting beetle cadavers with various cuticular extracts, the authors demonstrated a strong male preference for conspecific females according to species and sexual chemical specificity of their respective cuticular hydrocarbon profiles. This male mate choice reinforced sexual isolation.

Conclusion

Numerous populations or species can be recognized based on hydrocarbon patterns. However, even though hydrocarbon analysis may be informative and, sometimes, the only method available, it cannot be applied to all species or populations. In many cases, a combination of several different techniques is necessary for discrimination. Hydrocarbon composition can be affected by various external factors, e.g. food, temperature, and humidity, as well as social environment (Bagnères *et al.,* 2006b; Bonavita-Cougourdan *et al.,* 1996, 1997; Liang and Silverman, 2000) and physiology resulting in sexual dimorphisms (Mpuru *et al.,* 2001). Thus, although hydrocarbon analysis is a prodigious tool for recognizing and understanding insects, caution is necessary in interpreting hydrocarbon data. The technique should be used in combination with other analyses such as ethology, ecology, and genetics.

An original use for chemotaxonomy is forensic entomology. Some laboratories are using hydrocarbon analysis to identify insect larvae and to calculate developmental stage of necrophagous Dipteran larvae. Other original studies involved identification of cuticular compounds for comparison of glacially preserved acridian *Melanoplus* with modern species such as *M. sanguinipes* and *M. spretus* (Sutton *et al.,* 1996) and diverse insects, other arthropods and plants (Gupta *et al.,* 2006; Stankiewicz *et al.,*

1998). Specific cuticular mixtures have also been used in robot cockroach tests (Halloy *et al.,* 2007), and this application could be extended to various species for a futuristic approach to pest control.

We can surely embrace the future directions given by Symonds and Elgar (2007) in their review of the evolution of pheromone diversity. In particular initiating phylogenetic comparative studies of pheromone evolution should be a priority in order to remedy our poor understanding of it.

References

Alarie, Y., Joly, H. and Dennie, D. (1998). Cuticular hydrocarbon analysis of the aquatic beetle *Agabus anthracinus* Mannerheim (Coleoptera: Dytiscidae). *Can. Entomol.*, **130**, 615–629.

Akino, T. (2006). Cuticular hydrocarbons of *Formica truncorum* (Hymenoptera: Formicidae): Description of new very long chained hydrocarbon components. *Appl. Entomol. Zool.*, **41**, 667–677.

Aldrich, B. T. and Kambhampati, S. (2007). Population structure and colony composition of two *Zootermopsis nevadensis* subspecies. *Heredity*, **99**, 443–451.

Aldrich, B. T., Maghirang, B. B., Dowell, F. E. and Kambhampati, S. (2007). Identification of termite species and subspecies of the genus *Zootermopsis* using near-infrared reflectance spectroscopy. *J. Insect Sci.*, **7**, 18.

Anyanwu, G. I., Davies, D. H., Molyneux, D. H. and Phillips, A. (1997). Variation in cuticular hydrocarbons among strains of *Anopheles* (Cellia) *stephensi* Liston possibly related to prior insecticide exposure. *Ann. Trop. Med. Parasitol.*, **91**, 649–659.

Anyanwu, G. I., Davies, D. H., Molyneux, D. H., Phillips, A. and Milligan, P. J. (1993). Cuticular hydrocarbon discrimination/variation among strains of the mosquito, *Anopheles* (Cellia) *stephensi* Liston. *Ann. Trop. Med. Parasitol.*, **87**, 269–275.

Anyanwu, G. I., Davies, D. H., Molyneux, D. H. and Priestman, A. (2001). Cuticular hydrocarbon discrimination between *Anopheles gambiae* s.s. and *An. arabiensis* larval karyotypes. *Ann. Trop. Med. Parasitol.*, **95**, 843–852.

Anyanwu, G. I., Molyneux, D. H. and Phillips, A. (2000). Variation in cuticular hydrocarbons among strains of the *Anopheles gambiae* sensu stricto by analysis of cuticular hydrocarbons using gas liquid chromatography of larvae. *Mem. Inst. Oswaldo Cruz*, **95**, 295–300.

Anyanwu, G. I., Phillips, A. and Molyneux, D. H. (1994). Variation in the cuticular hydrocarbons of larvae of *Anopheles gambiae* and *A. arabiensis*. *Insect Sci. Applic.*, **15**, 117–122.

Austin, J. W., Bagnères, A.-G., Szalanski, A. L., Scheffrahn, R. H., Heintschel, B. P, Messenger, M. T., Clément, J.-L. and Gold, R. E. (2007). *Reticulitermes malletei* (Isoptera: Rhinotermitidae): a valid Nearctic subterranean termite from Eastern North America. *Zootaxa*, **1554**, 1–26.

Bagine, R. K. N., Brandl, R. and Kaib, M. (1994). Species delimitation in *Macrotermes* (Isoptera: Macrotermitidae): Evidence from epicuticular hydrocarbons, morphology, and ecology. *Ann. Entomol. Soc. Am.*, **87**, 498–506.

Bagnères, A. G. (1989). Les hydrocarbures cuticulaires des insects sociaux: Détermination et rôle dans la reconnaissance spécifique, coloniale et individuelle, PhD thesis, University Pierre and Marie Curie, Paris 6-UPMC, 151p.

Bagnères, A.-G., Clément, J.-L., Blum, M. S., Severson, R. F., Joulie, C and Lange, C. (1990b). Cuticular hydrocarbons and defensive compounds of *Reticulitermes flavipes* (Kollar) and *R. santonensis* (Feytaud): Polymorphism and chemotaxonomy. *J. Chem. Ecol.*, **16**, 3213–3244.

Bagnères, A.-G., Clément, J.-L., Lange, C. and Blum, M. S. (1990a). Cuticular compounds in *Reticulitermes* termites: species, caste and colonial signature. In *Social Insects and the Environment*, ed. G. K. Veeresh, B. Malik and C. A. Viraktamath. New Delhi: Oxford and IBH, p. 423.

Bagnères, A.-G., Clément J.-L., Lange, C. and Joulie, C. (1988). Les Hydrocarbures cuticulaires des *Reticulitermes* français: variations spécifiques et coloniales. *Actes. Coll. Insect Soc.*, **4**, 34–42.

Bagnères, A.-G., Huang, Z., Navajas, M., Salvy, M., Christides, J.-P., Zeng, Z. and Le Conte, Y. (2006b). Chemical mimicry by the ectoparasitic mite *Varroa destructor* infesting *Apis cerana* and *A. mellifera* broods. In *Proc. ISCE Meeting 2006, Barcelona*, p. 76.

Bagnères, A.-G., Killian, A., Clément, J.-L. and Lange, C. (1991b). Interspecific recognition among termites of the genus *Reticulitermes*: Evidence for a role for the hydrocarbons. *J. Chem. Ecol.*, **17**, 2397–2420.

Bagnères, A.-G., Lorenzi, M.-C., Dusticier, G., Turillazzi, S. and Clément, J.-L. (1996a). Chemical usurpation of a nest by paper wasp parasites. *Science*, **272**, 889–892.

Bagnères, A.-G. and Morgan, E. D. (1990). A simple method for analysis of insect cuticular hydrocarbons. *J. Chem. Ecol.*, **16**, 3263–3276.

Bagnères, A.-G and Morgan, E. D. (1991). The postpharyngeal glands and the cuticle of Formicidae contain the same characteristic hydrocarbons. *Experientia*, **47**, 106–111.

Bagnères, A.-G., Morgan, E. D. and Clément, J.-L. (1991). Species-specific secretions of the Dufour gland of three species of formicine ants (Hymenoptera: Formicidae). *Biochem. Syst. Ecol.*, **19**, 25–33.

Bagnères A.-G., Rivière, G. and Clément, J.-L. (1998). Artificial neural network modeling of caste odor discrimination based on cuticular hydrocarbons in termites. *Chemoecology*, **8**, 201–209.

Bagnères, A.-G., Uva, P. and Clément, J.-L. (2003). Description d'une nouvelle espèce de Termite : *Reticulitermes urbis* n.sp. (Isopt., Rhinotermitidae). *Bull. Soc. Entomol. Fr.*, **108**, 433–435.

Baker, G. L., Vroman, H. E. and Padmore, J. (1963). Hydrocarbons of the American cockroach. *Biochem. Biophys. Res. Comm.*, **13**, 360–365.

Bernier, U. R., Carlson, D. A. and Geden, C. J. (1998). Differentiating species of *Muscidifurax* by cuticular hydrocarbons identified by GC/MS. *J. Am. Soc. Mass Spectrom.*, **9**, 320–332.

Blomquist, G. J. and Dillwith, J. W. (1985). Cuticular lipids. In *Comprehensive Insect Physiology, Biochemistry and Pharmacology, Vol. 3, Integument, respiration and circulation*, ed. G. A. Kerkutand and L. Gilbert. Oxford: Pergamon, pp. 117–154.

Blomquist, G. J., Howard, R. W. and McDaniel, C. A. (1979). Structure of the cuticular hydrocarbons of the termite *Zootermopsis angusticullis* (Hagen). *Insect Biochem.*, **9**, 365–370.

Blomquist, G. J., Nelson, D. R. and deRenobales, M. (1987). Chemistry, biochemistry, and physiology of insect cuticular lipids. *Arch. Insect Biochem. Physiol.*, **6**, 227–265.

Blum, M.S. (1987). Specificity of pheromonal signals: A search for its recognitive bases in terms of a unified chemisociality. In *Chemistry and biology of social insects*, ed. J. Eder, and H. Rembold. Munich: Peperny.

Bonavita-Cougourdan, A., Bagnères, A.-G., Provost, E., Dusticier, G. and Clément, J.-L. (1997). Plasticity of the cuticular hydrocarbon profile of the slave-making ant *Polyergus rufescens* depending on the social environment. *Comp. Biochem. Physiol.*, **116B**, 287–302.

Bonavita-Cougourdan, A., Rivière, G., Provost, E., Bagnères, A.-G., Roux, M., Dusticier, G. and Clément, J.-L. (1996). Selective adaptation of the cuticular hydrocarbon profiles of the slave making ants *Polyergus rufescens* Latr. and their *Formica rufibarbis* and *F. cunicularia* Latr. slaves. *Comp. Biochem. Physiol.*, **113B**, 313–329.

Bonavita-Cougourdan, A., Theraulaz, G., Bagnères, A.-G., Roux, M., Pratte, M., Provost, E. and Clément, J.-L. (1991). Cuticular hydrocarbons, social organization and ovarian development in a Polistes Wasp: *Polistes dominulus* Christ. *Comp. Biochem. Physiol.*, **100B**, 667–680.

Brenner, R.J., Carlson, D.A., Rothe, L.M. and Paterson, R.S. (1993). Morphological and chemotaxonomic identification *of Blatella* cockroaches (Blattaria: Blattellidae) from Taiwan and selected Pacific basin locations. *Invertebrate taxonomy*, 7, 1205–1219.

Brown, W.V., Watson, J.A.L. and Lacey, M.J. (1994). The cuticular hydrocarbons of three Australian *Coptotermes* species, *C. michaelseni, C. brunneus,* and *C. dreghorni* (Isoptera: Rhinotermitidae). *Sociobiology*, **23**, 277–291.

Brown, W.V., Watson, J.A.L., Carter, F.L., Lacey, M.J., Barrett, R.A. and McDaniel, C.A. (1990). Preliminary examination of cuticular hydrocarbons of worker termites as chemotaxonomic characters for some Australian species of *Coptotermes* (Isoptera: Rhinotermitidae). *Sociobiology*, **16**, 181–197.

Brown, W.V., Watson, J.A.L. and Lacey, M.J. (1996a). A chemotaxonomic survey using cuticular hydrocarbons of some species of the Australian harvester termites genus *Drepanotermes* (Isoptera: Termitidae). *Sociobiology*, **27**, 199–221.

Brown, W.V., Watson, J.A.L., Lacey, M.J., Morton, R. and Miller, L.R. (1996b). Composition of cuticular hydrocarbons in the Australian harvester termite *Drepanotermes perniger* (Isoptera: Termitidae): Variation among individuals, castes, colonies and locations. *Sociobiology*, **27**, 181–197.

Broza, M., Nation, J.L., Milne, K. and Harrison, J. (2000). Cuticular hydrocarbons as a tool supporting recognition of *Gryllotalpa tali* and *G. marismortui* (Orthoptera: Gryllotalpidae) as distinct species in Israel. *Ann. Entomol. Soc. Am.*, **93**, 1022–1030.

Bruschini, C., Cervo, R., Dani, F.R. and Turillazzi, S. (2007). Can venom volatile be a taxonomic tool for *Polistes* wasps (Hymenoptera, Vespidae) *J. Zool. Syst. Evol. Res.*, **45**, 202–205.

Butts, D.P., Camann, M.A. and Espelie, K.E. (1993). Discriminant analysis of cuticular hydrocarbons of the baldfaced hornet, *Dolichovespula maculata* (Hymenoptera: Vespidae). *Sociobiology*, **21**, 193–201.

Caputo, B., Dani, F.R., Horne, G.L., N'Fale, S., Diabate, A., Turillazzi, S., Coluzzi, M., Costantini, C., Priestman, A.A., Petrarca, V. and della Torre, A. (2007). Comparative analysis of epicuticular lipid profiles of sympatric and allopatric field populations of *Anopheles gambiae* s.s. molecular forms and *An. arabiensis* from Burkina Faso (West Africa). *Insect Biochem. Mol. Biol.*, **37**, 389–398.

Carlson, D.A. (1988). Hydrocarbons for identification and phenetic comparisons: Cockroaches, honey bees and tsetse flies. *Florida Entomol.*, **71**, 333–345.

Carlson, D.A., Bernier, U.R., Hogsette, J.A. and Sutton, B.D. (2001). Distinctive hydrocarbons of the black dump fly, *Hydrotaea aenescens* (Diptera: Muscidae). *Arch. Insect Biochem. Physiol.*, **48**, 167–178.

Carlson, D.A. and Brenner, R.J. (1988). Hydrocarbon based discrimination of three North American *Blatella* cockroach species using gas chromatography. *Ann. Entomol. Soc. Am.*, **81**, 711–723.

Carlson, D.A., Langley, P.A. and Huyton, P. (1978). Sex pheromone of the tsetse fly: isolation, identification, and synthesis of contact aphrodisiacs. *Science*, **201**, 750–753.

Carlson, D.A., Reinert, J.F., Bernier, U.R., Sutton, B.D. and Seawright, J.A. (1997). Analysis of the cuticular hydrocarbons among species of the *Anopheles quadrimaculatus* complex (Diptera: Culicidae). *J. Am. Mosq. Control Assoc.*, **13** Suppl, 103–111.

Carlson, D.A. and Schlein, Y. (1991). Unusual polymethyl alkenes in tsetse flies acting as abstinon in *Glossina morsitans*. *J. Chem. Ecol.*, **17**, 267–284.

Carlson, D.A. and Service, M.W. (1979). Differentiation between species of the *Anopheles gambiae* Giles complex (Diptera: Culicidae) by analysis of cuticular hydrocarbons. *Ann. Trop. Med. Parasitol.*, **73**, 589–592.

Carlson, D.A. and Service, M.W. (1980). Identification of mosquitoes of *Anopheles gambiae* species complex A and B by analysis of cuticular components. *Science*, **207**, 1089–1091.

Chapman, R.F., Espelie, K.E. and Peck, S.B. (2000). Cuticular hydrocarbons of grass-hoppers from the Galapagos Islands, Ecuador. *Biochem. Syst. Ecol.*, **28**, 579–588.

Chapman, R.F., Espelie, K.E. and Sword, G.A. (1995). Use of cuticular lipids in grasshopper taxonomy: a study of variation in *Schistocerca shoshone* (Thomas). *Biochem. Syst. Ecol.*, **23**, 383–398.

Charlwood, J.D. and Jones, M.D.R. (1980). Mating in the mosquito *Anopheles gambiae* s.l. II. Swarming behavior. *Physiol. Entomol.*, **5**, 315–320.

Chertemps, T., Duportets, L., Labeur, C., Ueyama, M. and Wicker-Thomas, C. (2006). A female-specific desaturase gene responsible for diene hydrocarbon biosynthesis and courtship behaviour in *Drosophila melanogaster*. *Insect Mol. Biol.*, **15**, 465–473.

Chuah, C.-H. (2005). Interspecific variation in defense secretions of Malaysian termites from the genus *Bulbitermes*. *J. Chem. Ecol.*, **31**, 819–827.

Clément, J.-L., Bagnères, A.-G., Uva, P., Wilfert, L., Quintana, A., Reinhard, J. and Dronnet, S. (2001). Biosystematics of *Reticulitermes* termites in Europe. Morphological, chemical, molecular data. *Insect Soc.*, **48**, 202–215.

Clément, J.-L., Howard, R., Blum, M.S. and Lloyd, H. (1986). Isolement spécifique des termites du genre *Reticulitermes* du sud-est des Etats-unis. Mise en évidence grâce à la chimie et au comportement d'une espèce jumelle de *R. virginicus* : *R. malletei* sp. nov. et d'une semi-species de *R. flavipes*. *C. R. Acad. Sci.*, **302**, 67–70.

Clément, J.-L., Lange, C., Blum, M.S., Howard, R.W. and Lloyd, H. (1985). Chimiosystématique du genre *Reticulitermes* aux USA et en Europe. *Actes coll. Insect Soc.*, **2**, 123–131.

Clément, J.-L., Lemaire, M., Nagnan, P., Escoubas, P., Bagnères, A.-G and Joulie, C. (1988). Chemical ecology of European termites of the genus *Reticulitermes*: allomones, pheromones and kairomones. *Sociobiology*, **14**, 165–174.

Cobb, M. and Jallon, J.-M. (1990). Pheromones, mate recognition and courtship stimulation in the *Drosophila melanogaster* species sub-group. *Anim. Behav.*, **39**, 1058–1069.

Copren, K. A., Nelson, L. J., Vargo, E. L. and Haverty, M. I. (2005). Phylogenetic analyses of mtDNA sequences corroborate taxonomic designations based on cuticular hydrocarbons in subterranean termites. *Mol. Phyl. Evol.*, **35**, 689–700.

Coyne, J. A., Crittenden, A. P. and Mah, K. (1994). Genetics of a pheromonal difference contributing to reproductive isolation in *Drosophila. Science*, **265**, 1461–1464.

Coyne, J. A., Kim, S. Y., Chang, A. S., Lachaise, D. and Elwyn, S. (2002). Sexual isolation between two sibling species with overlapping ranges: *Drosophila santomea* and *Drosophila yakuba. Evolution Int. J. Org. Evolution*, **56**, 2424–2434.

Coyne, J. A. and Oyama, R. (1995). Localization of pheromonal sexual dimorphism in *Drosophila melanogaster* and its effect on sexual isolation. *Proc. Natl. Acad. Sci. USA,* **92**, 9505–9509.

Cvačka, J., Jiroš, P., Sobotník, J., Hanus, J. and Svatoš, A. (2006). Analysis of insect cuticular hydrocarbons using matrix-assisted laser desorption/ionization mass-spectrometry. *J. Chem. Ecol.*, **32**, 409–434.

Dahbi, A., Hefetz, A. and Lenoir, A. (2008). Chemotaxonomy of some *Cataglyphis* ants from Morocco and Burkina Faso. *Biochem. Syst. Ecol.*, **36**, 564–572.

Dahbi, A., Lenoir, A., Tinaud, A., Taghizadeh, T., Francke, W. and Hefetz, A. (1996). Chemistry of the postpharyngeal gland secretion and its implication for the phylogeny of the Iberian *Cataglyphis* species (Hymenoptera: Formicidae). *Chemoecology*, **7**, 163–171.

Dallerac, R., Labeur, C., Jallon, J.-M., Knipple, D. C., Roelofs, W. L. and Wicker-Thomas, C. (2000). A delta 9 desaturase gene with a different substrate specificity is responsible for the cuticular diene hydrocarbon polymorphism in *Drosophila melanogaster. Proc. Natl. Acad. Sci. USA*, **97**, 9449–9454.

Dani, F. (2006). Cuticular lipids as semiochemicals in paper wasps and other social insects. *Ann. Zool. Fennici*, **43**, 500–514.

Delphia, C. M., Copren, K. A. and Haverty, M. I. (2003). Agonistic behavior between individual worker termites from three cuticular hydrocarbon phenotypes of *Reticulitermes* (Isoptera: Rhinotermitidae) from northern California. *Ann. Entomol. Soc. Am.,* **96** 585–593.

Diabate, A., Dabire, R. K., Kengne, P., Brengues, C., Baldet, T., Ouari, A., Simard, F. and Lehmann, T. (2006). Mixed swarms of the molecular M and S forms of *Anopheles gambiae* (Diptera: Culicidae) in sympatric area from Burkina Faso. *J. Med. Entomol.*, **43**, 480–483.

Dronnet, S., Simon, X., Verhaeghe, J.-C., Rasmont, P. and Errard, C. (2005). Bumblebee inquilinism in *Bombus (Fernaldaepsithyrus) sylvestris* (Hymenoptera, Apidae): behavioural and chemical analyses of host-parasite interactions. *Apidologie*, **36**, 59–70.

Dronnet, S., Lohou, C., Christides, J.-P. and Bagnères, A.-G. (2006). Cuticular hydrocarbon composition reflects genetic relationship among colonies of the introduced termite *Reticulitermes santonensis* Feytaud. *J. Chem. Ecol.*, **32**, 1027–1042.

Elmes, G. W., Akino, T., Thomas, J. A., Clarke, R. T. and Knapp, J. J. (2002). Interspecific differences in cuticular hydrocarbon profiles of *Myrmica* ants are sufficiently consistent to explain host specificity by *Maculinea* (large blue) butterflies. *Oecologia*, **130**, 525–535.

Espelie, K. E., Berisford, C. W. and Dahlsten, D. L. (1990). Cuticular hydrocarbons of geographically isolated populations of *Rhopalicus pulchripennis* (Hymenoptera: Pteromalidae): evidence for two species. *Comp. Biochem. Physiol.*, **96B**, 305–308.

Estrada-Peña, A. and Dusbabek, F. (1993). Cuticular hydrocarbon gas chromatography analysis of *Argas vulgaris*, *A. polonicus*, and their hybrids. *Exp. Appl. Acarol.*, **17**, 365–376.

Estrada-Peña, A., Estrada-Peña, R. and Peiro, J. M. (1992a). Differentiation of *Rhipicephalus* ticks (Acari: Ixodidae) by gas chromatography of cuticular hydrocarbons. *J. Parasitol.*, **78**, 982–993.

Estrada-Peña, A., Gortazar, C. and Calvete, C. (1992b). Interspecific cuticular hydrocarbon variations and tentative hybrids of *Rhipicephalus sanguineus* and *R. pusillus* ticks (Acari: Ixodidae) in nature. *Ann. Parasitol. Hum. Comp.*, **67**, 197–201.

Etges, W. J. and Jackson, L. L. (2001). Epicuticular hydrocarbon variation in *Drosophila mojavensis* cluster species. *J. Chem. Ecol.*, **27**, 2125–2149.

Everaerts, C., Farine, J.-P. and Brossut, R. (1997). Changes of species cuticular hydrocarbons profiles in the cockroaches *Nauphoeta cinerea* and *Leucophaea maderae* reared in heterospecific groups. *Entomol. Exp. Applic.*, **85**, 145–150.

Everaerts, C., Maekawa, K., Farine, J.-P., Shimada K., Luykx, P., Brossut, R. and Nalepa, C. A. (2008). The *Crypotcercus punctatus* species complex (Dictyoptera: Cryptocercidae) in the eastern United States: comparison of cuticular hydrocarbons, chromosome number and DNA sequences. *Mol. Phyl. Evol.*, **47**, 950–959.

Fernandez, G. C., Juarez, M. P., Monroy, M. C., Menes, M., Bustamante, D. M. and Mijailovsky, S. (2005a). Intraspecific variability in *Triatoma dimidiata* (Hemiptera: Reduviidae) populations from Guatemala based on chemical and morphometric analyses. *J. Med. Entomol.*, **42**, 29–35.

Fernandez, G. C., Juarez, M. P., Ramsey, J., Salazar Schettino, P. M., Monroy, M. C., Ordonez, R. and Cabrera, M. (2005b). Cuticular hydrocarbon variability among *Triatoma dimidiata* (Hemiptera: Reduviidae) populations from Mexico and Guatemala. *J. Med. Entomol.*, **42**, 780–788.

Ferveur, J.-F., Cobb, M., Boukella, H. and Jallon, J.-M. (1996). World-wide variation in *Drosophila melanogaster* sex pheromone: behavioural effects, genetic bases and potential evolutionary consequences. *Genetica*, **97**, 73–80.

Ferveur, J.-F. and Jallon, J.-M. (1996). Genetic control of male cuticular hydrocarbons in *Drosophila melanogaster*. *Genet. Res.*, **67**, 211–218.

Gebre-Michael, T., Lane, R. P., Phillips, A., Milligan, P. and Molyneux, D. H. (1994). Contrast in the cuticular hydrocarbons of sympatric *Phlebotomus* (Synphlebotomus) females (Diptera: Phlebotominae). *Bull. Entomol. Res.*, **84**, 225–231.

Geden, C. J., Bernier, U. R., Carlson, D. A. and Sutton, B. D. (1998). Identification of *Muscidifurax* spp., parasitoids of muscoid flies, by composition patterns of cuticular hydrocarbons. *Biol. Control*, **12**, 200–207.

Gilby, A. R. and Cox, M. E. (1963). The cuticular lipids of the cockroach *Periplaneta americana* (L.). *J. Insect Physiol.*, **9**, 671–681.

Gökçen, O. A., Morgan, E. D., Dani, F. R., Agosti, D. and Wehner, R. (2002). Dufour gland contents of ants of the *Cataglyphis* bicolor group. *J. Chem. Ecol.*, **28**, 71–87.

Golden, K. L., Meinke, L. J. and Stanley-Samuelson, W. (1992). Cuticular hydrocarbon discrimination of *Diabrotica* (Coleoptera: Chrysomelidae) sibling species. *Ann. Entomol. Soc. Am.*, **85**, 561–570.

Grunshaw, J. P., Guermouche, H., Guermouche, S., Jago, N. D., Jullien, R., Knowles, E. and Perez, F. (1990). Chemical taxonomic studies of cuticular hydrocarbons in locusts of the *Schistocerca americana* complex (Acridae: Cyrtacanthacridinae): chemical relationships between New World and Old World species. *J. Chem. Ecol.*, **16**, 2835–2858.

Gupta, N. S., Michels, R., Briggs, D. E. G., Evershed, R. P. and Pancost, R. D. (2006). The organic preservation of fossil arthropods: an experimental study. *Proc. Royal Soc. B*, **273**, 2777–2783.

Hadley, N. F., Blomquist, G. J. and Lanham, U. N. (1981). Cuticular hydrocarbons of four species of Colorado hymenoptera. *Insect Biochem.*, **11**, 173–177.

Hadley, N. F. and Savill, A. (1989). Water loss in three subspecies of the New Zealand tiger beetle *Neocicindela perhispida*: correlation with cuticular hydrocarbons. *Comp. Biochem. Physiol.*, **94B**, 749–753.

Hadley, N. F. and Schultz, T. D. (1987). Water loss in three species of tiger beetles (*Cicindela*): correlations with epicuticular hydrocarbons. *J. Insect. Physiol.*, **33**, 677–682.

Halloy, J., Sempo, G., Caprari, G., Rivault, C., Asadpour, M., Tâche, F., Saïd, I., Durier, V., Canonge, S., Amé, J.-M., Detrain, C., Corell, N., Martinoli, A., Mondada, F., Siegwart, R. and Deneubourg, J.-L. (2007). Social integration of robots into groups of cockroaches to control self-organized choices. *Science*, **318**, 1155–1158.

Hamilton, J. G., Ward, R. D., Dougherty, M. J., Maignon, R., Ponce, C., Ponce, E., Noyes, H. and Zeledon, R. (1996). Comparison of the sex-pheromone components of *Lutzomyia longipalpis* (Diptera: Psychodidae) from areas of visceral and atypical cutaneous leishmaniasis in Honduras and Cost Rica. *Ann. Trop. Med. Parasitol.*, **90**, 533–541.

Hamilton, R. J. and Service, M. W. (1983). Value of cuticular and internal hydrocarbons for the identification of larvae of *Anopheles gambiae* Giles, *Anopheles arabiensis* Patton and *Anopheles melas* Theobald. *Ann. Trop. Med. Parasitol.*, **77**, 203–210.

Haverty, M. I., Collins, M. S., Nelson, L. J. and Thorne, B. L. (1997). Cuticular hydrocarbons of the termites of the British Virgin Islands. *J. Chem. Ecol.*, **23**: 927–964.

Haverty, M. I., Copren, K. A., Getty, G. M. and Lewis, V. R. (1999b). Agonistic behavior and cuticular hydrocarbon phenotypes of colonies of *Reticulitermes* (Isoptera: Rhinotermitidae) from northern California. *Ann. Entomol. Soc. Am.*, **92**, 269–277.

Haverty, M. I., Getty, G. M., Nelson, L. J. and Lewis, V. R. (2003). Flight phenology of sympatric populations of *Reticulitermes* (Isoptera: Rhinotermitidae) in northern California: Disparate flight intervals indicate reproductive isolation among cuticular hydrocarbon phenotypes. *Ann. Entomol. Soc. Am.*, **96**, 828–833.

Haverty, M. I., Grace, J. K., Nelson, L. J. and Yamamoto, R. T. (1996c). Intercaste, intercolony, and temporal variation in cuticular hydrocarbons of *Coptotermes formosanus* Shiraki (Isoptera: Rhinotermitidae). *J. Chem. Ecol.*, **22**, 1813–1834.

Haverty, M. I., Forschler, B. T. and Nelson, L. J. (1996a). An assessment of the taxonomy of *Reticulitermes* (Isoptera: Rhinotermitidae) from the southeastern United States based on cuticular hydrocarbons. *Sociobiology*, **28**, 287–318.

Haverty, M. I. and Nelson, L. J. (1997). Cuticular hydrocarbons of *Reticulitermes* (Isoptera: Rhinotermitidae) from northern California indicate undescribed species. *Comp. Biochem. Physiol.*, **118B**, 869–880.

Haverty, M. I. and Nelson, L. J. (2007). *Reticulitermes* (Isoptera: Rhinotermitidae) in Arizona: Multiple cuticular hydrocarbon phenotypes indicate additional taxa. *Ann. Entomol. Soc. Am.*, **100**, 206–221.

Haverty, M. I., Nelson, L. J. and Forschler, B. T. (1999a). New cuticular hydrocarbon phenotypes of *Reticulitermes* (Isoptera: Rhinotermitidae) from the United States. *Sociobiology*, **33**, 1–21.

Haverty, M. I., Nelson, L. J. and Page, M. (1990a). Cuticular hydrocarbons of four populations of *Coptotermes formosanus* Shiraki in the United States. Similarities and origins of introductions. *J. Chem. Ecol.*, **16**, 1635–1647.

Haverty, M. I., Nelson, L. J. and Page, M. (1991). Preliminary investigations of the cuticular hydrocarbons of *Reticulitermes* and *Coptotermes* (Isoptera: Rhinotermitidae) for chemosystematic studies. *Sociobiology*, **19**, 51–76.

Haverty, M. I., Page, M., Nelson, L. J. and Blomquist, G. J. (1988). Cuticular hydrocarbons of dampwood termites *Zootermopsis*: intra- and inercolony variation and potential as taxonomic characters. *J. Chem. Ecol.*, **14**, 1035–1058.

Haverty, M. I., Thorne, B. L. and Nelson, L. J. (1996b). Hydrocarbons of *Nasutitermes acajutlae* and a comparison of methodologies for sampling cuticular hydrocarbons of Carribean termites for taxonomic and ecological studies. *J. Chem. Ecol.*, **22**, 2081–2109.

Haverty, M. I., Thorne, B. L. and Page, M. (1990b). Surface hydrocarbon components of two species of *Nasutitermes* from Trinidad. *J. Chem. Ecol.*, **16**, 2441–2450.

Haverty, M. I., Woodrow, R. J., Nelson, L. J. and Grace, J. K. (2000). Cuticular hydro-carbons of the termites of the Hawaiian Islands. *J. Chem. Ecol.*, **26**, 1167–1191.

Haverty, M. I., Woodrow, R. J., Nelson, L. J. and Grace, J. K. (2005). Identification of termite species by the hydrocarbons in their feces. *J. Chem. Ecol.*, **31**, 2119–2151.

Hay-Roe, M. M., Lamas, G. and Nation, J. L. (2007). Pre- and post-zygotic isolation and Haldane rule effects in reciprocal crosses of *Danaus erippus* and *Danaus plexippus* (Lepidoptera: Danainae), supported by differentiation of cuticular hydrocarbons, establish their status as separate species. *Biol. J. Linn. Soc.*, **91**, 445–453.

Hirai, Y. and Kimura, M. T. (1997). Incipient reproductive isolation between two morphs of *Drosophila elegans* (Diptera: *Drosophilidae*). *Biol. J. Linn. Soc.*, **61**, 501–513.

Hollocher, H., Ting, C. T., Wu, M. L. and Wu, C. I. (1997). Incipient speciation by sexual isolation in *Drosophila melanogaster*: extensive genetic divergence without reinforcement. *Genetics*, **147**, 1191–1201.

Hoppe, K. L., Dillwith, J. W., Wright, R. E. and Szumlas, D. E. (1990). Identification of horse flies (Diptera: Tabanidae) by analysis of cuticular hydrocarbons. *J. Med. Entomol.*, **27**, 480–486.

Horne, G. L. and Priestman, A. A. (2002). The chemical characterization of the epicuticular hydrocarbons of *Aedes aegypti* (Diptera: Culicidae) *Bull. Entomol. Res.*, **92**, 287–294.

Howard, R. W. (1993). Cuticular hydrocarbons and chemical communication. In *Insect Lipids: Chemistry, Biochemistry and Biology*, ed. D. W. Stanley-Samuelson and D. R. Nelson. Lincoln, NE: University of Nebraska Press, pp. 179–226.

Howard, R. W. and Blomquist, G. J. (1982). Chemical ecology and biochemistry of insect hydrocarbons. *Annu. Rev. Entomol.*, **27**, 149–172.

Howard, R. W. and Blomquist, G. J. (2005). Ecological, behavioral, and biochemical aspects of insect hydrocarbons. *Annu. Rev. Entomol.*, **50**, 371–392.

Howard, R. W., McDaniel, C. A. and Blomquist, G. J. (1978). Cuticular hydrocarbons of the eastern subterranean termite, *Reticulitermes flavipes* (Kollar) (Isoptera: Rhinotermitidae). *J. Chem. Ecol.*, **4**, 233–245.

Howard, R. W., McDaniel, C. A., Nelson, D. R., Blomquist, G. J., Gelbaum, L. T. and Zalkow, L. H. (1982). Cuticular hydrocarbons of *Reticulitermes virginicus* (Banks) and their role as potential species- and caste-recognition cues. *J. Chem. Ecol.*, **8**, 1227–1239.

Inward, D., Beccaloni, G. and Eggleton, P. (2007). Death of an Order: a comprehensive molecular phylogenetic study confirms that termites are eusocial cockroaches. *Biol. Lett.*, **3**, 331–335.

Ishii, K., Hirai, Y., Katagiri, C. and Kimura, M. T. (2002). Sexual isolation and cuticular hydrocarbons in *Drosophila elegans*. *Heredity*, **87**, 392–399.

Jackson, L. L. (1970). Cuticular lipids of insects. II: Hydrocarbons of the cockroaches *Periplaneta australasiae*, *P. brunnea* and *P. fuliginosa*. *Lipids*, **5**, 38–41.

Jackson, L. L. (1972). Cuticular lipids of insects. IV: Hydrocarbons of the cockroaches *Periplaneta japonica* and *P. americana* compared to other cockroach hydrocarbons. *Comp. Biochem. Physiol.*, **41**, 331–336.

Jackson, L. L. and Bayer, G. L. (1970). Cuticular lipids of insects. *Lipids*, **5**, 239–246.

Jackson, L. L. and Blomquist, G. J. (1976). Insect waxes. In *Chemistry and Biochemistry of Natural Waxes*, ed. P. E. Kolattukudy. Amsterdam: Elsevier, pp. 201–233.

Jacob, J. and Hanssen, H. P. (1986). Distribution and variability of cuticular hydrocarbons within the Coleoptera. *Biochem. Syst. Ecol.*, **14**, 207–210.

Jallon, J.-M. (1984). A few chemical words exchanged by *Drosophila* during courtship and mating. *Behav. Genet.*, **14**, 441–478.

Jallon, J.-M. and David, J. (1987). Variations in cuticular hydrocarbons among the eight species of the *Drosophila melanogaster* subgroup. *Evolution*, **4**, 294–302.

Jallon, J.-M. and Péchiné, J.-M. (1989). Une autre race de chimique *Drosophila melanogaster* en Afrique. *C. R. Acad. Sci.*, **309**, 1551–1556.

Jallon, J.-M. and Wicker-Thomas, C. (2003). Genetic studies on pheromone production in Drosophila. Chapter 9 in *Insect Pheromone Biochemistry and Molecular Biology. Part 1. Biochemistry and Molecular Biology of Pheromone Production*, ed. G. J. Blomquist and R. G. Vogt, London: Elsevier, pp. 253–281.

Jenkins, T. M., Haverty, M. I., Basten, C. J., Nelson, L. J., Page, M. and Forschler, B. T. (2000). Correlation of mitochondrial haplotypes with cuticular hydrocarbon phenotypes of sympatric *Reticulitermes* species (Isoptera: Rhinotermitidae) from the southeastern United States. *J. Chem. Ecol.*, **26**, 1525–1542.

Jones, T. H., Moran, M. D. and Hurd, L. E. (1997). Cuticular extracts of five common mantids (Mantodea: Mantidae) of the Eastern United States. *Comp. Biochem. Physiol.*, **116B**, 419–422.

Juarez, M. P. and Blomquist, G. J. (1993). Cuticular hydrocarbons of *Triatoma infestans* and *T. mazzotti*. *Comp. Biochem. Physiol.*, **106B**, 667–674.

Juarez, M. P. and Fernandez, G. C. (2007). Cuticular hydrocarbons of triatomines. *Comp. Biochem. Physiol.*, **147A**, 711–730.

Kaib, M., Brandl, R. and Bagine, R. K. N. (1991). Cuticular hydrocarbon profiles: A valuable tool in termite taxonomy. *Naturwissenschaften*, **78**, 176–179.

Kamhawi, S., Molyneux, D. H., Killick-Kendrick, R., Milligan, P. J., Phillips, A., Wilkes, T. J. and Killick-Kendrick, M. (1987). Two populations of *Phlebotomus ariasi* in the Cevennes focus of leishmaniasis in the south of France revealed by analysis of cuticular hydrocarbons. *Med. Vet. Entomol.*, **1**, 97–102.

Keegans, S. J., Morgan, E. D., Agosti, D. and Wehner, R. (1992). What do the glands tell us about species? A chemical case study of *Cataphyphis* ants. *Biochem. Syst. Ecol.*, **20**, 559–572.

Kim, Y. K., Phillips, D. R., Chao, T. and Ehrman, L. (2004). Developmental isolation and subsequent adult behavior of *Drosophila paulistorum*. VI. Quantitative variation in cuticular hydrocarbons. *Behav. Genet.*, **34**, 385–394.

Kittayapong, P., Clark, J. M., Edman, J. D., Lavine, B. K., Marion, J. R. and Brooks, M. (1993). Survey of the *Anopheles maculatus* complex (Diptera: Culicidae) in peninsular Malaysia by analysis of cuticular lipids. *J. Med. Entomol.*, **30**, 969–974.

Kittayapong, P., Clark, J. M., Edman, J. D., Potter, T. L., Lavine, B. K., Marion, J. R. and Brooks, M. (1990). Cuticular lipid differences between the malaria vector and non-vector forms of the *Anopheles maculatus* complex. *Med. Vet. Entomol.*, **4**, 405–413.

Klochkov, S. C., Kozlovskii, V. I and Belyaeva, N. V. (2005). Caste and population specificity of termite cuticule hydrocarbons. *Chem. Nat. Compounds*, **41**, 1–6.

Korman, A. K., Pashley, D. P., Haverty, M. I. and LaFage, M. I. (1991). Allozymic relationships among cuticular hydrocarbon phenotypes of *Zootermopsis* species. *Ann. Entomol. Soc. Am.*, **84**, 1–9.

Kruger, E. L. and Pappas, C. D. (1993). Geographic variation of cuticular hydrocarbons among fourteen populations of *Aedes albopictus* (Diptera: Culicidae). *J. Med. Entomol.*, **30**, 544–548.

Kruger, E. L., Pappas, C. D. and Howard, R. W. (1991). Cuticular hydrocarbon geographic variation among seven North American populations of *Aedes albopictus* (Diptera: Culicidae). *J. Med. Entomol.*, **28**, 859–864.

Kutnik, M., Uva, P., Brinkworth, L. and Bagnères, A.-G. (2004). Phylogeography of two European *Reticulitermes* (Isoptera) species: the Iberian refugium. *Mol. Ecol.*, **13**, 3099–3113.

Lachaise, D., Harry, M., Solignac, M., Lemeunier, F., Benassi, V. and Cariou, M.-L. (2000). Evolutionary novelties in islands: *Drosophila santomea*, a new *melanogaster* sister species from Sao Tome. *Proc. Biol. Sci.*, **267**, 1487–1495.

Lange, C., Basselier, J.-J., Bagnères, A.-G., Escoubas, P., Lemaire, M., Lenoir, A., Clément, J.-L., Bonavita-Cougourdan, A., Trabalon, M. and Campan, M. (1989). Strategy for the analysis of cuticular hydrocarbon waxes from insects using gas chromatography with electron impact and chemical ionization. *Biomed. Environ. Mass. Spectrom.*, **18**, 787–800.

Langley, P. A., Huyton, P. M. and Carlson, D. A. (1987). Sex pheromone perception by males of the tsetse fly, *Glossina morsitans morsitans*. *J. Insect Physiol.*, **33**, 981–986.

Lavine, B. K. and Carlson, D. A. (1987). European bee or Africanized bee? *Analyt. Chem.*, **59**, 468–469.

Lavine, B. K., Carlson, D. A., Henry, D. and Jurs, P. C. (1988). Taxonomy based on chemical constitution: Differentiation of Africanized honey-bees from European honey-bees. *J. Chemometrics*, **2**, 29–37.

Lavine, B. K., Davidson, C., Vander Meer, R. K., Lahav, S., Soroker, V. and Hefetz, A. (2003). Genetic algorithms for deciphering the complex chemosensory code of social insects. *Chemometrics Intelligent Lab. Syst.*, **66**, 51–62.

Lavine, B. K. and Vora, M. N. (2005). Identification of Africanized honeybees. *J. Chromatogr.*, **1096A**, 69–75.

Lavine, B. K., Ward, A. J. I., Han, J. H., Smith, R.-K. and Taylor, O. R. (1990). Taxonomy based on chemical constitution: Differentiation of heavily Africanized honeybees from moderately Africanized honeybees. *Chemometrics Intelligent Lab. Syst.*, **8**, 239–243.

Lavine, B. K., Ward, A. J. I., Smith, R.-K. and Taylor, O. R. (1989). Application of gas chromatography/pattern recognition techniques to the problem of identifying Africanized honeybees. *Microchem. J.*, **39**, 308–316.

Lazzari, S. M. N., Swedenborg, P. D. and Jones, R. L. (1991). Characterization and discrimination of three *Rhopalosiphum* species (Homoptera: Acrididae) using cuticular hydrocarbons and cuticular surface patterns. *Comp. Biochem. Physiol.*, **100B**, 189–200.

Lefebvre, T., Châline, N. Limousin, D., Dupont, S. and Bagnères A.-G. (2008). From speciation to introgressive hybridization: the phylogeographic structure of an island subspecies of termite, *R. lucifugus corsicus*. *BMC Evolutionary Biology*, **8**, 38.

Legendre, A., Miao, X.-X., Da Lage, J.-L. and Wicker-Thomas, C. (2008). Evolution of a desaturase involved in female pheromonal cuticular hydrocarbon biosynthesis and courtship behavior in *Drosophila*. *Insect Biochem. Mol. Biol.*, **38**, 244–253.

Lemeunier, F., David, J.R., Tsacas, L. and Ashburner, M. (1986). The *melanogaster* species group. In *The Genetics and Biology of Drosophila,* Vol. 3, ed. M. Ashburner, H.L. Carson and J.N. Thompson Jr. London: Academic Press, pp. 147–256.

Leniaud, L., Pichon, A., Uva, P. and Bagnères, A-G. (2009a). Unicolonially in *Reticulitermes urbis*: a novel feature from a potentially invasive termite species. *Bull. Entomol. Res., ***99**, 1–10.

Leniaud, L., Dedeine, F., Pichon, A., Dupont, S. and Bagnères, A-G. (2009b) Geographical distribution, genetic diversity and social organization of a new European termite, *Reticulitermes urbis* (Isopetra: Rhinotermitidae). *Biological Invasions*. In press.

Liang, D., Blomquist, G.J. and Silverman J. (2001). Hydrocarbon-released nestmate aggression in the Argentine ant, *Linepithema humile*, following encounters with insect prey. *Comp. Biochem. Physiol.*, **129 B**, 871–882.

Liang, D. and Silverman, J. (2000). You are what you eat: Diet modifies cuticular hydrocarbons and nestmate recognition in the Argentine ant, *Linepithema humile*. *Naturwissenschaften*, **87**, 412–416.

Lo, N., Eldridge, R.H. and Lenz, M. (2006). Phylogeny of Australian *Coptotermes* (Isoptera, Rhinotermitidae) species inferred from mitochondrial COII sequences. *Bull. Entomol. Res.*, **96**, 433–437.

Lockey, K.H. (1976). Cuticular hydrocarbons of *Locusta, Schistocerca,* and *Periplaneta*, and their role in waterproofing. *Insect Biochem.*, **6**, 457–472.

Lockey, K.H. (1980). Insect cuticular hydrocarbons. *Comp. Biochem. Physiol.*, **65B**, 457–462.

Lockey, K.H. (1988). Lipids of the insect cuticle: Origin, composition and function. *Comp. Biochem. Physiol.*, **89B**, 595–645.

Lockey, K.H. (1991). Cuticular hydrocarbons of adult *Onymacris bicolor* (Haag) and *Onymacris boschimana* (Péringuey) (Coleoptera: Tenebrionidae). *Comp. Biochem. Physiol.*, **98B**, 151–163.

Lockey, K.H. (1992). Insect hydrocarbon chemotaxonomy: Cuticular hydrocarbons of adult and larval *Epiphysa* species (Blanchard) and adult *Onymacris unguicularis* (Haag) (Tenebrionidae: Coleoptera). *Comp. Biochem. Physiol.*, **102B**, 451–470.

Lockey, K.H. and Oraha, V.S. (1990). Cuticular lipids of adult *Locusta migratoria migratorioides* (R and F), *Schistocerca gregaria* (Forskal) (Acrididae) and other orthopteran species. II. Hydrocarbons. *Comp. Biochem. Physiol.*, **95B**, 721–744.

Lorenzi, M.-C., Bagnères, A.-G. and Clément J.-L. (1996). The role of cuticular hydrocarbons in social insects: Is it the same in paper wasps? In *Natural History and Evolution of Paper Wasps*, ed. S. Turillazzi and M.J. West-Eberhard. Oxford University Press, Chapter 10, pp. 178–189.

Lorenzi, M.-C., Bagnères, A-G., Clément, J.-L and Turillazzi, S. (1997). *Polistes biglumis bimaculatus* epicuticular hydrocarbons and nestmate recognition (Hymenoptera: Vespidae). *Insect Soc.*, **44**, 123–138.

Lucas, C., Fresneau, D., Kolmer, K., Heinze, J., Delabie, J.H. and Pho, D.B. (2002). A multidisciplinary approach to discriminating different taxa in the species complex *Pachycondyla villosa* (Formicidae). *Biol. J. Linn. Soc.*, **75**, 249–259.

Luyten, I. (1982). Variation intraspécifique et interspécifique des hydrocarbures cuticulaires chez *Drosophila simulans* et des espèces affines. *C. R. Acad. Sci. Paris*, **295**, 733–736.

Mafuyai, H. B., Phillips, A., Molyneux, D. H. and Milligan, P. (1994). Identification of the larvae of the *Simulium damnosum* complex from Nigeria by analysis of cuticular hydrocarbons. *Trop. Med. Parasitol.*, **45**, 130–132.

Mahamat, H. and Hassanali, A. (1998). Cuticular hydrocarbon composition analysis for taxonomic differentiation of phlebotomine sandfly species (Diptera: Psychodidae) in Kenya. *J. Med. Entomol.*, **35**, 778–781.

Marchand, R. P. (1984). Field observations on swarming and mating in *Anopheles gambiae* mosquitos in Tanzania. *Neth. J. Zool.*, **34**, 367–387.

Mas, F. and Jallon, J.-M. (2005). Sexual isolation and cuticular hydrocarbon differences between *Drosophila santomea* and *Drosophila yakuba*. *J. Chem. Ecol.*, **31**, 2747–2752.

McDaniel, C. A. (1990). Cuticular hydrocarbons of the Formosan termite *Coptotermes formosanus*. *Sociobiology*, **16**, 265–273.

McDaniel, C. A., Howard, R. W., Blomquist, G. J. and Collins, A. M. (1984). Hydrocarbons of the cuticle, sting apparatus, and sting shaft of *Apis mellifera* L. Identification and preliminary evaluation as chemotaxonomic characters. *Sociobiology*, **8**, 287–298.

Mpuru, S., Blomquist, G. J., Schal, C., Roux, M., Kuenzli, M., Dusticier, G., Clément, J.-L. and Bagnères, A.-G. (2001). Effect of age and sex on the production of internal and external hydrocarbons and pheromones in the Housefly, *Musca domestica*. *Insect Biochem. Mol. Biol.*, **31**, 139–155.

Mullen, S. P., Mendelson, T. C., Schal, C. and Shaw, K. L. (2007). Rapid evolution of cuticular hydrocarbons in a species radiation of acoustically diverse Hawaiian crickets (Gryllidae: Trigonidiinae: *Laupala*). *Evolution*, **61**, 223–231.

Mullen, S. P., Millar, J. G., Schal, C. and Shaw, K. L. (2008). Identification and characterization of cuticular hydrocarbons from a rapid species radiation of Hawaiian swordtailed crickets (Gryllidae: Trigonidiinae: *Laupala*). *J. Chem. Ecol.*, **34**, 198–204.

Neal, J. W., Leonhardt, B. A., Brown, J. K., Bentz, J. A. and Devilbiss, E. D. (1994). Cuticular lipids of greenhouse whitefly and sweetpotato whitefly type A and B (Homoptera: Aleyrodidae) pupal exuviae on the same hosts. *Ann. Entomol. Soc. Am.*, **87**, 609–618.

Nelson, D. R., Fatland, C. L., Howard, R. W., McDaniel, C. A. and Blomquist, G. J. (1980). Re-analysis of the cuticular methyl alkanes of *Solenopsis invicta* and *S. richteri*. *Insect Biochem.*, **10**, 409–418.

Nelson, L. J., Cool, L. G., Forschler, B. T. and Haverty, M. I. (2001). Correspondence of soldier defense secretion mixtures with cuticular hydrocarbon phenotypes for chemotaxonomy of the termite genus *Reticulitermes* in North America. *J. Chem. Ecol.*, **27**, 1449–1479.

Nelson, L. J., Cool, L. G., Solek, C. W. and Haverty, M. I. (2008). Cuticular hydrocarbons and soldier defense secretions of *Reticulitermes* in southern California: a critical analysis of the taxonomy of the genus in North America. *J. Chem. Ecol.*, **34**, 1452–1475.

Oldham, N. J., Billen, J. and Morgan, E. D. (1994). On the similarity of the Dufour gland secretion and the cuticular hydrocarbons of some bumblebees. *Physiol. Entomol.*, **19**, 115–123.

Oldham, N. J., Morgan, E. D., Agosti, D. and Wehner, R. (1999). Species recognition from postpharyngeal gland contents of ants of the *Cataglyphis bicolor* group. *J. Chem. Ecol.*, **25**, 1383–1393.

Ozaki, M., Wada-Katsumata, A., Fujikawa, K., Iwasaki, M., Yokohari, F., Satoji, Y., Nisimura, T. and Yamaoka, R. (2005). Ant nestmate and non-nestmate discrimination by a chemosensory sensillum. *Science*, **309**, 311–314.

Page, M., Nelson, L.J. Blomquist, G.J., and Seybold, S.J. (1997). Cuticular hydrocarbons as chemotaxonomic characters of pine engraver beetles (*Ips* spp.) in the *grandicollis* subgeneric group. *J. Chem. Ecol.*, **23**, 1053–1099.

Page, M., Nelson, L.J., Forschler, B.T. and Haverty, M.I. (2002). Cuticular hydrocarbons suggest three lineages in *Reticulitermes* (Isoptera: Rhinotermitidae) from North America. *Comp. Biochem. Physiol.*, **131B**, 305–324.

Page, M., Nelson, L.J., Haverty, M.I. and Blomquist, G.J. (1990a). Cuticular hydrocarbons of eight species of North American cone beetles, *Conophthorus* Hopkins. *J. Chem. Ecol.*, **16**, 1173–1197.

Page, M., Nelson, L.J., Haverty, M.I. and Blomquist, G.J. (1990b). Cuticular hydrocarbons as chemotaxonomic characters for bark beetles: *Dendroctonus ponderosae*, *D. jeffreyi*, *D. brevicomis*, and *D. frontalis* (Coleoptera: Scolytidae). *Ann. Entomol. Soc. Am.*, **83**, 892–901.

Pappas, C.D., Bricker, B.J., Christen, J.A. and Rumbaugh, S.A. (1994). Cuticular hydrocarbons of *Aedes hendersoni* Cockerell and *A. triseriatus* (Say). *J. Chem. Ecol.*, **20**, 1121–1136.

Park, Y.C., Kitade, O., Schwarz, M., Kim, J.P. and Kim, W. (2006). Intraspecific molecular phylogeny, genetic variation and phylogeography of *Reticulitermes speratus* (Isoptera: Rhinotermitidae). *Mol. Cells*, **21**, 89–103.

Peterson, M.A., Dobler, S., Larson, E.L., Juarez, D., Schlarbaum, T., Monsen, K.J. and Francke, W. (2007). Profiles of cuticular hydrocarbons mediate male mate choice and sexual isolation between hybridising *Chrysochus* (Coleoptera: Chrysomelidae). *Chemoecology*, **17**, 87–96.

Phillips, A., Le Pont, F., Desjeux, P., Broomfield, G. and Molyneux, D.H. (1990b). Separation of *Psychodopygus carrerai carrerai* and *P. yucumensis* (Diptera: Psychodidae) by gas chromatography of cuticular hydrocarbons. *Acta Trop.*, **47**, 145–149.

Phillips, A., Sabatini, A., Milligan, P.J.M., Boccolini, D., Broomsfield, G. and Molyneux, D.H. (1990a). The *Anopheles maculipennis* complex (Diptera: Culicidae): comparison of the hydrocarbon profiles determined in adults of five palaeartic species. *Bull. Entomol. Res.*, **80**, 459–464.

Phillips, A., Walsh, J.F., Garms, R., Molyneux, D.H., Milligan, P. and Ibrahim, G. (1985). Identification of adults of the *Simulium damnosum* complex using hydrocarbon analysis. *Trop. Med. Parasitol.*, **36**, 97–101.

Pianaro, A., Flach, A., Patricio, E.F., Nogueira-Neto, P. and Marsaioli, A.J. (2007). Chemical changes associated with the invasion of a *Melipona scutellaris* colony by *Melipona rufiventris* workers. *J. Chem. Ecol.*, **33**, 971–984.

Provost, E. (1994). Fermeture de la société et mécanismes chimiques de la reconnaissance chez diverses espèces de fourmis. In *Thèse de doctoral d'Etat*, University of Marseille, pp. 43–47.

Provost, E., Blight, O., Tirard, A. and Renucci, M. (2008). Hydrocarbons and Insects' Social Physiology. In *Insect Physiology: New Research*, ed. P. Rayan. Maes, Hauppauge, NY: Nova Science.

Raboudi, F., Mezghani, M., Makni, H., Marrakchi, M., Rouault, J.D. and Makni, M. (2005). Aphid species identification using cuticular hydrocarbons and cytochrome b gene sequences *J. Appl. Entomol.*, **129**, 75–80.

Reynolds, T. (2007). The evolution of chemosystematics. *Phytochemistry*, **68**, 2887–2895.

Richard, F.-J., Poulsen, M., Drijfhout, F., Jones, G. and Boomsma, J.J. (2007). Specificity in chemical profiles of workers, brood and mutualistic fungi in *Atta, Acromyrmex*, and *Sericomyrmex* fungus-growing ants. *J. Chem. Ecol.*, **33**, 2281–2292.

Rosa-Freitas, M.G., Broomfield, G., Priestman, A., Milligan, P.J., Momen, H. and Molyneux, D.H. (1992). Cuticular hydrocarbons, isoenzymes and behavior of three populations of *Anopheles darlingi* from Brazil. *J. Am. Mosq. Control Assoc.*, **8**, 357–366.

Rouault, J., Capy, P. and Jallon, J.-M. (2001). Variations of male cuticular hydrocarbons with geoclimatic variables: an adaptative mechanism in *Drosophila melanogaster*? *Genetica*, **110**, 117–130.

Rouault, J. D., Marican, C., Wicker-Thomas, C. and Jallon, J.-M. (2004). Relations between cuticular hydrocarbon (HC) polymorphism, resistance against desiccation and breeding temperature; a model for HC evolution in *D. melanogaster* and *D. simulans*. *Genetica*, **120**, 195–212.

Roux, O., Gers, C. and Legal, L. (2006). When, during ontogeny, waxes in the blowfly (Calliphoridae) cuticle can act as phylogenetic markers. *Biochem. Syst. Ecol.*, **34**, 406–416.

Ryan, L., Phillips, A., Milligan, P., Lainson, R., Molyneux, D. H. and Shaw, J. J. (1986). Separation of female *Psychodopygus wellcomei* and *P. complexus* (Diptera: Psychodidae) by cuticular hydrocarbon analysis. *Acta Trop.*, **43**, 85–89.

Saïd, I., Costagliola, G., Leoncini, I. and Rivault, C. (2005). Cuticular hydrocarbon profiles and aggregation in four *Periplaneta* species (Insecta: Dictyoptera). *J. Insect Physiol.*, **51**, 995–1003.

Savarit, F., Sureau, G., Cobb, M. and Ferveur, J.-F. (1999). Genetic elimination of known pheromones reveals the fundamental chemical bases of mating and isolation in *Drosophila*. *Proc. Natl. Acad. Sci. USA*, **96**, 9015–9020.

Sillam-Dussès, D., Sémon, E., Lacey, M. L., Robert, A., Lenz, M. and Bordereau, C. (2007). Trail-following pheromones in basal termites, with special references to *Mastotermes darwiniensis*. *J. Chem. Ecol.*, **33**, 1960–1977.

Singer, T. L. (1998). Roles of hydrocarbons in the recognition systems of insects. *Am. Zool.*, **38**, 394–405.

Stankiewicz, B. A., Scott, A. C., Collinson, M. E., Finch, P., Mösle, B., Briggs, D. E. G. and Evershed, R. P. (1998). Molecular taphonomy of arthropod and plant cuticles from the Carboniferous of North America: implications for the origin of kerogen. *J. Geol. Soc. London*, **155**, 453–462.

Steiner, F. M., Schilck-Steiner, B. C., Nikiforov, A., Kalb, R. and Mistrik, R. (2002). Cuticular hydrocarbons of *Tetramorium* ants from Central Europe: Analysis of GC-MS data with self-organizing maps (SOM) and implications for systematics. *J. Chem. Ecol.*, **28**, 2569–2584.

Sutton, B. D. and Carlson, D. A. (1997a). Cuticular hydrocarbon variation in the Tabanidae (Diptera): *Tabanus nigrovittatus* complex of the North American Atlantic coast. *Ann. Entomol. Soc. Am.*, **90**, 545–549.

Sutton, B. D. and Carlson, D. A. (1997b). Cuticular hydrocarbons of *Glossina*. III: subgenera *Glossina* and *Nemorhina*. *J. Chem. Ecol.*, **23**, 1291–1320.

Sutton, B. D., Carlson, D. A., Lockwood, J. K. and Nunamaker, R. A. (1996). Cuticular hydrocarbons of glacially-preserved *Melanoplus* (Orthoptera: Acrididae): Identification and comparision with hydrocarbons of *M. sanguinipes* and *M. spretus*. *J. Orth. Res.*, **5**, 1–12.

Sutton, B. D. and Steck, G. J. (1994). Discrimination of Caribbean and Mediterranean fruit fly larvae (Diptera: Tephritidae) by cuticular hydrocarbon analysis. *Florida Entomol.*, **77**, 232–237.

Sutton, B. D. and Steck, G. J. (2005). An annotated checklist of the Tephritidae (Diptera) of Florida. *Insecta Mundi*, **19**, 227–245.

Symonds, M. R. E. and Elgar, M. A. (2007). The evolution of pheromone diversity. *Trends in Ecol. Evol.*, **23**, 220–228.

Takahashi, A., Tsaur, S.C., Coyne, J.A. and Wu, C.I. (2001). The nucleotide changes governing cuticular hydrocarbon variation and their evolution in *Drosophila melanogaster*. *Proc. Natl. Acad. Sci. USA*, **98**, 3920–3925.

Takematsu, Y. and Yamaoka, R. (1999). Cuticular hydrocarbons of *Reticulitermes* (Isoptera: Rhinotermitidae) in Japan and neighboring countries as chemotaxonomic characters. *Appl. Entomol. Zool.*, **34**,179–188.

Tartivita, K. and Jackson, L.L. (1970). Cuticular lipids of insects. I. Hydrocarbons of *Leucophaea maderae* and *Blatta orientalis*. *Lipids*, **5**, 35–37.

Thorne, B.L., Haverty, M.I. and Collins, M.S. (1994). Taxonomy and biogeography of *Nasutitermes acajutlae* and *N. nigriceps* (Isoptera: Termitidae) in the Caribbean and Central America. *Ann. Entomol. Soc. Am.*, **87**, 762–770.

Thorne, B.L., Haverty, M.I., Page, M. and Nutting, W.L. (1993). Distribution and biogeography of the North American termite genus *Zootermopsis* (Isoptera: Termopsidae). *Ann. Entomol. Soc. Am.*, **86**, 532–544.

Tompkins, L., McRobert, S.P. and Kaneshiro, K.Y. (1993). Chemical communication in Hawaiian *Drosophila*. *Evolution*, **47**, 1407–1419.

Torres, C.W., Brandt, M. and Tsutsui, N.D. (2007). The role of cuticular hydrocarbons as chemical cues for nestmate recognition in the invasive Argentine ant (*Linepithema humile*). *Insect Soc.*, **54**, 363–373.

Urech, R., Brown, G.W., Moore, C.J. and Green, P.E. (2005). Cuticular hydrocarbons of buffalo fly, *Haematobia exigua*, and chemotaxonomic differentiation from horn fly, *H. irritans*. *J. Chem. Ecol.*, **31**, 2451–2461.

Uva, P., Clément, J.-L., Austin, J.W., Aubert, J., Zaffagnini, V., Quintana, A. and Bagnères, A.-G. (2004b). Origin of a new *Reticulitermes* termite (Isoptera, Rhinotermitidae) inferred from mitochondrial and nuclear DNA data. *Mol. Phylogenet. Evol.*, **30**, 344–353.

Uva, P., Clément, J.-L. and Bagnères, A.-G. (2004a). Colonial and geographic variations in behaviour, cuticular hydrocarbons and mtDNA of Italian populations of *Reticulitermes lucifugus* (Isoptera, Rhinotermitidae). *Insect Soc.*, **51**, 163–170.

Vander Meer, R.K. and Lofgren, C.S. (1989). Biochemical and behavioral evidence for hybridization between fire ants, *Solenopsis invicta* and *Solenopsis richteri* (Hymenoptera, Formicidae). *J. Chem. Ecol.*, **15**, 1757–1765.

Vander Meer, R.K., Lofgren, C.S. and Alvarez, F.M. (1985). Biochemical evidence for hybridization in fire ants. *Florida Entomol.*, **68**, 501–506.

Watson, J.A.L., Brown, W.V., Miller, L.R., Carter, F.L. and Lacey, M.J. (1989). Taxonomy of *Heterotermes* (Isoptera: Rhinotermitidae) in south-eastern Australia: cuticular hydrocarbons of workers and soldier and alate morphology. *Syst. Entomol.*, **14**, 299–325.

Wu, C.I., Hollocher, H., Begun, D.J., Aquadro, C.F., Xu, Y. and Wu, M.L. (1995). Sexual isolation in *Drosophila melanogaster*: a possible case of incipient speciation. *Proc. Natl. Acad. Sci. USA*, **92**, 2519–2523.

Wu, C.I. and Ting, C.T. (2004). Genes and speciation. *Nat. Rev. Genet.*, **5**, 114–122.

Ye, G., Li, K., Zhu, J., Guanghui, H. and Hu, C. (2007). Cuticular hydrocarbon composition in pupal exuviae for taxonomic differentiation of six necrophagous flies. *J. Med. Entomol.*, **44**, 450–456.

Youngsteadt, E., Fan, Y., Stay, B. and Schal, C. (2005). Cuticular hydrocarbon synthesis and its maternal provisioning to embryos in the viviparous cockroach *Diploptera punctata*. *J. Insect Physiol.*, **51**, 803–809.

8

Chemical synthesis of insect cuticular hydrocarbons

Jocelyn G. Millar

The external cuticle of insects is covered by a waxy layer composed of mixtures of hydrophobic lipids that include long-chain alkanes, alkenes, wax esters, fatty acids, alcohols, aldehydes, and sterols. The primary purpose of this layer is to maintain water balance and prevent desiccation, as described in Chapter 6, but many of the cuticular lipid components have important secondary roles as intraspecific contact chemical signals (pheromones). These roles include species and sex recognition during reproductive interactions, and nestmate recognition and other colony organization functions in social insects. Thus, these compounds are essential mediators of insect behaviors. Cuticular compounds are also exploited by parasitoids and predators as interspecific contact cues (kairomones) to aid in host location.

Despite their critical importance in mediating insect behaviors, we still know very little about the detailed roles and mechanisms of insect cuticular lipids as signal molecules, in part because it is not immediately obvious how such signals could be manipulated and exploited for insect management, as is done with the more well-known volatile sex attractant pheromones. Even for the comparatively few species that have been examined in any detail, there are large gaps in our knowledge, for several reasons.

First, almost all studies of insect cuticular lipids have used gas chromatography (GC) to analyze lipid extracts, using standard GC conditions that only allow compounds under ~C_{40} to be detected. More specialized GC equipment that can extend this range to >C_{60}, in particular, columns that can withstand high temperatures (>400°C) are now available. However, they have not yet found routine use for cuticular lipid analysis, despite recent studies that have clearly demonstrated that cuticular lipids do indeed contain hydrocarbons, waxes, and other compounds with molecular weights above 500 daltons (Cvačka et al., 2006). Thus, to date, many studies may only have examined subsets of the cuticular components.

Second, the majority of studies of cuticular lipids have simply listed the compounds present in cuticular extracts, and in some cases compared the lipid profiles of different life stages, populations, or species. Relatively few studies have fractionated extracts to isolate and identify the specific biologically active contact pheromone components from the mixtures of lipids in extracts, and then followed up with reconstruction of active pheromone blends from synthesized compounds to verify the bioactivity. This trend towards comparative studies listing hydrocarbons present without proper determination of their biological activities and functions has been abetted by the comparative ease with which cuticular hydrocarbons can be identified. Several seminal studies have provided the chromatographic retention time and mass spectral interpretation underpinnings that render analysis of most insect cuticular hydrocarbons quite straightforward (see Chapter 2).

Third, several contact pheromone chemicals have been identified as methyl-branched hydrocarbons. A hydrocarbon with a single methyl branch can exist in one of two enantiomeric forms, whereas hydrocarbons with two, three, or four methyl branches can have 4, 8, or 16 different stereoisomeric forms. However, almost all studies of cuticular hydrocarbons have ignored the issue of stereochemistry, despite the fact that insect receptors for volatile pheromones clearly discriminate between enantiomeric forms, and in many cases, behaviors are strongly antagonized by the incorrect stereoisomer (reviewed in Mori, 2007). This is in part due to the fact that, with rare exceptions, it has not been possible to determine which enantiomer an insect produces by analytical chemistry means, because currently available chiral stationary phase chromatography columns are incapable of resolving long-chain methyl-branched hydrocarbons. Furthermore, most insect cuticular compounds are available only in relatively small amounts, and chiral long-chain methyl-branched hydrocarbons often have small optical rotations (e.g., Kuwahara and Mori, 1983), so it may be impractical to attempt to determine which enantiomer an insect produces by measuring the optical rotation of a purified compound. The situation is further clouded by the fact that in a few of the cases in which synthetic enantiomers have been prepared and bioassayed, test insects have responded to both enantiomers (Fukaya *et al.*, 1997; Mori, 2008a). Thus, almost all work with synthetic compounds has been done with racemic compounds or mixtures of diastereomers, and luckily, these mixtures of stereoisomers have generally proven to be at least partially biologically active. Nevertheless, the role of chirality in contact chemical signals remains an area desperately in need of more detailed research. This is particularly true in light of the fact that insects are sensitive to other, often minor, changes in the structures of contact semiochemicals, such as chain length, the position and geometry of double bonds in long-chain alkenes, or the position of one or more methyl groups in a long hydrocarbon chain (e.g., Uebel *et al.*, 1975; Fukaya *et al.*, 1997).

Overall, given the relatively simple structures of many insect cuticular hydrocarbons and the ease with which they can be synthesized, it is remarkable that most studies of cuticular hydrocarbons have not been carried out in greater depth. Many papers have simply listed compounds present in cuticular extracts, rather than conducting more comprehensive studies in which the hydrocarbons have been synthesized, followed by methodical reconstruction of biologically active contact pheromone blends. Thus, one of the primary goals of this chapter is to review some of the straightforward methods by which cuticular hydrocarbons can be made and purified. In addition, examples of more advanced methods of stereoselective synthesis of compounds with multiple methyl branches will be described.

Preamble: convenient methods of purification of synthesized cuticular hydrocarbons

Purification of synthesized cuticular hydrocarbon products from starting materials and reaction by-products can be the most difficult and time-consuming part of a synthesis. Specifically, low-pressure liquid chromatography (= flash chromatography) on silica gel will remove any unreacted starting materials or by-products that contain polar functional groups, but this method is of limited or no use in separating hydrocarbons from each other because they have no polar functional groups to adsorb to the silica. Thus, hydrocarbons essentially pass right through a silica column, even with the most nonpolar solvents. It may be possible to achieve limited separation of saturated hydrocarbons from alkenes or acetylenes by activation of the silica gel in a drying oven at >100°C for several hours. The activated silica then is used immediately before it can readsorb water from the air, using hexane for elution. However, this method will be of no use in separating saturated alkanes from each other, or alkenes from each other. Instead, a variety of other separation methods can be used, with the choice of method(s) depending on the relative amounts of undesired impurities, differences in boiling points between the impurities and the product, and differences in degree of unsaturation (i.e., number and type of double bonds). These methods include the following:

1. *Recrystallization at 0 or –20°C.* Many cuticular hydrocarbons have melting points in the range of ~0–40°C, and many can be purified by recrystallization from acetone, a polar solvent, or nonpolar hexane, using ~10–25 ml solvent per gram of hydrocarbon. Any polar impurities should be removed before recrystallization is attempted by passing a hexane solution of the crude hydrocarbon product through a short (~5 cm) bed of silica gel, then concentrating again. For recrystallization to be successful, it is usually necessary to start with a product which is ~80% or more pure, particularly if the impurities are similar in size or structure to the desired product. When using acetone, it is usually necessary to heat the solution to dissolve the hydrocarbon, which tends to form an oily layer at room temperature, whereas heating may not be required with hexane. For either solvent, simply chilling the solution overnight at –20°C in a freezer is usually sufficient to induce crystallization. Because the solids are relatively low melting, the product should be filtered off with vacuum filtration in a cold room, precooling the filter apparatus, and rinsing the filter cake with clean solvent also precooled to –20°C. As soon as the filtration is complete, it is advisable to transfer the crystalline product to a container, in case it begins to melt as it warms up. Should one cycle of recrystallization fail to achieve the desired purity, the recrystallization can be repeated one or more times. Because of its ease of use, its efficacy, and the fact that recrystallization can be done in one batch on amounts from milligrams to tons, we now purify nearly all of our synthesized hydrocarbons by this method as a matter of course.
2. *Vacuum distillation.* Long-chain hydrocarbons are usually assembled by coupling two smaller fragments, one or both of which may be much more volatile than the product, and/or the by-products of the reaction may be more volatile than the product. As a first step in purification, these semivolatile compounds can be distilled off under vacuum (0.1 mm Hg or less), using even quite high temperatures because the desired hydrocarbon products are generally quite stable to heat, and have boiling points well above 100°C, even at pressures of ~0.1 mm Hg.

3. *Reverse phase liquid chromatography.* It is astonishing that reverse phase chromatography has not been more widely used in the purification of hydrocarbons, because in our hands, it appears to be ideally suited to such separations. Specifically, reverse phase chromatography separates compounds largely on the basis of lipophilicity, with shorter-chain hydrocarbons eluting before longer-chain ones. When working with small amounts of compounds on reverse phase HPLC, hydrocarbons can be eluted with non-aqueous solvent mixtures of methanol or acetonitrile with less polar solvents such as ethyl acetate, acetone, or dichloromethane (e.g., Metzger *et al.*, 1985; Peschke and Metzger, 1987; Sugeno *et al.*, 2006). In our hands, preparative scale separations on reverse phase flash chromatography can be done with stepwise gradients of increasing percentages of ethyl acetate in methanol, ending with pure ethyl acetate. Once the desired product has eluted, the column can be stripped with methylene chloride or even hexane, then reequilibrated in the starting solvent mixture for reuse. Reuse of the columns is desirable because of the considerable expense of reverse phase packing material.

4. *Removal of straight-chain impurities by absorbtion with molecular sieves.* Coupling reactions to produce branched-chain hydrocarbons frequently result in straight-chain impurities of similar chain length from dimerization of organometallic precursors. These straight-chain precursors can be readily removed by stirring an isooctane solution of the mixture with pre-dried, powdered 5Å molecular sieves, at room temperature or at reflux temperature if necessary (O'Connor *et al.*, 1962). The pore size is such that straight-chain compounds can enter molecular sieve pores, whereas branched-chain compounds are excluded. After the straight-chain compounds have been absorbed, the molecular sieves are filtered off, leaving a solution containing only the branched-chain compound(s). Note that it is essential to use the branched-chain solvent isooctane with this method; if hexane or other unbranched solvents were used, they would simply be absorbed by the sieves in place of the straight-chain reaction side-products.

5. *Purification of alkenes on silica gel impregnated with silver nitrate.* This classic method of purification of alkenes can be used in one pass for (a) removal of traces of alkanes from alkenes, (b) separation of minor amounts of *E*-isomers from the desired *Z*-isomer product, and (c) separation of minor amounts of terminal or internal alkynes from alkenes. Some or all of these impurities may be present in reaction mixtures. For example, an alkene made by catalytic reduction of an internal alkyne with hydrogen, which in turn was generated from coupling a terminal alkyne with an alkyl halide (see below) may be contaminated with unreacted alkyl halide, alkanes from reduction of the unreacted terminal alkyne starting material and/or from over-reduction of the internal alkyne intermediate, small amounts of the internal alkyne, and traces of the *E*-isomer of the desired alkene. Bulk $AgNO_3$-impregnated silica is available from Aldrich Chemical Co. for preparative separations by flash chromatography, or for small-scale separations, it is available preloaded in solid phase extraction cartridges from Supelco (Discovery Ag-Ion cartridges), or in HPLC columns (e.g. Chromspher Lipids™ from Varian-Chrompack). For flash chromatography, the crude hydrocarbon is loaded onto the column as a hexane solution, and the column is then eluted with hexane (to remove any saturated hydrocarbons or alkyl halides) followed by elution with solvent mixtures containing hexane with increasing amounts of benzene, cyclohexene, or ether, to sequentially elute (*E*)- and then (*Z*)-isomers (e.g., Uebel *et al.*, 1978; Oguma *et al.*, 1992; Akino *et al.*, 2004). The preparation of silver ion packing materials and columns is reviewed and described at www.lipidlibrary.co.uk/index.html and references therein.

In our experience, one or more of these methods is sufficient to purify any synthetic hydrocarbon to high levels of chemical and isomeric purity, be it a straight-chain, branched-chain, or unsaturated compound.

Methods of synthesizing various classes of cuticular hydrocarbons

The methods used for synthesizing straight-chain, branched-chain, or unsaturated hydro-carbons often use the same basic coupling reactions for assembling the carbon skeleton, and a variety of different coupling reactions and synthetic routes are possible. What fol-lows are a few of the most common and robust methods, but it should be borne in mind that numerous carbon–carbon bond-forming reactions can be used to assemble the carbon chains of cuticular hydrocarbons. When choosing between various routes, the deciding factors are usually the availability of appropriate starting materials, the efficiency of the coupling reaction to join the synthons together, and the anticipated by-products, i.e., the ease of purification of the desired product, which can be laborious as described above. In that context, it is worth considering that hydrocarbon impurities are typical by-products from organometallic reagents such as Grignard reagents, organolithiums, and organocu-prates that are generated from alkyl halides. These unwanted hydrocarbons are formed by dimerization of the organometallic reagents during the reaction or work-up, and if they are of similar chain length to the desired hydrocarbon products, they may be difficult to remove cleanly. In such cases, it may be more efficient to use a route which includes a polar inter-mediate that can be readily separated from nonpolar contaminants. For example, reaction of a Grignard reagent with a ketone produces an alcohol intermediate which is readily sep-arable from hydrocarbon by-products. Once separated from these by-products, the alcohol can then be manipulated further as required to form the desired hydrocarbon.

Straight-chain alkanes. Most straight-chain alkanes found in insect cuticular lipids are commercially available, either as individual compounds or as components of kits assem-bled for GC standards. Should it be necessary to synthesize them, even-numbered chains can be made by dimerization of alkylmagnesium halides (Grignard reagents, prepared from the corresponding alkyl halides) with divalent copper, chromium, or cobalt halides or by treatment with monovalent silver salts (Figure 8.1A) (compilations of methods, Larock 1999; Lehmler *et al.*, 2002). Odd-numbered chains can be made by coupling of Grignard reagents with alkyl halides, tosylates, or mesylates with Li_2CuCl_4 catalysis (Figure 8.1B) (examples, Larock 1999, p. 119; Herber and Breit, 2007). Alternatively, Wittig reactions of alkyltriphenylphosphoranes (from treatment of alkyltriphenylphosphonium halide salts with strong bases in THF or ether) and aldehydes (Figure 8.1C) (e.g. Brooke *et al.*, 1996), or coupling of terminal alkynes with alkyl halides in THF (Figure 8.1D) (Buck and Chong 2001, and references therein), produce alkenes and alkynes respectively. Unreacted ter-minal alkyne can be easily removed by extraction with aqueous $AgNO_3$ solution during the work-up (e.g., Klewer *et al.*, 2007). The resulting alkenes or alkynes are then reduced to the corresponding alkanes by catalytic reduction with Pd on carbon and hydrogen, with hexane or other solvents. These reductions are very simple, requiring only a balloon of hydrogen attached to a septum-sealed, well-stirred flask. Overall, a compilation of methods of syn-thesizing straight-chain alkanes can be found in Lehmler *et al.* (2002).

Longer-chain alkyl halides may not be commercially available, but they are readily made in one step from the corresponding alcohols (Larock, 1999), as are tosylates and mesylates. Similarly, longer-chain terminal alkynes are not commercially available, but can be readily made by reaction of alkyl halides with lithium acetylide–ethylene diamine complex in dry

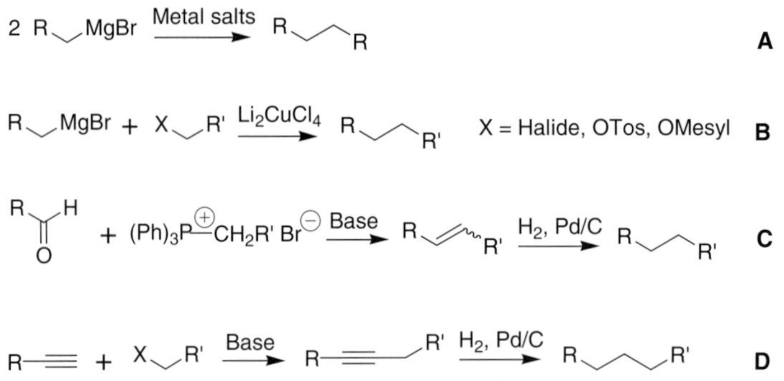

Figure 8.1 Synthetic routes to straight-chain alkanes.

DMSO with NaI catalysis (Sonnet and Heath, 1980), by bromination of a terminal alkene followed by double elimination of HBr (Hoye *et al.*, 1999), or by isomerization of an internal alkyne with the acetylene zipper reaction (Abrams, 1984).

Branched-chain alkanes. The vast majority of branched-chain hydrocarbons found in insect cuticular lipids are methyl-branched, and so this review will focus on these types of compound. However, exactly the same reactions and strategies can be used for making ethyl- or other alkyl-branched compounds by using, for example, an ethyl-branched instead of a methyl-branched fragment in the reaction.

The introduction of branches also makes it possible to have stereoisomers. Compounds with a single methyl branch at any position other than carbon 2 or the exact center of the chain can exist as one of two possible enantiomers, whereas compounds with two or more branches have a number of different stereoisomers (e.g., enantiomers, *meso* isomers, or diastereomers). Generic reactions that produce racemic mixtures or mixtures of stereoisomers will be discussed first, followed by descriptions of methods used to make individual stereoisomers.

The choice of which route to use to synthesize racemic monomethylalkanes depends to some extent on the availability of precursors. Thus, if a synthon with a terminal halide or alcohol (readily converted to a halide) and a methyl group in the correct position is commercially available, the target hydrocarbon can be synthesized by reaction of the branched alkyl halide with a Grignard reagent prepared from the appropriate straight-chain alkyl halide using Li_2CuCl_4 catalysis (Figure 8.2A). The reaction can of course be conducted the other way around, i.e., coupling a branched chain Grignard reagent with a straight-chain alkyl halide, tosylate, or mesylate (e.g., Ginzel *et al.*, 2006a). Methyl-branched fragments of any desired length and position of the methyl group can be made by the malonic ester synthesis, either by coupling a 2-haloalkane with the anion from diethylmalonate (Figure 8.2B) (Hwang *et al.*, 1974, 1978), or by coupling a straight-chain alkyl halide with the anion of diethyl methylmalonate. Hydrolysis, decarboxylation, and reduction yield a methyl-branched alcohol, ready for tosylation or conversion to an alkyl halide.

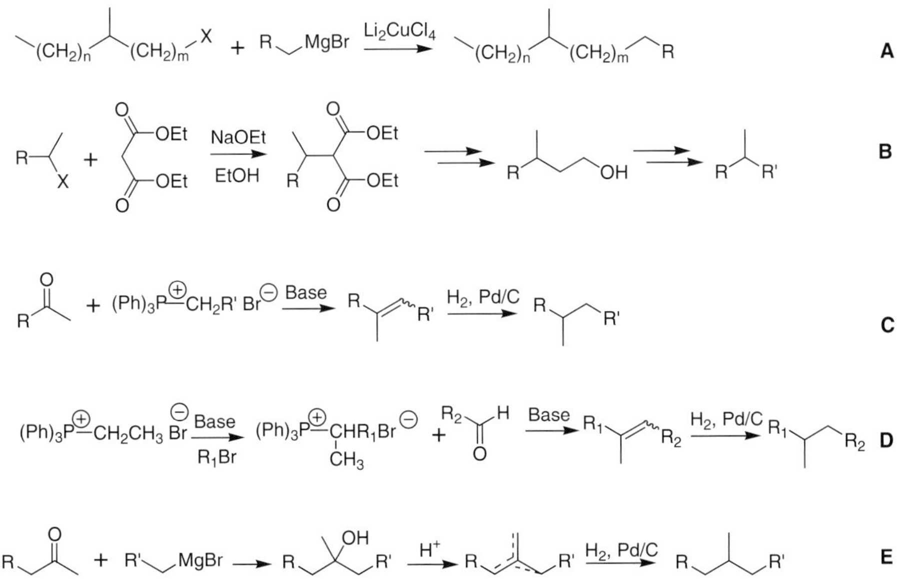

Figure 8.2 Synthetic routes to racemic alkanes with a single methyl branch.

If the appropriate fragments are not readily available for the reactions discussed above, then it may be more expedient to react a 2-alkanone with a Wittig reagent, followed by hydrogenation of the resulting trisubstituted alkene (Figure 8.2C) (Naoshima and Mukaidani, 1987; Carrière *et al.*, 1988; Ginzel *et al.*, 2006). In an efficient and clever variation that places a methyl anywhere in a chain, Sonnet (1976) first alkylated ethyl-triphenylphosphorane with a primary alkyl halide, and then reacted the resulting Wittig reagent (now containing a methyl and an alkyl chain) with an aldehyde, followed by catalytic reduction to the methylalkane (Figure 8.2D). This *in situ* formation of the secondary alkyltriphenylphosphonium salt was reportedly cleaner and more efficient than direct formation of the salt from triphenylphosphine and a secondary alkyl halide (Sonnet *et al.*, 1977; D'Ettorre *et al.*, 2004).

In another variation, a 2-alkanone can be reacted with a Grignard reagent, followed by acid-catalyzed dehydration to a mixture of alkenes, and catalytic reduction (Figure 8.2E) (Hwang *et al.*, 1976; Suzuki, 1981; Lacey *et al.*, 2008). Other methods have also been used. For example, in some of the pioneering work on insect cuticular hydrocarbons and their analyses, Pomonis and co-workers prepared a large number of monomethyl and dimethylalkanes from 2,5-disubstituted thiophenes (e.g., Pomonis *et al.*, 1978). However, this route cannot be recommended because it requires as many as 12 steps per hydrocarbon.

Chiral monomethylalkanes have usually been synthesized from a double-ended chiral synthon, with two different functional groups to which two different chains can be sequentially attached. Common synthons used are those based on citronellol or its analogs, whereby

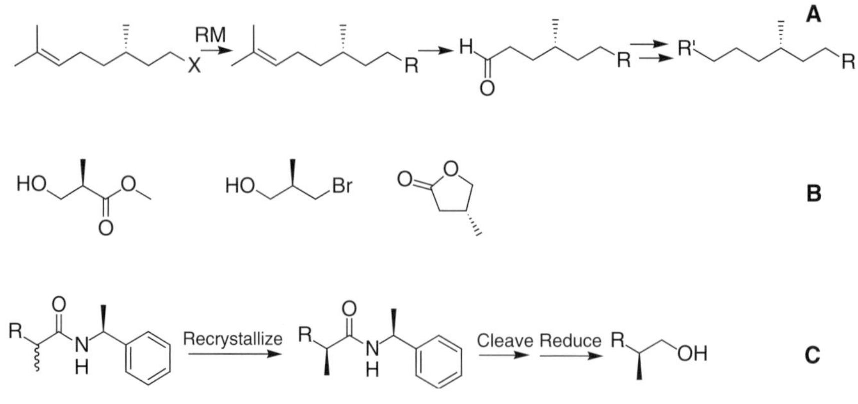

Figure 8.3 Synthetic routes and useful chiral synthons for synthesis of the enantiomers of alkanes with a single methyl branch.

the alcohol function is converted to a leaving group, and the chain is extended by alkylation as described above (Figure 8.3A) (e.g., Naoshima and Mukaidani, 1987; Mori, 2008a). The double bond is then oxidatively cleaved to an aldehyde, and the chain is extended by a Wittig reaction, followed by reduction of the alkene to give the saturated methylalkane. Marukawa *et al.* (2001) described a variation using reaction of a Grignard reagent with a chiral methyl-branched aldehyde prepared from citronellol, followed by elimination of water from the resulting alcohol, and reduction to the alkane. This route was used because the polar alcohol intermediate was easy to separate from nonpolar by-products, in contrast to preparation of a hydrocarbon via Wittig reaction of the same aldehyde precursor. The same authors generated chiral 3-methylalkanes by an analogous method, reacting chiral 3-methylalkylmagnesium bromide with a long straight-chain aldehyde to give an alcohol, followed by removal of the alcohol to give the desired chiral methylalkanes. An analogous strategy of using intermediates with polar functional groups to avoid problems with separation of by-products was used by Mori and Wu (1991a, b) in syntheses of all four stereoisomers of 3,7-dimethylnonadecane and the (5*S*,9*S*)-enantiomers of 5, 9-dimethylheptadecane and 5, 9-dimethyloctadecane respectively.

Other useful double-ended chiral synthons can be readily derived from derivatives of commercial methyl 3-hydroxy-2-methylpropanoate enantiomers (Figure 8.3B). Even more convenient, both enantiomers of 3-bromo-2-methyl-1-propanol (Figure 8.3B) are commercial products that can be protected at the alcohol function, chain extended on the other side, then deprotected and the resulting alcohol converted to a halide or sulfonate for a second alkylation (e.g., Fukaya *et al.*, 1997; Schlamp *et al.*, 2005). We have used these synthons to efficiently generate both enantiomers of three long-chain methylalkanes (J.G.M., unpublished data). Another useful double-ended methyl-branched synthon can be readily derived from enantiomers of 3-methylbutyrolactone (Figure 8.3B) (Mori, 2008a).

Sonnet (1984) developed another general method of synthesizing chiral methylalkanes, based on fractional crystallization of the mixture of diastereomeric amides generated from a racemic 2-methylalkanoic acid and one of the enantiomers of α-methylbenzylamine (Figure 8.3C). After separation of the diastereomers, each was cleaved to give the enantiomerically pure acids, which were then reduced to the chiral alcohols, followed by chain extension in the usual way.

Nonstereoselective syntheses of mixtures of stereoisomers of alkanes with two or more methyl branches can be carried out by coupling two synthons of the appropriate lengths and methyl branch positions, using for example Wittig (1,3, 1,4, and 1,5-dimethyl motifs, Doolittle *et al.*, 1995; Carrière *et al.*, 1988) or acetylene-based coupling reactions as described above. Alternatively, two (or more) methyl branches can be inserted at any desired positions by the malonic ester synthesis, coupling the anion of diethyl methylmalonate with a methyl-branched alkyl halide. The resulting product is decarboxylated and reduced to a dimethylalkanol, which can then be chain extended (Zambelli *et al.*, 1978).

For particular methyl branching motifs, it may be possible to take advantage of naturally occurring compounds which have two or more methyl branches appropriately placed. For example, 1,5-dimethyl-branched alkanes are common, and this branching pattern is found in citronellol and other monoterpenes. Citronellol and geraniol (or their acetates) are readily oxidized at one of the allylic methyl groups (Figure 8.4A), providing a double-ended, dimethyl synthon for attachment of the appropriate hydrocarbon chains (Shibata *et al.*, 2002). These authors extended this strategy to tetramethylalkanes by coupling two of the dimethyl synthons together. This concept also can be extended to 1,5,9-trimethyl or 1,5,9,13-tetramethyl branching motifs by use of farnesyl acetate and geranylgeranyl acetate as the starting synthons respectively (Figure 8.4B) (Rowland *et al.*, 1982). In a variation, Hoshino and Mori (1980) oxidatively cleaved the terminal isopropylidene group of a geranylgeraniol derivative to generate the symmetrical 4,8,12-trimethylpentadecanedial with a 1,5,9-trimethyl motif, which was then converted to the symmetrical 15,19,23-trimethylheptatriacontane (a sex pheromone of the female tsetse fly) by a double Wittig reaction

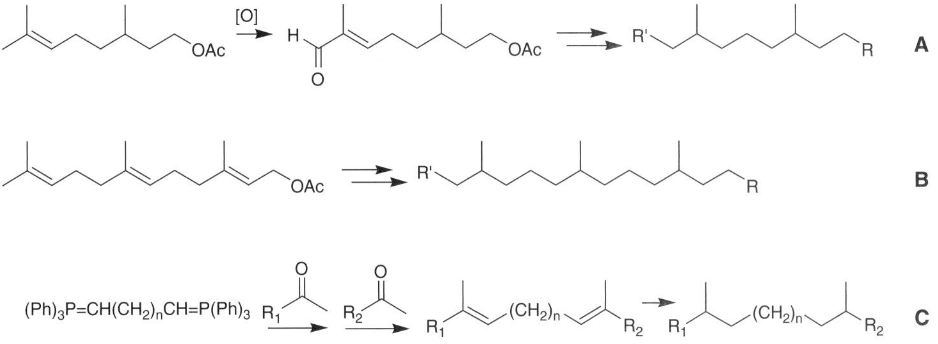

Figure 8.4 Synthetic routes and synthons for nonstereoselective syntheses of alkanes with two or more methyl branches.

followed by reduction of the diene to the saturated hydrocarbon. A 1,5,9-trimethyl alkane also has been generated by double alkylation of diethyl 3-oxoglutarate with two methyl-branched fragments, decarboxylation, and reaction of the resulting ketodiene with methyl-magnesium bromide to add the third methyl (Naoshima *et al.*, 1986).

For more general syntheses that place methyls anywhere along a chain, double-ended Wittig reagents can be reacted sequentially in the same pot with two different methylke-tones (Figure 8.4C) (Zarbin *et al.*, 2004). The resulting diene is then catalytically reduced to the dimethylalkane. The reaction partners can also be interchanged, i.e., a dimethyl-diketone can be reacted sequentially with two different straight-chain Wittig reagents or Grignard reagents, giving a diene or diol respectively, both of which are easily converted to the dimethylalkane (Jocelyn G. Millar and F. Soriano, unpublished data). Overall, a sequential series of reactions involving protection of one of the ketones and reaction of the other, followed by deprotection and reaction of the first, may provide greater control and fewer problems with purification (Matsuyama and Mori, 1994). Alkanes with three or more methyl branches can be made by simply incorporating additional methyls into the diketone, or into one or both of the Wittig or Grignard fragments (Jocelyn G. Millar and F. Soriano, unpublished data).

Individual stereoisomers of alkanes with two or more methyl branches can be obtained by coupling two chiral fragments by the reactions described above (e.g., Pempo *et al.*, 1996; Takikawa *et al.*, 1997; Nakamura and Mori, 2000; Pempo *et al.*, 2000; Masuda and Mori, 2002; Mori, 2008a, b). The chiral fragments are usually generated from citronellol or methyl 3-hydroxy-2-methylpropanoate enantiomers as described above for chiral methylalkanes. Mori and Wu (1991a) reported a nice example which took advantage of both of these chiral synthons to generate all four stereoisomers of 3,7-dimethylnonadecane (Figure 8.5A). They first prepared the enantiomers of 2-methylbutyl iodide from the enantiomers of methyl 3-hydroxy-2-methylpropanoate. They then prepared each of the enantiomers of 3-methyl-1-(phenylsulfonyl)pentadecane from (*R*)-citronellic acid by conducting different series of reactions on each end of this double-ended synthon. Straightforward coupling reactions between the anions of the 3-methyl-1-(phenylsulfonyl)pentadecane enantiomers and the 2-methylbutyl iodide enantiomers, followed by removal of the sulfonyl activating group, completed the syntheses. A similar strategy was used to produce (5*S*,9*S*)-dimeth-ylheptadecane and (5*S*,9*S*)-dimethyloctadecane (Mori and Wu, 1991b; Tamagawa *et al.*, 1999). Chiral 1,5-dimethyl motifs such as these can also be obtained from double-ended synthons with two chiral methyl groups, available from (-)-isopulegol and (-)-neoisopule-gol (Moreira and Correa, 2003).

As an alternative to coupling fragments by using Wittig, cuprate, or acetylene chemistry, Kuwahara and Mori (1983) prepared two chiral fragments from (*R*)-citronellic acid, and then coupled them sequentially to methyl acetoacetate at the γ and α positions respectively (Figure 8.5B). The resulting ketoester was decarboxylated, and the ketone was removed with a Wolff–Kishner reduction to give chiral 13,23-dimethylpentatriacontanes. Because of symmetry, this compound only has three stereoisomers (13*R*,23*R*, 13*S*,23*S*, and *meso*).

Alternatively, the chiral centers can be generated de novo by induction of asymmetry in achiral precursors, using enantioselective or diastereoselective reactions. Some useful

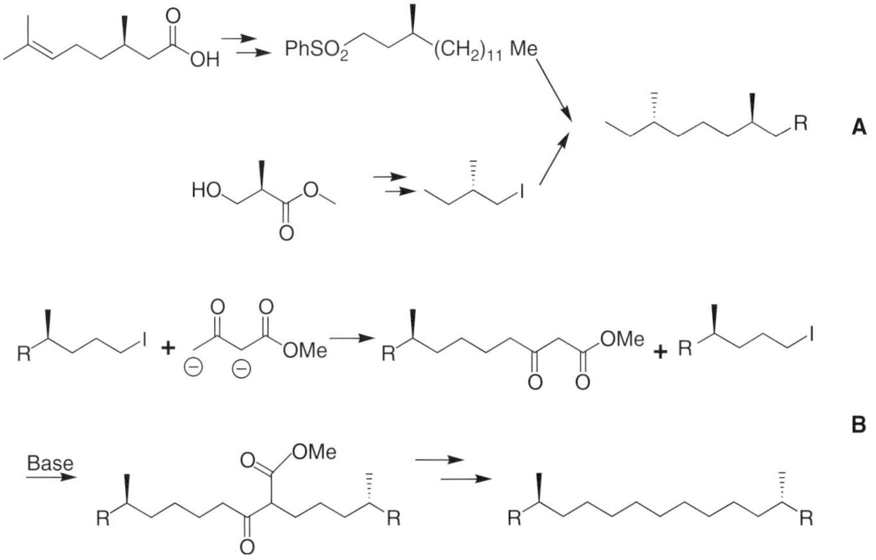

Figure 8.5 Syntheses of specific enantiomers of dimethylalkanes with 1,5- and 1,11-branching motifs.

representative examples are described here, but numerous other methods can be found in the literature.

In a nice example of controlling the relative stereochemistry of 1,4 (Figure 8.6) or 1,5 (Figure 8.7) dimethyl motifs, Heathcock *et al.* (1988) used aldol reactions of α,β-unsaturated aldehydes with preformed (Z)- or (E)-enolates to prepare *anti* and *syn* allylic alcohol products respectively. These in turn served as substrates for stereoselective Eschenmoser–Claisen or Claisen rearrangements which produced double-ended synthons containing two chiral methyl groups. These were elaborated into lipids containing 2, 4, 8, or even 16 methyl branches, each with known relative configuration.

Van Summeren *et al.* (2005) reported a convenient method of generating double-ended synthons for making any desired stereoisomer of dimethylalkanes with a 1,5-motif by sequential conjugate additions of Me₂Zn to 2,7-cyclooctadienone, with a chiral catalyst (Figure 8.8). Simply switching between the two enantiomeric forms of the catalyst allowed each methyl group to be placed with complete stereocontrol.

The identification of 4,6,8,10,16-pentamethyldocosane and 4,6,8,10,16,18-hexamethyl-docosane in the cuticular lipids of Australian cane beetles (Fletcher *et al.*, 2003, 2008) presented both a substantial challenge for identification (particularly the relative and absolute stereochemistries of the two compounds), and for their syntheses. Determination of the relative stereochemistries was carried out in tandem with the first diastereoselective syntheses, because very careful comparison of NMR data with those of known standards was the only feasible way to determine the stereochemical relationships between the multiple methyl groups. Thus, in the first syntheses, 1,3,5-methyl triads with known relative

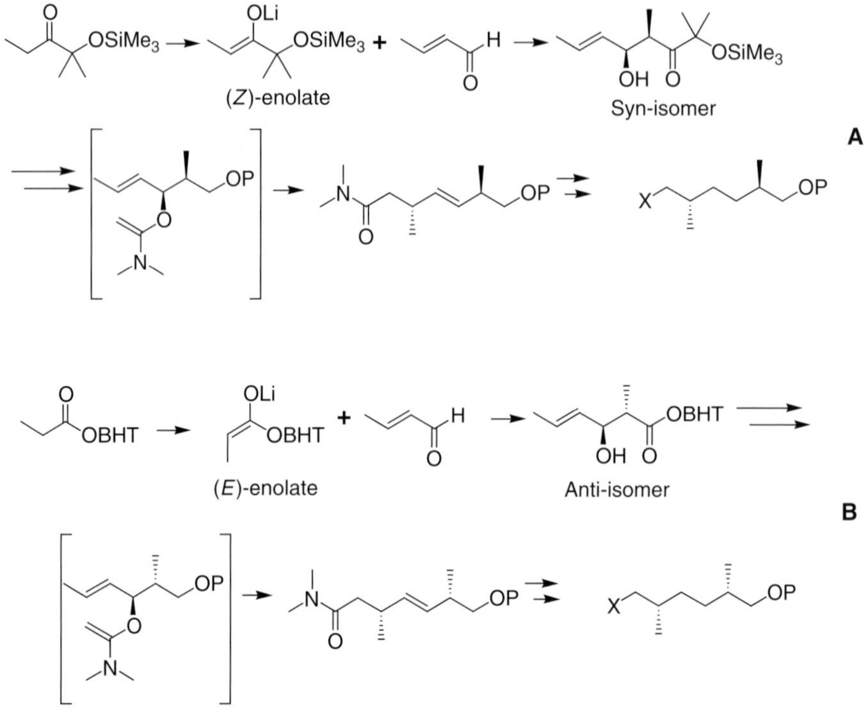

Figure 8.6 Preparation of enantiomeric forms of dimethylalkanes with 1,4-branching patterns, via stereoselective Eschenmoser-Claisen rearrangements.

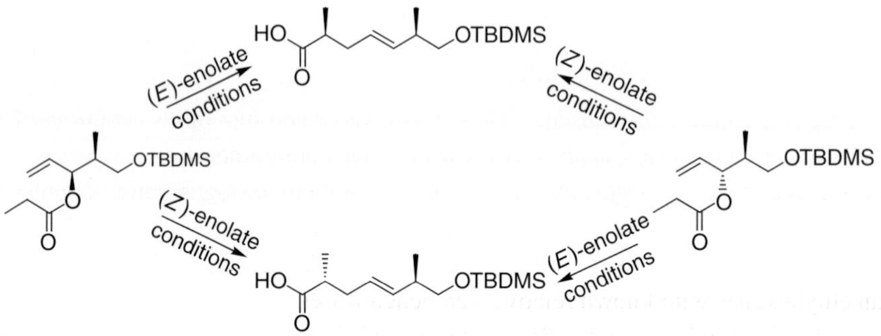

Figure 8.7 Stereoselective preparation of key intermediates in the syntheses of *syn*- or *anti*-1, 5-dimethylalkanes, by control of the geometry of enolate precursors.

stereochemistries were constructed by reduction of 2,4,6-trimethylphenol with a rhodium catalyst and high pressure H_2, producing all-*cis* 2,4,6-trimethylcyclohexanol (Figure 8.9) (Fletcher *et al.*, 2003; Chow *et al.*, 2005). This was elaborated to an all-*cis*, double-ended trimethyl synthon by oxidation to the cyclohexanone, Baeyer–Williger oxidation, and lactone opening with NaOMe. Furthermore, one or both of the outermost methyl groups

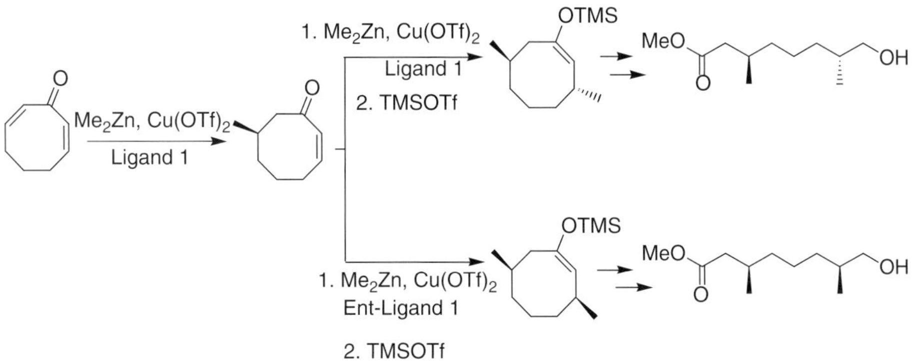

Figure 8.8 Synthesis of chiral 1,5-dimethylalkane motifs via stereo- and enantioselective Michael addition of methyl groups to 2,7-cyclooctadienone, with a chiral catalyst.

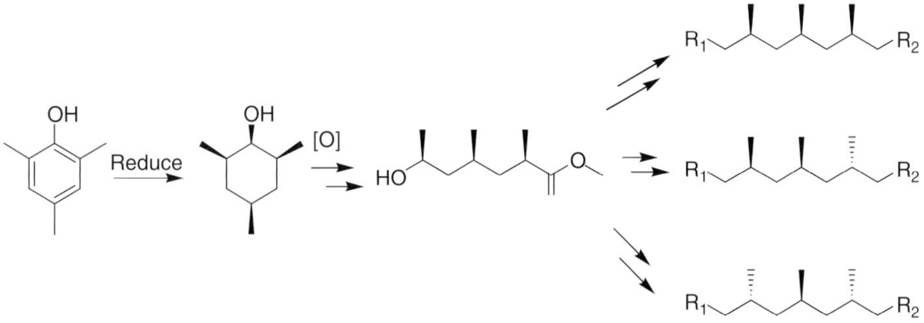

Figure 8.9 Stereoselective syntheses of 1,3,5-trimethyl motifs of any desired relative configuration from all-*cis* 2,4,6-trimethylcyclohexanol.

could be inverted, effectively providing access to all four diastereomers of the methyl triad. Analogous methodology was applied to *cis-* or *trans-*3,5-dimethylcyclohexanone to generate 1,3-dimethyl synthons with fixed relative configurations. Coupling of the appropriate dimethyl and trimethyl synthons then allowed elaboration to the pentamethyl- or hexamethylalkanes with known relative stereochemistries.

Several fully chiral syntheses of these compounds have now been published, using different synthetic strategies. In the first, the chiral allylic alcohol in an early stage synthon, in combination with a reagent system that was completely chemo-, regio-, and stereoselective, allowed the iterative assembly of 1,3,5... methyl-branched carbon chains (Breit and Herber, 2004; Herber and Breit, 2005, 2007). The key feature was the iterative use of *o*-diphenylphosphanyl benzoate ester synthons generated from chiral allylic alcohols as the reagent-directing leaving groups in S_N2' copper-mediated allylic substitution reactions, resulting in 1,3 chirality transfer with complete stereochemical control (Figure 8.10). The resulting alkene was then cleaved to an aldehyde, reduced to an alcohol, and converted

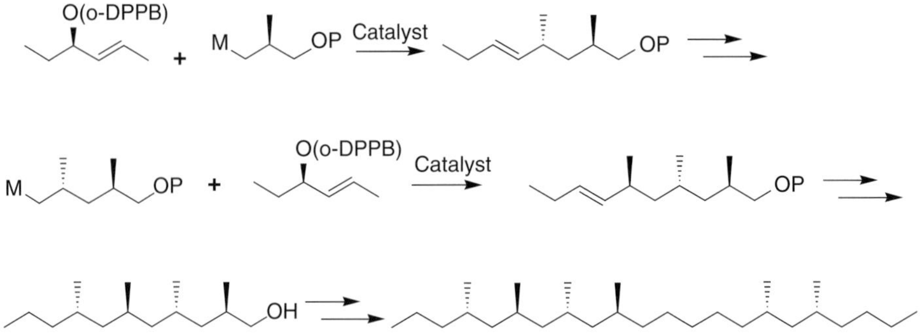

Figure 8.10 Stereoselective synthesis of (S,R,R,S,R,S)-4,6,8,10,16,18-hexamethyldocosane via reagent-directing *o*-diphenylphosphanyl benzoate ester leaving groups.

to the alkyl iodide precursor of the organometallic reagent required for the next iteration. Addition of this organometallic to either the (R)- or the (S)-enantiomer of another *o*-diphenylphosphanyl benzoate ester synthon gave complete control of the stereochemistry of the new methyl group, so that the multiply branched chain could be iteratively assembled with any desired relative and absolute stereochemistry.

The second approach used catalyst-controlled diastereoselective hydrogenation of alkene precursors as the key methodology (Figure 8.11) (Zhou and Burgess, 2007; Zhou *et al.*, 2007). Reduction of a chiral (E)-α,ß-unsaturated ester precursor with the (L)-enantiomer of the catalyst gave a 23:1 ratio of the *anti*-isomer, whereas reduction of the (Z)-isomer of the corresponding alcohol with the (D)-enantiomer of the catalyst gave predominantly the *syn*-isomer (34:1). In either case, the product of reduction was a double-ended synthon, which in a few steps could be elaborated to a new alkene ready for another diastereoselective reduction to set the next methyl group. Once all the methyls were installed, straightforward reactions added a terminal chain to one piece, and coupled the resulting two chiral intermediates. Because the hydrogenations were diastereoselective rather than diastereospecific, purification of intermediates was required to remove small amounts of stereoisomeric impurities.

A third approach used enantioface-selective zirconium-catalyzed asymmetric carboalumination (ZACA) as the key step for systematic installation of methyls in a 1,3,5… motif, with the *syn* or *anti* relationship between the newly installed methyl and the previous methyl group being determined by which enantiomer of the catalyst was used (Figure 8.12) (Zhu *et al.*, 2008). (S)-Citronellal provided the first chiral methyl group, and each additional chiral methyl was introduced with the required stereochemistry by the ZACA methodology. Coupling of the fragment containing a methyl tetrad with a second fragment containing a methyl dyad delivered the final product with six chiral methyl groups in, remarkably, only 11 consecutive steps from (S)-citronellal and with an overall 11% yield.

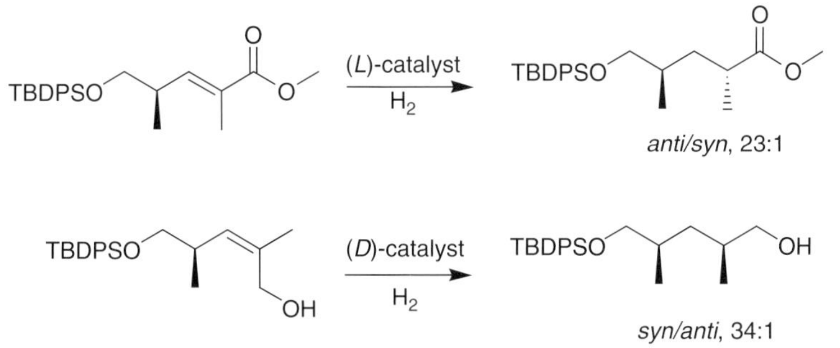

Figure 8.11 Syntheses of 1,3-dimethyl motifs via stereoselective reductions of α,ß-unsaturated esters with a chiral catalyst and H_2.

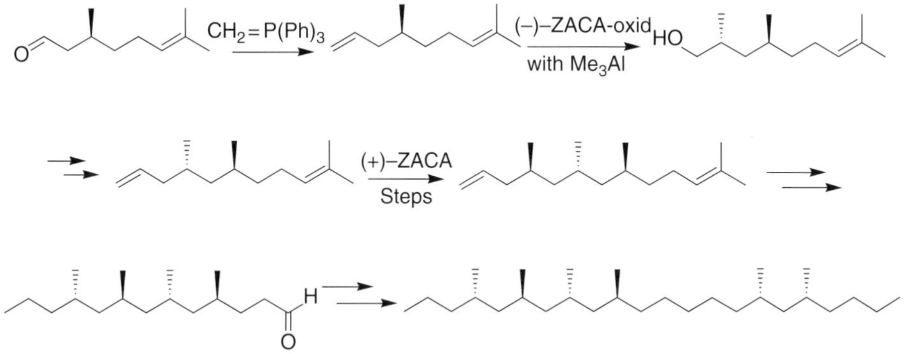

Figure 8.12 Stereo- and enantioselective synthesis of (*S,R,R,S,R,S*)-4,6,8,10,16,18-hexa-methyldocosane via iterative zirconium-catalyzed asymmetric carboalumination of alkenes (ZACA reaction).

Alkenes. Almost all alkenes found in insect cuticular lipids have the (Z) configuration. These can be easily made in two ways. First, internal alkynes, from reaction of a metallated terminal alkyne with an alkyl halide or tosylate as described above, can be reduced to (Z)-alkenes with hydrogen and Lindlar's catalyst (Pd on calcium carbonate treated with lead) or Pd on $BaSO_4$ poisoned with quinoline (Buck and Chong, 2001, and references therein; Masuda and Mori, 2002; Zhang *et al.*, 2003; Akino *et al.*, 2004; Klewer *et al.*, 2007) (Figure 8.13A). However, these catalytic reductions can be idiosyncratic because the catalyst is easily poisoned by impurities which may slow or stop the reduction completely. Even with an active catalyst, it may be difficult to get a clean reduction, and so the product is invariably contaminated with overreduced by-products, unreacted starting material, or traces of the (E)-isomer. Generally these impurities can be removed by chromatography and/or recrystallization, but if necessary, stoichiometric reduction with dicyclohexylborane or thexylborane may provide cleaner reductions.

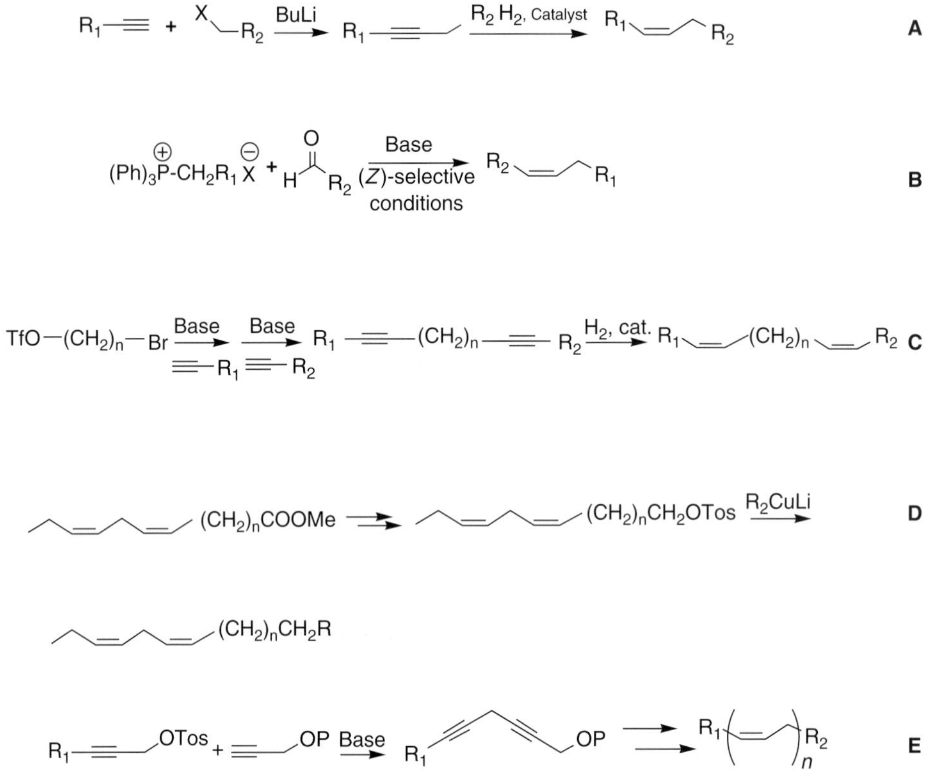

Figure 8.13 Stereoselective syntheses of (Z)-alkenes, and all-(Z)-dienes and polyenes.

Second, (Z)-alkenes can be made by Wittig reactions under conditions that favor the production of the (Z)-isomer (Figure 8.13B) (e.g., Ginzel *et al.*, 2006). These include the use of so-called "salt-free" conditions, in which sodium or potassium hexamethyl-disilazide is used to deprotonate an alkyltriphenylphosphonium halide in a relatively non-polar solvent like toluene (e.g., Zanoni *et al.*, 2007), and reaction of the resulting anion at low temperature with an aldehyde. After work-up, most of the triphenylphosphine oxide by-product can be precipitated out of a hexane solution of the crude product, and the remainder, along with any other nonhydrocarbon by-products, can be removed by passage of the hexane solution of the alkene through a short plug of silica gel. Further purification, including removal of small amounts of the (E)-isomer, can be accomplished by chroma-tography on silica–silver nitrate and/or recrystallization. Overall, the Wittig reaction is very sensitive to the base, solvent, and reaction conditions. For example, Uebel *et al.* (1975) reported that reaction of an alkylidenetriphenylphosphorane (generated by action of strong base on a corresponding alkyltriphenylphosphonium halide) with an aldehyde gave 94–96% (Z)-isomer in THF/HMPA solvent, whereas the same reaction conducted in toluene gave about 60% (E)-isomer. As an example of the utility of recrystallization of these long-chain hydrocarbons, these authors were able to purify the (*E*)-isomer from the 60:40 mixture of isomers solely by recrystallization.

In general, should (*E*)-isomers be required as standards to prove that a particular insect alkene does indeed have the (*Z*)-configuration by comparison of GC retention times, (*Z*)-isomers can be readily scrambled to mixtures of (*Z*) and (*E*), for example by treatment with thiophenol, or even with exposure to UV or sunlight. Isomerization methods are summarized in Attygalle (1998). Also, should (*E*)- and (*Z*)-isomers prove difficult to separate by GC, as is often the case with double bonds in the middle of a chain, alkenes are readily converted to the corresponding epoxides by treatment with *m*-chloroperbenzoic acid in methylene chloride for an hour or so at room temperature, followed by dilution with hexane, and washing with aqueous NaOH to remove excess reagent and byproducts (Ginzel *et al.*, 2006). After drying over anhydrous Na_2SO_4, the resulting solution of epoxides is usually clean enough to inject directly onto GC. On a nonpolar DB-5 column, and probably on other columns as well, the resulting *cis*- and *trans*-epoxides are separated to baseline. Thus, this represents a useful method for determining not only which isomer the insect produces, but also the stereoisomeric purity of the synthesized alkene.

Nonconjugated dienes can be made by analogous routes. For example, dienes can be made by sequential reaction of double-ended, ω-bromoalkyl triflates (readily available from diols or bromoalcohols) with two terminal alkynes (Figure 8.13C) (Armstrong-Chong *et al.*, 2004). The resulting diyne is then reduced as described above for monoalkynes. With care, the reaction can be done in one pot, or broken into two steps to completely avoid any double alkylation with either of the two alkynes, which might complicate purification. This is a more efficient version of a longer route, in which a protected haloalcohol would be coupled with the first alkyne, followed by removal of the protecting group and conversion of the resulting alcohol to a sulfonate or halide leaving group, followed by coupling with the second alkyne (e.g., Mant *et al.*, 2005). In a nice example of using available substrates, Nemoto *et al.* (1994) made (7Z,27Z)-tritriacontadiene by Li_2CuCl_4-catalyzed coupling of (11Z)-hexadecenylmagnesium bromide (readily prepared from (11Z)-hexadecen-1-ol, a commercially available lepidopteran pheromone) and (11Z)-heptadecenyl bromide. In another clever example employing commercially available synthons, Carlson and Mackley (1985) used 1,7,13-tetradecatriyne in very short syntheses of trienes with a 1,7,13-triene motif. Analogous syntheses of dienes can be done with double-ended terminal alkynes, a number of which are available, or which can be easily made from α,ω-dihaloalkanes.

Nonconjugated dienes can also be made by sequential reaction of double-ended Wittig reagents with two different aldehydes in one reaction sequence (Pohnert and Boland, 2000; Zarbin *et al.*, 2004). Again, this is a more efficient version of a longer route in which the Wittig reactions are conducted in two different steps after various protections, deprotections, and manipulations of functional groups.

Dienes or polyenes with a single methylene between two sequential (*Z*)-double bonds (so-called skipped methylene polyenes) represent a ubiquitous motif in nature, being found in fatty acids and numerous fatty acid derivatives. If the double-bond placement is correct, they can be synthesized from the appropriate fatty acids in a few steps by chain extension

(Figure 8.13D) (e.g., Conner *et al.*, 1980; Heath *et al.*, 1983; Davies *et al.*, 2007). Those in which the double-bond placement does not match available fatty acid precursors can be synthesized by iterative coupling of three-carbon propargyl units (Figure 8.13E) (Millar *et al.*, 1987).

Conjugated dienes are much less common in cuticular lipids, and there are a large number of ways in which any desired isomer can be made with high stereoselectivity. The interested reader is referred to the compilation of methods in Mori (1992) as a starting point. A convenient route to any one or all four stereoisomers of a conjugated diene from two acetylenic precursors has been described by Svatoš and Saman (1997).

Polyenes can be made via the same methods that are used for monoenes and dienes, using iterative sequences of coupling reactions to build up a chain with each unsaturation placed correctly. Thus, a coupling reaction between a terminal alkyne and a synthon containing a leaving group on one end and a protected alcohol on the other places the first latent double bond. After deprotection of the alcohol and conversion to a leaving group, coupling to a second terminal alkyne with a protected alcohol at its other terminus places the next latent double bond. The cycle of deprotection, conversion to a leaving group, and coupling is repeated as many times as necessary. Once the complete carbon chain is assembled, all the alkynes can be reduced simultaneously. Alternatively, an iterative series of Wittig reactions can be used. However, because Wittig reactions are not completely stereospecific, the resulting products will be contaminated with variable amounts of other stereoisomers which may prove more or less difficult to remove.

Allenes. To date, allenes have been found in the cuticular lipids of only a single group of insects, comprised of some Australian scarab beetles in the tribe Melolonthini (McGrath *et al.*, 2003). These authors readily synthesized the racemic allenes by thermal decomposition of hydrazines produced in a Mitsunobu-like reaction by treating appropriate secondary propargyllic alcohols with *o*-nitrobenzenesulfonylhydrazine, triphenylphosphine, and diethylazodicarboxylate. Alternatively, racemic internal allenes were formed by deprotonation and alkylation of terminal allenes in THF/HMPA with primary alkyl halides. Because the thermal elimination of the hydrazine was stereospecific, the corresponding chiral allenes could be generated simply by using chiral propargyl alcohol starting materials, available in one step from the corresponding propargyl ketones. Alternatively, *trans*-addition of tributyltin hydride to the triple bond of the chiral propargyl alcohol, acetylation of the alcohol, and treatment of the resulting acetoxystannane with tetrabutylammonium fluoride gave the desired chiral allenes. In both cases, the enantiomeric purity of the products was determined largely by the enantiomeric excess of the intermediate propargyl alcohols.

Acetylenes. Acetylenes have only been reported from cuticular lipids of two insect species to date, the cerambycid beetle *Callidiellum rufipenne* (9-pentacosyne and 9-heptacosyne; Rutledge *et al.*, 2009) and the ant *Platythrea punctata* (Hartmann *et al.*, 2005). However, acetylenes have been found in several insect pheromones (see *The Pherolist*, http://www.nysaes. cornell.edu/pheronet/). Thus, it is highly likely that more examples will turn up. Overall, acetylenes are easy to make, as described above, and have the advantage that there is no stereochemistry to worry about.

Summary. This chapter describes straightforward and commonly used methods of synthesizing insect cuticular hydrocarbons, with the goal of encouraging more research groups to try making their own synthetic standards. Such compounds can be used both for proper verification of the identities of insect-produced compounds, and to provide materials for bioassays so that biological activities of hydrocarbons can be assessed. I hope that I have shown that synthesis of many of these compounds is not difficult, even for chiral compounds. Furthermore, by describing a number of easy methods of purifying synthesized compounds, I hope to have minimized or eliminated a significant stumbling block in these types of synthesis. As a final word of encouragement, it should be pointed out that most hydrocarbons are very stable (with the exception of polyenes), so that these compounds, once made and purified carefully, can form part of a library of standards that will last indefinitely without degrading.

References

Abrams, S. R. (1984). Alkyne isomerization reagents: mixed alkali metal amides. *Can. J. Chem.*, **62**, 1333–1334.

Akino, T., Yamamura, K., Wakamura, S. and Yamaoka, R. (2004). Direct behavioral evidence for hydrocarbons as nestmate recognition cues in *Formica japonica* (Hymenoptera: Formicidae). *Appl. Entomol. Zool.*, **39**, 381–387.

Armstrong-Chong, R. J., Matthews, K. and Chong, J. M. (2004). Sequential alkynylation of ω-bromoalkyl triflates: facile access to unsymmetrical non-conjugated diynes including precursors to diene pheromones. *Tetrahedron*, **60**, 10239–10244.

Attygalle, A. (1998). Microchemical techniques. In *Methods in Chemical Ecology, Vol. 1, Chemical Methods*, ed. J. G. Millar and K. F. Haynes, New York: Chapman and Hall.

Breit, B. and Herber, C. (2004). Iterative deoxypropionate synthesis based on a copper-mediated directed allylic substitution. *Angew. Chem. Int. Ed.*, **43**, 3790–3792.

Brooke, G. M., Burnett, S., Mohammed, S., Proctor, D. and Whiting, M. C. (1996). A versatile process for the syntheses of very long chain alkanes, functionalized derivatives, and some branched chain hydrocarbons. *J. Chem. Soc., Perkin Trans.*, **1**, 1635–1645.

Buck, M. and Chong, J. M. (2001). Alkylation of 1-alkynes in THF. *Tetrahedron Lett.*, **42**, 5825–5827.

Carlson, D. A. and Mackley, J. W. (1985). Polyunsaturated hydrocarbons in the stable fly. *J. Chem. Ecol.*, **11**, 1485–1496.

Carrière, Y., Millar, J. G., McNeil, J. N., Miller, D. and Underhill, E. W. (1988). Identification of female sex pheromone in alfalfa blotch leafminer, *Agromyza frontella* (Rondani)(Diptera: Agromyzidae). *J. Chem. Ecol.*, **14**, 947–956.

Chow, S., Fletcher, M. T., Lambert, L. K., Gallagher, O. P., Moore, C. J., Cribb, B. W., Allsopp, P. G. and Kitching, W. (2005). Novel cuticular hydrocarbons from the cane beetle *Antitrogus parvulus* – 4,6,8,10,16-penta- and 4,6,8,10,16,18-hexamethyldocosanes – unprecedented anti-anti-anti-stereochemistry in the 4,6,8, 10-methyltetrad. *J. Org. Chem.*, **70**, 1808–1827.

Conner, W. E., Eisner, T., Vander Meer, R. K., Guerrero, A., Ghiringelli, D. and Meinwald, J. (1980). Sex attractant of an arctiid moth (*Utethesa ornatrix*): a pulsed chemical signal. *Behav. Ecol. Sociobiol.*, **7**, 55–63.

Cvačka, J., Jiroš, P., Sobotník, J., Hanus, R. and Svatoš, A. (2006). Analysis of insect cuticular hydrocarbons using matrix-assisted laser desorption/ionization mass spectrometry. *J. Chem. Ecol.*, **32**, 409–434.

Davies, N. W., Meredith, G., Molesworth, P. P. and Smith, J. A. (2007). Use of the antioxidant butylated hydroxytoluene *in situ* for the synthesis of readily oxidized compounds: application to the synthesis of the moth pheromone (*Z,Z,Z*)-3,6,9-nonadecatriene. *Aus. J. Chem.*, **60**, 848–849.

D'Ettorre, P., Heinze, J., Schulz, C., Francke, W. and Ayasse, M. (2004). Does she smell like a queen? Chemoreception of a cuticular hydrocarbon signal in the ant *Pachycondyla inversa*. *J. Exp. Biol.*, **207**, 1085–1091.

Doolittle, R. E., Proveaux, A. T., Alborn, H. T. and Heath, R. R. (1995). Quadrupole storage mass spectrometry of mono- and dimethylalkanes. *J. Chem. Ecol.*, **21**, 1677–1695.

Fletcher, M. T., Allsopp, P. G., McGrath, M. J., Chow, S., Gallagher, O. P., Hull, C., Cribb, B. W., Moore, C. J. and Kitching, W. (2008). Diverse cuticular hydrocarbons from Australian cane beetles (Coleoptera: Scarabaeidae). *Austral. J. Entomol.*, **47**, 153–159.

Fletcher, M. T., Chow, S., Lambert, L. K., Gallagher, O. P., Cribb, B. W., Allsopp, P. G., Moore, C. J. and Kitching, W. (2003). 4,6,8,10,16-Penta- and 4,6,8,10,16,18-hexamethyldocosanes from the cane beetle *Antitrogus parvulus* – cuticular hydrocarbons with unprecedented structure and stereochemistry. *Org. Lett.*, **5**, 5083–5086.

Fukaya, M., Wakamura, S., Yasuda, T., Senda, S., Omata, T. and Fukusaki, E. (1997). Sex pheromonal activity of geometric and optical isomers of synthetic contact pheromone to males of the yellow-spotted longicorn beetles, *Psacothea hilaris* (Pascoe) (Coleoptera: Cerambycidae). *Appl. Entomol. Zool.*, **32**, 654–656.

Ginzel, M. D., Moreira, J. A., Ray, A. M., Millar, J. G. and Hanks, L. M. (2006). (9*Z*)-Nonacosene – major component of the contact sex pheromone of the beetle *Megacyllene caryae*. *J. Chem. Ecol.*, **32**, 435–451.

Hartmann, A., d'Etorre, P., Jones, G. R., and Heinze, J. (2005). Fertility signaling – the proximate method of worker policing in a clonal ant. *Naturwissen.* **92**, 282–286.

Heath, R. R., Tumlinson, J. H., Leppla, N. C., McLaughlin, J. R., Dueben, B., Dundulis, E. and Guy, R. H. (1983). Identification of a sex pheromone produced by female velvetbean caterpillar moth. *J. Chem. Ecol.*, **9**, 645–656.

Heathcock, C. H., Finkelstein, B. L., Jarvi, E. T., Radel, P. A. and Hadley, C. R. (1988). 1,4- and 1,5-Stereoselection by sequential Aldol addition to α,ß-unstaurated aldehydes followed by Claisen rearrangement. Application to total synthesis of the vitamin E side chain and the archaebacterial C40 diol. *J. Org. Chem.*, **53**, 1922–1942.

Herber, C. and Breit, B. (2005). Enantioselective total synthesis and determination of the absolute configuration of the 4,6,8,10,16,18-hexamethyldocosane from *Antitrogus parvulus*. *Angew. Chem. Int. Ed.*, **44**, 5267–5269.

Herber, C. and Breit, B. (2007). Enantioselective total synthesis and determination of the absolute configuration of the 4,6,8,10,16,18-hexamethyldocosane from *Antitrogus parvulus*. *Eur. J. Org. Chem.*, **21**, 3512–3519.

Hoshino, C. and Mori, K. (1980). Pheromone synthesis. Part XL. Synthesis of a diastereomeric mixture of 15,19,23-trimethylheptatriacontane, the most active component of the sex pheromone of the female tsetse fly, *Glossina morsitans morsitans*. *Agric. Biol. Chem.*, **44**, 3007–3009.

Hoye, R. C., Baigorria, A. S., Danielson, M. E., Pragman, A. A. and Rajapakse, H. A. (1999). Synthesis of elenic acid, an inhibitor of topoisomerase II. *J. Org. Chem.*, **64**, 2450–2453.

Hwang, Y.-S., Mulla, M. S. and Arias, J. R. (1974). Overcrowding factors of mosquito larvae. V. Synthesis and evaluation of some branched-chain fatty acids against mosquito larvae. *J. Agr. Food Chem.*, **22**, 400–403.

Hwang, Y.-S., Mulla, M. S., Arias, J. R. and Majori, G. (1976). Overcrowding factors of mosquito larvae. VII. Preparation and biological activity of methyloctadecanes and methylnonadecanes against mosquito larvae. *J. Agr. Food Chem.*, **24**, 160–163.

Hwang, Y.-S., Navvab-Gojrati, H. A. and Mulla, M. S. (1978). Overcrowding factors of mosquito larvae. 10. Structure-activity relationship of 3-methylalkanoic acids and their esters against mosquito larvae. *J. Agr. Food Chem.*, **26**, 557–560.

Klewer, N., Ruzicka, Z. and Schulz, S. (2007). (Z)-Pentacos-12-ene, an oviposition-deterring pheromone of *Cheilomenes sexmaculata*. *J. Chem. Ecol.*, **33**, 2167–2170.

Kuwahara, S. and Mori, K. (1983). Pheromone synthesis. Part LXII. Synthesis of all of the three possible stereoisomers of 13,23-dimethylpentatriacontane, a sex pheromone of the tsetse fly, *Glossina pallidipes*. *Agric. Biol. Chem.*, **47**, 2599–2606.

Lacey, E. S., Ginzel, M. D., Millar, J. G. and Hanks, L. M. (2008). 7-Methylheptacosane is a major component of the contact sex pheromone of the cerambycid beetle *Neoclytus acuminatus acuminatus*. *Physiol. Entomol.*, **33**, 209–216.

Larock, R. C. (1999). *Comprehensive Functional Group Transformations*, 2nd edn. New York: Wiley.

Lehmler, H.-J., Bergosh, R. G., Meier, M. S. and Carlson, R. M. K. (2002). A novel synthesis of branched high-molecular-weight (C40+) long-chain alkanes. *Biosci. Biotech. Biochem.*, **66**, 523–531.

Mant, J., Brändli, C., Vereecken, N. J., Schulz, C. M., Francke, W. and Schiestl, F. P. (2005). Cuticular hydrocarbons as sex pheromone of the bee *Colletes cunicularius* and the key to its mimicry by the sexually deceptive orchid *Ophrys exaltata*. *J. Chem. Ecol.*, **31**, 1765–1787.

Marukawa, K., Takikawa, H. and Mori, K. (2001). Pheromone Synthesis. Part 207. Synthesis of the enantiomers of some methyl-branched cuticular hydrocarbons of the ant, *Diacamma* sp. *Biosci. Biotech. Biochem.*, **65**, 305–314.

Masuda, Y. and Mori, K. (2002). Pheromone synthesis. Part 215. Synthesis of the four stereoisomers of 3,12-dimethylheptacosane, (Z)-9-pentacosene and (Z)-9-heptacosene, the cuticular hydrocarbons of the ant, *Diacamma* sp. *Biosci. Biotech. Biochem.*, **66**, 1032–1038.

Matsuyama, K. and Mori, K. (1994). Pheromone synthesis. Part 159. Synthesis of a stereoisomeric mixture of 13,25-, 11,21- and 11,23-dimethylheptatriacontane, the contact sex pheromone of the tsetse fly, *Glossina tachinoides*. *Biosci. Biotech. Biochem.*, **58**, 539–543.

McGrath, M. J., Fletcher, M. T., König, W. A., Moore, C. J., Cribb, B. W., Allsopp, P. G. and Kitching, W. (2003). A suite of novel allenes from Australian melolonthine scarab beetles. Structure, synthesis, and stereochemistry. *J. Org. Chem.*, **68**, 3739–3748.

Metzger, P., Casadevall, E., Pouet, M. J. and Pouet, Y. (1985). Structures of some botryococcenes: branched hydrocarbons from the B-race of the green alga *Botryococcus braunii*. *Phytochemistry*, **24**, 2995–3002.

Millar, J. G., Underhill, E. W., Giblin, M. and Barton, D. (1987). Sex pheromone components of three species of *Semiothisa* (Geometridae), (Z,Z,Z)-3,6,9-heptadecatriene and two monoepoxide analogs. *J. Chem. Ecol.*, **13**, 1271–1283.

Moreira, J.A. and Corrêa, A.G. (2003). Enantioselective synthesis of three stereoisomers of 5,9-dimethylpentadecane, sex pheromone component of *Leucoptera coffeella*, from (-)-isopulegol. *Tetrahedron Asymm.*, **14**, 3787–3795.

Mori, K. (1992). The synthesis of insect pheromones, 1979–1989. In *The Total Synthesis of Natural Products*, Vol. 9, ed. J. Apsimon. New York: Wiley pp. 60–63.

Mori, K. (2007). The significance of chirality in pheromone science. *Biorg. Med. Chem.*, **15**, 7505–7523.

Mori, K. (2008a). Synthesis of the (5*S*,9*R*)-isomer of 5,9-dimethylpentadecane, the major component of the female sex pheromone of the coffee leaf miner, *Leucoptera coffeella*. *Tetrahedron: Asymm.*, **19**, 857–861.

Mori, K. (2008b). Synthesis of all the six components of the female-produced contact sex pheromone of the German cockroach, *Blattella germanica* (L.). *Tetrahedron*, **64**, 4060–4071.

Mori, K. and Wu, J. (1991a). Pheromone synthesis. CXXV. Synthesis of the four possible stereoisomers of 3,7-dimethylnonadecane, the female sex pheromone of *Agromyza frontella* Rondani. *Liebigs Ann. Chem.*, **1991**, 213–217.

Mori, K. and Wu, J. (1991b). Synthesis of the (5*S*,9*S*)-isomers of 5,9-dimethylheptadecane and 5,9-dimethyloctadecane, the major and the minor components of the sex pheromone of *Leucoptera malifoliella* Costa. *Liebigs Ann. Chem.*, **1991**, 439–443.

Nakamura, Y. and Mori, K. (2000). Pheromone synthesis. Part 204. Synthesis of the enantiomers of anti-2,6-dimethylheptane-1,7-diol monotetrahydropyranyl ether and their conversion into the enantiomers of the sex pheromone components of the apple leafminer, *Lyonetia prunifoliella*. *Eur. J. Org. Chem.*, **15**, 2745–2753.

Naoshima, Y. and Mukaidani, H. (1987). Synthesis of racemate and enantiomers of 15-methyltritriacontane, sex-stimulant pheromone of stable fly *Stomoxys calcitrans* L. *J. Chem. Ecol.*, **13**, 325–333.

Naoshima, Y., Mukaidani, H., Shibayama, S. and Murata, T. (1986). Selective alkylation of diethyl 3-oxoglutarate. Part XII. Synthesis of diastereomeric mixture of 15,19, 23-trimethylheptatriacontane, contact sex pheromone of tsetse fly, *Glossina morsitans morsitans* Westwood. *J. Chem. Ecol.*, **12**, 127–133.

Nemoto, T., Doi, M., Oshio, K., Matsubayashi, H., Oguma, Y., Suzuki, T. and Kuwahara, Y. (1994). (Z,Z)-5,27-Tritriacontadiene: major sex pheromone of *Drosophila pallidosa* (Diptera: Drosophilidae). *J. Chem. Ecol.*, **20**, 3029–3037.

O'Connor, J.G., Burow, F.H. and Norris, M.S. (1962). Determination of normal paraffins in C20 to C32 paraffin waxes by molecular sieve adsorbtion. *Analyt. Chem.*, **34**, 82–86.

Oguma, Y., Nemoto, T. and Kuwahara, Y. (1992). (Z)-11-Pentacosene is the major sex pheromone component in *Drosophila virilis* (Diptera). *Chemoecology*, **3**, 60–64.

Pempo, D., Cintrat, J.-C., Parrain, J.-L. and Santelli, M. (2000). Synthesis of [^{3}H$_2$]-(11*S*,17*R*)-11,17-dimethylhentriacontane: a useful tool for the study of the internalisation of communication pheromones of ant *Camponotus vagus*. *Tetrahedron*, **56**, 5493–5497.

Pempo, D., Viala, J., Parrain, J.-L. and Santelli, M. (1996). Synthesis of (11*R*,17*S*)-11, 17-dimethylhentriacontane: a communication pheromone of ant *Camponotus vagus*. *Tetrahedron: Asymm.*, **7**, 1951–1956.

Peschke, K. and Metzger, M. (1987). Cuticular hydrocarbons and female sex pheromones of the rove beetle, *Aleochara curtula* (Goeze) (Coleoptera: Staphylinidae). *Insect Biochem.*, **17**, 167–178.

Pohnert, G. and Boland, W. (2000). Highly efficient one-pot double-Wittig approach to unsymmetrical (1Z,4Z,7Z)-homoconjugated trienes. *Eur. J. Org. Chem.*, **2000** 1821–1826.

Pomonis, J. G., Fatland, C. L., Nelson, D. R. and Zaylskie, R. G. (1978). Insect hydrocarbons. Corroboration of structure by synthesis and mass spectrometry of mono- and dimethylalkanes. *J. Chem. Ecol.*, **4**, 2319–2333.

Rowland, S. J., Lamb, N. A., Wilkinson, C. F. and Maxwell, J. R. (1982). Confirmation of 2,6,10,15,19-pentamethyleicosane in methanogenic bacteria and sediments. *Terahedron Lett.*, **23**, 101–104.

Rutledge, C. E., Millar, J. G., Romero, C. M., and Hanks, L. M. (2009). Identification of an important component of the contact sex pheromone of *Callidiellum rufipenne* (Coleoptera: Cerambycidae). *Environ. Entomol.*, **38**, 1267–1275.

Schlamp, K. K., Gries, R., Khaskin, G., Brown, K., Khaskin, E., Judd, G. J. R. and Gries, G. (2005). Pheromone components from body scales of female *Anarsia lineatella* induce contacts by conspecific males. *J. Chem. Ecol.*, **31**, 2897–2911.

Shibata, C., Furukawa, A. and Mori, K. (2002). Syntheses of racemic and diastereomeric mixtures of 3,7,11,15-tetramethylhentriacontane and 4,8,12, 16-tetramethyldotriacontane, the cuticular tetramethylalkanes of the tsetse fly, *Glossina brevipalpis. Biosci. Biotech. Biochem.*, **66**, 582–587.

Sonnet, P. E. (1976). Synthesis of 1,5-dimethylalkanes, components of insect hydrocarbons. *J. Am. Oil Chem. Soc.*, **53**, 57–59.

Sonnet, P. E. (1984). General approach to synthesis of chiral branched hydrocarbons in high configurational purity. *J. Chem. Ecol.*, **10**, 771–781.

Sonnet, P. E. and Heath, R. R. (1980). Stereospecific synthesis of (*Z,Z*)-11, 13-hexadecadienal, a female sex pheromone of the navel orangeworm, *Amyelois transitella* (Lepidoptera: Pyralidae). *J. Chem. Ecol.*, **6**, 221–228.

Sonnet, P. E., Uebel, E. C., Harris, R. L. and Miller, R. W. (1977). Sex pheromone of the stable fly: evaluation of methyl- and 1,5-dimethylalkanes as mating stimulants. *J. Chem. Ecol.*, **3**, 245–249.

Sugeno, W., Hori, M. and Matsuda, K. (2006). Identification of the contact sex pheromone of *Gastrophysa atrocyanea* (Coleoptera: Chrysomelidae). *Appl. Entomol. Zool.*, **41**, 269–276.

Suzuki, T. (1981). Identification of the aggregation pheromone of flour beetles *Tribolium castaneum* and *T. confusum* (Coleoptera: Tenebrionidae). *Agric. Biol. Chem.*, **45**, 1357–1363.

Svatoš, A. and Saman, D. (1997). Efficient stereoselective synthesis of all geometrical isomers of heptadeca-11,13-dienes. *Collect. Czech. Chem. Commun.*, **62**, 1457–1467.

Takikawa, H., Fujita, K, and Mori, K. (1997). Synthesis of (3*S*,11*S*)-3,11-dimethyl-2-heptacosanone, a new component of the female sex pheromone of the German cockroach. *Liebigs Ann. Chem.*, **1997**, 815–820.

Tamagawa, H., Takikawa, H. and Mori, K. (1999). Pheromone synthesis. Part 192. Synthesis of all the stereoisomers of 10,14-dimethyloctadec-1-ene, 5,9-dimethyloctadecane, and 5,9-dimethylheptadecane, the sex pheromone components of the apple leafminer, *Lyonetia prunifoliella. Eur. J. Org. Chem.*, **5**, 973–978.

Uebel, E. C., Schwarz, M., Miller, R. W. and Menzer, R. E. (1978). Mating stimulant pheromone and cuticular lipid constituents of *Fannia femoralis* (Stein) (Diptera: Muscidae). *J. Chem. Ecol.*, **4**, 83–93.

Uebel, E. C., Sonnet, P. E., Miller, R. W. and Beroza, M. (1975). Sex pheromone of the face fly *Musca autumnalis* (Diptera: Muscidae). *J. Chem. Ecol.*, **1**, 195–202.

Van Summeren, R. P., Reijmer, S. J. W., Feringa, B. A. and Minnard, A. J. (2005). Catalytic asymmetric synthesis of enantiopure isoprenoid building blocks: application in the synthesis of apple leafminer pheromones. *Chem. Comm.*, **2005**, 1387–1389.

Zambelli, A., Bajo, G. and Rigamonti, E. (1978). Model compounds and carbon-13 NMR investigation of isolated ethylene units in ethylene/propene copolymers. *Makromolek. Chem.*, **179**, 1249–1259.

Zanoni, G., Brunoldi, E. M., Porta, A. and Vidari, G. (2007). Asymmetric synthesis of 14-A4t-neuroprostane: hunting for a suitable biomarker for neurodegenerative diseases. *J. Org. Chem.*, **72**, 9698–9703.

Zarbin, P. H. G., Princival, J. L., de Lima, E. R., dos Santos, A. A., Ambrogio, B. G. and de Oliveira, A. R. M. (2004). Unsymmetrical double Wittig olefination in the syntheses of insect pheromones. Part 1: synthesis of 5,9-dimethylpentadecane, the sexual pheromone of *Leucoptera coffeella*. *Tetrahedron Lett.*, **45**, 239–241.

Zhang, A., Oliver, J. E., Chauhan, K., Zhao, B., Xia, L. and Xu, Z. (2003). Evidence for contact sex recognition pheromone of the Asian longhorned beetle, *Anoplophora glabripennis* (Coleoptera: Cerambycidae). *Naturwissen.*, **90**, 410–413.

Zhou, J. and Burgess, K. (2007). α,ω-Functionalized 2,4-dimethylpentane dyads and 2,4,6-trimethylheptane triads through asymmetric hydrogenation. *Angew. Chem. Int. Ed.*, **46**, 1129–1131.

Zhou, J., Zhu, Y. and Burgess, K. (2007). Synthesis of (*S,R,R,S,R,S*)-4,6,8,10,16, 18-hexamethyldocosane from *Antitrogus parvulus* via diastereoselective hydrogenations. *Org. Lett.*, **9**, 1391–1393.

Zhu, G., Liang, B. and Negishi, E.-I. (2008). Efficient and selective synthesis of (*S,R,R,S,R,S*)-4,6,8,10,16,18-hexamethyldocosane via Zr-catalyzed asymmetric carboalumination of alkenes (ZACA reaction). *Org. Lett.*, **10**, 1099–1101.

9

Oxygenated derivatives of hydrocarbons

James S. Buckner

Arthropod waxes (complex mixtures of long-chain-carbon compounds) include straight-chain saturated and unsaturated hydrocarbons, methyl-branched hydrocarbons and more polar lipids that contain one or more oxygen functional groups on long aliphatic carbon chains that include wax esters, sterol esters, ketones, alcohols, aldehydes and acids for Insecta (Lockey, 1988; Buckner, 1993; Nelson and Blomquist, 1995) and Arachnida (Chapters 7 and 16, this book). As a major lipid class, hydrocarbons are present in the cuticular extracts of virtually all insects studied and there have been previous reviews on their occurrence, identification, biosynthesis and function (Jackson and Blomquist, 1976; Blomquist and Jackson, 1979; Blomquist and Dillwith, 1985; Blomquist *et al.*, 1987; Lockey, 1988; Blomquist *et al.*, 1993; Howard, 1993; Nelson and Blomquist, 1995; Howard and Blomquist, 2005). This chapter focuses on the occurrence, structural identification and function of those hydrocarbons that possess one or more oxygenated functional groups that include: ethers, epoxides, ketones, secondary alcohols, and their esters. These lipid classes represent hydrocarbon derivatives in which oxygen was biosynthetically introduced onto a non-terminal carbon(s) of long-chain hydrocarbons. Other lipid classes included in this review are long-chain methyl-branched primary alcohols and those wax esters with either ketone groups or methyl branches on their acid and/or alcohol moieties. This review does not include those oxygenated insect lipids not derived from pre-formed hydrocarbon constituents: long-chain fatty acids, primary alcohols, aldehydes, wax esters (esters of long-chain alcohols and long-chain acids), sterols, sterol esters, and mono-, di- and triacylglycerols.

Occurrence and structural identification

Secondary alcohols

Long-carbon-chain secondary alcohols (*sec*-alkanols) are not common constituents of insect cuticular lipids (Buckner, 1993). Espelie and Bernays (1989) reported that the cuticular lipids of *Manduca sexta* larvae reared on tomato or potato foliage contained

6–7% free C_{29} and C_{27} secondary alcohols. For tomato-reared larvae, the C_{29} carbon chain secondary alcohols were a mixture of positional isomers with the hydroxyl group at either C_8 (51%), C_9 (33%), C_{10} (10%), or C_7 (6%). The distribution for the C_{27} isomers was C_7 (52%), C_6 (28%), and C_8 (19%). The Dufour glandular secretion of *Myrmecocystus mexicanus* worker ants contained the three 2-alkanols, 2-tridecanol, 2-pentadecanol and 2-heptadecanol (Lloyd *et al.*, 1989). Approximately 2% of the cuticular lipids of the cabbage seedpod weevil, *Ceutorrhynchus assimilis*, are secondary alcohols with even and odd carbon-numbers ranging from 26 to 30 (Richter and Krain, 1980). The major components, secondary nonacosanol (95%) and hexacosanol (2%), were the same prominent secondary alcohols that are present in the host plant of this insect. The C_{21} secondary alcohol alkene enantiomers (6Z, 9Z, 11S)-6,9-heneicosadien-11-ol and (6Z, 9Z, 11R)-6-,9-heneicosadien-11-ol were identified as major sex pheromone components of female tussock moths, *Orgyia detrita* (Gries *et al.*, 2003) and the same C_{21} secondary alcohol di-alkene was identified as a major sex pheromone of the painted apple moth, *Teia anartoides* (El-Sayed *et al.*, 2005).

Secondary alcohol esters

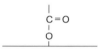

The occurrence of esters of secondary alcohols has been reported (Blomquist *et al.*, 1972; Warthen and Uebel, 1980; Pomonis *et al.*, 1993; Finidori-Logli *et al.*, 1996; Howard and Baker, 2003) and reviewed (Buckner, 1993; Nelson and Blomquist, 1995). Long-chain secondary alcohols are major constituents of the wax ester fraction from the cuticular lipids of six species of migratory grasshoppers, genus *Melanoplus*. These odd carbon-numbered C_{37}–C_{45} esters comprise 18 and 28% of the surface lipid of *M. sanguinipes* and *M. packardii*, respectively (Blomquist *et al.*, 1972). The secondary alcohol moieties of *M. sanguinipes* esters were C_{21} to C_{27} odd carbon-numbered compounds with the C_{23} compounds as major moieties (59%), and the major isomer as tricosan-11-ol, with smaller amounts of tricosan-10-ol. In the nymphs of *M. differentialis*, tricosanyl octadecanoate (C_{41}) and pentacosanyl hexadecanoate (C_{41}) are the major wax esters (Warthen and Uebel, 1980). The alcohol moieties are isomeric mixtures of 11- and 12-tricosanol and 8- and 9-pentacosanol. Adult *M. differentialis* have a different wax ester composition: tricosanyl octadecanoate (C_{41}) and pentacosanyl octadecanoate (C_{43}) with isomeric mixtures of 10-, 11- and 12-tricosanol and 11- and 12-pentacosanol respectively, as the secondary alcohol moieties (Warthen and Uebel, 1980). Odd-chain wax esters (C_{31}–C_{41}) with secondary alcohols as alcohol moieties are major components (18–30%) of the cuticular lipids of the migratory grasshoppers, *M. bivittatus*, *M. femurrubrum* and *M. dawsoni* (Jackson, 1981). Six unbranched and seven methyl-branched acetate ester homologs of long-chain (C_{29}) secondary alcohols (3-hydroxy to 8-hydroxy) were identified in the cuticular lipids of the female screwworm fly, *Cochliomyia hominivorax* (Pomonis *et al.*, 1993). The cuticular lipids of the adult male little housefly, *Fannia cannicularis*, contained 27% of the acetate ester of the secondary alcohol, 8-heneicosanol (Uebel *et al.*, 1977). The secondary alcohol acetate ester, 2*S*,12*Z*-2-acetoxy-12-heptadecene, was identified as the major sex pheromone component of the pistachio twig borer, *Kermania pistaciella* (Gries *et al.*, 2006). Small quantities of novel esters of long-chain secondary alcohols (C_{25}–C_{32}) and short-chain acids (C_2–C_4) were reported in hexane extracts of the male antenna and forelegs of *Helicoverpa zea* and *Heliothis virescens* (Böröczky *et al.*, 2008). The C_{25} and C_{27} secondary alcohol moieties for both species were identified as 7- and 8-pentacosanol and 8- and 9-heptacosanol respectively.

Diols

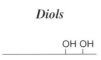

Diols have been rarely observed in insect cuticular lipids (Buckner, 1993). Odd-carbon-number diols (C_{23}–C_{29}) were the major lipid class (55%) of the larval cuticular lipids from the flour beetle, *Tenebrio molitor* (Bursell and Clements, 1967). The major diol constituent was 8, 9-pentacosanediol. For the cuticular lipids of *M. sexta* larvae, very small amounts (<1%) of 7,8- and 8,9-C_{27} diols and 8,9- and 9,10-C_{29} diols were identified (Espelie and Bernays, 1989). Hydroxy *n*-alkanols are diols with a hydroxyl functional group on the C_1 position (terminal) of the alkyl chain, but are technically not alcohol derivatives of hydrocarbons. There are a few reports of the occurrence of insect hydroxy *n*-alkanols (Buckner, 1993; Nelson and Blomquist, 1995; Buckner *et al.*, 1996). In a structure analysis study of beeswax, the major alcohol moieties of the diester fraction were identified as 1,23-tetracosanediol (42.2%), 1,27-octacosanediol (26.0%) and 1,25-hexacosanediol (20.2%) (Tulloch, 1971). The hydroxy *n*-alkanols comprised 16% of the cuticular lipids of *H. zea* pupae and were identified as C_{30}–C_{36} even-chain *n*-alcohols with hydroxyl groups on carbon numbers 11, 12, 13, 14, or 15 (Buckner *et al.*, 1996). Mass spectral analysis indicated the presence of unsaturation in the alkyl chain of the major diol components.

Methyl-branched alcohols

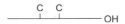

Very-long-chain methyl-branched alcohols (C_{38} to >C_{44}) and their esters with short-chain acids (C_2 to C_5) represent a novel class of long-chain internal lipids which mainly occur during insect metamorphosis (Nelson, 1993). The very-long-chain methyl-branched alcohols were first characterized in the internal lipids of developing pupae (pharate adults)

of the cabbage looper, *Trichoplusia ni* (Nelson *et al.,* 1989). The major alcohols had carbon backbones of 38, 40 and 42 carbons with 1, 2 or 3 methyl branches and occurred in 4 homologous series. The major alcohol components of each of the series respectively, were 24-methyltetracontan-1-ol, 24,28-dimethyltetracontan-1-ol, 24,36-dimethyltetracontan-1-ol, and 24,28,36-trimethyltetracontan-1-ol. These constituents existed in the internal lipids of the pupa both as free alcohols and as ester components. The acid moiety of the very-long-chain methyl-branched alcohol esters was identified as acetic acid (Nelson and Blomquist, 1990). Neither the free alcohols nor their esters were found in the surface lipid of any stage of the cabbage looper.

The internal lipids of developing pupae of *M. sexta* contained long-chain methyl-branched alcohols (C_{25} to C_{32}) (Nelson *et al.,* 1990). These alcohols occurred as mixtures of monomethyl- and dimethyl-branched isomers in which the methyl branches were closer to the hydroxyl end of the molecule than to the alkyl end of the molecule, the opposite of that observed for the very-long-chain methyl-branched alcohols (C_{38} to >C_{44}) (Nelson and Fatland, 1992). The long-chain methyl-branched alcohols occurred mainly as acetate esters (Nelson *et al.,* 1990) with minor amounts of propionate esters (Nelson and Fatland, 1992). The propionate esters have only been characterized in *M. sexta*. In addition to their presence in developing pupae, small amounts of the alcohols and their acetate esters were identified in the internal lipids of larvae and adults of *M. sexta*. The very-long-chain methyl-branched alcohols (C_{35} to >C_{46}) and their acetate esters have also been identified in the internal lipids of pupae of the southwestern corn borer, *Diatraea grandiosella* (Nelson and Blomquist, 1990; Nelson and Fatland, 1997), the southern armyworm, *Spodoptera eridania* (Guo *et al.,* 1992), and in the tobacco budworm, *H. virescens*, the corn earworm, *H. zea*, the sunflower moth, *Homoeosoma electellum* and the banded sunflower moth, *Cochylis hospes* (Nelson and Fatland, 1997). Minor amounts of long-chain methyl-branched alcohols (C_{25}–C_{34}) were also found mainly in *H. virescens* and *H. zea* (Nelson and Fatland, 1997).

Ethers

Aliphatic ethers have been observed in cuticular lipids from a few insect species. The surface lipids of the locust, *L. m. cinerascens* contained 4–5% aliphatic ethers (Génin *et al.,* 1987). The major ethers were C_{29}, C_{31} and C_{33} compounds with the alkyl moieties ranging in size from 11 to 20 carbons. The locust showed dimorphism: solitary locusts had a majority of the longer-carbon-chain ethers while the gregarious locusts had a majority of shorter-carbon-chain ethers. The surface lipids of the red-shouldered leaf beetle, *Monolepta australis*, contained a series of 7-octadecenyl alkyl ethers, the major constituent being 7-octadecenyl pentadecyl ether (Southwell and Stiff, 1989).

Epoxides

Epoxy derivatives of n-alkanes

The occurrence of hydrocarbons (usually mono- and di-alkenes) with an epoxide function group have been reported usually as sex attractants. The sex attractant of the female gypsy moth, *Lymantria dispar*, was identified as the C_{18} 2-methyl alkane derivative *cis*-7,8-epoxy-2-methyloctadecane (Bierl *et al.*, 1972). For the housefly, *M. domestica*, a major sex pheromone component is the C_{23} *n*-alkane epoxide *cis*-9,10-epoxytricosane (Uebel *et al.*, 1978) with a lesser quantity of 9,10-epoxyheptacosane (Mpuru *et al.*, 2001).

Epoxy derivatives of mono-alkenes

C_{19} mono-alkenes with an epoxide functional group include the lepidopteran female sex pheromones 9-*cis*-(Z)-6,7-epoxy-nonadecene and 6-*cis*-(Z)-9,10-epoxy-nonadecene of the geometrid moths, *Biston robustum* (Yamamoto *et al.*, 2000) and the common forest looper, *Pseudocoremia suavis* (Gibb *et al.*, 2006), respectively. The first epoxide of a C_{21} mono-alkene, (6Z)-*cis*-9,10-epoxyheneicosene, was first described by Rollin and Pougny (1986) as a pheromone component of the ruby tiger moth, *Phragmatobia fuliginosa*. A novel C_{21} di-epoxide pheromone component, 3Z-*cis*-6,7-*cis*-9,10-diepoxyheneicosene was first identified from the lepidopteran, *Leucoma salicis* (Gries *et al.*, 1997) and more recently, (3R,4S,6S,7R, 9Z)-3,4–6,7-diepoxyheneicosene was identified in the pheromone gland of the clear-winged tussock moth, *Perina nuda* (Wakamura *et al.*, 2002).

Epoxy derivatives of di-alkenes

For C_{19} epoxy di-alkenes, identified sex attractants for geometrid moths include 3Z,9Z-6-,7-epoxy-nonadecadiene, *Eufidonia convergaria* (Millar *et al.*, 1990a) and *B. robustum* (Yamamoto *et al.*, 2000); 6Z,9Z-3,4-epoxy-nonadecadiene, *Probole amicaria* (Millar *et al.*, 1990b) and *Milionia basalis pryeri* (Yasui *et al.*, 2005); and 3Z,6Z-9,10-epoxy-nonadecadiene, *P. suavis* (Gibb *et al.*, 2006). For the noctuid moth, *Rivula propinqualis*, 3Z,9Z-6,7-epoxy-nonadecadiene was identified in the female pheromone-gland extract (Millar *et al.*, 1990a). Components of the tiger moth (*Arctia caja*) sex pheromone were identified as the C_{20} and C_{21} epoxides, (3Z, 6Z)-*cis*-9,10-epoxyeicosadiene and 3Z,6Z-*cis*-9,10-epoxyheneicosadiene (Bestmann *et al.*, 1992). The same 9,10-epoxy C_{21} di-alkene was a sex pheromone component of the saltmarsh caterpillar moth, *Estigmene acrea* (Hill and Roelofs, 1981) and ruby tiger moth, *P. fuliginosa* (Rollin and Pougny, 1986). The pheromone glands of the geometrid moth species, *Semiothisa signaria dispuncta*, *Caenurgina distincta* and *Euclidea cuspidea*, produced pheromone blends that contained the C_{17}, C_{20} and C_{21} epoxy alkenes, 6Z,9Z-cis-3,4-epoxyheptadecadiene (Millar *et al.*, 1987), 3Z,6Z-*cis*-9,10-epoxyeicosadiene and 3Z,6Z-cis-9,10-epoxyheneicosadiene, respectively (Millar *et al.*, 1991).

Ketones

$$\underset{\displaystyle \parallel}{\overset{\displaystyle O}{\rule{4cm}{0.4pt}}}$$

Ketones as cuticular lipids

Like secondary alcohols, ketones are not common constituents of the cuticular lipids of insects (Lockey, 1988). The cuticular lipid of the female housefly, *M. domestica*, contains 6% of an unsaturated ketone, (Z)-14-tricosen-10-one (Uebel *et al.*, 1978) and lesser amounts of tricosan-10-one and heptacosen-12-one (Mpuru *et al.*, 2001). The cuticular lipids of several species of *Drosophila* contain C_{13}–C_{17} saturated and unsaturated ketones, including 2-tridecanone and 2-pentadecanone in *Drosophila hydei* (Moats *et al.*, 1987), 10-heptadecen-2-one in *D. mulleri* (Bartelt *et al.*, 1989), and 2-pentadecanone in *D. busckii* (Schaner *et al.*, 1989). Odd-chain ketones (2-nonadecanone, 2-heneicosanone and 2-tricosanone) comprise 1% and 3% of adult male and female cuticular lipids, respectively, of the pecan weevil, *Curculio caryae* (Espelie and Payne, 1991). The cuticular lipids of mature screwworm females, *C. hominivorax*, contained small quantities of two C_{31} ketones: the symmetrical ketone, 16-hentriacontanone and the methyl-branched ketone, 21-methyl-7-hentriacontanone (Pomonis *et al.*, 1993).

Ketones as glandular lipids

Long-chain ketones have been reported as semiochemicals in lipid mixtures of specialized glands. (Z)-10-Heptadecen-2-one was identified in surface lipids and the ejaculatory bulb of *D. martensis*, *D. buzzatii* and *D. serido* (Schaner and Jackson, 1992). Both (Z)-10-heptadecen-2-one and 2-tridecanone were synthesized by the microsomal fraction of the ejaculatory bulbs from mature male *D. buzzatii* (Skiba and Jackson, 1993). Identified keto-hydrocarbons from moth pheromone gland components include: the C_{21} keto monoenes, (Z)-6-heneicosen-11-one of the Douglas-fir tussock moth, *Orgyia pseudotsugata* (Smith *et al.*, 1975) and (Z)-6-heneicosen-9-one, *O. thyellina* (Gries *et al.*, 1999); the C_{21} keto di-alkene, (Z,E)-6,8-heneicosadien-11-one of the Douglas-fir tussock moth, *O. pseudotsugata* (Gries *et al.*, 1997) and the western tussock moth *O. vetusta* (Gries *et al.*, 2005). Lipids from Dufour gland secretions of ants in the genus *Myrmecocystus* included a homologous series of the odd-chain ketones 2-tridecanone, 2-pentadecanone and 2-heptadecanone (Lloyd *et al.*, 1989). Two keto mono-alkenes, 10-nonacosen-2-one and 16-pentacosen-2-one, were identified in the postpharyngeal gland lipids of the male solitary European beewolf wasp, *Philanthus triangulum* (Schmidt *et al.*, 1990; Herzner *et al.*, 2007a). The lipids of the postpharyngeal gland secretion of female *P. triangulum* that is used to embalm paralyzed prey include the two keto mono-alkenes Δ-16-pentacosen-8-one and Δ-18-heptacosen-8-one (Herzner *et al.*, 2007b).

Methyl-branched ketones

The major cuticular lipids of the German cockroach, *Blattella germanica*, have been identified as homologous series of straight chain and methyl-branched C_{27} and C_{29} alkanes (Augustynowicz *et al.*, 1987; Jurenka *et al.*, 1989). The cuticular lipids of *B. germanica* also contained long-chain ketones that were characterized as contact (nonvolatile) sex pheromones (Nishida and Fukami, 1983; Jurenka *et al.*, 1989). The major component is the dimethyl ketone 3,11-dimethylnonacosan-2-one, with lesser amounts of 29-hydroxy-3,11-dimethylnonacosan-2-one, 29-oxo-3,11-dimethylnonacosan-2-one and the C27 isomer, 3,11-dimethylheptacosan-2-one (Nishida and Fukami, 1983; Jurenka *et al.*, 1989). Recently, two other oxygenated derivatives of the C_{27} dimethyl-branched alkanes, 27-oxo-3,11-dimethylheptacosan-2-one and 27-hydroxy-3,11-dimethylheptacosan-2-one, were identified as components of the German cockroach contact pheromone (Eliyahu *et al.*, 2008).

Keto-alcohols, keto-aldehydes and keto-wax esters

In addition to the occurrence of the ketone functional group on *n*-alkanes and methyl-branched alkanes of insect lipids, they have been reported as functional groups on long-chain aldehydes, *n*-alcohols and acids (Buckner, 1993). For the lepidopteran pupae of *M. sexta* and *H. zea*, the major pupal cuticular lipids were identified as long-chain oxoaldehydes and oxoalcohols (Buckner *et al.*, 1984b; Buckner *et al.*, 1996) and for *M. sexta*, short-chain acid esters of oxoalcohols (Buckner *et al.*, 1984a). Surface lipid of diapausing *M. sexta* pupae consisted mainly of 11- and 12-oxooctacosanol, most of which (35–45%) was esterified to 3-oxobutyric acid (Buckner *et al.*, 1984a), and long-chain oxoaldehydes (30–35%) with the major constituents as 11- and 12-oxooctacosanal (Buckner *et al.*, 1984b). Oxoalcohols (mainly 12-oxotriacontanol) and oxoaldehydes (mainly 12-oxotriacontanal) were reported as minor components of the cuticular lipid of *H. zea* (Buckner *et al.*, 1996). Novel wax esters containing ketone groups on both very-long-chain acid and alcohol moieties have been identified in several homopteran insects. The surface wax of the cochineal insect, *Coccus cacti* and the woolly alder aphid, *Prociphilus tessellatus*, contain mainly the C_{66} ester, 15-oxotetratriacontanyl 13-oxodotriacontanoate (Chibnall *et al.*, 1934; Meinwald *et al.*, 1975). The cuticular lipid of the white pine chermes aphid, *Adelges (Pineus) strobi* includes a C_{66} wax ester composed of 17-oxohexatriacontan-1-ol and 11-oxotriacontanoic acid (Blount *et al.*, 1937). The major oxoalcohol and oxoacid moieties of the diketo esters of the woolly apple aphid, *Eriosoma lanigerum*, are 15-oxotetratriacontan-1-ol and 13-oxodotriacontanoic acid (Cameron and Drake, 1976). The C_{64} ester, 15-oxo-tetratriacontanyl 11-oxo-triacontanoate, is the main constituent in the wax of the cochineal insect, *Dactylopius confusus* (Meinwald *et al.*, 1975), the lantern bug, *Cerogenes auricoma* and the related fulgorid species, *Fulgora castresii* (Mason *et al.*, 1989). The wax of another fulgorid species, *F. lampetis*, has 17-oxo-hexatriacontanyl 11-oxo-triacontanoate (C_{66}) as the major keto wax ester, but it also contains minor amounts of C_{44}–C_{54} normal carbon chain wax esters (Mason *et al.*, 1989).

Function of oxygenated hydrocarbons

The cuticular or surface lipids of arthropods are necessary for survival and among the many insect species cuticular lipids play a major role in minimizing water loss, in chemical communication, and in providing a wide range of other functions (Hadley, 1981; Blomquist and Dillwith, 1985; Noble-Nesbitt, 1991; St. Leger, 1991; Buckner, 1993; Nelson and Blomquist, 1995; Howard and Blomquist, 2005; see relevant subject chapters, this book). The functions of oxygenated derivatives of hydrocarbons have mainly been studied and reported in regard to their effects on protecting insects against desiccation and their role(s) in insect chemical communication.

Protective water barrier

The main function of cuticular lipid on terrestrial arthropods is to minimize the loss of water by transpiration across the integument; the lipids provide a better waterproofing barrier when they are in a solid rather than fluid state (Gibbs, 1998; Rourke and Gibbs, 1999; Rourke, 2000; Gibbs, 2002; Gibbs and Rajpuhorit, Chapter 6, this book). The introduction of a functional group (i.e., ester linkage, methyl branch, double bond, ketone, secondary alcohol) to the long, hydrophobic alkyl chain can introduce a kink in the lipid chain and disrupt lipid packing. A kink can result in increased fluidity and decrease in the melting temperature. Methyl branching gives lipid increased fluidity over a range of temperatures, and unsaturated hydrocarbons melt at a much lower temperature than their corresponding *n*-alkanes (Lockey, 1988; Gibbs and Pomonis, 1995). The presence of an ester linkage in long-chain wax esters has been shown to substantially lower the melting temperature relative to hydrocarbons containing the same number of carbon atoms (Gibbs and Pomonis, 1995; Patel *et al.*, 2001). Wax esters may interact with hydrocarbons to affect the properties of the overall lipid mixture (Riederer and Schneider, 1990; Gibbs, 1995; Dodd and Afzal-Rafii, 2000). Experimental data on the interactions between wax esters and *n*-alkanes showed a slight reduction in lipid melting temperature ($\leq 5°$ C) (Patel *et al.*, 2001).

In addition to wax esters, the inclusion of a mid-chain keto or alcohol functional group to a long-chain hydrocarbon would cause a kink in the lipid chain and result in cuticular lipids with lower melting temperatures. An example is provided by the wax esters of secondary alcohols, which occur on melanopline grasshoppers (Blomquist *et al.*, 1972; see above, this chapter). Structurally, these esters of secondary alcohols are T-shaped molecules that pack less closely than esters of primary alcohols, and Patel *et al.* (2001) showed that the secondary wax esters of grasshoppers (*M. sanguinipes*) melted >60°C below primary esters of the same molecular weight. The quantity of lipid on the cuticular surfaces of diapausing pupae of the tobacco hornworm, *M. sexta* was three times that of non-diapausing pupae (Bell *et al.*, 1975) and those lipids include large amounts of oxoaldehydes and oxoalcohols esterified to 3-oxobutyric and 3-hydroxybutyric acids (Buckner *et al.*, 1984a, b). The presence of mid-chain ketone groups could act to increase the fluidity of the mixture of highly oxygenated cuticular lipids during deposition.

For the long- and very-long-methyl-branched alcohols and their esters that occur internally in the mid-pupal stages of Lepidoptera (Nelson, 1993; see above, this chapter), the fate and function of these alcohols have not been established. They are not synthesized and are at low levels or undetectable in larvae and adults, and at the beginning and end of the pupal stage (Dwyer *et al.*, 1986; de Renobales *et al.*, 1989; Nelson *et al.*, 1990; Guo *et al.*, 1992; Nelson, 1993). Nelson and Fatland (1997) suggest that their presence in mid-stage pupae suggests a role in metamorphosis, but further research would be required to establish a definitive physiological role for the methyl-branched alcohols and their acetate esters in lepidopteran pupae.

Chemical communication

The majority of the known cuticular lipid constituents that function in chemical communication processes are hydrocarbons (Howard and Blomquist, 1982; Blomquist and Dillwith, 1985; Howard, 1993; Nelson and Blomquist, 1995). Most oxygenated lipids that function as pheromones are not part of the surface lipids but are secretory products of specialized glands (Tamaki, 1985; Arn *et al.*, 1986; Morgan and Mandava, 1988). Many of the compounds in pheromones are short-chain unsaturated aldehydes, ketones and acetate esters of short-chain (C_{10}–C_{14}) unsaturated alcohols. Those short-chain lipids (<C_{16}) with biological activity have been reviewed elsewhere (Blomquist and Dillwith, 1985; Tamaki, 1985; Lockey, 1988; Blomquist *et al.*, 1993; Jurenka and Roelofs, 1993; Howard, 1993).

Ketones and secondary alcohols

An unsaturated ketone (Z)-14-tricosen-10-one, a prominent (6%) constituent of the cuticular lipid of the female housefly, *M. domestica*, is the major component of the sex pheromone mixture (Uebel *et al.*, 1978). For several species of *Drosophila*, the cuticular lipids contain saturated and unsaturated ketones (C_{13}–C_{17}) which were characterized as components of aggregation pheromones. Identified structures included 2-tridecanone and 2-pentadecanone in *D. hydei* (Moats *et al.*, 1987), (Z)-10-heptadecen-2-one in *D. mulleri* (Bartelt *et al.*, 1989), and 2-pentadecanone in *D. busckii* (Schaner *et al.*, 1989). The ejaculatory bulb and surface lipids of three *Drosophila* species (*D. martensis, D. buzzatii* and *D. serido*) had the compound (Z)-10-heptadecen-2-one that functions as the major component of the aggregation pheromone of those insects (Schaner and Jackson, 1992). Both (Z)-10-heptadecen-2-one and 2-tridecanone (an inhibitor of aggregation) were synthesized by the microsomal fraction of the ejaculatory bulbs from mature male *D. buzzatii* (Skiba and Jackson, 1993). The C_{19} keto alkenes 10-nonacosen-2-one and 16-pentacosen-2-one, from the postpharyngeal gland of the male solitary wasp, european beewolf (*P. triangulum*), were identified as marking pheromones to attract females (Schmidt *et al.,* 1990; Kroiss *et al.,* 2006; Herzner *et al.,* 2007a).

Two acetate esters of mono-methyl-branched secondary alcohols were identified as major components of the contact mating pheromone of the New World screwworm *C. hominivorax*. Five homologous acetate derivatives of long-chain secondary alcohols and a related ketone, as identified by Pomonis *et al.* (1993) (see above in this chapter), were chemically synthesized (Furukawa *et al.,* 2002) and the two acetate esters of C_{29} secondary alcohols racemic 6-acetoxy-19-methylnonacosane and 7-acetoxy-15-methylnonacosane were characterized as sex stimulant pheromones (Carlson *et al.,* 2007). These two acetate derivatives were the first sex pheromones identified in a calliphorid fly. In the hymenopteran parasitoid *Diglyphus isaea*, long-chain secondary alcohol esters present on the females had an aphrodisiac effect on males. The major components were C_{21}–C_{25} 11-hydroxy esters of C_8 and C_{10} fatty acids (Finidori-Logli *et al.,* 1996). Secondary alcohols as pheromone components were reported for the first time in a ditrysian lepidopteran species. (6Z,9Z,11S)-6, 9-Heneicosadiene-11-ol and (6Z,9Z,11R)-6,9-heneicosadiene-11-ol were identified as the major sex pheromone components of female tussock moths, *O. detrita* (Gries *et al.,* 2003). These two C_{21} *sec*-alcohol di-alkenes, in combination but not singly, attracted significant numbers of male moths.

Methyl-branched ketones (German cockroach pheromone)

As discussed earlier in this chapter, the occurrence of dimethyl-branched ketone components of the surface lipids of the German cockroach, *B. germanica*, has been reported and shown to have contact sex pheromone activity (Nishida and Fukami, 1983; Jurenka *et al.,* 1989; Blomquist, Chapter 3, this book). Structurally, the mixture of pheromone components all have methyl-branched positions at carbons 3 and 11 and the keto functional group at the carbon 2 position. Biosynthetic studies demonstrated that the methyl ketone pheromone component is produced by a sex-specific hydroxylation of 3,11-dimethylnonacosane to the corresponding 3,11-dimethylnonacosan-2-ol (by females only), which is then oxidized to the di-methyl ketone (by both females and males) (Chase *et al.,* 1992; Blomquist, Chapter 3, this book). Recently, two additional C_{27} components of the contact pheromone mixture of the German cockroach were identified: 27-oxo-3,11-dimethylheptacosan-2-one and 27-hydroxy-3,11-dimethylheptacosan-2-one (Eliyahu *et al.,* 2008). Even though it is not a hydrocarbon derivative, a volatile female sex pheromone that attracts *B. germanica* males over some distance was discovered (Liang and Schal, 1993) and has been identified as gentisyl quinone isovalerate, a short alkyl chain ester of an alcohol derivative (gentisyl quinone) of *p*-benzoquinone (Nojima *et al.,* 2005).

Epoxides

Most epoxides that have been characterized are epoxy derivatives of *n*-alkanes, mono-alkenes and di-alkenes that function as pheromone and sex attractant components, and the occurrence of many of them is reviewed above in this chapter. Epoxides are usually bio-synthetically derived from monoene and polyene long-chain hydrocarbons and many are used as pheromone components and sex attractants by four macrolepidopteran families: the Geometridae, Noctuidae, Arctiidae, and Lymantriidae (Millar, 2000; Millar, Chapter 18, this book). As of 2009, sex pheromones or attractants have been identified for more than 120 geometrids (El-Sayed, 2009).

Pheromone databases

Recently, comprehensive World Wide Web (Internet) databases have been established on insect pheromones and semiochemicals: "The Pherolist", a database of chemicals identified from sex pheromone glands of female lepidopteran insects and other chemicals attractive to male moths (Arn *et al.*, 1999); and "The Pherobase", a database of pheromones and semiochemicals for Lepidoptera and other insect orders (El-Sayed, 2006). These large databases on behavior modifying chemicals have extensive cross-linkages for animal taxa, indexes of compounds and source (reference) indexes. The indexes include those compounds cited in this chapter and many more with pheromone and semiochemical function: acetate esters, diols, epoxides, ethers, ketones and secondary alcohols. For example, "The Pherolist" reports approximately 90 epoxy derivatives of C_{17}–C_{23} of *n*-alkanes, mono-alkenes and di-alkenes as insect semiochemicals.

References

Arn, H. Tóth, M. and Priesner, E. (1986). List of sex pheromones of Lepidoptera and related attractants. *OILB-SROP/IOBC-WPRS, ISBN* 92–9067–002–9-Paris.

Arn, H., Tóth M. and Priesner, E. (1999). *The Pherolist*. http://www.nysaes.cornell.edu/ pheronet/

Augustynowicz, M., Malinski, E., Warnke, Z., Szafranek, J. and Nawrot, J. (1987). Cuticular hydrocarbons of the German cockroach, *Blattella germanica* L. *Comp. Biochem. Physiol.*, **86B**, 519–523.

Bartelt, R. J., Schaner, A. M. and Jackson, L. L. (1989). Aggregation pheromone components in *Drosophila mulleri*. A chiral ester and an unsaturated ketone. *J. Chem. Ecol.*, **15**, 399–411.

Bell, R. A., Nelson, D. R., Borg, T. K. and Cardwell, D. L. (1975). Wax secretion in non-diapausing and diapausing pupae of the tobacco hornworm, *Manduca sexta*. *J. Insect Physiol.*, **21**, 1725–1729.

Bestmann, H. J., Kern, F., Mineif, A., Platz, H. and Vostrowsky, O. (1992). Pheromone 84[1]. Der sexualpheromonkomplex des, "Braunen Bären" *Arctia caja* (Lepidoptera: *Arctiidae*). *Z. Naturforsch.*, **47**, 132–135.

Bierl, B. A., Beroza, M., and Collier, C. W. (1972). Isolation, identification, and synthesis of the gypsy moth sex attractant. *J. Econ. Entomol.*, **65**, 659–664.

Blomquist, G. J. and Dillwith, J. W. (1985). Cuticular Lipids. In *Comprehensive Insect Physiology, Biochemistry and Pharmacology, Vol. 3: Integument, Respiration and Circulation*, ed. G. A. Kerkut and L. I. Gilbert. Oxford: Pergamon, pp. 117–154.

Blomquist, G. J. and Jackson, L. L. (1979). Chemistry and biochemistry of insect waxes. *Prog. Lipid Res.*, **17**, 319–345.

Blomquist, G. J., Nelson, D. R. and de Renobales, M. (1987). Chemistry, biochemistry, and physiology of insect cuticular lipids. *Arch. Insect Biochem. Physiol.*, **6**, 227–265.

Blomquist, G. J., Soliday, C. L., Byers, B. A., Brakke, J. W. and Jackson, L. L. (1972). Cuticular lipids of insects: cuticular wax esters of secondary alcohols from the grasshoppers *Melanoplus packardii* and *Melanoplus sanguinipes*. *Lipids*, **7**, 356–362.

Blomquist, G. J., Tillman-Wall, J. A., Guo, L., Quilici, D. R., Gu, P. and Schal, C. (1993). Hydrocarbons and hydrocarbon derived sex pheromones in insects: biochemistry and endocrine regulation. In *Insect Lipids: Chemistry, Biochemistry and Biology*, ed. D. W. Stanley-Samuelson and D. R. Nelson. Lincoln, NE: University of Nebraska Press, pp. 317–351.

Blount, B. K., Chibnall, A. C. and Mangouri, H. A. (1937). CLXXI. The wax of the white pine chermes. *Biochem. J.*, **31**, 1375–1378.

Böröczky, K., Park, K. C., Minard, R. D., Jones, T. H., Baker, T. C. and Tumlinson, J. H. (2008). Differences in cuticular lipid composition of the antennae of *Heliocoverpa zea, Heliothis virescens,* and *Manduca sexta*. *J. Insect Physiol.*, **54**, 1385–1391.

Buckner, J. S. (1993). Polar cuticular lipids. In *Insect Lipids: Chemistry, Biochemistry and Biology*, ed. D. W. Stanley-Samuelson and D. R. Nelson. Lincoln, NE: University of Nebraska Press, pp. 227–270.

Buckner, J. S., Mardaus, M. C. and Nelson, D. R. (1996). Cuticular lipid composition of *Heliothis virescens* and *Helicoverpa zea* pupae. *Comp. Biochem. Physiol.*, **114B**, 207–216.

Buckner, J. S., Nelson, D. R., Fatland, C. L., Hakk, H. and Pomonis, J. G. (1984a). Novel surface lipids of diapausing *Manduca sexta* pupae: Long chain oxoalcohol esters of acetoacetic, hydroxybutyric, and acetic acids. *J. Biol. Chem.*, **259**, 8461–8470.

Buckner, J. S., Nelson, D. R., Hakk, H. and Pomonis, J. G. (1984b). Long chain oxoaldehydes and oxoalcohols from esters as major constituents of the surface lipids of *Manduca sexta* pupae in diapause. *J. Biol. Chem.*, **259**, 8452–8460.

Bursell, E. and Clements, A. N. (1967). The cuticular lipids of the larva of *Tenebrio molitor* L. (Coleoptera). *J. Insect Physiol.*, **13**, 1671–1678.

Cameron, D. W. and Drake, C. B. (1976). Colouring matters of the Aphidoidea. XL. The external wax of the woolly apple aphid *Eriosoma lanigerum* (Hemiptera: Insecta). *Aust. J. Chem.*, **29**, 2713–2721.

Carlson, D. A., Berkebile, D. R., Skoda, S. R., Mori, K., and Mihok, S. (2007). Candidate sex pheromones of the New World screwworm *Cochliomyia hominivorax*. *Med. Vet. Entomol.*, **21**, 93–96.

Chase, J., Touhara, K., Prestwich, G. D., Schal, C. and Blomquist, G. J. (1992). Biosynthesis and endocrine control of the production of the German cockroach sex pheromone, 3,11-dimethynonacosan-2-one. *Proc. Natl. Acad. Sci. USA*, **89**, 6050–6054.

Chibnall, A. C., Latner, A. L., Williams, E. F. and Ayre, C. A. (1934). The constitution of coccerin. *Biochem. J.*, **28**, 313–325.

de Renobales, M., Nelson, D. R., Zamboni, A. C., Mackay, M. E., Dwyer, L. A., Theisen, M. O. and Blomquist, G. J. (1989). Biosynthesis of very long-chain methyl-branched alcohols during pupal development in the cabbage looper, *Trichoplusia ni*. *Insect Biochem.*, **19**, 209–214.

Dodd, R. S. and Afzal-Rafii, Z. (2000). Habitat-related adaptive properties of plant cuticular lipids. *Evolution*, **54**, 1438–1444.

Dwyer, L. A., Zamboni, A. C. and Blomquist, G. J. (1986). Hydrocarbon accumulation and lipid biosynthesis during larval development in the cabbage looper, *Trichoplusia ni*. *Insect Biochem.*, **16**, 463–469.

Eliyahu, D., Nojima, S., Capracotta, S. S., Comins, D. L. and Schal, C. (2008). Identification of cuticular lipids eliciting interspecific courtship in the German cockroach, *Blattella germanica*. *Naturwissenschaften*, **95**, 403–412.

El-Sayed, A. M. (2009). The Pherobase: Database of Insect Pheromones and Semiochemicals. http://www.pherobase.com.

El-Sayed, A. M., Gibb, A. R., Suckling, D. M., Bunn, B., Fielder, S., Comeskey, D., Manning, L. A., Foster, S. P., Morris, B. D., Ando, T., and Mori, K. (2005). Identification of sex pheromone components of the painted apple moth: a tussock moth with a thermally labile pheromone component. *J. Chem. Ecol.*, **31**, 621–646.

Espelie, K. E. and Bernays, E. A. (1989). Diet-related differences in the cuticular lipids of *Manduca sexta* larvae. *J. Chem. Ecol.*, **15**, 2003–2017.

Espelie, K. E. and Payne, J. A. (1991). Characterization of the cuticular lipids of the larvae and adults of the pecan weevil, *Curculio caryae*. *Biochem. Syst. Ecol.*, **19**, 127–132.

Finidori-Logli, V., Bagnères, A.-G., Erdmann, D., Francke, W. and Clément, J.-L. (1996). Sex recognition in *Diglyphus isaea* Walker (Hymenoptera: Eulophidae): role of an uncommon family of behaviorally active compounds. *J. Chem. Ecol.*, **22**, 2063–2079.

Furukawa, A., Shibata, C., and Mori, K. (2002). Syntheses of four methyl-branched secondary acetates and a methyl-branched ketone as possible candidates for the female pheromone of the screwworm fly, *Cochliomyia hominivorax*. *Biosci. Biotechn. Biochem.*, **66**, 1164–1169.

Génin, E., Jullien, R. and Fuzeau-Braesch, S. (1987). New natural aliphatic ethers in cuticular waxes of gregarious and solitary locusts *Locusta migratoria cinerascens* (II). *J. Chem. Ecol.*, **13**, 265–282.

Gibb, A. R., Comeskey, D., Berndt, L., Brockerhoff, E. G., El-Sayed, A. M., Jactel, H. and Suckling, D. M. (2006). Identification of sex pheromone components of a New Zealand geometrid moth, the common forest looper *Pseudocoremia suavis*, reveals a possible species complex. *J. Chem. Ecol.*, **32**, 865–879.

Gibbs, A. (1995). Physical properties of insect cuticular hydrocarbons: model mixtures and lipid interactions. *Comp. Biochem. Physiol.*, **112**, 667–672.

Gibbs, A. (1998). Water-proofing properties of cuticular lipids. *Am. Zool.*, **38**, 471–482.

Gibbs, A. (2002). Lipid melting and cuticular permeability: new insights into an old problem. *J. Insect Physiol.*, **48**, 391–400.

Gibbs, A. and Pomonis, J. G. (1995). Physical properties of insect cuticular hydrocarbons: The effects of chain length, methyl-branching and unsaturation. *Comp. Biochem. Physiol.*, **112B**, 243–249.

Gries, G., Clearwater, J., Gries, R., Khaskin, G., King, S. and Schaefer, P. (1999). Synergistic sex pheromone components of white-spotted tussock moth, *Orgyia thyellina*. *J. Chem. Ecol.*, **25**, 1091–1104.

Gries, R., Gries, G., Schäfer, P. W., Yoo, H. J. S., and Greaves, M. (2005). (Z,E)-6, 8-Heneicosadien-11-one: Major sex-pheromone component of *Orgyia vetusta* (Lepidoptera: Lymantriidae). *Can. Entomol.*, **137**, 471–475.

Gries, R., Khaskin, G., Daroogheh, H., Mart, C., Karadag, S., Er, M. K., Britton, R., and Gries, G. (2006). (2S,12Z)-2-Acetoxy-12-heptadecene: major sex pheromone component of pistachio twig borer, *Kermania pistaciella*. *J. Chem. Ecol.*, **32**, 2667–2677.

Gries, R., Khaskin, G., Khaskin, E., Foltz, J. L., Schaefer, P. W. and Gries, G. (2003). Enantiomers of (Z,Z)-6,9-heneicosadien-11-ol: sex pheromone components of *Orgyia detrita*. *J. Chem. Ecol.*, **29**, 2201–2212.

Gries, G., Slessor, K. N., Gries, R., Khaskin, G., Wimalaratne, P. D. C., Gray, T. G., Grant, G. G., Tracey, A. S. and Hulme, M. (1997). (Z)-6,(E)-8-Heneicosadiene-11-one: Synergistic sex pheromone component of Douglas-fir tussock moth, *Orgyia pseudotsugata* (McDunnough) (Lepidoptera: Lymantriidae). *J. Chem. Ecol.*, **23**, 19–34.

Guo, L., Nelson, D. R., Fatland, C. L. and Blomquist, G. J. (1992). Very long-chain methyl-branched alcohols and their acetate esters in pupae of the southern armyworm, *Spodoptera eridania*: Identification and biosynthesis. *Insect Biochem. Mol. Biol.*, **22**, 277–283.

Hadley, N. F. (1981). Cuticular lipids of terrestrial plants and arthropods, a comparison of their structure, composition and waterproofing function. *Biol. Rev.*, **56**, 23–47.

Herzner, G., Goettler, W., Kroiss, J., Purea, A., Webb, A. G., Jakob, P. M., Rössler, W., and Strohm, E. (2007a). Males of a solitary wasp possess a postpharyngeal gland. *Arth. Struct. Dev.*, **36**, 123–133.

Herzner, G., Schmitt, T., Peschke, K., Hilpert, A. and Strohm, E. (2007b). Food wrapping with the post pharyngeal gland secretion by females of the European beewolf *Philanthus triangulum*. *J. Chem. Ecol.*, **33**, 849–859.

Hill, A. S. and Roelofs, W. L. (1981). Sex pheromone of the saltmarsh caterpillar moth, *Estigmene acrea*. *J. Chem. Ecol.*, **7**, 655–668.

Howard, R. W. (1993). Cuticular hydrocarbons and chemical communication. In *Insect Lipids: Chemistry, Biochemistry and Biology*, ed. D. W. Stanley-Samuelson and D. R. Nelson. Lincoln, Nebraska: University of Nebraska Press, pp. 179–226.

Howard, R. W. and Baker, J. E. (2003). Cuticular hydrocarbons and wax esters of the ectoparasitoid *Habrobracon hebetor*: ontogenetic, reproductive, and nutritional effects. *Arch. Insect Biochem. Physiol.*, **53**, 1–18.

Howard, R. W. and Blomquist, G. J. (1982). Chemical ecology and biochemistry of insect hydrocarbons. *Annu. Rev. Entomol.*, **27**, 149–172.

Howard, R. W. and Blomquist, G. J. (2005). Ecological, behavioral and biochemical aspects of insect hydrocarbons. *Annu. Rev. Entomol.*, **50**, 371–393.

Jackson, L. L. (1981). Cuticular lipids of insects – IX. Surface lipids of the grasshoppers *Melanoplus bivittatus*, *Melanoplus femurrubrum* and *Melanoplus dawsoni*. *Comp. Biochem. Physiol.*, **70B**, 441–445.

Jackson, L. L. and Blomquist, G. J. (1976). Insect Waxes. In *Chemistry and Biochemistry of Natural Waxes*, ed. P. E. Kolattukudy. New York: Elsevier, pp. 201–233.

Jurenka, R. A. and Roelofs, W. L. (1993). Biosynthesis and endocrine regulation of fatty acid derived sex pheromones in moths. In *Insect Lipids: Chemistry, Biochemistry and Biology*, ed. D. W. Stanley-Samuelson and D. R. Nelson. Lincoln, NE: University of Nebraska Press, pp. 353–388.

Jurenka, R. A., Schal, C., Burns, E., Chase, J. and Blomquist, G. J. (1989). Structural correlation between cuticular hydrocarbons and female contact sex pheromone of German cockroach *Blattella germanica* (L.). *J. Chem. Ecol.*, **15**, 939–949.

Kroiss, J., Schmitt, T., Schreier, P., Strohm, E. and Herzner, G. (2006). A selfish function of a "social" gland? A postpharyngeal gland functions as a sex pheromone reservoir in males of the solitary wasp *Philanthus triangulum*. *J. Chem. Ecol.*, **32**, 2763–2776.

Liang, D. and Schal, C. (1993). Volatile sex pheromone in the female German cockroach. *Experientia*, **49**, 324–328.

Lloyd, H.A., Blum, M.S., Snelling, R.R. and Evans, S.L. (1989). Chemistry of mandibular and Dufour's gland secretions of ants in genus *Myrmecocystus. J. Chem. Ecol.*, **15**, 2589–2599.

Lockey, K.H. (1988). Review – Lipids of the insect cuticle: origin, composition and function. *Comp. Biochem. Physiol.*, **89B**, 595–645.

Mason, R.T., Fales, H.M., Jones, T.H., O'Brien, L.B., Taylor, T.W., Hogue, C.L. and Blum, M.S. (1989). Characterization of fulgorid waxes (Homoptera:Fulgoridae:Insecta). *Insect Biochem.*, **19**, 737–740.

Meinwald, J., Smolanoff, J., Chibnall, A.C. and Eisner, T. (1975). Characterization and synthesis of waxes from homopterous insects. *J. Chem. Ecol.*, **1**, 269–274.

Millar, J.G. (2000). Polyene hydrocarbons and epoxides: a second major class of lepidopteran sex attractant pheromones. *Annu. Rev. Entomol.*, **45**, 575–604.

Millar, J.G., Giblin, M., Barton, D., and Underhill, E.W. (1990a). (3Z,6Z,9Z)-Nonadecatriene and enantiomers of (3Z,9Z)-cis-6,7-epoxy-nonadecadiene as sex attractants for two geometrid and one noctuid moth species. *J. Chem. Ecol.*, **16**, 2153–2166.

Millar, J.G., Giblin, M., Barton, D., and Underhill, E.W. (1990b). 3Z,6Z,9Z-trienes and unsaturated epoxides as sex attractants for geometrid moths. *J. Chem. Ecol.*, **16**, 2307–2316.

Millar, J.G., Giblin, M., Barton, D., Wong, J.W. and Underhill, E.W. (1991). Sex attractants and sex pheromone components of noctuid moths *Euclidea cuspidea, Caenurgina distincta*, and geometrid moth *Eupithecia annulata. J. Chem. Ecol.*, **17**, 2095–2111.

Millar, J.G., Underhill, E.W., Giblin, M. and Barton, D. (1987). Sex pheromone components of three species of Semiothisa (Geometridae), (Z,Z,Z)-3,6,9-heptadecatriene and two monoepoxydiene analogs. *J. Chem. Ecol.*, **13**, 1371–1383.

Moats, R.A., Bartelt, R.J., Jackson, L.L. and Schaner, A.M. (1987). Ester and ketone components of the aggregation pheromone of *Drosophila hydei* (Diptera: Drosophilidae). *J. Chem. Ecol.*, **13**, 451–462.

Morgan, E.D. and Mandava, N.B. (1988). Pheromones. In *Handbook of Natural Pesticides*,Vol IV, part B, ed. N.B. Mandava. Boca Raton, FL: CRC Press.

Mpuru, S., Blomquist, G.J., Schal, C., Roux, M., Kuenzli, M., Dusticier, G., Clément, J.-L. and Bagnères, A.-G. (2001). Effect of age and sex on the production of internal and external hydrocarbons and pheromones in the housefly, *Musca domestica. Insect Biochem. Mol. Biol.*, **31**, 139–155.

Nelson, D.R. (1993). Methyl-branched lipids in insects. In *Insect Lipids: Chemistry, Biochemistry and Biology*, ed. D.W. Stanley-Samuelson and D.R. Nelson. Lincoln, Nebraska: University of Nebraska Press, pp. 271–315.

Nelson, D.R. and Blomquist, G.J. (1990). Acetate esters of very long-chain methyl-branched alcohols in lepidopteran pupae. *FASEB J.*, **4**, A1827.

Nelson, D.R. and Blomquist, G.J. (1995). Insect waxes. In *Waxes: Chemistry, Molecular Biology and Functions*, ed. R.J. Hamilton. Dundee, Scotland: Oily Press, pp. 1–90.

Nelson, D.R. and Fatland, C.L. (1992). Verification of structure and direction of biosynthesis of long-chain methyl-branched alcohols and very long-chain methyl-branched alcohols in pupae of the tobacco hornworm, *Manduca sexta. Insect Biochem. Mol. Biol.*, **22**, 111–123.

Nelson, D. R. and Fatland, C. F. (1997). Very long-chain methyl-branched alcohols
    and their acetate esters in the internal lipids of lepidopteran pupae: *Cochylis
    hospes, Diatraea grandiosella, Homoeosoma electellum, Heliothis virescens* and
    Helicoverpa zea. Comp. Biochem. Physiol., **116B**, 243–256.

Nelson, D. R., Fatland, C. L., Buckner, J. S., de Renobales, M. and Blomquist,
    G. J. (1990). Long-chain and very long-chain methyl-branched alcohols and their
    acetate esters in pupae of the tobacco hornworm, *Manduca sexta. Insect Biochem.*,
    20, 809–819.

Nelson, D. R., de Renobales, M., Dwyer, L. A., Zamboni, A. C., Fatland, C. L. and
    Blomquist, G. J. (1989). Novel very long-chain methyl-branched alcohols and their
    esters, and methyl-branched alkanes in pupae of the cabbage looper, *Trichoplusia ni*
    (Hubner). *Insect Biochem.*, **19**, 197–208.

Nishida, R. and Fukami, H. (1983). Female sex pheromone of the German cockroach,
    Blattella germanica. Mem. Coll. Agric. Kyoto Univ., **122**, 1–24.

Noble-Nesbitt, J. (1991). Cuticular permeability and its control. In *Physiology of
    the Insect Epidermis*, ed. K. Binnington and A. Retnakaran. East Melbourne,
    Australia: CSIRO Publications, pp. 252–283.

Nojima, S., Schal, C., Webster, F. X., Santangelo, R. G. and Roelofs, W. L. (2005).
    Identification of the sex pheromone of the German Cockroach, *Blattella germanica.
    Science*, **307**, 1104–1106.

Patel, S., Nelson, D. R. and Gibbs, A. G. (2001). Chemical and physical analyses of wax
    ester properties. *J. Insect Sci.*, **1, 4**.

Pomonis, J. G., Hammack, L. and Hakk, H. (1993). Identification of compounds in an
    HPLC fraction from female extracts that elicit mating responses in male screwworm
    flies, *Cochliomyia hominivorax. J. Chem. Ecol.*, **19**, 985–1008.

Riederer, M. and Schneider, G. (1990). The effect of the environment on the permeability
    and composition of *Citrus* leaf cuticles. 2. Composition of soluble cuticular lipids
    and correlation with transport properties. *Planta*, **180**, 154–165.

Richter, I. and Krain, H. (1980). Cuticular lipid constituents of cabbage seedpod weevils
    and host plant oviposition sites as potential pheromones. *Lipids*, **15**, 580–586.

Rollin, P. and Pougny, J. R. (1986). Synthesis of (6Z,9S,10R)-6-*cis*-9,10-
    epoxyheneicosene, a compound of the ruby tiger moth pheromone. *Tetrahedron*, **42**,
    3479–3490.

Rourke, B. C. (2000). Geographic and altitudinal variation in water balance and metabolic
    rate in a California grasshopper, *Melanoplus sanguinipes. J. Exp. Biol.*, **203**,
    2699–2712.

Rourke, B. C. and Gibbs, A. G. (1999). Effects of lipid phase transitions on the cuticular
    permeability: model membrane and *in situ* studies. *J. Exp. Biol.*, **202**, 3255–3262.

Schaner, A. M. and Jackson, L. L. (1992). (Z)-10-Heptadecen-2-one and other 2-ketones
    in the aggregation pheromone blend of *Drosophila martensis, D. buzzatii*, and
    D. serido. J. Chem. Ecol., **18**, 53–64.

Schaner, A. M., Tanico-Hogan, L. D. and Jackson, L. L. (1989). (*S*)-2-Pentadecyl acetate
    and 2-pentadecanone: Components of aggregation pheromone of *Drosophila busckii.
    J. Chem. Ecol.*, **15**, 2577–2588.

Schmidt, J. O., McDaniel, C. A., and Simon Thomas, R. T. (1990). Chemistry of
    male mandibular gland secretions of *Philanthus triangulum. J. Chem. Ecol.*, **16**,
    2135–2143.

Skiba, P.J. and Jackson, L.L. (1993). (Z)-10-Heptadecen-2-one and 2-tridecanone biosynthesis from [1–14C]acetate by *Drosophila buzzatii*. *Insect Biochem. Mol. Biol.*, **23**, 375–380.

Smith, R.G., Daterman, G.E. and Daites, G.D. (1975). Douglas-fir tussock moth: Sex pheromone identification and synthesis. *Science*, **188**, 63–64.

Southwell, I.A. and Stiff, I.A. (1989). Presence of long-chain dialkyl ethers in cuticular wax of the Australian chrysomelid beetle *Monolepta australis*. *J. Chem. Ecol.*, **15**, 255–263.

St. Leger, R.J. (1991). Integument as a barrier to microbial infections. In *Physiology of the Insect Epidermis*, ed. K. Binnington and A. Retnakaran. East Melbourne, Victoria, Australia: CSIRO Publications, pp. 284–306.

Tamaki, Y. (1985). Sex Pheromones. In *Comprehensive Insect Physiology, Biochemistry and Pharmacology*, Vol. 9: Behavior, ed. G.A. Kerkut and L.I. Gilbert. Oxford: Pergamon, pp. 145–191.

Tulloch, A.P. (1971). Beeswax: Structure of the esters and their component hydroxy acids and diols. *Chem. Phys. Lipids*, **6**, 235–265.

Uebel, E.C., Schwarz, M., Lusby, W.R., Miller, R.W. and Sonnet, P.E. (1978). Cuticular non-hydrocarbons of the female house fly and their evaluation as mating stimulants. *Lloydia*, **41**, 63–67.

Uebel, E.C., Sonnet, P.E., Menzer, R.E., Miller, R.W. and Lusby, W.R. (1977). Mating-stimulant pheromone and cuticular lipid constituents of the little housefly, *Fannia canicularis* (L.). *J. Chem. Ecol.*, **3**, 269–278.

Wakamura, S., Arakaki, N., Yamazawa, H., Nakajima, N., Yamamoto, M. and Ando, T. (2002). Identification of epoxyhenicosadiene and novel diepoxy derivatives as sex pheromone components of the clear-winged tussock moth *Perina nuda*. *J. Chem. Ecol.*, **28**, 449–467.

Warthen, J.D. and Uebel, E.C. (1980). Differences in the amounts of two major cuticular esters in males, females and nymphs of *Melanoplus differentialis* (Thomas) (Orthoptera: Acrididae). *Acrida*, **9**, 101–106.

Yamamoto, M., Kiso, M., Yamazawa, H., Takeuchi, J. and Ando, T. (2000). Identification of chiral sex pheromone secreted by giant geometrid moth, *Biston robustum* Butler. *J. Chem. Ecol.*, **26**, 2579–2590.

Yasui, H., Wakamura, S., Arakaki, N., Irei, H., Kiyuna, C., Ono, H., Yamazawa, H. and Ando, T. (2005). Identification of a sex pheromone component of the geometrid moth *Milionia basalis pryeri*. *J. Chem. Ecol.*, **31**, 647–656.

Part II
Chemical Communication

Part II

10

Perception and olfaction of cuticular compounds

Mamiko Ozaki and Ayako Wada-Katsumata

When humans communicate with each other, we rely on an arsenal of acoustic sounds and signals, as well as words and body language. Just as the simple words themselves tell only part of the story for humans, so, too, in the insect world, species-specific single- or few-component chemical messaging (i.e. sex pheromones) do not convey all the needed information. Our intonation, tone, intensity, gesturing and posture all combine to allow our fellow humans greater perception and analysis of the larger meaning we are trying to convey. In a comparable way, the chemosensory systems of insects also release mixtures of multiple compounds forming larger patterns for con-specifics to interpret. The communication among insects, like that of humans, has the ability to receive both selectively and collectively. Furthermore, the chemical information, when sent to the central nervous system, is integrated with other information or referred to memorized information, until, finally, it affects behavior. In this chapter, we will focus on the sensory system in cuticular hydrocarbon (CHC) perception.

We can find a variety of chemosensory organs on insect antenna, mouth and other body parts, such as tarsus and wing. In *Heterotermes tenuis* and *Periplaneta americana*, some antennal sensory organs have been demonstrated to respond to CHCs by electroantennogram (EAG) or gas chromatography-electroantennographic detection (GC-EAD) (Batista-Pereira *et al.*, 2004; Saïd *et al.*, 2005). In *Schistocerca gregaria*, it was suggested that stimulation of antennae by CHCs induced intracellular IP_3 synthesis via some chemoreceptors (Heifetz *et al.*, 1997). Honeybees can discriminate between different CHCs and learn them through chemical perception in the antennae (Getz and Smith, 1987; Châline *et al.*, 2005). Nevertheless, it took a long time for the CHC sensitive sensilla to be identified, because various types of chemosensilla were observed on insect antennae.

Insect chemosensory organs have been differentially developed for taste and olfactory sensing. The contact and the distant chemosensory sensilla are responsible for non-volatile and volatile chemical reception, respectively. The CHCs with long carbon chains are non-volatile, and therefore thought to be received by taste sensilla (Ebbs and Amrein, 2007). However, because of their insolubility in water, it was very difficult to obtain response recordings to them from taste sensilla. Success was recently obtained, however, in *Drosophila melanogaster*, where a male-specific CHC as a sex-pheromone inhibiting male–male courtship was found to stimulate the bitter taste receptor neuron within the taste sensillum (Lacaille *et al.*, 2007). Before that, in 2005, Ozaki had discovered in the blowfly *Phormia regina* that a water-insoluble lipophilic material stimulated the bitter taste receptor neuron with the help of an odorant-binding protein (OBP) distributed in the taste sensillar lymph (Ozaki *et al.*, 2003). Recently, Park *et al.* (2006) showed that a novel small protein, CheB42a, not related to OBPs but secreted in small subsets of taste sensilla on male forelegs, is required for a female-specific CHC to modulate male courtship behavior. Thus, the sex-pheromone CHCs consisting of species-specific single or a few CHC components appear to stimulate the taste sensillum.

Generally, the taste sensillum contains only small numbers of receptor neurons for fundamental tastes. Hence, the taste sensillum might not be suitable for perception of CHC pheromones that contain many components. For such multi-component pheromone perception, olfactory sensilla with many receptor neurons might be more suitable, even though CHC contact pheromones are non-volatile in most cases.

Tip-recording method for electrophysiological study of chemosensory sensilla

In order to record the action potentials or impulses from insect contact chemosensory sensilla, the tip-recording method, first developed by Hodgson and Roeder, is very popular (1956). The tip-recording method was originally developed for the fly taste hairs that are hair-shaped contact chemosensory sensilla having a single top pore at the tip. When a neutral electrode is inserted at the basement of the sensillum and a recording electrode containing electrolyte plus stimulant is connected with the sensillar tip, receptor stimulation and impulse recording simultaneously start. When the receptor membrane of the dendritic outer segment of a receptor neuron within the sensillum is depolarized by chemical stimulation, a receptor current occurs at the dendritic outer segment (Ozaki and Tominaga, 1999). Thus, the induced inward current is driven by the transepithelial voltage to the impulse generating site. The transepithelial voltage is generated by an electrogenic pump, which the cell membranes of auxiliary cells facing the receptor lymph cavity have. It is believed that the impulses are generated by outward current, which would occur around the ciliary segment between the outer and the inner segments of the receptor neuron (Ozaki and Tominaga, 1999). Actually, the receptor membrane depolarization is one of the earliest electrical events, and following that, impulse generation occurs. However, as long as the impulse frequency is lineally related with the receptor potential (Ozaki and Amakawa, 1990), impulse frequency, which can easily be measured by the tip-recording method, is used as a quantitative indicator of neuronal activity of the stimulated receptor neuron.

In order to make precise kinetic measurements of the relationship between the strength of stimulus and the magnitude of response in each receptor neuron, it is necessary to use adequate stimuli for the targeting receptor neuron. For example, a taste sensillum of flies houses four functionally differentiated chemoreceptor neurons corresponding to insect fundamental tastes: sugar, salt, water and bitter taste receptor neurons. These four receptor neurons, when stimulated by adequate stimuli, generate distinguishable impulses by their amplitudes (Ozaki and Tominaga, 1999). By the tip-recording method, all these receptor neurons in the investigated sensillum are potentially stimulated. Nevertheless, if one can use an adequate stimulus for each neuron, the impulse of which is identified by its particular amplitude, single-receptor recording is possible. Moreover, the adaptation rate or magnitude of adaptation is also estimated by considering gradual decreases in impulse frequency (Ozaki and Amakawa, 1990).

Thus, using these techniques, it has been suggested that a sex-pheromone CHC of *Drosophila* stimulates the bitter taste receptor neuron in a CHC concentration-dependent manner. Moreover, the cross adaptation test between the sex-pheromone CHC and an ordinary bitter substance supported the idea that the sex-pheromone CHC tastes bitter (Lacaille *et al.*, 2007).

Functional searching for the ant CHC sensillum by electrophysiological method

In addition to the Drosophila sex pheromone CHCs mentioned above, it is well known that CHCs are used by ants for various chemical communications: nestmate recognition, caste discrimination, etc. (see related chapters). In various ant species, chemical analysis of the body surface materials suggested that the colony-specific blends of a multi-component CHC mixture act as the nestmate-discriminative pheromone (Bonavita-Cougourdan *et al.*, 1987; Yamaoka, 1990; Howard, 1993; Vander Meer and Morel, 1998; Howard and Blomquist, 2005). In a Japanese carpenter ant, *Camponotus japonicus*, the CHC pheromone, consisting of 18 CHC components of 20–40 carbons, is used as a chemical cue for nestmate and non-nestmate discrimination (Yamaoka, 1990). Because of the antennation behavior for inspecting encountered ants, the CHC-sensitive sensillum was expected to be discovered on the antenna.

Insect CHCs are too lipophilic to be dissolved in an aquatic electrolyte solution. For functional searching of the CHC sensillum by the electrophysiological procedure, this is one of the biggest problems in stimulus preparation with CHCs. In order to solve this problem, the stimulus CHCs were dissolved in the electrolyte with a mild detergent in early experiments, and then latterly dissolved with a chemosensory protein (CSP). The CjapCSP was discovered in the CHC sensillum of *Camponotus japonicus* as a lipophilic substance carrier protein (Ozaki *et al.*, 2005). It was suggested that the CjapCSP transports CHCs to the receptor membranes across the aquatic environment of receptor lymph in the CHC sensillum. When a recording electrode filled with the stimulus CHC solution contacted the target CHC sensillum, vigorous impulse discharge with a variety of amplitudes was observed. Therefore, it was presumed that the single CHC sensillum possessed multiple receptor neurons, the electrophysiological properties of which were different from each other.

The antennal contact chemosensory sensillum sensitive to CHCs was 15 μm long and 4 μm in diameter, and morphologically identified not as a taste sensillum, but as an olfactory sensillum (Figure 10.1). As a result, the tip-recording method, which was originally developed for taste response recording, was also applicable to the CHC sensillum without a gustatory top pore. According to scanning electron microscopic observation, the cuticle wall of the CHC sensillum has numerous tiny olfactory pores, through which the stimulus CHCs

Figure 10.1 Aggressive behavior of *Camponotus japonicus* and the CHC sensillum on the antennal surface. Left: Photograph of *Camponotus japonicus* encountering a non-nestmate worker ant. Right: Scanning electron micrograph (SEM) of the antennal surface. The non-nestmate-CHC-sensitive sensillum is indicated by an arrow.

may be able to penetrate into the inside of the sensillum and stimulate the receptor neurons (see Figure 10.2). By transmission electron microscopic observation, about 130 receptor neurons are seen in a single CHC sensillum (Ozaki, unpublished data). Vosshall (2001) and Couto *et al.* (2005) proposed that one receptor neuron responds to multiple odorants and one odorant stimulates multiple receptor neurons. In accordance with this principle, even a single CHC component might induce impulse generation from several receptor neurons within a CHC sensillum. When a CHC sensillum is stimulated with a CHC mixture extracted from the body surface of ants, multiple receptor neurons should generate impulses. As expected, in electrophysiological experiments with various combinations of colonies, to which the test ant and the CHC donor ant belonged, vigorous impulse discharge from the stimulated CHC sensillum was observed sometimes, but not always. Surprisingly, the CHC sensillum did discriminate between the non-nestmate CHC mixture and the nestmate CHC mixture. Indeed, the CHC sensillum selectively responded to the non-nestmate CHC mixtures dissolved in the stimulus solution containing CSP (Figure 10.3).

Perireceptor events and role of lipophilic ligand-carrier proteins

In the chemosensory sensillum, even lipophilic pheromones and odorants have to reach the receptor membranes across the receptor lymph. The receptor lymph surrounding the receptor membranes is rich in small amphipathic proteins secreted from

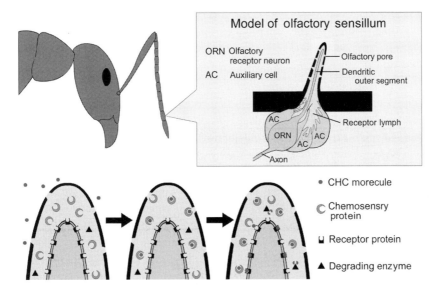

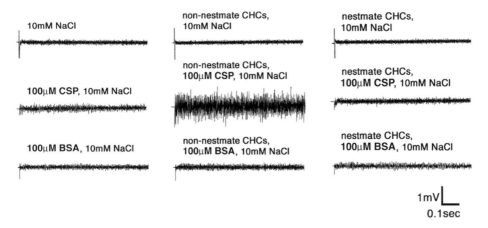

Figure 10.2 Schematic drawing of the perireceptor model in an ant CHC sensillum catching CHC molecules.

Figure 10.3 Electrophysiological response of the CHC sensillum to nestmate or non-nestmate CHCs with or without CjapCSP. Left column: Recordings to stimulus solutions of 10 mM NaCl, 10 mM NaCl plus CSP and 10 mM NaCl plus bovine serum albumin (BSA). Middle column: Recordings to non-nestmate CHCs dissolved in the same stimulus solutions. Right column: Recordings to nestmate CHCs dissolved in the same stimulus solutions.

auxiliary cells. Such proteins were predicted to work as carriers of lipophilic ligands (Vogt and Riddiford, 1981; Vogt *et al.*, 1990; Pelosi, 1998; Steinbrecht, 1998; Jacquin-Joly *et al.*, 2001; Calvello *et al.*, 2003; Ozaki *et al.*, 2005). Two classes of the carrier proteins, odorant-binding proteins (OBPs) and CSPs, share some characteristics: small size (11–18 kDa), high water solubility and reversible binding activity for small lipophilic molecules. There is, however, little similarity among their amino acid sequences (Pelosi *et al.*, 2006). OBPs and CSPs commonly contain α-helical domains, but are folded differently.

The OBPs are widely differentiated and distributed across and within species, with around 10–15% identical amino acid residues (Pelosi *et al.*, 2006). Their signature is represented by a pattern of six cysteines in conserved positions that in the native protein form three interlocking disulphide bridges. These disulfide bridges create a rigid structure so that the helices cannot move freely. Instead, only the N- and C-terminal parts of the protein are flexible and can move to open and close the ligand-binding cavity (Prestwich *et al.*, 1995; Leal *et al.*, 1999; Scaloni *et al.*, 1999). OBPs are further classified into pheromone-binding proteins (PBPs) and general odorant-binding proteins (GOBPs), which are thought to be tuned to pheromone binding and binding of general odorants, respectively (Steinbrecht *et al.*, 1995; Laue *et al.*, 1994; Krieger *et al.*, 1996; Zhang *et al.*, 2001).

The GOBPs mainly found in the female moth antennae seem to carry odorant molecules found in plant volatiles (Steinbrecht *et al.*, 1995). In the honeybee, *Apis mellifera*, ASP2 (AmelOBP2) expressed in the antennae of workers and drones can bind odorant molecules of floral scents such as 1,8-cineol, 2-isobutyl-3-methoxypyrazine, 2-heptanone and isoamyl-acetate but do not bind pheromone components (Danty *et al.*, 1997, 1999; Briand *et al.*, 2001).

The PBPs in the male moth antennae show considerable binding selectivity for the major constituent of the female sex pheromones in each species (Leal *et al.*, 2005). The three-dimensional folding of PBP of *Bombyx mori* (BmorPBP) has been resolved. The position of its ligand, bombykol, was identified inside the binding cavity, which is located at the core of the protein (Shandler *et al.*, 2000). Thus, the lipophilic ligand in the binding pocket can be perfectly separated from the hydrophilic receptor lymph (Shandler *et al.*, 2000). The C-terminal region of BmorPBP, which has no definite structure at neutral pH, folds into an α-helical segment and fits into the ligand-binding cavity at low pH (Wojtasek and Leal, 1999). This pH-dependent C-terminal change can occur in the local acidic environment at the proximity of the dendritic membrane, so that the α-helical C-terminal of BmorPBP may push the pheromone molecule out of the ligand-binding cavity and pass the pheromone to the receptor protein on the dendritic membrane (Leal, 2005).

On the other hand, the CSPs are conserved with 40–50% amino acid identity even between phylogenetically distant species (Pelosi *et al.*, 2006). Their signature consists of four cysteines connected by disulphide bonds between adjacent cysteine residues, resulting in the formation of two small loops of eight and four amino acids. Thus, one can imagine that CSPs do not have as rigid a structure as OBPs. The ligand-binding pocket of CSPs is a sort of hydrophobic tunnel, which can accommodate a long-carbon-chain molecule (Lartigue *et al.*, 2002; Mosbah *et al.*, 2003; Campanacci *et al.*, 2003). It is thought that the ligand-binding cavities of CSPs are more flexible than those of OBPs (Picimbon, 2003).

Originally, OBPs were discovered in Lepidoptera antennae and considered to be involved in perireceptor events around odorant receptor membranes, while CSPs were found not only on antennae but also on other parts where chemosensory sensilla are located (Steinbrecht, 1998; Shanbhag *et al.*, 2001), yet undefined in their function. However, antennal specificity of OBPs has not been verified, since an OBP was found in the labellar taste sensilla of the blowfly *Phormia regina* (Ozaki *et al.*, 1995). In some social hymenopterans, OBP is expressed in body parts other than the antennae, such as legs and wings (Calvello *et al.*, 2003, 2005). In Formicidae, it is reported that the Argentine ant, the Japanese carpenter ant and the red fire ant have antenna-specific CSPs, which may carry lipophilic CHC molecules to the receptor membranes across the aquatic receptor lymph (Figure 10.2) (Ishida *et al.*, 2002; Ozaki *et al.*, 2005; Leal and Ishida, 2008).

Ligand binding of lipophilic ligand carrier proteins

Ligand-binding studies of OBPs and CSPs have been carried out in different insect species which led to different results (Pelosi *et al.*, 2006). In the social wasp, *Polistes dominulus*, both OBP and CSP have good affinity for differently classified long-chain compounds, such as acid, ester and amine (Calvello *et al.*, 2003). The OBP is commonly found in the antennae, head, wings and legs of all castes and ages, and the CSP is mainly expressed in the antennae of workers and males, while in females (except for workers), it is also expressed in the legs (Calvello *et al.*, 2003, 2005). Studies to determine the binding activity for linear saturated amides, alcohols, and carboxylic acids with 12–18 carbons, showed that CSP exhibited a broader binding spectrum than OBP did for saturated amides with 12–16 carbons and saturated straight-chain alcohols of 12–18 carbons. In the honeybee, *Apis mellifera*, ASP3c was found in the antennae, wings and legs and is proposed to be a general lipid carrier protein. It binds several fatty acids and fatty acid methyl esters: myristic acid, palmitic acid, stearic acid, C16-Me and C18-Me, the latter two of which are components in brood pheromone blend (Briand *et al.*, 2002). In the locust *Schistocerca gregaria* CSPs are found in the taste sensilla rather than in the olfactory sensilla (Angeli *et al.*, 1999). CSP-sg4 binds various plant volatiles, carboxylic acids and linear alcohols with 12, 14 and 18 carbon atoms, but does not bind aggregation pheromones (Ban *et al.*, 2002). In *Mamestra brassicae*, CSPMbraA6 found in the antenna and pheromone gland binds volatile pheromone components and brominated alkyl alcohols or fatty acids with 12–18 carbons (Jacquin-Joly *et al.*, 2001; Lartigue *et al.*, 2002).

At the present time CSPMbraA6 of *Mamestra brassicae* is the only structurally characterized CSP (Lartigue *et al.*, 2002; Mosbah *et al.*, 2003; Campanacci *et al.*, 2003). The molecular structure was analyzed in the binding of a pheromone component, Z11–16:Ac, and showed large conformational changes after ligand binding. The binding pocket is reported to bind up to three 12-bromo-dodecanol molecules (Lartigue *et al.*, 2002). It is as yet unclear whether the CSP binds ligands with a positive cooperative kinetics, but precise investigation of the three-dimensional structure of CSPs in various insect species would suggest more about the function of this lipophilic ligand carrier protein family.

Recently, in *Drosophila melanogaster*, CheB42a was found in the taste sensilla on male front legs (Xu *et al.*, 2002; Starostina *et al.*, 2008). In this species, female-specific CHC pheromones play essential roles in triggering and modulating mating behavior. CheB42a is secreted into the inner lumen of the particular pheromone-sensing taste hairs (Starostina *et al*., 2008). The CheB proteins, including CheB42a, are similar to MD-like proteins (MD, myeloid differentiation protein). MD-like proteins, which are found in all eukaryotes, form the ML family. The ML family is a different lipid-binding protein family from the OBP or the CSP family. These proteins are soluble proteins of 150–200 amino acids with two or more disulfide bonds and have signal peptides at the N-terminal. Although not much has been reported about the molecular mechanisms of MD-like proteins for hydrocarbon pheromone detection, it has been suggested that the protein is concerned with the perception of hydrocarbons used as contact sex pheromones (Starostina *et al*., 2008).

In the Japanese carpenter ant, *Camponotus japonicus*, CjapCSP found in a particular type of antennal chemosensory sensilla is involved in the perception of CHCs. For nestmate recognition, the worker ants distinguish the colony-specific CHC profiles consisting of over 18 compounds by the antennal chemosensory sensilla sensitive to CHCs (Ozaki *et al*., 2005). In order to prove that CjapCSP functions as a CHC carrier protein in *Camponotus japonicus*, CHCs extracted from the ant body surface were adsorbed on the glass wall of a test tube by evaporating organic solvents, and subsequently the buffer solution containing CjapCSP was added to the tube, so that the CHCs in the CjapCSP-binding form were dissolved in the buffer. Then, the CHCs were extracted with a small amount of organic solvent and quantitatively analyzed by gas chromatography. Thus, it was found that CjapCSP nonspecifically bound every CHC component. When an aquatic buffer containing CjapCSP was used instead of intrinsic receptor lymph, the same CHC mixture pattern as that on the ant body surface could be reproduced in a test tube, regardless of colony specificity of the CHC pattern (Figure 10.4). The buffer containing bovine serum albumin (BSA) instead of CjapCSP dissolved very little CHC and did not reproduce the original CHC pattern. Hence, by using a CjapCSP-containing electrolyte solution, electrophysiological stimulus solutions with various colonies' CHC mixtures could be prepared, and the impulses to those CHC mixtures were recorded from the CHC-sensilla. It can therefore be determined that nestmate CHCs never stimulate the CHC-sensilla but non-nestmate CHCs do stimulate the CHC sensilla (Figure 10.3). This implies that nestmate and non-nestmate discrimination primarily occurs at the sensillar level before the brain.

Chemoreception of cuticular hydrocarbons and aggressive behavior expression

The body surface chemicals of insects contain a large proportion of CHCs, including straight-chain, methyl-, dimethyl-, trimethyl-branched and unsaturated components. They function as important components in chemical communication (Howard, 1993). Ant species, especially, rely much on CHCs as chemical communication to inform the membership of colonies, caste, sex and other physiological status. The CHC components are different among species (see related chapters), and the mixture profiles of CHCs are characteristically different among colonies, but rather resemble the individual worker ants in each colony. Since colony members usually share similar CHC profiles in any ant species,

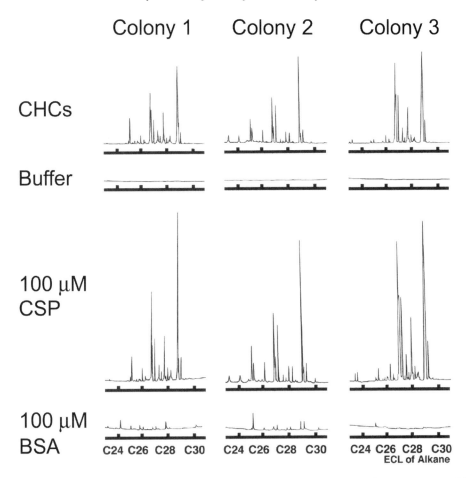

Figure 10.4 Introduction of CHCs derived from ant body surface into aquatic environment with CjapCSP. A. Gas chromatograms for the original CHC profiles of three different colonies. B. CHC profiles dissolved in plane buffer solution or in buffer solution with CjapCSP or bovine serum albumin (BSA), or in plane buffer.

CHCs were proposed as the chemical sign for nestmate recognition (Bonavita-Cougourdan *et al.*, 1987; Yamaoka, 1990; Howard, 1993; Vander Meer and Morel, 1998; Howard and Blomquist, 2005). The colony-specific CHC profiles could be detected even by a single sweep of their antennae (Wilson, 1971; Tanner, 2008). Thus, based on the chemical signatures of CHC mixtures, nestmate recognition underlies various social behaviors of ants. The hypothesis that CHCs play essential roles in the social behaviors of ants has been supported by experimental evidence through a variety of behavioral bioassays or correlation studies in which cuticular compounds are removed and reapplied via solvent extraction (Bonavita-Cougourdan *et al.*, 1987; Howard, 1993; Vander Meer and Morel, 1998; Breed, 1998; Martin *et al.*, 2008a, b; Guerreri and d'Ettore, 2008). This hypothesis was further

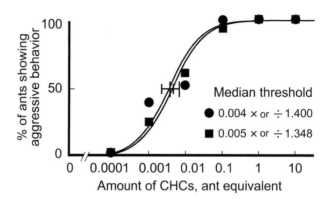

Figure 10.5 CHC amount–aggressive behavior relationship for stimulation with non-nestmate CHCs. The percentage of ants showing aggressive behaviors is plotted against the logarithmically scaled amount of non-nestmate CHCs treated on glass bead dummies. Aggressive behavior is defined as heavier expression of aggressiveness than biting. Amount of CHC is relatively indicated by ant equivalent amount. Each curve is based on a serial behavioral experiment using 30 workers from a single colony. The experiment was repeated twice, using different worker populations from independent colonies. The median threshold is obtained as the CHC amount giving half the maximum rate of the ants exhibiting aggressive behavior.

supported by comparison of the pheromone activity between natural and artificial hydro-carbon blends using synthetic hydrocarbon components (Lahav *et al.*, 1999; Akino *et al.*, 2002; 2004; Ozaki *et al.*, 2005; Martin *et al.*, 2008a).

The Japanese carpenter ant, *Camponotus japonicus,* is one of the ant species in which the CHCs have been directly shown to serve as nestmate recognition cues (Ozaki *et al.*, 2005). The chemosensory sensilla on the worker antennae only responded to non-nest-mate CHC profiles. Nestmate recognition can be evaluated by assessing the aggressive-ness of workers against non-nestmate workers. Such colony discriminative aggression was observed against not only living non-nestmate workers but also glass bead dummies onto which non-nestmate CHCs had been deposited. By using glass beads as surrogate ants, stimulation for triggering aggressive behaviors was well assessed not only qualitatively, but also quantitatively. Aggressive behavior toward non-nestmates is regulated by the CHC profiles consisting of over 18 compounds. The ants can distinguish the CHC profiles of non-nestmates from that of nestmates by the CHC sensilla on the antennae.

Figure 10.5 shows a CHC-dose-dependent elevation of aggressiveness in *Camponotus japonicus.* Preparing a population of workers derived from the same colony, aggressive behavior was induced by using different amounts of non-nestmate CHCs on the glass beads, and the aggressive threshold concentration, which triggers biting, charging or spraying for-mic acid, was individually determined. Thus, the number of ants showing aggressive behav-iors indicated a sigmoidal change against the amount of non-nestmate CHCs expressed in the logarithmic scale. The aggression threshold of individuals broadly varies over three log-units from 10^{-4} to 10^{-1} ant equivalent of non-nestmate CHCs, although the median thresholds of different ant populations, even from different colonies, are similarly defined to be around 0.005 ant equivalent. Thus, considering the broad aggression threshold of individuals, they might have various CHC sensitivities at the sensillar level or at higher neuronal levels. It should be ascertained at what level and how the aggressive behavior expression toward non-nestmates is regulated. As shown in Figure 10.3, when a single sensillum is stimulated with a non-nestmate CHC blend of multiple CHC components, multiple receptor neurons respond with multi-components of impulses. By measuring the magnitude of the response, the sensillar sensitivity to CHCs can be evaluated. When the antenna sweeps the body surface of a non-nestmate, multiple sensilla are stimulated. Thus, the generated impulses from the stimulated sensilla are at once summed into a primary olfactory center (Nishikawa *et al.*, 2008) and then the aggression-inducing neuronal signal is sent to the projection neurons (Kirschner *et al.*, 2006). If the neuronal activity is recorded from the projection neurons, a higher neural activity, which might regulate the aggression threshold, could be evaluated.

Another unsolved problem is the mechanism of a sort of filter function of the CHC sensillum, which cuts off the chemical information of nestmates and passes that of the non-nestmates. Moreover, the CHC sensillum of *Camponotus japonicus*, housing about 130 receptor neurons, could potentially distinguish 2^{130} different patterns by an on/off combination of 130 receptor neurons. This number of receptor neurons in a sensillum seems too large to recognize CHC blends of 18 components, although the sensillum might need to respond to CHCs not only of its own species, but also of other species.

Much is still unknown about the use of the CHC sensillum. Using natural and synthetic hydrocarbons, more precise studies should be done at both neuronal and behavioral levels.

References

Akino, T., Terayama, M., Wakamura, S. and Yamaoka, R. (2002). Intraspecific variation of cuticular hydrocarbon composition in *Formica japonica* Motschoulsky (Hymenoptera: Formicidae). *Zool. Sci.*, **19**, 1155–1165.

Akino, T., Yamamura, K., Wakamura, S. and Yamaoka, R. (2004). Direct behavioral evidence for hydrocarbons as nestmate recognition cues in *Formica japonica* (Hymenoptera: Formicidae). *Appl. Entomol. Zool.*, **39**, 381–387.

Angeli, S., Ceron, F., Scaloni, A., Monti, M., Monteforti, G., Minnocci, A., Petacchi, R. and Pelosi, P. (1999). Purification, structural characterization, cloning and immunocytochemical localization of chemoreception proteins from *Schistocerca gregaria*. *Eur. J. Biochem.*, **262**, 745–754.

Ban, L., Zhang, L., Yan, Y. and Pelosi, P. (2002). Binding properties of a locust's chemosensory protein. *Biochem. Biophys. Res. Comm.*, **293**, 50–54.

Batista-Pereira, L. G., dos Santos, M. G., Corrêa, A. G., Fernandes, J. B., Arab, A., Costa-Leonardo, A.- M., Dietrich, C. R. R. C., Pereira, D. A. and Bueno O. C. (2004). Cuticular hydrocarbons of *Heterotermes tenuis* (Isoptera: Rhinotermitidae): Analyses and electrophysiological studies. *Z. Naturforsch.*, **59c**, 135–139.

Bonavita-Cougourdan, A., Clément, J.-L. and Lange, C. (1987). Nestmate recognition: The role of cuticular hydrocarbons in the ant *Camponotus vagus* Scop. *J. Entomol. Sci.*, **22**, 1–10.

Breed, M. D. (1998). Chemical cues in kin recognition: criteria for identification, experiental approaches, and the honey bee as an example. In *Pheromone Communication in Social Insects: Ants, Wasps, Bees and Termites*, ed. R. K. Vander Meer, M. D. Breed, K. E. Espelie and M. L. Winston. Boulder CO: Westview Press, pp. 57–78.

Briand, L., Nespoulous, C., Huet, J.-C., Takahashi, M. and Pernollet, J.-C. (2001). Ligand binding and physico-chemical properties of ASP2, a recombinant odorant-binding protein from honeybee (*Apis mellifera* L.). *Eur. J. Biochem.*, **268**, 752–760.

Briand, L., Nespoulous, C., Huet, J.-C., Takahashi, M. and Pernollet, J.-C. (2002). Characterization of a chemosensory protein (ASP3c) from honeybee (*Apis mellifera* L.) as a brood pheromone carrier. *Eur. J. Biochem.*, **269**, 4586–4596.

Calvello, M., Brandazza, A., Navarrini, A., Dani, F. R., Turillazzi, S., Felicioli, A. and Pelosi, P. (2005). Expression of odorant-binding proteins and chemosensory proteins in some Hymenoptera. *Insect Biochem. Mol. Biol.*, **35**, 297–307.

Calvello, M., Guerra, N., Brandazza, A., D'Ambrosio, C., Scaloni, A., Dani, F. R., Turillazzi, S. and Pelosi, P. (2003). Soluble proteins of chemical communication in the social wasp *Polistes dominulus*. *Cell. Mol. Life Sci.*, **60**, 1933–1943.

Campanacci, V., Lartigue, A., Hällberg, B. M., Jones, T. A., Giudici-Orticoni, M-T., Tegoni, M. and Cambillau, C. (2003). Moth chemosensory protein exhibits drastic conformational changes and cooperativity on ligand binding. *Proc. Natl. Acad. Sci. USA*, **100**, 5069–5074.

Châline, N., Sandoz J.-C., Martin S. J., Ratnieks, F. L. W. and Jones, G. R. (2005). Learning and discrimination of individual cuticular hydrocarbons by honeybees (*Apis mellifera*). *Chem. Senses*, **30**, 327–335.

Couto A., Alenius, M. and Dickson, B. J. (2005). Molecular, anatomical, and functional organization of the *Drosophila* olfactory system. *Curr. Biol.*, **15**, 1535–1547.

Danty, E., Briand, L., Michard-Vanhée, C., Perez, V., Arnold, G., Gaudemer, O., Huet, D., Huet, J.-C., Ouali, C., Masson, C. and Pernollet, J.-C. (1999). Cloning and expression of a queen pheromone-binding protein in the honeybee: an olfactory-specfic, developmentally regulated protein. *J. Neurosci.*, **19**, 7468–7475.

Danty, E., Michard-Vanhée, C., Huet, J. C., Genecque, E., Pernollet, J.-C. and Masson, C. (1997). Biochemical characterization, molecular cloning and localization of a putative odorant binding protein in the honeybee *Apis mellifera* L. *FEBS Lett.*, **414**, 595–598.

Ebbs, M. L. and Amrein, H. (2007). Taste and pheromone perception in the fruit fly, *Drosophila melanogaster*. *Pflugers Arch – Eur. J. Physiol.*, **454**, 735–747.

Getz, W. M. and Smith, K. B. (1987). Olfactory sensitivity and discrimination of mixtures in the honeybee *Apis mellifera*. *J. Comp. Physiol. A*, **160**, 239–245.

Guerreri, F. J. and d'Ettore, P. (2008). The mandible opening response: quantifying aggression elicited by chemical cues in ants. *J. Experimental Biol.*, **211**, 1109–1113.

Heifetz, Y., Boekhoff, I., Breer, H. and Appplebaum, S. W. (1997). Cuticular hydrocarbons control behavioural phase transition in *Schistocerca gregaria* nymphs and elicit biochemical responses in antennae. *Insect Biochem. Mol. Biol.*, **27**, 563–568.

Hodgson, E. S. and Roeder, K. D. (1956). Electrophysiological studies of arthropod chemoreception. I. General properties of the labellar chemoreceptors of Diptera. *J. Cell. Comp. Physiol.*, **48**, 41–74.

Howard, R. W. (1993). Cuticular hydrocarbons and chemical communication. In *Insect Lipids: Chemistry, biochemistry and biology*, ed. D. W. Stanley-Samuelson and D. R. Nelson. Lincoln, NE: University of Nebraska Press, pp. 179–226.

Howard, R. W. and Blomquist, G. J. (2005). Ecological, behavioral, and biochemical aspects of insect hydrocarbons. *Annu. Rev. Entomol.*, **50**, 371–393.

Ishida, Y., Chiang, V. and Leal, W. S. (2002). Protein that makes senses in the Argentine ant. *Naturwissenschaften*, **89**, 505–507.

Jacquin-Joly, E., Vogt, R. G., Francois, M.-C. and Nagnan-Le Meillour, P. (2001). Functional and expression pattern analysis of chemosensory proteins expressed in antennae and pheromonal gland of *Mamestra brassicae*. *Chem. Senses*, **26**, 833–844.

Kirschner, S., Kleineidam, C. J., Zube, C., Rybal, J., Grunewald, B. and Rossler, W. (2006). Dual olfactory pathway in the honeybee, *Apis mellifera*. *J. Comp. Neurol.*, **499**, 933–952.

Krieger, J. von Nickisch-Rosenegk, E., Mameli, M., Pelosi, P. and Breer, H. (1996). Binding proteins from the antennae of *Bombyx mori*. *Insect Biochem. Mol. Biol.*, **3**, 297–307.

Lacaille, F., Hiroi, M., Twele, R., Inoshita, T., Umemoto, D., Manière, G., Marion-Poll, F., Ozaki, M., Francke, W., Cobb, M., Everaerts, C., Tanimura, T. and Ferveur, J.-F. (2007). An inhibitory sex pheromone tastes bitter for *Drosophila* males. *PLoS ONE*, **2**, e661.

Lahav, S., Soroker, V., Hefetz, A. and Vander Meer, R. K. (1999). Direct behavioral evidence for hydrocarbons as ant recognition discriminators. *Naturwissenschaften*, **86**, 246–249.

Lartigue, A., Campanacci, V., Roussel, A., Larsson, A. M., Jones, T. A., Tegoni, M. and Cambillau, C. (2002). X-ray structure and ligand binding study of a moth chemosensory protein. *J. Biol. Chem.*, **277**, 32094–32098.

Laue, M., Steinbrecht, R. A. and Ziegelberger, G. (1994). Immunocytochemical localization of general odorant binding protein in olfactory sensilla of the silkworm *Antheraea polyphemus*. *Natuurwissenschaften*, **81**, 178–180.

Leal, W. S. (2005). Pheromone reception. *Top. Curr. Chem.*, **240**, 1–36.

Leal, W. S., Chen, A. M. and Erickson, M. L. (2005). Selective and pH-dependent binding of a moth pheromone to a pheromone-binding protein. *J. Chem. Ecol.*, **31**, 2493–2499.

Leal, W. S. and Ishida, Y. (2008). GP-9s are ubiquitous proteins unlikely involved in olfactory mediation of social organization in the red imported fire ant, *Solenopsis invicta*. *PLoS ONE*, **3**, e3762.

Leal, W. S., Nikonova, L. and Peng, G. (1999). Disulfide structure of the pheromone binding protein from the silkworm moth, *Bombyx mori*. *FEBS Lett.*, **464**, 85–90.

Martin, S. J., Helanterä, H. and Drijfhout, F. P. (2008a). Colony-specific hydrocarbons identify nest mates in two species of *Formica* ant. *J. Chem. Ecol.*, **34**, 1072–1080.

Martin, S. J., Vitikainen, E., Helanterä, H. and Drijfhout, F. P. (2008b). Chemical basis of nest-mate discrimination in the ant *Formica exsecta*. *Proc. R. Soc. B*, **270**, 153–158.

Mosbah, A., Campanacci, V., Lartigue, A., Tegoni, M., Cambillau, C. and Darbon, H. (2003). Solution structure of a chemosensory protein from the moth *Mamestra brassicae*. *Biochem. J.*, **369**, 39–44.

Nishikawa, M., Nishino, H., Misaka, Y., Kubita, M., Tsuji, E., Satoji, Y., Ozaki, M. and Yokohari, F. (2008). Sexual dimorphism in the antennal lobe structure of the ant, *Camponotus japonicus*. *Zool. Sci.*, **25**, 195–204.

Ozaki, M. and Amakawa, T. (1990). Adaptation-promoting effects of IP3, Ca^{2+} and phorbol ester on the sugar taste receptor cell of the blowfly, *Phormia regina*. *J. Gen. Physiol.*, **100**, 867–879.

Ozaki, M., Morisaki, K., Idei, W., Ozaki, K and Tokunaga, F. (1995). A putative lipophilic stimulant carrier protein commonly found in the taste and olfactory system:

A unique member of the pheromone-binding protein superfamily. *Eur. J. Biochem.*, **230**, 298–308.

Ozaki, M., Takahara, T., Kawahara, Y., Wada-Katsumata, A., Seno, K., Amakawa, T., Yamaoka, R. and Nakamura, T. (2003). Perception of noxious compounds by contact chemoreceptors of the blowfly, *Phormia regina*: Putative role of an odorant-binding protein. *Chem. Senses*, **28**, 349–359.

Ozaki, M. and Tominaga, Y. (1999). Contact Chemoreceptors. In *Atlas of Arthropod Sensory Receptors*, ed. E. Eguchi and Y. Tominaga. Tokyo: Springer, pp. 143–154.

Ozaki M., Wada-Katsumata, A., Fujikawa, K., Iwasaki, M., Yokohari, F., Satoji, Y., Nishikawa, T. and Yamaoka, R. (2005). Ant nestmate/non-nestmate discrimination by a chemosensory sensillum. *Science*, **309**, 311–314.

Park, S. K., Mann, K. J., Lin, H., Starostina, E., Kolski-Andreaco, A. and Pikielny, C. W. (2006). A Drosophila protein specific to pheromone-sensing gustatory hairs delays males' copulation attempts. *Curr. Biol.*, **16**, 1154–1159.

Pelosi, P. (1998). Odorant-binding proteins: Structural aspects. *Ann. N.Y. Acad. Sci.*, **855**, 281–293.

Pelosi, P., Zhou, J.-J., Ban, L. P. and Calvello, M. (2006). Soluble proteins in insect chemical communication. *Cell. Mol. Life Sci.*, **63**, 1658–1676.

Picimbon, J.-F. (2003). Biochemistry and evolution of OSD and OBP proteins. In *Pheromone Biochemistry and Molecular Biology*, ed. G. J. Blomquist and R. G. Vogt. New York: Academic, pp. 539–566.

Prestwich, G. D., Du, G. and LaForest, S. (1995). How is pheromone specificity encoded in proteins? *Chem. Senses*, **20**, 461–469.

Saïd, I., Gaertner, C., Renou, M. and Rivault, C. (2005). Perception of cuticular hydrocarbons by the olfactory organs in *Periplaneta americana* (L.) (Insecta, Disctyoptera). *J. Insect Physiol.*, **51**, 1384–1389.

Scaloni, A., Monti, M., Angeli, S. and Pelosi, P. (1999). Structural analysis and disulfide-bridge pairing of two odorant-binding proteins from *Bombyx mori*. *Biochem. Biophys. Res. Comm.*, **266**, 386–391.

Shanbhag, S. R., Hekmat-Scafe, D., Kim, M. S., Park, S. K., Carlson, J. R., Pikielny, C., Smith, D. P. and Seinbrecht, R. A. (2001). Expression mosaic of odorant-binding proteins in *Drosophila* olfactory organs. *Microsc. Res. Tech.*, **55**, 297–306.

Shandler, B. H., Nikonova, L., Leal, W. S. and Clardy, J. (2000). Sexual attraction in the silkworm moth: structure of the pheromone-binding-protein-bombykol complex. *Chem. Biol.*, **7**, 143–151.

Starostina, E., Xu, A., Lin, H. and Pikielny, C. W. (2008). A *Drosophila* protein family implicated in pheromone perception is related to Tay–Sachs GM2-activator protein. *J. Biol. Chem.*, **284**, 585–594.

Steinbrecht, R. A. (1998). Odorant binding proteins: expression and function. *Ann. N. Y. Acad. Sci.*, **855**, 323–332.

Steinbrecht, R. A., Laue M. and Ziegelberger, G. (1995). Immunolocalization of pheromone binding protein and general odorant-binding protein in olfactory sensilla of the silk moths *Antheraea* and *Bombyx*. *Cell Tissue Res.*, **282**, 203–217.

Tanner, C. J. (2008). Aggressive group behaviour in the ant *Formica xerophila* is coordinated by direct nestmate contact. *Anim. Behav.*, **76**, 1335–1341.

Vander Meer, R. K. and Morel, L. (1998). Nestmate recognition in ants. In *Pheromone Communication in Social Insects: Ants, Wasps, Bees and Termites*, ed. R. K. Vander Meer, M. D. Breed, K. E. Espelie and M. L. Winston. Boulder, CO: Westview Press, pp. 79–103.

Vosshall, L. B. (2001). The molecular logic of olfaction in *Drosophila*. *Chem. Senses*, **26**, 207–213.

Vogt, R. G. and Riddiford, L. M. (1981). Pheromone binding and interaction by moth antenna. *Nature*, **293**, 161–163.

Vogt, R. G., Rybczynshi, R. and Lerner, M. R. (1990). The biochemistry of odorant reception and transduction. In *Chemosensory Information Processing*, ed. D. Schild. Berlin: Springer, pp. 33–76.

Wilson, E. O. (1971). *The Insect Societies*. Cambridge, MA: Belknap Press of Harvard University Press.

Wojtasek, H. and Leal, W. S. (1999). Conformational change in pheromone-binding protein from *Bombyx mori* induced by pH and by interaction with membranes. *J. Biol. Chem.*, **274**, 30950–30956.

Xu, A., Park, S. K., D'Mello, S., Kim, E., Wang Q. and Pikielny, C. W. (2002). Novel genes expressed in subsets of chemosensory sensilla on the front legs of male *Drosophila melanogaster*. *Cell Tissue Res.*, **307**, 381–392.

Yamaoka, R. (1990) Chemical approach to understanding interaction among organisms. *Physiol. Ecol. Japan*, **27**, 31–52.

Zhang, S., Maida, R. and Steinbrecht, R. A. (2001). Immunolocalization of odorant-binding-proteins in Noctuid moths (Insecta, Lepidoptera). *Chem. Senses*, **26**, 885–896.

11

Nestmate recognition in social insects and the role of hydrocarbons

Jelle S. van Zweden and Patrizia d'Ettorre

One of the conditions favoring the evolution and maintenance of sociality is the ability to discriminate between kin and non-kin, because it allows altruistic acts to be directed to recipients of high relatedness (Hamilton, 1987). Nestmate recognition is the process whereby social insects recognize individuals belonging to their own colony or an alien colony, and accordingly allow or prohibit entry to their nest. Nestmate recognition often results in the discrimination of kin from non-kin, but in social insects there is a distinction to be made between nestmate and kin recognition. Where kin recognition is the assessment of the degree of relatedness towards another individual, nestmate recognition is the binary recognition of group membership (Arnold *et al.*, 1996; Lenoir *et al.*, 1999). In ants, wasps and termites, the blend of hydrocarbons present on the cuticle appears to comprise the essential compounds that serve as nestmate recognition cues (Howard and Blomquist, 2005), although in honeybees fatty acids and esters may also play an important role (Breed, 1998a; but see Dani *et al.*, 2005; Châline *et al.*, 2005). At the end of the 1990s, the notion that cuticular hydrocarbons act as recognition cues was supported mainly by correlative evidence (Singer, 1998; Vander Meer and Morel, 1998; Lenoir *et al.*, 1999), but significant progress has been made over the last decade. In recent years, not only have some studies provided the direct evidence needed, but they have also shown the differential importance of structural groups of hydrocarbons (i.e. linear and methyl-branched alkanes and alkenes) in the discrimination between nestmate and non-nestmate. On the other hand, it is still largely unknown which physiological, perceptual and cognitive processes are responsible for nestmate recognition, from encoding of cues in the hydrocarbon profile to the detection and processing of these cues in the antennae and further up in the neural pathway (see Chapter 10). Only a few studies have started to experimentally address these proximate mechanisms underlying behavioral decision rules. In this essay we will first summarize the evidence that cuticular hydrocarbons serve as recognition cues, with a particular focus on which hydrocarbons are most informative. We will then explore the mechanism behind nestmate recognition, i.e. how the perception of these hydrocarbons leads to the rejection of non-nestmates and the acceptance of nestmates.

Nestmate recognition is believed to involve the matching of a label, the so-called colony odor (i.e. the chemical profile containing the nestmate cues), with a template (the neural representation of the colony odor stored in the memory), and depending on the similarity (or dissimilarity) between the two, a conspecific will be accepted or rejected (Hölldobler and Michener, 1980; Vander Meer and Morel, 1998). The expression of the label can be influenced by multiple sources, namely heritable cues of the individuals in a nest, cues derived from the environment (e.g. food, nest material) or cues from the queen (which could also include heritable and environmental factors). For example, *Temnothorax* (*Leptothorax*) *ambiguus* and *T.(L.) longispinosus* ant workers that had eclosed in isolation were attacked significantly less by nestmate adults than by non-nestmate adults, suggesting that the recognition cues produced by the callow worker were similar to those of nestmates, and hence a genetic source is supported (Stuart, 1988). When, however, workers were allowed to eclose in a non-nestmate colony and were subsequently introduced to their parental colony, they were attacked just as often as control non-nestmates, suggesting that the genetic odor is transferred among individuals in the colony (Stuart, 1988). Similarly, honeybee workers reared and kept under laboratory conditions use genetic cues to distinguish between nestmates and non-nestmates (Breed, 1983). However, under more natural conditions, the exposure to comb wax overrides these genetic cues, so that the bees come to rely on environmental recognition cues (Breed *et al.*, 1988; Downs and Ratnieks, 1999). Evidence for queen-derived recognition cues comes from the ant *Camponotus floridanus*, where intercolonial aggression between unrelated workers whose adopted queens are sisters is lower than between sister workers whose adopted queens are unrelated, suggesting that the odors of the queen are the dominant recognition cues (Carlin and Hölldobler, 1988). The relative influences of genetic, environmental and queen odors may differ per species, and may thus depend on general ecology of the species, colony kin-structure and (social) parasite pressure. In ants, recognition cues are mixed throughout the colony by means of trophallaxis (transfer of liquid food between individuals) and allogrooming, in combination with the use of the postpharyngeal gland (PPG), thereby establishing and continuously updating the colony odor (Soroker *et al.*, 1994, 1995). In wasps and bees, the nest paper and comb wax, respectively, seem to have this role as colony odor homogenizers (Espelie *et al.*, 1990; Singer and Espelie, 1996; Breed *et al.*, 1988). For termites, the dynamics of homogenization of recognition cues are less well understood. Here caste odor profiles generally appear to be more distinct than colony-specific odor profiles, but differences in colony odor profiles correlate with aggression and genetic differences, indicating nestmate recognition based on these cues (Kaib *et al.*, 2004; Dronnet *et al.*, 2006; Liebig *et al.*, 2009; Weil *et al.*, 2009).

The manifestation of the template is less well understood. The observation that ant brood and callow workers are less aggressive towards non-nestmates than are adults suggests that there is an early learning period for template acquisition (Jaisson, 1991; Errard, 1994a). There is, nevertheless, evidence that individuals constantly need to reinforce and fine-tune their template with nestmate odors (e.g. Errard and Hefetz, 1997; Lenoir *et al.*, 2001a; Breed *et al.*, 2004; Leonhardt *et al.*, 2007). The degree of mismatch between label and template generally translates into a graded response, from complete acceptance to rejection, i.e. from investigation to threat to overt aggression (e.g. Carlin and Hölldobler, 1986; Obin and Vander Meer, 1988; Errard and Hefetz, 1997).

Cuticular hydrocarbon profiles of the same species typically have the same qualitative make-up, i.e. the same set of compounds (e.g. Nowbahari *et al.*, 1990; see Chapter 7), and as a result social insects should rely on the quantitative variation of hydrocarbons within the profile (different concentrations and/or ratios) for nestmate recognition (e.g. Bonavita-Cougourdan *et al.*, 1987; Espelie *et al.*, 1990; Vander Meer *et al.*, 1989; Martin *et al.*, 2008a). Consistent with this observation, cuticular hydrocarbon profiles have a significant heritable component (e.g. Ross *et al.*, 1987), although environmental components have been implicated as well (e.g. Liang and Silverman, 2000) and the profiles may change over time (Vander Meer *et al.*, 1989; Provost *et al.*, 1993; Lenoir *et al.*, 2001a). Hence, both label and template are not fixed, but are continuously changing. This naturally leads to the question of what kind of template can be both restrictive and flexible, something we will address in the second part of this essay. First, we will go through the evidence that leads us to believe that hydrocarbons are indeed the key nestmate recognition cues.

Evidence for hydrocarbons as nestmate recognition cues

Social parasites

Some of the indirect evidence for the use of hydrocarbons in nestmate recognition comes from social parasites. Since these are able to break the code and integrate into the normally closed colony of a social insect, the recognition cues responsible for their success can be found by studying their chemical profiles. We will give a few examples that serve the purpose of illustration; a full review of social parasitism and hydrocarbons is given elsewhere (Lenoir *et al.*, 2001b; Nash and Boomsma, 2008; see Chapter 14).

Social parasite species can use either chemical camouflage (recognition cues are acquired from its host) or chemical mimicry (recognition cues are synthesized by the parasite), or a combination of both (Lenoir *et al.*, 2001b). However, the result is the same: they overcome detection as non-nestmates. The staphylinid termitophile beetles *Trichopsenius frosti*, *T. depressus*, *Xenistusa hexagonalis*, and *Philotermes howardi* are guests in termite nests and were found to have hydrocarbon profiles similar to those of their respective hosts (Howard *et al.*, 1980, 1982). Similarly, the paper wasp *Polistes atrimandibularis* is an obligate social parasite of another paper wasp, *P. biglumis bimaculatus*. It was shown that, at the point of the colony life cycle that the parasite needs to integrate into its host colony, the hydrocarbon profile of the parasite changes from one that is characterized by unsaturated hydrocarbons into one that matches the host's profile: characterized by the same saturated methyl-branched hydrocarbons, without any unsaturated hydrocarbons (Bagnères *et al.*, 1996). Another paper wasp, *P. sulcifer*, was found to adopt a colony-specific hydrocarbon profile of its host, *P. dominulus*. When presented to a non-nestmate parasitized colony, or when treated with the hydrocarbons of the parasite of a non-nestmate parasitized colony, the *P. sulcifer* individuals were aggressively rejected (Sledge *et al.*, 2001). Also, it was shown that the ant *Polyergus rufescens* can achieve social integration in nests of multiple hosts of the genus *Formica* (subgenus *Serviformica*) by adopting a hydrocarbon profile that is similar to that of their host, even if the experimental host is not a natural host, like *Formica selysi* (d'Ettorre *et al.*, 2002).

Box 11.1. Diagnostic power (*DP*)

One of the problems in investigating the evolutionary chemical ecology of nestmate recognition is that researchers are somewhat impaired in manipulating the chemical profile, due to the lack of synthetic hydrocarbons. There is no commercial availability of methyl-branched alkanes at the present moment (although linear alkanes are and some alkenes may be available) and synthesizing these is a costly and time-consuming process. The search for likely nestmate recognition cues therefore often relies on correlative evidence, i.e. stepwise discriminant analysis (DA) between colonies based on cuticular hydrocarbons. The downside to this analysis is that it requires large sample sizes, unless preceded by principal component analysis (PCA). However, this procedure can complicate things, as this often results in one or two of the principal components giving good separation between colonies, with the putative recognition cues associated to these principal components to some degree.

We propose that the diagnostic power for each separate hydrocarbon can be calculated, by dividing the standard deviation among sampled individuals by the pooled standard deviation within nests (Christensen *et al.*, 2005). The rationale behind this is that likely nestmate recognition cues should be most variable between colonies, but should be consistent within colonies due to mixing of odors and/or high relatedness). Therefore, the diagnostic power (DP) of a compound is calculated as:

$$DP_i = \frac{SD_i}{\sqrt{\dfrac{\sum\limits_{j=1}^{N} SD_{ij}^{2} \times \left(n_j - 1\right)}{\sum\limits_{j=1}^{N} \left(n_j - 1\right)}}}$$

in which SD_i is the standard deviation of the normalized peak area of the i'th compound over all individuals of N colonies, SD_{ij} is the standard deviation of the normalized peak area of the i'th compound in the j'th colony, and n_j is the number of individuals in the j'th colony. Compounds can then be ranked according to their DP and the most likely candidates for nestmate recognition cues can be found.

In *Formica rufibarbis* and *Camponotus aethiops*, for example, we found the linear alkanes to be almost as variable within colonies as between colonies, whereas some methyl-branched hydrocarbons were more than six times as variable between colonies as within colonies (J.S. van Zweden *et al.*, in preparation; van Zwedan *et al.*, 2009). This is a similar approach as used by Martin and others (2008a) in their search for colony-specific hydrocarbons within the profiles of *F. exsecta* and *F. fusca*. They meas-ured the correlation (r^2) between the relative abundance of compounds at the species and at the colony levels and subsequently excluded those compounds that had either low or high correlation at both levels, but included those that had low correlation at the species level and high at the colony level, which would be those with diagnostic power (*DP*) a high.

Correlative evidence and removal-and-replacement experiments

Evidence for identification of nestmate recognition cues can constitute either correlative evidence, i.e. when cues are shown to be colony specific, or can come from removal-and-replacement experiments, i.e. extracting the putative recognition cues from an individual and supplementing with the extract from another individual, followed by behavioral exper-iments (Breed, 1998b). Although the latter may appear to be a rather direct approach, the problem is that the extract may also contain other compounds that remain undetectable, even by gas-chromatography with mass-spectrometry (GC–MS) (Howard and Blomquist, 2005). These types of evidence for the use of hydrocarbons in nestmate recognition in ants and wasps come from a number of studies (e.g. Vander Meer *et al.*, 1989; Espelie and Hermann, 1990; Butts *et al.*, 1991; Gamboa *et al.*, 1996; Lorenzi *et al.*, 1997; Lenoir *et al.*, 1999; Hefetz, 2007), but we shall restrict ourselves to a few examples.

Workers of the ant *Camponotus vagus* showed significantly more aggression towards freshly killed nestmates when their chemical profiles had been removed and their bod-ies covered by an extract of a non-nestmate worker. Only hydrocarbons were found in a GC–MS analysis of these extracts, giving a good indication that these are the compounds responsible for the reaction (Bonavita-Cougourdan *et al.*, 1987). Adults of the paper wasp *Polistes metricus* were found to be unable to recognize their own nest next to an alien nest when these had been washed in hexane, but the recognition was restored when the extracts were reapplied to the nests (Espelie *et al.*, 1990). Using multivariate statistics (e.g. factor analysis), it was also shown that the hydrocarbon profiles of adult workers and nest paper were colony-specific. Again, hydrocarbons were found to be the major compounds in the cuticular washes, which suggests their use in recognition. For the ant *Cataglyphis cursor* it was shown that there is a correlation between the adoption of workers into alien colonies and the similarity of the hydrocarbon profiles of the original and the adopting colony, as calculated by the Nei index (Nowbahari *et al.*, 1990). These authors observed equally low aggression towards freshly killed nestmates and non-nestmates when these had been washed in the solvent, but aggression could be restored by applying the extract of a non-nestmate.

More recently, some evidence for the use of hydrocarbons in recognition in semi-social insects has been found. In the burying beetles *Nicrophorus vespilloides* and *N. orbicollis* a male and female bury a small vertebrate carcass together and defend it against intruders, in order to utilize it for their reproduction (Müller *et al.*, 2003). They will attack any conspecific adult, unless it is in the same breeding stage as themselves and their partner. This progressed breeding stage is associated with a change in their hydrocarbon profiles, which are likely to be the cues used in recognition (Steiger *et al.*, 2007, 2008; Scott *et al.*, 2008). Even though this is not nestmate recognition *sensu stricto* and in *N. vespilloides* a more volatile substance seems to be involved in the recognition system as well (S. Steiger, personal communication), this example does illustrate the more widespread use of hydrocarbons in nest defense.

Volatile chemicals have also been proposed as nestmate recognition cues in highly eusocial insects. For several species of ants, volatile components of alarm pheromones were suggested, because the freeze-drying of ants (during which volatiles can evaporate) removed differential behavior towards nestmates and non-nestmates (Jaffe, 1987). However, to our knowledge, no compelling evidence has been presented that the actual variation (quantitative or qualitative) in these secretions, instead of the mere presence, is the cause of differential aggression. On the other hand, volatile chemicals may also be involved in the nestmate recognition system of the ant *Camponotus fellah*. Workers that were isolated with a double mesh within the nest showed significantly different hydrocarbon profiles from non-isolated workers after 21 days, but aggression was not significantly elevated, suggesting the involvement of volatile nest odors (Katzav-Gozansky *et al.*, 2004).

Direct evidence for cuticular hydrocarbons

Direct evidence for the use of hydrocarbons in nestmate recognition in social insects has been accumulating over the last fifteen years. When workers of the subterranean termites *Reticulitermes speratus* and *Coptotermes formosanus* were topically supplemented with the purified hydrocarbon profile of the other species, this was followed by an increase of aggression by nestmate soldiers (Takahashi and Gassa, 1995). Similarly, the topical application of (Z)-9-$C_{23:1}$ onto the cuticle of workers of the ant *Camponotus vagus* resulted in increased antennation and threat in the form of mandible opening (Meskali *et al.*, 1995b). These examples, however, concern qualitatively different hydrocarbons, and thus not variation in abundance that is typical for colonies of the same species.

The ant *Cataglyphis niger* was among the first for which conspecific cuticular hydrocarbons were tested (Lahav *et al.*, 1999). The postpharyngeal gland (PPG) contains the same mixture of hydrocarbons as is found on the cuticle of this species (Soroker *et al.*, 1994; see Chapter 5). Extracts obtained from the PPG were applied onto ant workers and the aggression these workers elicited in nestmates in a neutral arena was observed. When the purified hydrocarbon fraction of the PPG extract (or the entire PPG extract) of a non-nestmate was applied, significantly more aggression was observed than in the sham control, whereas this was not the case for the non-hydrocarbon fraction (Figure 11.1). This shows that it is, indeed, the hydrocarbons that are the key to the nestmate recognition code. Similarly, when the hydrocarbon fraction of a nestmate was applied onto a non-nestmate, the aggression was reduced, even though the effect was not as strong.

The differential importance of alkenes, and linear and methyl-branched alkanes in the nestmate recognition system of the paper wasp *Polistes dominulus* was tested using

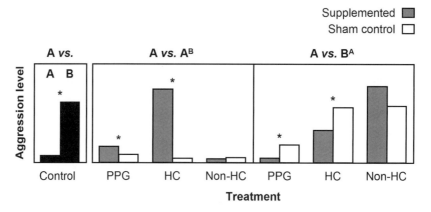

Figure 11.1 *Cataglyphis niger* workers were supplemented with extracts of the postpharyngeal gland (PPG), the hydrocarbon fraction of this extract (HC) or the non-hydrocarbon lipid fraction of this extract (Non-HC). Subsequently, untreated workers of a colony A were tested against untreated nestmates (A) or untreated non-nestmates (B) (Control), against nestmates supplemented with non-nestmate extract (A^B), or against non-nestmates supplemented with nestmate extract (B^A), and their aggression was observed. Asterisks depict significant statistically differences. Modified from Lahav *et al.* (1999).

synthetic hydrocarbons (Dani *et al.*, 2001). Synthetic hydrocarbons (n-C_{21}, n-C_{23}, n-C_{25}, n-C_{27}, n-C_{29}, n-C_{31}, 11-MeC_{29}, 7-MeC_{29}, (Z)-11-$C_{29:1}$, (Z)-9-$C_{29:1}$, and (Z)-11-$C_{31:1}$) were topically applied onto anesthetized workers one at a time, after which these were reintroduced to their nest and the reaction of their nestmates was observed. Linear alkanes were never found to have any aggression-eliciting effect on nestmates, whereas alkenes and methyl-branched alkanes did have this effect. Especially when the authors applied an unnaturally high quantity (200µg), the addition of a single hydrocarbon (but not linear alkanes) disrupted the nestmate signal and elicited aggressive reactions. These were all hydrocarbons that naturally occur on the cuticles of *P. dominulus* (although the alkenes are only found in traces), so it is the *increase in relative abundance* of a single hydrocarbon that elicits the aggression.

Similarly, in honeybees, *Apis mellifera*, alkenes were found to be more important in disrupting the nestmate signal than alkanes (Dani *et al.*, 2005). On the other hand, hydrocarbons may not be the only nestmate recognition cues in honeybees. Comb wax, used in nest construction, seems to be the crucial medium for the acquisition of nestmate recognition cues among nestmates (Breed *et al.*, 1988; d'Ettorre *et al.*, 2006; Couvillon *et al.*, 2007). The major compounds found in this wax are free fatty acids (12% of the total wax content) and hydrocarbons (14%) (Breed, 1998a). In supplementation experiments, the increase in relative abundance of several hydrocarbons, free fatty acids and esters was found to lower the acceptability of worker bees to nestmates (Breed and Stiller, 1992; summarized in Buchwald and Breed, 2005). The honeybee is, therefore, one of the few examples where the evidence suggests that hydrocarbons are only part of the nestmate recognition cues.

It may be that this applies to more bees, since other species in the families Halictidae and Apidae have been found to have fatty acids and esters on their cuticles (*Lasioglossum zephyrum*: Smith *et al.*, 1985; *Trigona fulviventris*: Buchwald and Breed, 2005; *Frieseomelitta varia, Lestrimelitta limao*: Nunes *et al.*, 2008; *Megalopta genalis*: J. S. van Zweden and W. T. Wcislo, unpublished results).

A recent study on *Formica japonica* (Akino *et al.*, 2004) showed that both alkenes and linear alkanes are necessary for nestmate recognition in this species. When the entire CHC-profile (consisting of (Z)-9-alkenes and linear alkanes from 25 to 33 carbon atoms) of a non-nestmate was imitated on an oval-shaped glass dummy, it was attacked. On the other hand, when only the alkene or only the alkane fraction of the profile was applied there was no significant difference between the reactions to nestmate and non-nestmate profiles. Hence, the presence of linear alkanes was necessary to elicit aggressive reactions between non-nestmates. These results are consistent with the results reported by Dani *et al.* (2001, 2005), because linear alkanes were present on the cuticle of the *Polistes* wasps. A similar study has been performed with *Formica exsecta* (Martin *et al.*, 2008b), but here linear alkanes do not seem to be necessary for nestmate recognition. The hydrocarbon profile of these ants also consists of (Z)-9-alkenes and linear alkanes, ranging from 23 to 31 carbon atoms. When workers were presented with glass beads coated with the nestmate profile, they exhibited no aggression. When the alkene fraction was missing or was altered, they attacked the glass beads significantly more, whereas the alkane fraction did not show this effect, either when it was missing, or when it was altered.

Unlike *P. dominulus*, *A. mellifera* and *F. exsecta*, the European hornet, *Vespa crabro*, and the Argentine ant, *Linepithema humile*, show some support for linear alkanes as nestmate recognition cues. Workers of *V. crabro*, when leaving their nest, were significantly more aggressive to dead nestmates when n-C_{21}, n-C_{23} or (Z)-9-$C_{23:1}$ had been increased in the hydrocarbon profile of these nestmates (Ruther *et al.*, 2002). Workers of *L. humile* also responded aggressively to an increase in multiple linear alkanes (n-C_{23} to n-C_{30} and n-C_{33}) in the profile of nestmates, but when the mixture of linear alkanes or the nestmate profile was presented alone, no significant increase in aggression was observed compared with a blank control. Hence, alteration of the relative concentrations of the linear alkanes only elicited any effect when methyl-branched alkanes and alkenes were present as well, which shows that multiple structural classes of hydrocarbons are necessary for the nestmate signal in this species (Greene and Gordon, 2007).

Altogether these studies show that hydrocarbons are essential for nestmate recognition in ant and wasp species, but they are, perhaps, not the only cues in bees. For termites, direct evidence for hydrocarbons as nestmate recognition cues is still lacking, although correlative data suggest its use (Kaib *et al.*, 2004; Dronnet *et al.*, 2006). The differential importance of structural classes – linear alkanes, methyl-branched alkanes and alkenes – may also differ among species, although linear alkanes have so far received the least support for a role in nestmate recognition. This may be attributed partly to the fact that linear alkanes have only the length of the chain of carbon atoms as a discriminative feature, whereas methyl-branched and unsaturated hydrocarbons also have the position of the methyl group or the double bond (cf. Châline *et al.*, 2005). A non-exclusive alternative hypothesis is that the relative abundance of linear alkanes is mostly related to (recognition of) castes, tasks, and social status within the colony (Wagner *et al.*, 2001). The variation in linear alkanes is then equally high within colonies and between colonies and, therefore, the supplementation of one or a few of these compounds will not alter nestmate-recognition-related aggression (see Chapter 12).

Models of decision rules in nestmate recognition

The early models

Crozier and Dix (1979) quantitatively analyzed two genetic models formalizing how nestmate recognition could work (Figure 11.2A). Under their "individualistic" model, odors are not transferred between individuals within the nest and individuals only recognize each other if they share at least one recognition allele on the nest odor loci. Under the "Gestalt" model, odors mix within the nest and individuals recognize each other only if they share all recognition alleles. In a similar analysis, Getz (1982) extended the possible models to include a "genotype recognition" model, in which individuals only accept each other as nestmates when their exact recognition allele genotypes are present in the nest (Figure 11.2A). Furthermore, he considers a "foreign-label rejection" model, in which an individual is rejected when it carries a foreign recognition allele. Lastly, he proposed an "habituated-label acceptance" model, which assumes the acceptance of nestmates when they share at least one recognition allele regardless of the other recognition allele and is, therefore, even less restrictive than the Crozier–Dix "individualistic" model (Figure 11.2A).

The acceptance threshold

In the models of Crozier and Dix (1979) and Getz (1982), nestmate recognition is based on the matching of alleles at recognition loci. However, directly comparing genotypes is impossible, and the matching of alleles would thus have to be done at the phenotypic level, e.g. the hydrocarbon pattern (Waldman, 1987; Liebert and Starks, 2004). And, as mentioned before, nestmate recognition in social insects most probably relies on quantitative differences rather than qualitative differences (Hölldobler and Michener, 1980).

Sherman *et al.* (1997) mention two models regarding how the detection of cues leads to acceptance of nestmates. These are similar to the earlier models proposed by Crozier and Dix (1979) and Getz (1982), yet explicitly on the phenotypic level. Under the desirable-present model (D-present), individuals are accepted when they possess desirable cues, i.e. cues that are possessed by nearly all desirable individuals (nestmates), but also by some undesirable individuals (non-nestmates), due to overlapping distribution of cues (Figure 11.2B). Under the undesirable-absent (U-absent) model, individuals are accepted when they lack undesirable cues, i.e. cues that are possessed by undesirable individuals, but also by some desirable individuals. These two fundamentally different models are implicitly also different in their

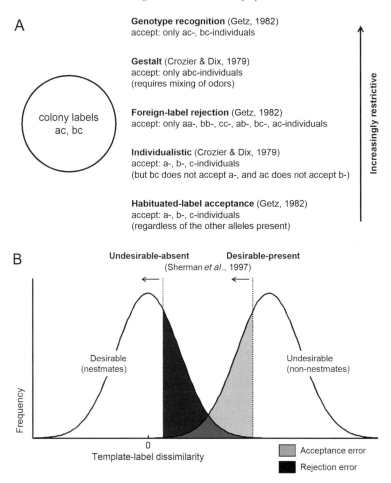

Figure 11.2 Schemes of nestmate recognition models. (A) Models based on matching of alleles. Haplodiploidy is assumed, as in Hymenopteran societies. (B) Models based on phenotype matching. Arrows indicate which phenotypes will be accepted under the respective model. Acceptance errors will be made more frequently when the desirable-present model applies, whereas more rejection errors will be made under the undesirable-absent model (Sherman *et al.*, 1997).

restrictiveness: the D-present model allows for some undesirable (additional) cues to be present and still be accepted, thus leading to more acceptance errors, whereas the U-absent model does not allow this deviation and thus may lead to more rejection errors.

Some support for the U-absent model comes from a study that was designed to test if hydrocarbons are the essential nestmate recognition cues in the ant *Cataglyphis niger* (Lahav *et al.*, 1999; Figure 11.1). When nestmates were treated with the hydrocarbon fraction of a non-nestmate (A versus A^B), aggression towards them increased dramatically to the level of an untreated non-nestmate, whereas when non-nestmates were treated with nestmate hydrocarbon extract (A versus B^A), aggression was only lowered slightly (though significantly), and not down to the level of an untreated nestmate. This pattern indicates that it is easier for *C. niger* workers to detect differences than similarities (Lenoir *et al.*, 1999; see also Gamboa *et al.*, 1986b).

More recently, Couvillon and Ratnieks (2008) performed a study using the stingless bee *Frieseomelitta varia*. Using a simple protocol with unwashed collecting vials, workers were treated with non-nestmate odors and subsequently introduced either into their own colony or into the non-nestmate colony from which the odor had come. It was found that nestmates were significantly less accepted when "contaminated" with non-nestmate odor, whereas non-nestmates were not better accepted when they were "contaminated" with nest-mate odor. This shows that the presence of non-nestmate odor on nestmates led to rejection, whereas the presence of nestmate odor on non-nestmates did not lead to better acceptance, which favors the U-absent model over the D-present model for *F. varia*. Similar experi-ments in the honeybee, *Apis mellifera*, yield the same outcome (F. L. W. Ratnieks, personal communication; see Couvillon *et al.*, 2007), which together with experiments showing equally high rejection of a number of harmless insects (Kärcher and Ratnieks, in prepara-tion), support a more widespread occurrence of a U-absent recognition system.

Plasticity of the acceptance threshold

Reeve (1989) proposed an adaptively shifting acceptance threshold that minimizes the chance of making an acceptance or rejection error. His mathematical model showed that an acceptance threshold that maximizes fitness depends on (1) the fitness consequences of accepting or rejecting desirable (nestmates) and undesirable individuals (non-nestmates), and (2) the frequency of interaction with these classes of individuals. A critical test of this adaptive acceptance threshold in *Polistes dominulus* confirmed that discrimination depended on the presence of nest material or a nestmate that may indicate vicinity to the nest and therefore increased fitness loss of accepting non-nestmates (Starks *et al.*, 1998). In *Apis mellifera* an increasing acceptance of non-nestmates was found to follow an increase in nectar availability or extra feeding. Since higher nectar availability results in reduced robbing of food by non-nestmates, and therefore reduced fitness loss upon acceptance of non-nestmates, this fits the predictions of the adaptive threshold model (Downs and Ratnieks, 2000; Downs *et al.*, 2001). Evidence for the frequency of interaction comes from a study using the ant *Formica rufibarbis* and its social parasite, *Polyergus rufescens*. In an unparasitized population, the aggression exhibited by host workers towards parasite work-ers was always high, whereas the parasitized population (which had encountered the social parasite before) only showed high aggression in the raiding season of the social parasite when the frequency of encounters increases (d'Ettorre *et al.*, 2004).

The referents and gestalt odor

The accuracy and flexibility of the nestmate recognition system may depend on the refer-ents used for the template against which incoming individuals will be compared (Lacy and

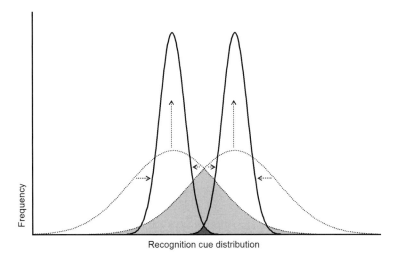

Figure 11.3 The effect of mixing odors within colonies. The distribution of recognition cues is narrowed, which reduces the probability of acceptance and rejection errors (shaded area).

Sherman, 1983). Self-referent phenotype matching would likely result in the most accurate assessment of kin, but, to our knowledge, there is no evidence for this mechanism in social insects. Nestmate-referent phenotype matching, on the other hand, has been proposed in a number of species (e.g. Buckle and Greenberg, 1981; Gamboa *et al.*, 1986a; Errard, 1994a). The idea of nestmates as referents is that the average quantity of each recognition cue among individuals in the nest is learned, and individuals are accepted as long as they do not deviate more than a certain threshold from this mean ("mean template" model; Breed and Bennett, 1987). The efficiency of learning this mean odor is significantly increased when all individuals in the colony indeed also carry the mean odor, i.e. by creating a Gestalt. Evidence for a Gestalt comes from the mixing of odors (Soroker *et al.*, 1994; Meskali *et al.*, 1995a; Lenoir *et al.*, 2001a), cross-fostering within species (Stuart, 1988; Downs and Ratnieks, 1999; Harano and Sasaki, 2006; J. S. van Zweden *et al.*, unpublished results) and cross-fostering between species (Errard, 1994b; Errard and Hefetz, 1997).

An additional feature of creating a Gestalt is that, while reducing the variation in cues, it also results in nestmates and non-nestmates becoming more dissimilar (less cue overlapping). This, in turn, reduces the possibility of rejection and acceptance errors, compared to unmixed odors (Figure 11.3). This can thus be viewed as an example of an evolutionary shift in recognition cues to optimize the balance between rejection and acceptance errors that can operate next to a shifting rejection threshold (Sherman *et al.*, 1997).

The neural mechanism and a new model

The traditionally accepted model for nestmate recognition of label-template matching involving long-term memory and thus higher integration centers in the brain, has recently been challenged by the hypothesis that more peripheral neuronal structures might be used (Ozaki *et al.*, 2005). A sensillum in the antennae of the ant *Camponotus japonicus* gave spikes when it perceived the hydrocarbon profile of a non-nestmate, but not when a nest-mate profile was presented. This implies that the receptor should be adapted (desensitized) to the multi-component nestmate odor, and hence some integration of the concentrations and the ratios between the different components would occur already at the level of the antenna (see Chapter 10).

Evidence against a template at the antennal level comes from a related ant, *Camponotus floridanus*. By treating the antennae of restrained workers (so that these could not clean their antennae) with the PPG extract of either nestmates or non-nestmates, it was shown that a constant exposure to a non-nestmate odor can induce a change in the neural tem-plate leading to reduced aggression towards these non-nestmates. However, the effect was only observed after 15h of exposure, and not after 2h, so the authors argue that receptor adaptation is unlikely to play a role since this process works on a much shorter time scale (Leonhardt *et al.*, 2007).

The integration of information from different receptor neurons takes place in the anten-nal lobes, the first synaptic relay of the central nervous system. Hence, habituation (the most basic form of learning) to nestmate odor could occur at the level of the antennal lobes and therefore the nestmate odor signal would not be passed on to higher brain centers. In this habituation state, a reaction would follow only the perception of non-nestmate odor (Dalton, 2000). In the experiments with *C. floridanus* (Leonhardt *et al.*, 2007), associative learning can likely be excluded, since treated workers were isolated and the perception of hydrocarbons was therefore not reinforced by any stimulus. Also, the hydrocarbons were permanently present on the antennae, not having the predictive value needed for associa-tive learning, thereby arguing against a template being stored in the long-term memory. Habituation is more parsimonious, however, since it also involves neuronal reformation in the central nervous system (but does not involve long-term memory) and operates on the right time scale to explain the results obtained in *C. floridanus* (Leonhardt *et al.*, 2007).

Since individuals within the colony are almost constantly encountering their nestmates, it would not be adaptive to continuously have to compare their profiles with an internal template in the long-term memory. The alternative would be to filter out all this input early in the neural structures and only react to hydrocarbon profiles that are dissimilar to the nest-mate ones. With either desensitized antennae or habituated antennal lobes, individuals in the social insect colony would not detect or respond to the hydrocarbon profiles of nestmates (D-present), nor would they actively detect that non-nestmate hydrocarbon patterns are absent (U-absent) when accepting an encountered individual, as proposed by Sherman *et al.* (1997). Instead, they would only detect cues present on non-nestmates, since only these trigger the antennae or the antennal lobe to respond, and on this basis reject the encoun-tered individual. Social insects would thus use the following as a simple rule of thumb: if the receptor neurons are not spiking and/or the pattern of activation in the antennal lobes is not altered (no differences in nestmate recognition cues are detected), the encountered indi-vidual is accepted as a nestmate. This results in the same acceptance and rejection outcome

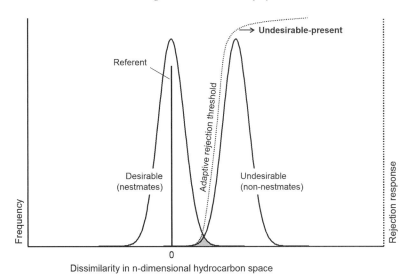

Figure 11.4   The Gestalt non-nestmate recognition model. Individuals are habituated (or desensitized) to the referent phenotype and will thus only detect dissimilarities from this phenotype. On the right-hand side of the rejection threshold (dashed line) the dissimilarity is large enough, so that the detection of undesirable phenotypes leads to rejection. The shaded area depicts rejection errors.

as under the long-term-memory model, but the neural structures involved are different. We thus propose not a new *nestmate* recognition model, but a *non-nestmate* recognition model: the "U-present" model (Figure 11.4).

If nestmate recognition occurs early in the olfactory system, other more sophisticated levels of recognition are likely to require higher integration structures in the brain. This is the case for within-colony recognition of worker castes (Greene and Gordon, 2003; see Chapter 12), queen fertility signal (Monnin, 2006; Hefetz, 2007; Heinze and d'Ettorre, 2009; see Chapter 13), genetic lineages (Hannonen and Sundström, 2003; but see Holzer *et al.*, 2006), cheating workers (Ratnieks *et al.*, 2006, and references therein) and recognition of individual identity (d'Ettorre and Heinze, 2005; Dreier *et al.*, 2007).

Synthesis

In order to synthesize the ideas that have been proposed and supported regarding hydrocarbons and nestmate recognition, we suggest that most evidence is pointing towards a model for nestmate recognition which is (1) based on quantitative variation among hydrocarbons that are homogeneous among nestmates, (2) based on a graded response to increasingly different concentrations and ratios, but perhaps not a linear response, so that we could still refer to the existences of a "soft" response threshold, (3) taking a context-dependent rejection threshold into account, (4) using nestmates as referents for comparison of phenotypic characters (at whatever level of the nervous system this matching occurs) and (5) working according to an "undesirable-present" principle (Figure 11.4). Below, we will briefly summarize why we believe these points are supported.

The increase in the abundance of specific hydrocarbons (or also fatty acids, in the case of the honeybee) on the cuticle of a social insect worker can increase the aggression that it receives, which implies that the worker is now perceived as a non-nestmate (Lahav *et al.*, 1999; Dani *et al.*, 2001; Breed *et al.*, 2004; Dani *et al.*, 2005; Guerrieri *et al.*, 2009). Also, the mere observation that workers can perceive each other as non-nestmates, even though they have the same set of hydrocarbons on their cuticles, adds to the evidence in support of the first point of this model. A social insect worker will thus (at the level of the antennae or antennal lobes) compare the dissimilarity of the cuticular hydrocarbon profile of an encountered individual with its nestmate profile, which will give a graded function and hence a graded rejection response. Graded responses, from acceptance to complete rejection, have been found in ants (e.g. Carlin and Hölldobler, 1986; Obin and Vander Meer, 1988; Crosland, 1990; Provost, 1991; Errard and Hefetz, 1997; Lenoir *et al.*, 2001a), wasps (e.g. Gamboa, 1988; Bura and Gamboa, 1994) and bees (e.g. Dani *et al.*, 2005; Harano and Sasaki, 2006). However, the relationship between dissimilarity and rejection response does not have to be a linear one. We propose that an S-curve will fit the experimental data best (Figure 11.4): a mild rejection response (prolonged investigation followed by extra grooming and trophallaxis) towards individuals with low dissimilarity (shaded area in Figure 11.4), followed by a steep increase in rejection (from mandible opening, to biting, to stinging or gaster flexing), after which the rejection response will remain approximately at the same level with increasing dissimilarity in hydrocarbon profiles. The slope and intercept of this S-shaped function may depend on the frequency of interaction between desirable and undesirable individuals and on the fitness consequences of rejecting these classes of individuals (Reeve, 1989). Shifting rejection responses have at least been found in *Polistes dominulus* (Starks *et al.*, 1998), *Apis mellifera* (Downs *et al.*, 2001), and *Formica rufibarbis* (d'Ettorre *et al.*, 2004).

Implicit to the U-present model is that a social insect is habituated (or desensitized) to its colony odor: to the odor of its nestmates. The observation that cross-fostered individuals are less aggressive to colonies related to their foster colony, as compared to control individuals, shows that social insects generally use nestmates as referents (e.g. Buckle and Greenberg, 1981; Errard and Hefetz, 1997; Downs and Ratnieks, 1999; Harano and Sasaki, 2006). Because labels change over time, the template must be continuously updated and thus continuously reinforced (Breed and Bennett, 1987; Vander Meer *et al.*, 1989; Lenoir *et al.*, 2001a). This seems to argue against a template stored in the long-term memory. Habituation in the antennal lobe is more flexible and can easily explain the pattern that social insects detect differences rather than similarities (Lahav *et al.*, 1999; Couvillon and Ratnieks, 2008). It can also explain how exposure to non-nestmate odor through non-associative learning can lead to a reduction in aggression towards these non-nestmates after >15h (Leonhardt *et al.*, 2007). Yet, it still leads to the essential discrimination between nestmates and non-nestmates. Furthermore, inside the colony, it seems maladaptive to continuously have to compare nestmates with a template stored in the long-term memory, whereas filtering this information early in the nervous system and only passing specific sensory stimuli would avoid this overload (Wehner, 1987). We propose habituation to nestmate odor and subsequent detection of undesirable hydrocarbon profiles as a parsimonious explanation for the pattern of recognition of non-nestmates observed among social insects. In our view, acceptance of nestmates would be a simple consequence of non-rejection.

Acknowledgments

We would like to thank the editors for inviting us to contribute to this book. Furthermore, we are grateful to J. J. Boomsma, N. Bos, K. R. Foster, L. Holman, and V. Nehring for comments and discussion and to all members of the Copenhagen *Centre for Social Evolution* for providing a stimulating working environment. The work was supported by the EU Marie Curie Excellence Grant CODICES-EXT-CT-2004–014202 assigned to PdE.

References

Akino, T., Yamamura, K., Wakamura, S. and Yamaoka, R. (2004). Direct behavioral evidence for hydrocarbons as nestmate recognition cues in *Formica japonica* (Hymenoptera: Formicidae). *Appl. Entomol. Zool.*, **39**, 381–387.

Arnold, G., Quenet, B., Cornuet, J.-M., Masson, C., De Schepper, B., Estoup, A. and Gasqui, P. (1996). Kin recognition in honeybees. *Nature*, **379**, pp. 498.

Bagnères, A.-G., Lorenzi, M. C., Dusticier, G., Turillazzi, S. and Clément, J.-L. (1996). Chemical usurpation of a nest by paper wasp parasites. *Science*, **272**, 889–892.

Bonavita-Cougourdan, A., Clément, J.-L. and Lange, C. (1987). Nestmate recognition: the role of cuticular hydrocarbons in the ant *Camponotus vagus* Scop. *J. Entomol. Sci.*, **22**, 1–10.

Breed, M. D. (1983). Nestmate recognition in honey bees. *Anim. Behav.*, **31**, 86–91.

Breed, M. D. (1998a). Recognition pheromones of the honey bee. *BioScience*, **48**, 463–470.

Breed, M. D. (1998b). Chemical cues in kin recognition: criteria for identification, experimental approaches, and the honey bee as an example. In *Pheromone Communication in Social Insects: Ants, Wasps, Bees, and Termites*, ed. R. K. Vander Meer, M. D. Breed, K. E. Espelie and M. L. Winston. Boulder, CO: Westview Press, pp. 57–78.

Breed, M. D. and Bennett, B. (1987). Kin recognition in highly eusocial insects. In *Kin Recognition in Animals*, ed. D. J. C. Fletcher and C. D. Michener. Chichester: Wiley, pp. 243–285.

Breed, M. D., Diaz, P. H. and Lucero, K. D. (2004). Olfactory information processing in honeybee, *Apis mellifera*, nestmate recognition. *Anim. Behav.*, **68**, 921–928.

Breed, M. D. and Stiller, T. M. (1992). Honey bee, *Apis mellifera*, nestmate discrimination: hydrocarbon effects and the evolutionary implications of comb choice. *Anim. Behav.*, **43**, 875–883.

Breed, M. D., Williams, K. R. and Fewell, J. H. (1988). Comb wax mediates the acquisition of nest-mate recognition cues in honey bees. *Proc. Natl. Acad. Sci. USA*, **85**, 8766–8769.

Buchwald, R. and Breed, M. D. (2005). Nestmate recognition cues in a stingless bee, *Trigona fulviventris*. *Anim. Behav.*, **70**, 1331–1337.

Buckle, G. R. and Greenberg, L. (1981). Nestmate recognition in sweat bees (*Lasioglossum zephyrum*): does an individual recognize its own odour or only odours of its nestmates? *Anim. Behav.*, **29**, 802–809.

Bura, E. A. and Gamboa, G. J. (1994). Kin recognition by social wasps: asymmetric tolerance between aunts and nieces. *Anim. Behav.*, **47**, 977–979.

Butts, D. P., Espelie, K. E. and Hermann, H. R. (1991). Cuticular hydrocarbons of four species of social wasps in the subfamily vespinae: *Vespa crabro* (L.), *Dolichovespula maculata* (L.), *Vespula squamosa* (Drury), and *Vespula maculifrons* (Buysson). *Comp. Biochem. Physiol. B*, **99**, 87–91.

Carlin, N. and Hölldobler, B. (1988). Influence of virgin queens on kin recognition in the carpenter ant *Camponotus floridanus* (Hymenoptera: Formicidae). *Insectes Soc.*, **35**, 191–197.

Carlin, N. F. and Hölldobler, B. (1986). The kin recognition system of carpenter ants (*Camponotus* spp.) I: hierarchical cues in small colonies. *Behav. Ecol. Sociobiol.*, **19**, 123–134.

Châline, N., Sandoz, J.-C., Martin, S. J., Ratnieks, F. L. W. and Jones, G. R. (2005). Learning and discrimination of individual cuticular hydrocarbons by honeybees (*Apis mellifera*). *Chem. Senses*, **30**, 327–335.

Christensen, J. H., Hansen, A. B., Karlson, U., Mortensen, J. and Andersen, O. (2005). Multivariate statistical methods for evaluating biodegradation of mineral oil. *J. Chromatogr. A*, **1090**, 133–145.

Couvillon, M. and Ratnieks, F. L. W. (2008). Odour transfer in stingless bee marmelada (*Frieseomelitta varia*) demonstrates that entrance guards use an "undesirable–absent" recognition system. *Behav. Ecol. Sociobiol.*, **62**, 1099–1105.

Couvillon, M. J., Caple, J. P., Endsor, S. L., Kärcher, M., Russell, T. E., Storey, D. E. and Ratnieks, F. L. W. (2007). Nest-mate recognition template of guard honeybees (*Apis mellifera*) is modified by wax comb transfer. *Biol. Lett.*, **3**, 228–230.

Crosland, M. W. J. (1990). Variation in ant aggression and kin discrimination ability within and between colonies. *J. Insect Behav.*, **3**, 359–379.

Crozier, R. H. and Dix, M. W. (1979). Analysis of two genetic models for the innate components of colony odor in social Hymenoptera. *Behav. Ecol. Sociobiol.*, **4**, 217–224.

d'Ettorre, P., Brunner, E., Wenseleers, T. and Heinze, J. (2004). Knowing your enemies: seasonal dynamics of host–social parasite recognition. *Naturwissenschaften*, **91**, 594–597.

d'Ettorre, P. and Heinze, J. (2005). Individual recognition in ant queens. *Curr. Biol.*, **15**, 2170–2174.

d'Ettorre, P., Mondi, N., Lenoir, A. and Errard, C. (2002). Blending in with the crowd: social parasites integrate into their host colonies using a flexible chemical signature. *Proc. R. Soc. B*, **269**, 1911–1918.

d'Ettorre, P., Wenseleers, T., Dawson, J., Hutchinson, S., Boswell, T. and Ratnieks, F. L. W. (2006). Wax combs mediate nestmate recognition by guard honeybees. *Anim. Behav.*, **71**, 773–779.

Dalton, P. (2000). Psychophysical and behavioral characteristics of olfactory adaptation. *Chem. Senses*, **25**, 487–492.

Dani, F. R., Jones, G. R., Corsi, S., Beard, R., Pradella, D. and Turillazzi, S. (2005). Nestmate recognition cues in the honey bee: differential importance of cuticular alkanes and alkenes. *Chem. Senses*, **30**, 477–489.

Dani, F. R., Jones, G. R., Destri, S., Spencer, S. H. and Turillazzi, S. (2001). Deciphering the recognition signature within the cuticular chemical profile of paper wasps. *Anim. Behav.*, **62**, 165–171.

Downs, S. G. and Ratnieks, F. L. W. (1999). Recognition of conspecifics by honeybee guards uses nonheritable cues acquired in the adult stage. *Anim. Behav.*, **58**, 643–648.

Downs, S. G. and Ratnieks, F. L. W. (2000). Adaptive shifts in honey bee (*Apis mellifera* L.) guarding behavior support predictions of the acceptance threshold model. *Behav. Ecol.*, **11**, 326–333.

Downs, S. G., Ratnieks, F. L. W., Badcock, N. S. and Mynott, A. (2001). Honeybee guards do not use food-derived odors to recognize non-nest mates: a test of the odor convergence hypothesis. *Behav. Ecol.*, **12**, 47–50.

Dreier, S., van Zweden, J. S. and D'Ettorre, P. (2007). Long-term memory of individual identity. *Biol. Lett.*, **3**, 459–462.

Dronnet, S., Lohou, C., Christides, J.-P. and Bagnères, A.-G. (2006). Cuticular hydrocarbon composition reflects genetic relationship among colonies of the introduced termite *Reticulitermes santonensis Feytaud*. *J. Chem. Ecol.*, **32**, 1027–1042.

Errard, C. (1994a). Development of interspecific recognition behavior in the ants *Manica rubida* and *Formica selysi* (Hymenoptera: Formicidae) reared in mixed-species groups. *J. Insect Behav.*, **7**, 83–99.

Errard, C. (1994b). Long-term memory involved in nestmate recognition in ants. *Anim. Behav.*, **48**, 263–271.

Errard, C. and Hefetz, A. (1997). Label familiarity and discriminatory ability of ants reared in mixed groups. *Insectes Soc.*, **44**, 189–198.

Espelie, K. E. and Hermann, H. R. (1990). Surface lipids of the social wasp *Polistes annularis* (L.) and its nest and nest pedicel. *J. Chem. Ecol.*, **16**, 1841–1852.

Espelie, K. E., Wenzel, J. W. and Chang, G. (1990). Surface lipids of social wasp *Polistes metricus* Say and its nest and nest pedicel and their relation to nestmate recognition. *J. Chem. Ecol.*, **16**, 2229–2241.

Gamboa, G. J. (1988). Sister, aunt-niece, and cousin recognition by social wasps. *Behav. Genet.*, **18**, 409–423.

Gamboa, G. J., Grudzien, T. A., Espelie, K. E. and Bura, E. A. (1996). Kin recognition pheromones in social wasps: combining chemical and behavioural evidence. *Anim. Behav.*, **51**, 625–629.

Gamboa, G. J., Reeve, H. K., Ferguson, I. D. and Wacker, T. L. (1986a). Nestmate recognition in social wasps: the origin and acquisition of recognition odours. *Anim. Behav.*, **34**, 685–695.

Gamboa, G. J., Reeve, H. K. and Pfennig, D. W. (1986b). The evolution and ontogeny of nestmate recognition in social wasps. *Annu. Rev. Entomol.*, **31**, 431–454.

Getz, W. M. (1982). An analysis of learned kin recognition in Hymenoptera. *J. Theor. Biol.*, **99**, 585–597.

Greene, M. J. and Gordon, D. M. (2003). Cuticular hydrocarbons inform task decisions. *Nature*, **423**, pp. 32.

Greene, M. J. and Gordon, D. M. (2007). Structural complexity of chemical recognition cues affects the perception of group membership in the ants *Linepithema humile* and *Aphaenogaster cockerelli*. *J. Exp. Biol.*, **210**, 897–905.

Guerrieri, F. J., Nehring, V., Jørgensen, C. G., Nielsen, J., Galizia, C. G., and d'Ettorre, P. (2009). Ants recognize foes and not friends. *Proc. R. Soc. B*, **276**, 2461–2468

Hamilton, W. D. (1987). Discrimination nepotism: expectable, common, overlooked.

In *Kin Recognition in Animals* ed. D. J. C. Fletcher and C. D. Michener. New York: Wiley, pp. 417–437.

Hannonen, M. and Sundström, L. (2003). Sociobiology: Worker nepotism among polygynous ants. *Nature*, **421**, 910.

Harano, K.-I. and Sasaki, M. (2006). Renewal process of nestmate recognition template in European honeybee *Apis melliffera* L. (Hymenoptera: Apidae). *Appl. Entomol. Zool.*, **41**, 325–330.

Hefetz, A. (2007). The evolution of hydrocarbon pheromone parsimony in ants (Hymenoptera: Formicidae) – interplay of colony odor uniformity and odor idiosyncrasy: a review. *Myrmecol. News*, **10**, 59–68.

Heinze, J. and d'Ettorre, P. (2009). Honest and dishonest communication in social Hymenoptera. *J. Exp. Biol.*, **212**, 1775–1779.

Hölldobler, B. and Michener, C. D. (1980). Mechanisms of identification and discrimination in social Hymenoptera. In *Evolution of Social Behavior: Hypotheses and Empirical Tests*, ed. H. Markl. Weinheim: Chemie Verlag, pp. 35–58.

Holzer, B., Kümmerli, R., Keller, L. and Chapuisat, M. (2006). Sham nepotism as a result of intrinsic differences in brood viability in ants. *Proc. R. Soc. B*, **273**, 2049–2052.

Howard, R. W. and Blomquist, G. J. (2005). Ecological, behavioral, and biochemical aspects of insect hydrocarbons. *Annu. Rev. Entomol.*, **50**, 371–393.

Howard, R. W., McDaniel, C. A. and Blomquist, G. J. (1980). Chemical mimicry as an integrating mechanism: cuticular hydrocarbons of a termitophile and its host. *Science*, **210**, 431–433.

Howard, R. W., McDaniel, C. A. and Blomquist, G. J. (1982). Chemical mimicry as an integrating mechanism for three termitophiles associated with *Reticulitermes virginicus* (Banks). *Psyche*, **89**, 157–168.

Jaffe, K. (1987). Evolution of territoriality and nestmate recognition in ants. *Experientia Supplementum*, **54**, 295–311.

Jaisson, P. (1991). Kinship and fellowship in ants and social wasps. In *Kin Recognition*, ed. P. G. Hepper. Cambridge: Cambridge University Press, pp. 60–93.

Kaib, M., Jmhasly, P., Wilfert, L., Durka, W., Franke, S., Francke, W., Leuthold, R. H. and Brandl, R. (2004). Cuticular hydrocarbons and aggression in the termite *Macrotermes subhyalinus*. *J. Chem. Ecol.*, **30**, 365–385.

Katzav-Gozansky, T., Boulay, R., Vander Meer, R. and Hefetz, A. (2004). In-nest environment modulates nestmate recognition in the ant *Camponotus fellah*. *Naturwissenschaften*, **91**, 186–190.

Lacy, R. C. and Sherman, P. W. (1983). Kin recognition by phenotype matching. *Am. Nat.*, **121**, 489–512.

Lahav, S., Soroker, V., Hefetz, A. and Vander Meer, R. K. (1999). Direct behavioral evidence for hydrocarbons as ant recognition discriminators. *Naturwissenschaften*, **86**, 246–249.

Lenoir, A., Cuisset, D. and Hefetz, A. (2001a). Effects of social isolation on hydrocarbon pattern and nestmate recognition in the ant *Aphaenogaster senilis* (Hymenoptera, Formicidae). *Insectes Soc.*, **48**, 101–109.

Lenoir, A., D'Ettorre, P., Errard, C. and Hefetz, A. (2001b). Chemical ecology and social parasitism in ants. *Annu. Rev. Entomol.*, **46**, 573–599.

Lenoir, A., Fresneau, D., Errard, C. and Hefetz, A. (1999). Individuality and colonial identity in ants: the emergence of the social representation concept. In *Information Processing in Social Insects,* ed. C. Detrain, J.-L. Deneubourg and J. M. Pasteels. Basel: Birkhäuser, pp. 219–237.

Leonhardt, S., Brandstaetter, A. and Kleineidam, C. (2007). Reformation process of the neuronal template for nestmate-recognition cues in the carpenter ant *Camponotus floridanus*. *J. Comp. Physiol. A*, **193**, 993–1000.

Liang, D. and Silverman, J. (2000). "You are what you eat": diet modifies cuticular hydrocarbons and nestmate recognition in the Argentine ant, *Linepithema humile*. *Naturwissenschaften*, **87**, 412–416.

Liebert, A. E. and Starks, P. T. (2004). The action component of recognition systems: a focus on the response. *Ann. Zool. Fennici*, **41**, 747–764.

Liebig. J., Eliyahu, D. and Brent, C. S. (2009). Cuticular hydrocarbon profiles indicates reproductive status in the termite *Zootermopsis nevedensis. Behav. Ecol. Sociobiol.*, in press.

Lorenzi, M. C., Bagnères, A. G., Clément, J. L. and Turillazzi, S. (1997). *Polistes biglumis bimaculatus* epicuticular hydrocarbons and nestmate recognition (Hymenoptera, Vespidae). *Insectes Soc.*, **44**, 123–138.

Martin, S., Helanterä, H. and Drijfhout, F. (2008a). Colony-specific hydrocarbons identify nest mates in two species of *Formica* ant. *J. Chem. Ecol.*, **34**, 1072–1080.

Martin, S. J., Vitikainen, E., Helanterä, H. and Drijfhout, F. P. (2008b). Chemical basis of nest-mate discrimination in the ant *Formica exsecta. Proc. R. Soc. B*, **275**, 1271–1278.

Meskali, M., Bonavita-Cougourdan, A., Provost, E., Bagnères, A.-G., Dusticier, G. and Clément, J.-L. (1995a). Mechanism underlying cuticular hydrocarbon homogeneity in the ant *Camponotus vagus* (SCOP.) (Hymenoptera: Formicidae): Role of postpharyngeal glands. *J. Chem. Ecol.*, **21**, 1127–1148.

Meskali, M., Provost, E., Bonavita-Cougourdan, A. and Clément, J.-L. (1995b). Behavioural effects of an experimental change in the chemical signature of the ant *Camponotus vagus* (Scop.). *Insectes Soc.*, **42**, 347–358.

Monnin, T. (2006). Chemical recognition of reproductive status in social insects. *Ann. Zool. Fennici*, **43**, 531–549.

Müller, J. K., Eggert, A.-K. and Elsner, T. (2003). Nestmate recognition in burying beetles: the "breeder's badge" as a cue used by females to distinguish their mates from male intruders. *Behav. Ecol.*, **14**, 212–220.

Nash, D. R. and Boomsma, J. J. (2008). Communication between hosts and social parasites. In *Sociobiology of Communication: an interdisciplinary perspective,* ed. P. d'Ettorre and D. P. Hughes. Oxford: Oxford University Press, pp. 55–79.

Nowbahari, E., Lenoir, A., Clément, J. L., Lange, C., Bagnères, A. G. and Joulie, C. (1990). Individual, geographical and experimental variation of cuticular hydrocarbons of the ant *Cataglyphis cursor* (Hymenoptera: Formicidae): Their use in nest and subspecies recognition. *Biochem. Syst. Ecol.*, **18**, 63–73.

Nunes, T. M., Nascimento, F. S., Turatti, I. C., Lopes, N. P. and Zucchi, R. (2008). Nestmate recognition in a stingless bee: does the similarity of chemical cues determine guard acceptance? *Anim. Behav.*, **75**, 1165–1171.

Obin, M. S. and Vander Meer, R. K. (1988). Sources of nestmate recognition cues in the imported fire ant *Solenopsis invicta* Buren (Hymenoptera: Formicidae). *Anim. Behav.*, **36**, 1361–1370.

Ozaki, M., Wada-Katsumata, A., Fujikawa, K., Iwasaki, M., Yokohari, F., Satoji, Y., Nisimura, T. and Yamaoka, R. (2005). Ant nestmate and non-nestmate discrimination by a chemosensory sensillum. *Science*, **309**, 311–314.

Provost, E. (1991). Nonnestmate kin recognition in the ant *Leptothorax lichtensteini*: evidence that genetic factors regulate colony recognition. *Behav. Genet.*, **21**, 151–167.

Provost, E., Rivière, G., Roux, M., Morgan, E. D. and Bagnères, A.-G. (1993). Change in the chemical signature of the ant *Leptothorax lichtensteini* Bondroit with time. *Insect Bioch. Mol. Biol.*, **23**, 945–957.

Ratnieks, F. L. W., Foster, K. R. and Wenseleers, T. (2006). Conflict resolution in insect societies. *Annu. Rev. Entomol.*, **51**, 581–608.

Reeve, H. K. (1989). The evolution of conspecific acceptance thresholds. *Am. Nat.*, **133**, 407–435.

Ross, K. G., Meer, R. K. V., Fletcher, D. J. C. and Vargo, E. L. (1987). Biochemical phenotypic and genetic studies of two introduced fire ants and their hybrid (Hymenoptera: Formicidae). *Evolution*, **41**, 280–293.

Ruther, J., Sieben, S. and Schricker, B. (2002). Nestmate recognition in social wasps: manipulation of hydrocarbon profiles induces aggression in the European hornet. *Naturwissenschaften*, **89**, 111–114.

Scott, M. P., Madjid, K. and Orians, C. M. (2008). Breeding alters cuticular hydrocarbons and mediates partner recognition by burying beetles. *Anim. Behav.*, **76**, 507–513.

Sherman, P. W., Reeve, H. K. and Pfennig, D. W. (1997). Recognition systems. In *Behavioural Ecology: An Evolutionary Approach*, ed. J. R. Krebs and N. B. Davies. Oxford: Blackwell, pp. 69–96.

Singer, T. and Espelie, K. (1996). Nest surface hydrocarbons facilitate nestmate recognition for the social wasp, *Polistes metricus* Say (Hymenoptera: Vespidae). *J. Insect Behav.*, **9**, 857–870.

Singer, T. L. (1998). Roles of hydrocarbons in the recognition systems of insects. *Am. Zool.*, **38**, 394–405.

Sledge, M. F., Dani, F. R., Cervo, R., Dapporto, L. and Turillazzi, S. (2001). Recognition of social parasites as nest-mates: adoption of colony-specific host cuticular odours by the paper wasp parasite *Polistes sulcifer*. *Proc. R. Soc. B*, **268**, 2253–2260.

Smith, B. H., Carlson, R. G. and Frazier, J. (1985). Identification and bioassay of macrocyclic lactone sex pheromone of the halictine bee *Lasioglossum zephyrum*. *J. Chem. Ecol.*, **11**, 1447–1456.

Soroker, V., Vienne, C. and Hefetz, A. (1995). Hydrocarbon dynamics within and between nestmates in *Cataglyphis niger* (Hymenoptera: Formicidae). *J. Chem. Ecol.*, **21**, 365–378.

Soroker, V., Vienne, C., Hefetz, A. and Nowbahari, E. (1994). The postpharyngeal gland as a "gestalt" organ for nestmate recognition in the ant *Cataglyphis niger*. *Naturwissenschaften*, **81**, 510–513.

Starks, P. T., Fischer, D. J., Watson, R. E., Melikian, G. L. and Nath, S. D. (1998). Context-dependent nestmate discrimination in the paper wasp, *Polistes dominulus*: a critical test of the optimal acceptance threshold model. *Anim. Behav.*, **56**, 449–458.

Steiger, S., Peschke, K., Francke, W. and Müller, J. K. (2007). The smell of parents: breeding status influences cuticular hydrocarbon pattern in the burying beetle *Nicrophorus vespilloides*. *Proc. R. Soc. B*, **274**, 2211–2220.

Steiger, S., Peschke, K. and Müller, J. (2008). Correlated changes in breeding status and polyunsaturated cuticular hydrocarbons: the chemical basis of nestmate recognition in the burying beetle *Nicrophorus vespilloides*? *Behav. Ecol. Sociobiol.*, **62**, 1053–1060.

Stuart, R. J. (1988). Collective cues as a basis for nestmate recognition in polygynous leptothoracine ants. *Proc. Natl. Acad. Sci. USA*, **85**, 4572–4575.

Takahashi, S. and Gassa, A. (1995). Roles of cuticular hydrocarbons in intra-and interspecific recognition behavior of two Rhinotermitidae species. *J. Chem. Ecol.*, **21**, 1837–1845.

van Zwenden, J. S., Dreier, S. and d'Ettorre, P. (2009). Disentangling environmental and heritable nestmate recognition cues in a carpenter ant. *J. Insect Physiol.*, **55**, 158–163.

Vander Meer, R. K. and Morel, L. (1998). Nestmate recognition in ants. In *Pheromone Communication in Social Insects*, ed. R. K. Vander Meer, M. D. Breed, M. L. Winston and K. E. Espelie. Boulder, CO: Westview Press, pp. 79–103.

Vander Meer, R. K., Saliwanchik, D. and Lavine, B. (1989). Temporal changes in colony cuticular hydrocarbon patterns of *Solenopsis invicta*: implications for nestmate recognition. *J. Chem. Ecol.*, **15**, 2115–2125.

Wagner, D., Tissot, M. and Gordon, D. (2001). Task-related environment alters the cuticular hydrocarbon composition of harvester ants. *J. Chem. Ecol.*, **27**, 1805–1819.

Waldman, B. (1987). Mechanisms of kin recognition. *J. Theor. Biol.*, **128**, 159–185.

Wehner, R. (1987). 'Matched filters' – neural models of the external world. *J. Comp. Physiol. A*, **161**, 511–531.

Weil, T., Hoffmann, K., Kroiss, J., Strohm, E. and Korb, J. (2009). Scent of a queen – cuticular hydrocarbons specific for female reproductives in lower termites. *Naturwissenschaften*, **96**, 315–319.

12

Cuticular hydrocarbon cues in the formation and maintenance of insect social groups

Michael Greene

Insect social groups are formed and maintained by the many interactions among the members of the system (Gordon, 1996; Deneubourg *et al.*, 2002; O'Donnell and Bulova, 2007). Some solitary insect species are gregarious, forming self-organized aggregations associated with protection, reproduction, and feeding (Wertheim *et al.*, 2005). Eusocial insects live in societies with a division of labor between a reproductive caste and sterile workers in which the activity of workers is regulated in a non-hierarchical manner (Gordon, 1996). Patterns of social interaction can inform individual behavioral decisions that, in the aggregate, lead to changes in group dynamics. Many insect species use information coded in cuticular hydrocarbons to recognize other individuals during social interactions.

Aggregations of adult and nymph desert locusts (*Schistocerca gregaria*) are triggered by plant chemical attractants and aggregation pheromones (Heifetz *et al.*, 1997). This increase in population density results in an increase in direct contact and close-range chemical interactions among individuals. The increase in density subsequently induces a behavioral transition from a solitary phase to a gregarious migratory phase of behavior in which locusts become more active and interact with each other more (Heifetz *et al.*, 1996). Behavioral and physiological data show that cuticular lipids, specifically the hydrocarbon fraction, are responsible for the transition to the gregarious phase (Heifetz *et al.*, 1997, 1998). As locusts interact at high density, cues in cuticular hydrocarbons are detected by the antennae. Hydrocarbons from *S. gregaria* have been shown to interact with antennal receptors which affect levels of the second messenger inositol triphosphate (Heifetz *et al.*, 1997).

Cockroaches are gregarious insects that form aggregations in which individuals use communal shelters during rest (Rivault *et al.*, 1998). Gregarious behavior facilitates cooperation among cockroaches (Dambach and Goehlen, 1999). Cuticular hydrocarbons are used by cockroaches to organize the formation of aggregations though direct antennal interactions or close-range chemical detection of hydrocarbons; cockroaches are attracted to cuticular hydrocarbon odors during interactions (Rivault *et al.*, 1998; Saïd *et al.*, 2005). Species-specific cuticular hydrocarbon extracts from nymphs of *Periplaneta americana*, *P. brunnea*, *P. fuliginosa*, and *P. australasiae* induced an aggregation response when tested in a binary choice test in which filter paper was treated with hydrocarbons or a solvent control (Saïd *et al.*, 2005). More nymphs were found to aggregate near relevant species-specific cuticular hydrocarbons compared to controls. Each of the four species possessed a unique hydrocarbon profile ranging from 19 to 25 hydrocarbons that ranged in chain length from 21 to 43 carbons (Saïd *et al.*, 2005).

Cuticular hydrocarbons also act in the organization of aggregations of *Blattella germanica*. Rivault *et al.* (1998) tested the proportion of larvae attracted to filter papers conditioned with controls or solvent extracts of larvae. The highest response was towards filter papers treated with extracts containing only cuticular hydrocarbons extracted from any area of larval bodies. Furthermore, nymph and adult *B. germanica* discriminate siblings from non-siblings using quantitative differences in cuticular hydrocarbons (Lihoreau and Rivault, 2008). In this example of kin recognition, discrimination of siblings occurred without any previous social experience and was a context-dependent response: siblings were preferentially used as social partners while non-siblings were preferentially used as mating partners (Lihoreau and Rivault, 2008).

The aggregation of cockroaches under shelters is a self-organizing process (Deneubourg *et al.*, 2002; Halloy *et al.*, 2007). Chemical information present in cuticular hydrocarbons is assessed during direct antennal contacts or at close-range and informs behavioral decisions to aggregate. Theoretical models combined with empirical data show that the collective decisions that lead to the self-organization of cockroach aggregations depend on the resting time of individuals under a shelter (Deneubourg *et al.*, 2002; Halloy *et al.*, 2007). This process is organized so that cockroaches that are attracted to each other by information present in cuticular hydrocarbons gather together under shelters.

Eusocial insects, such as ants, bees, wasps, and termites have experienced tremendous evolutionary and ecological success as reflected by their high abundance, high species diversity, and broad distribution across most geographic regions of the world (Boomsma and Franks, 2006). The success of eusocial insects can be in part attributed to their social organization in which there exists a well-regulated division of labor (Smith *et al.*, 2008). A queen, or in some cases multiple queens, specialize in reproduction while sterile workers perform the basic functions necessary for colony survival including colony defense, foraging, brood tending, and nest construction (Gordon, 1996, 1999).

In colonies, the recognition of castes, including queens, workers, eggs, and brood occurs through cues in cuticular hydrocarbons. For example, differences between castes of the termites *Reticulitermes flavipes* and *R. virginicus* occur in the relative abundances of shared hydrocarbon molecules (Howard *et al.*, 1978, 1982; Haverty *et al.*, 1996; Kaib *et al.*, 2002). Caste differences in hydrocarbon profiles were also found in *Zootermopsis nevadensis* (Sevala *et al.*, 2000). Cuticular hydrocarbon cues allow discrimination of reproductive ants versus non-reproductive ants (Bonavita-Cougourdan *et al.*, 1993; Dietemann *et al.*, 2003; Chapter 13 of this book).

The regulation of worker activity in social insect colonies is non-hierarchical (Gordon, 1996). No one entity, including the queen, has the ability to control the activity of the other colony members. Instead, individual workers make behavioral decisions informed by the assessment of local information cues (Seeley and Tovey, 1994; Biesmeijer *et al.*, 1998; Greene and Gordon, 2003; Pratt, 2005; Detrain and Deneubourg, 2006; Greene and Gordon, 2007a). Colony-wide changes in behavior occur because of the collective decision making of workers (Gordon, 2002). Local informational cues must inform workers' behavioral decisions so that colony behavior changes appropriately in response to changing colony needs.

Task allocation is the process by which social insect colonies regulate the number of workers performing various tasks in a manner that appropriately meets changing colony needs (Gordon, 1996). Workers perform a variety of tasks including tending to the queen and brood, nest construction and maintenance, patrolling or scouting nest areas, and foraging for food (Gordon, 2002). Which individuals perform a task and the number of individuals performing a task are regulated during the process (Gordon, 1996, 2002).

Local cues relevant to task allocation are often found in the pattern of interactions among workers within a colony. For example, interactions between returning foragers and inactive foragers stimulate foraging activity in honey bees in response to differing food reward rates (Fernandez *et al.*, 2003). Honey bee workers that receive vibration signals produced by other workers in the nest respond with an increase in activity (Lewis *et al.,* 2002). The time elapsed for a honey bee to unload its nectar influences its decision to leave the hive on another foraging trip (Seeley and Tovey, 1994). Paper wasps are stimulated to forage in response to biting interactions among workers (O'Donnell, 2001). The time interval elapsed between two loads of water brought to the nest was found to influence nest construction behavior in the social wasp *Polybia occidentalis* (Jeanne and Nordheim, 1996). Workers of the ant *Temnothorax albipennis* use encounter rate to assess nestmate density while locating suitable nest sites (Pratt, 2005).

Task allocation in the red harvester ant (*Pogonomyrmex barbatus*) is a well-studied system that is regulated by patterns of interactions among workers on different tasks that are identified by cues present in the mixture of hydrocarbons on the worker cuticle (Greene and Gordon, 2003, 2007a). The cues are detected by ants during direct antennal contacts with the cuticle of another worker (Greene and Gordon, 2003). *Pogonomyrmex barbatus* is a seed-eating ant native to the southwestern USA and central Mexico (Gordon, 1996). Colonies contain a single queen and thousands of workers that perform various tasks including patrolling, foraging, midden work and nest maintenance work (Gordon, 1986, 1989).

Harvester ant workers doing different tasks do not differ in shape and size; however, they can easily be distinguished by their behavior on the nest mound (Gordon, 1986, 1996). Nest maintenance workers transition between inside the nest and outside, carrying pieces of soil, seed husks, corpses, and other colony garbage from the nest. Midden workers organize materials on the nest mound including materials dropped by nest maintenance workers and the small pebbles that cover the nest mound. Patrollers are the first task to emerge each morning, stimulating foragers to emerge from the nest and setting the foraging directions for the day (Gordon, 1989, 1991; Greene and Gordon, 2007b). Foragers exit the nest entrance on foraging trips which last approximately 20 minutes during which they search for seeds buried in soil. Workers perform different tasks at different points in their life. Workers switch tasks according to changes in colony needs, and task-activity levels are inter-related (Gordon, 1986, 1989; Gordon and Mehdiabadi, 1999).

Harvester ant workers possess task-specific differences in their cuticular hydrocarbon profiles. Thirty-four cuticular hydrocarbon molecules have been identified in the harvester ant profile (Wagner *et al.*, 1998). Hydrocarbons range in chain-length from 23 carbons to 35 carbons with compounds representing *n*-alkane, mono-methylalkane, dimethyl-alkane, and *n*-alkene structural classes (Wagner *et al.*, 1998, 2001). Workers of different colonies possess most of the same hydrocarbon molecules on their cuticle; however, ants of different colonies differ in the relative abundance of shared hydrocarbon compounds (peak area from gas chromatogram divided by sum of all peak areas). Within colonies, foragers and patrollers, the tasks that spend the longest duration times outside of the nest, have a higher relative abundance of *n*-alkane hydrocarbons compared to nest maintenance workers of the same colony (Wagner *et al.*, 1998). Exposure to a combination of high temperature and low relative humidity, as is experienced by patrollers and foragers while working outside of the nest, leads to increases in the relative abundance of *n*-alkanes on the cuticle (Wagner *et al.*, 2001). Ultraviolet wave exposure has no effect on *n*-alkane abundance. Evidence in *Myrmicaria eumenoides* demonstrates that juvenile hormone III titers can affect task-specific hydrocarbon cues in ants (Lengyel *et al.*, 2007).

Cuticular hydrocarbons contain cues that allow harvester ants to distinguish nestmate ants from conspecific non-nestmate ants (Wagner *et al.*, 2000). Other non-hydrocarbon surface lipids on harvester cuticles, such as wax esters and long-chain fatty acids, have been shown to play no meaningful communicative role in harvester ants (Wagner *et al.*, 1998, 2000; Nelson *et al.*, 2001; Greene and Gordon, 2003). The antennae are the organs of chemical perception and it is during direct antennal contact between workers that one ant can assess the cuticular hydrocarbon profile of another (Greene and Gordon, 2003; Ozaki *et al.*, 2005; Chapter 10 of this book).

A field study was conducted that exploited the relationship between patrollers and foragers (Greene and Gordon, 2003). If patrollers are prevented from returning to the nest, foragers will not leave the nest to collect seeds. When patrollers enter the nest, they mix with other foragers in a chamber just below the entrance where interactions occur among workers. Patrollers were removed from nest mounds as they exited the nest entrance, thus inhibiting foraging in focal colonies. After a period of inactivity inside the nest entrance, ant mimics were returned to the nest entrance at a rate of 1 bead every 10 seconds, a return rate of patrollers that was observed to occur at colonies prior to the exit of foragers from the nest. The number of foragers exiting nests was counted and data were normalized in order to control for absolute differences in forager numbers among colonies. Ant mimics were created by coating 5 mm diameter glass beads with (a) hydrocarbons extracted from patrollers, (b) hydrocarbons extracted from nest maintenance workers or (c) whole surface lipids from patrollers, which include hydrocarbons, wax esters, and fatty acids. Blank control beads were washed in pentane solvent. Live patrollers were also returned to the nest mound as a "positive" control. Using ant mimics was advantageous because the return rate into the nest entrance could be precisely controlled and all information that might indicate task – such as behavior – except for chemical information could be removed. Significantly more foragers emerged from nests in response to the return of live patrollers and ant

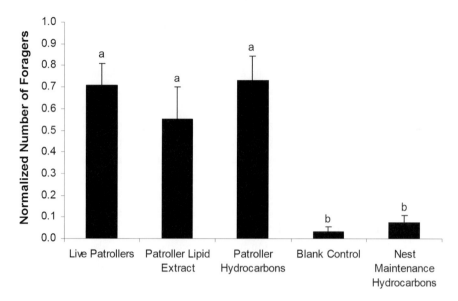

Figure 12.1 Harvester ant foragers are stimulated to leave the nest in search of seeds in response to cues present in the task-specific cuticular hydrocarbon profile of patrollers. After inhibition of foragers at harvester ant colonies by removal of patrollers, colony foraging behavior was rescued by the addition of ant mimics, glass beads coated with task-specific cuticular hydrocarbons from patrollers, to the nest entrance at a rate of 1 bead every 10 seconds. Data were normalized to account for differences in the absolute number of foragers active each day. The same letters above bars denote statistical significance using a Tukey's post-hoc test. From Greene and Gordon, 2003.

mimics (beads) coated with patroller surface lipids and patroller cuticular hydrocarbons than to blank control beads and beads coated with nest-maintenance worker hydrocarbons (Figure 12.1). Thus, task-specific cues in cuticular hydrocarbons were sufficient to elicit a significant foraging response in comparison to controls (Greene and Gordon, 2003).

However, it is not the cue in and of itself that stimulates forager activity in harvester ants. Instead, it is the pattern of social interactions that occur among workers in the colony that informs worker behavioral decisions. Task allocation occurs despite the inability of individual workers to assess the total number of workers performing a task because of a reliance on the detection of meaningful local information cues (Gordon and Mehdiabadi, 1999; Greene and Gordon, 2007a). A study using captive harvester ants showed that workers had a greater tendency to perform midden work when their rate of encounter with midden workers was high (Gordon and Mehdiabadi, 1999). Also, the time workers spent performing midden work was positively correlated to the number of midden workers they met while away from the midden.

A field experiment demonstrated that foragers assess the return rate of patrollers in order to decide to emerge from the nest to collect seeds each day (Greene and Gordon, 2007a). This experiment also took advantage of the relationship between foragers and patrollers. As they emerged, patrollers were collected so they could not return to the nest entrance.

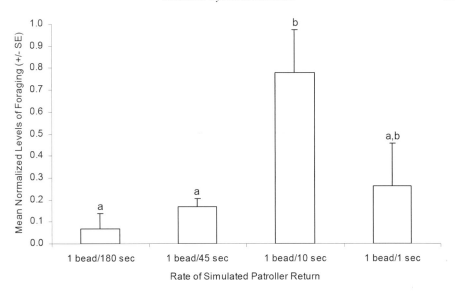

Figure 12.2 Foraging activity at harvester ant colonies was affected by the simulated return rate of patrollers back to the colony. Letters above bars denote differences in statistical significance among treatments (LSD, $p < 0.05$). Error bars denote standard error of the mean. From Greene and Gordon, 2007a.

After a period of inactivity, glass beads coated with patroller hydrocarbons were dropped into the nest entrance at different rates: (a) 1 bead every 180 seconds, (b) 1 bead every 45 seconds, (c) 1 bead every 10 seconds and (d) 1 bead every second. Colonies responded with the highest levels of foraging to the 1 bead every 10 seconds treatment, a rate used in the prior experiment (Greene and Gordon, 2003) and a rate observed prior to foraging in non-manipulated colonies (Figure 12.2; Greene and Gordon 2007a). Furthermore, when patroller-hydrocarbon-coated ant mimics were used to boost the return rate of patrollers back to the nest, foragers emerged faster than when blank control beads were added to colonies (Greene and Gordon, 2007b).

Harvester ant patrollers set the direction that foragers take each day by depositing glandular secretions from Dufour glands onto the nest mound (Greene and Gordon, 2007b). Dufour gland secretions of harvester ants are mainly composed of species-specific blends of hydrocarbons (Hölldobler *et al.*, 2004). Each day foragers exit the nest and travel along up to 8 trunk trails to forage for seeds. Unlike many other ant species, harvester ants very rarely use trail recruitment pheromones to recruit foragers to food sources. When trails were blocked to patrollers, there was a decrease in the number of foragers using the trails after the barrier was removed. Foraging along blocked trails was rescued by application of Dufour gland extracts but was not rescued by application of poison gland extracts or a blank control (Greene and Gordon, 2007b; Figure 12.3). Thus, cues are present in harvester ant Dufour secretions, composed mainly of hydrocarbons, which influence the direction foragers take to collect seeds each day.

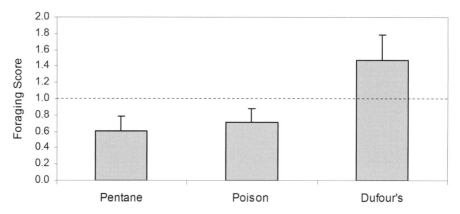

Figure 12.3 Foraging along trails on which patrolling was blocked was rescued by the application of extracts of Dufour's gland secretions but not by application of poison gland extracts. The foraging score indicates the measured response in comparison to an expected value if all trails at a nest were used equally. A foraging score of greater than one indicates a greater foraging response along trails than would be expected by random trail choice by foragers. From Greene and Gordon, 2007b (© The University of Chicago).

Increasing attention is being paid to the self-organization of insect social groups and to how insect societies are maintained and regulated in a non-hierarchical manner (Gordon, 1996; Deneubourg *et al.*, 2002; Detrain and Deneubourg, 2006; O'Donnell and Bulova, 2007). Members of groups typically assess local information cues to inform behavioral decisions which, in the aggregate, affect the dynamics of the group. As reviewed in this chapter, the perception of information coded in cuticular hydrocarbons along with patterns of social interactions can act as information cues that inform the decisions of group members for some insect species. This is an emerging area of research and future work will undoubtedly reveal additional examples. Furthermore, future work will continue to operate on two fronts: (1) studying the dynamics of how interactions affect patterns of activity in insect social groups while (2) studying the biosynthesis, expression and perception of cuticular hydrocarbon-based cues so that the rules by which individuals make behavioral decisions can be understood.

References

Biesmeijer, J. C, van Nieuwstadt, M. G. L., Lukács, S. and Sommeijer, M. J. (1998). The role of internal and external information in foraging decisions of *Melipona* workers (Hymenoptera: Meliponinae). *Behav. Ecol. Sociobiol.*, **42**, 107–116.

Bonavita-Cougourdan, A. Clément, J. L. and Lange, C. (1993). Functional subcaste discrimination (foragers and brood-tenders) in the ant *Camponotus vagus* Scop: polymorphism of cuticular hydrocarbon patterns. *J. Chem. Ecol.*, **19**, 1461–1477.

Boomsma, J. J. and Franks, N. R. (2006). Social Insects: From Selfish Genes to Self Organization and Beyond. *Trends Ecol. Evol.* **21**, 303–308.

Dambach, M. and Goehlen, B. (1999). Aggregation density and longevity correlate with humidity in first-instar nymphs of the cockroach (*Blattella germanica* L. Dictyoptera). *J. Insect Physiol.*, **45**, 423–429.

Deneubourg, J.-L., Lioni, A. and Detrain, C. (2002). Dynamics of aggregation and emergence of cooperation. *Biol. Bull.*, **202**, 262–267.

Detrain, C. and Deneubourg, J.-L. (2006). Self-organized structures in a superorganism: do ants "behave" like molecules? *Phys. Life Rev.*, **3**, 162–187.

Dietemann, V., Peeters, C., Liebig, J., Thivet, V. and Hölldobler, B. (2003). Cuticular hydrocarbons mediate discrimination of reproductives and nonreproductives in the ant *Myrmecia gulosa*. *Proc. Natl. Acad. Sci. USA*, **100**, 10341–10346.

Fernandez, P. C., Gil, M. and Farina, W. M. (2003). Reward rate and forager activation in honeybees: recruiting mechanisms and temporal distribution of arrivals. *Behav. Ecol. Sociobiol.*, **54**, 80–87.

Gordon, D. M. (1986). The dynamics of the daily round of the harvester ant colony. *Anim. Behav.*, **34**, 1402–1419.

Gordon, D. M. (1989). Dynamics of task switching in harvester ants. *Anim. Behav.*, **38**, 194–204.

Gordon, D. M. (1991). Behavioral flexibility and the foraging ecology of seed-eating ants. *Am. Nat.*, **138**, 379–411.

Gordon, D. M. (1996). The organization of work in social insect colonies. *Nature*, **380**, 121–124.

Gordon, D. M. (1999). Interaction patterns and task allocation in ant colonies. In *Information Processing in Social Insects*, ed. C. Detrain, J. M. Pasteels and J. L. Deneubourg. Basel, Switzerland: Birkhäuser, pp. 51–67.

Gordon, D. M. (2002). The regulation of foraging activity in red harvester ant colonies. *Am. Nat.*, **159**, 509–518.

Gordon, D. M. and Mehdiabadi, N. (1999). Encounter rate and task allocation in harvester ants. *Behav. Ecol. Sociobiol.*, **45**, 370–377.

Greene, M. J. and Gordon, D. M. (2003). Social insects: Cuticular hydrocarbons inform task decisions. *Nature*, **423**, 32.

Greene, M. J. and Gordon, D. M. (2007a). Interaction rate informs harvester ant task decisions. *Behav. Ecol.*, **18**, 451–455.

Greene, M. J. and Gordon, D. M. (2007b). How patrollers set foraging direction in harvester ants. *Am. Nat.*, **179**, 943–948.

Halloy, J., Sempo, G., Caprari, G., Rivault, C., Asadpour, M., Tâche, F., Saïd, I., Durier, V., Canonge, S., Amé, J.-M., Detrain, C., Corell, N., Martinoli, A., Mondada, F., Siegwart, R. and Deneubourg, J.-L. (2007). Social integration of robots into groups of cockroaches to control self-organized choices. *Science*, **318**, 1155–1158.

Haverty, M. I., Grace, J. K., Nelson, L. J. and Yamamoto, R. T. (1996). Intercaste, intercolony, and temporal variation in cuticular hydrocarbons of *Coptotermes formosanus* Shiraki (Isoptera: Rhinotermitidae). *J. Chem. Ecol.*, **22**, 1813–1834.

Heifetz, Y., Boekhoff, I., Breer, H. and Applebaum, S. W. (1997). Cuticular hydrocarbons control behavioral phase transition in *Schistocerca gergaria* nymphs and elicit biochemical responses in antennae. *J. Biochem. Mol. Biol.*, **27**, 563–568.

Heifetz, Y., Miloslavski, I., Aizenshtat, Z. and Applebaum, S. W. (1998). Cuticular surface hydrocarbons of desert locust nymphs, *Schistocerca gregaria*, and their effect on phase behavior. *J. Chem. Ecol.*, **24**, 1033–1046.

Heifetz, Y., Voet, H. and Appelbaum, S.W. (1996). Factors affecting behavioral phase transition in the desert locust, *Schistocerca gregaria* (Forskal) (Orthoptera: Acrididiae). *J. Chem. Ecol.*, **22**, 1717–1734.

Hölldobler, B., Morgan, E.D., Oldham, N.J., Liebig, J. and Liu, Y. (2004). Dufour gland secretion in the harvester ant genus *Pogonomyrmex*. *Chemoecology*, **14**, 101–106.

Howard, R.W., McDaniel, C.A. and Blomquist, G.J. (1978). Cuticular hydrocarbons of the eastern subterranean termite, *Reticulitermes flavipes* (Kollar) (Isoptera: Rhinotermitidae). *J. Chem. Ecol.*, **4**, 233–245.

Howard, R.W., McDaniel, C.A., Nelson, D.R., Blomquist, G.J., Gelbaum, L.T. and Zalkow, L.H. (1982). Cuticular hydrocarbons of *Reticulitermes virginicus* (Banks) and their role as potential species- and caste-recognition cues. *J. Chem. Ecol.*, **8**, 1227–1239.

Jeanne, R. and Nordheim, E. (1996). Productivity in a social wasp: per capita output increases with swarm size. *Behav. Ecol.*, **7**, 43–48.

Kaib, M., Franke, S., Francke, W. and Brandl, R. (2002). Cuticular hydrocarbons in a termite: phenotypes and a neighbour-stranger effect. *Physiol. Entomol.*, **27**, 189–198.

Lengyel, F., Westerlund, S.A. and Kaib, M. (2007). Juvenile hormone III influences task-specific cuticular hydrocarbon profile changes in the ant *Myrmicaria eumenoides*. *J. Chem. Ecol.*, **33**, 167–181.

Lewis, L.A., Schneider, S.S. and Degrandi-Hoffman, G. (2002). Factors influencing the selection of recipients by workers performing vibration signals in colonies of the honeybee, *Apis mellifera*. *Anim. Behav.*, **63**, 361–367.

Lihoreau, M. and Rivault, C. (2008). Kin recognition via cuticular hydrocarbons shapes cockroach social life. *Behav. Ecol.*, **20**, 46–53.

Nelson, D.R., Tissot, M., Nelson, L.J., Fatland, C.L. and Gordon, D.M. (2001). Novel wax esters and hydrocarbons in the cuticular surface lipids of the red harvester ant, *Pogonomyrmex barbatus*. *Comp. Biochem. Physiol. B*, **128**, 575–595.

O'Donnell, S. (2001). Worker biting interactions and task performance in a swarm-founding eusocial wasp (*Polybia occidentalis*, Hymenoptera: Vespidae). *Behav. Ecol.*, **12**, 353–359.

O'Donnell, S. and Bulova, S.J. (2007). Worker connectivity: a review of the design of worker communication systems and their effects on task performance in insect societies. *Insect Soc.*, **54**, 203–210.

Ozaki, M., Wada-Katsumata, A., Fujikawa, K., Iwasaki, M., Yokohari, F., Satoji, Y., Nisimura, T., and Yamaoka, R. (2005). Ant Nestmate and Non–Nestmate Discrimination by a Chemosensory Sensillum. *Science*, **309**, 311–314.

Pratt, S. (2005). Quorum sensing by encounter rates in the ant *Temnothorax albipennis*. *Behav. Ecol.*, **10**, 488–496.

Rivault, C., Cloarec, A. and Streng, L. (1998). Cuticular extracts inducing aggregation in the German cockroach, *Blatella germanica* (L.). *J. Insect Physiol.*, **44**, 909–918.

Saïd, I., Costagliola, G., Leoncini, I. and Rivault, C. (2005). Cuticular hydrocarbon profiles and aggregation in four *Periplaneta* species (Insecta: Dictyoptera). *J. Insect Physiol.*, **51**, 995–1003.

Seeley, T.D. and Tovey, C.V. (1994). Why search time to find a food-storer bee accurately indicates the relative rates of nectar collecting and nectar processing in honey bee colonies. *Anim. Behav.*, **47**, 311–316.

Sevala, V., Bagnères, A.-G., Kuenzli, M., Blomquist, G.J. and Schal, C. (2000). Cuticular hydrocarbons of the dampwood termite, *Zootermopsis nevadensis*: Caste differences and role of lipophorin in transport of hydrocarbon and hydrocarbon metabolites. *J. Chem. Ecol.*, **26**, 765–789.

Smith, C. R., Anderson, R. T., Tillberg, C. V., Gadau, J. and Suarez, A. V. (2008). Caste Determination in a Polymorphic Social Insect: Nutritional, Social, and Genetic Factors. *Am. Nat.,* **172**, 497–507.

Wagner, D., Brown, M. J. F., Broun, P., Cuevas, W., Moses, L. E., Chao, D. L. and Gordon, D. M. (1998). Task-related differences in the cuticular hydrocarbon composition of harvester ants, *Pogonomyrmex barbatus. J. Chem. Ecol.*, **24**, 2021–2037.

Wagner, D., Tissot, M., Cuevas, W. and Gordon, D. M. (2000). Harvester ants utilize cuticular hydrocarbons in nestmate recognition. *J. Chem. Ecol.*, **26**, 2245–2257.

Wagner, D., Tissot, M. and Gordon, D. M. (2001). Task-related environment alters the cuticular hydrocarbon composition of harvester ants. *J. Chem. Ecol.*, **27**, 1805–1819.

Wertheim, B., van Baalen, E.-J. A., Dicke, M. and Vet, L. E. M. (2005). Pheromone-mediated aggregation in nonsocial arthropods: an evolutionary ecological perspective. *Annu. Rev. Entomol.*, **50**, 321–346.

13

Hydrocarbon profiles indicate fertility and dominance status in ant, bee, and wasp colonies

Jürgen Liebig

The social organization of insect colonies indicates the importance of information that is usually not needed in solitary insects. Information about the presence and fertility of a queen strongly affects worker behavior and colony organization. Reproductive competition in colonies requires the correct assessment of each others' rank. All of this information about fertility status and/or dominance status can be encoded in the cuticular hydrocarbon profile of members of ant, wasp, and bee colonies. Understanding variations in these hydrocarbon profiles, their composition, and relation to fertility is key to the further understanding of the major property of eusocial insects, reproductive division of labor.

Cuticular hydrocarbons are part of the lipid layer of the insect cuticle that protects from desiccation (Lockey, 1988) and are thus present in basically every social insect (see Chapter 6). Insects have the sensory apparatus to detect these profiles. So it is not surprising that they utilize variations in hydrocarbon profiles between individuals within and between species to detect various properties in other individuals, such as species identity, gender, colony membership (Howard and Blomquist, 1982, 2005; and various chapters in Part II of this book). In this chapter I will review the evidence indicating that hydrocarbon profiles are also used in colonies of ants, bees, and wasps for the regulation of reproduction. I will especially focus on patterns of variation in hydrocarbon profiles on the cuticle and the eggs in relation to fertility differences, which has not been done in such detail in previous reviews (Heinze, 2004; Monnin, 2006; Hefetz, 2007; Le Conte and Hefetz, 2008; Peeters and Liebig, 2009).

The importance of chemical communication in the regulation of reproduction in social insects

In a typical hymenopteran society, a reproductively specialized individual, the queen, lays the eggs, whereas the reproductively degenerated workers raise them to adulthood. The level of reproductive degeneration varies between groups and species. In very few species, workers are totally sterile. In the vast majority of hymenopteran species, however, workers have retained their ovaries and are thus still able to lay fertile eggs (Bourke, 1988; Choe, 1988). Due to the sex-determination system in Hymenopterans, they can lay unfertilized haploid eggs that develop into males. In a relatively small number of ants, workers also maintain the ability to mate and thus are capable also of producing diploid eggs which develop into workers or queens. In some bees and wasps, there is no clear dimorphism between the foundress of the colony and her helpers, i.e. all colony members could potentially become a principal egg-layer (Ross and Matthews, 1991). This also applies to a few ant species that secondarily lost the queen caste and now consist entirely of workers. In such species, workers that are mated and lay viable eggs are called gamergates (Peeters, 1993; Peeters and Ito, 2001). These gamergates are totipotent, except that they do not found colonies in the wild, although they may be able to do so (Liebig *et al.*, 1998).

Despite their reproductive potential, workers do not normally activate their ovaries in the presence of the established reproductives. Early observations already indicated the involvement of pheromones in the lack of worker reproduction (Fletcher and Ross, 1985; Hölldobler and Bartz, 1985). These pheromones most likely do not directly act on the reproductive physiology of workers, which would be a type of parental manipulation. They are more likely signals that indicate the presence of a healthy and fertile queen to the workforce (Seeley, 1985; Keller and Nonacs, 1993). Workers benefit in this situation from raising offspring rather than from their own reproduction. Nevertheless some workers may activate their ovaries, which is not in the interest of other colony members (Bourke and Franks, 1995). Kin selection theory predicts that either the principal reproductive or other workers should prevent their reproduction (Ratnieks, 1988; Ratnieks *et al.*, 2006). A necessary requirement for this interference is that incipient subordinate egg-layers or their eggs are identified.

Conflicts over reproduction can also arise when a reproductive position becomes vacant after the death of a principal reproductive. In this case, potential successors may fight with each other for successorship. They may form a hierarchy or they may already have a hierarchy in place, from which the highest ranking individual moves up to the reproductive position. In any case, the competitors and other workers need to identify and evaluate the status of each other. Non-reproductive workers that do not line up for successorship may play an important part in the final outcome, as they may attack inferior candidates. All the information needed for status assessment and individual recognition can also be encoded in the cuticular hydrocarbon profiles of the individuals involved.

Communication of reproductive status in social insects

When a queen dies or becomes senescent workers either benefit from starting to lay male-destined eggs, from fighting for the vacant reproductive position and potentially from raising a sexual brood. Accurate information about the health and reproductive status of the established reproductives is thus crucial for them.

Despite the importance of such pheromones for the understanding of insect sociality, very little is yet known about them (Le Conte and Hefetz, 2008). So far, in only one species (i.e. the honeybee), has it been shown that a pheromone emitted by a reproductive prevents worker reproduction. In the honeybee, *Apis mellifera*, it has been suggested that the queen mandibular pheromone (QMP) causes worker ovarian inhibition (Butler and Fairey, 1963).

Although this pheromone has been synthesized and was available for bioassays, various tests produced ambiguous and contradictory results (Slessor *et al.*, 1998). It was only four decades later that its function was finally proven (Hoover *et al.*, 2003). This indicates how difficult it is in general to prove such an inhibitory effect of these pheromones. The successful bioassay with this pheromone is also a special case. The QMP is originally used for mate attraction and is thus very specific to honeybees. It may apply to bumblebees (Röseler *et al.*, 1981) but certainly cannot have a general function in social insects.

The only other study that shows an inhibitory effect on nestmates has been performed in fire ants. However, the effect is restricted to alate queens, since workers are sterile. Poison gland extracts have been shown to prevent female alates from dealation, which would finally result in egg-laying (Vargo, 1997, 1998). However, another experiment suggests that other pheromones may be involved as well (Vargo and Hulsey, 2000). The inhibitory effect is most likely due to the piperidines present in the poison gland. These compounds are typical for fire ants, but do not generally occur in ants or other social insects. This means that this inhibitory effect is again restricted to this taxonomic group.

The question is whether there is a general mechanism that explains the chemical nature of the reproductive regulation in social insects. Cuticular hydrocarbons may be the basis of such a general mechanism. They fulfill all the requirements needed for such a system, as I will describe in the following paragraphs.

Hydrocarbon profiles and reproductive status

In many social Hymenoptera, reproductive individuals differ in their profile from individuals who either do not lay eggs or only produce trophic eggs. In addition, individuals that activate their ovaries often change their cuticular hydrocarbon profile even though they may not have the reproductive potential of the established egg-layer.

Ants

Ants are the best-studied group in this respect. So far, differences in cuticular hydrocarbon profiles relating to reproduction have been found in 21 ant species representing 16 genera. The most common differences are between established queens or gamergates and non-fertile workers providing the basis for the identification of a healthy and reproductive queen.

Species with workers having a limited reproductive potential

Camponotus floridanus and *Linepithema humile*, both species with very distinct queen–worker dimorphism, demonstrate how strong differences in the cuticular hydrocarbon profile between reproductive queens and workers can be (de Biseau *et al.*, 2004; Endler *et al.*, 2004, 2006). In *C. floridanus*, on average 50% of the relative amount of the cuticular hydrocarbons of highly fertile queens was not present in the profile of workers (Figure 13.1). Founding queens slowly develop the profile typical of highly fertile queens. When they lay about one egg per day during the colony founding stage their

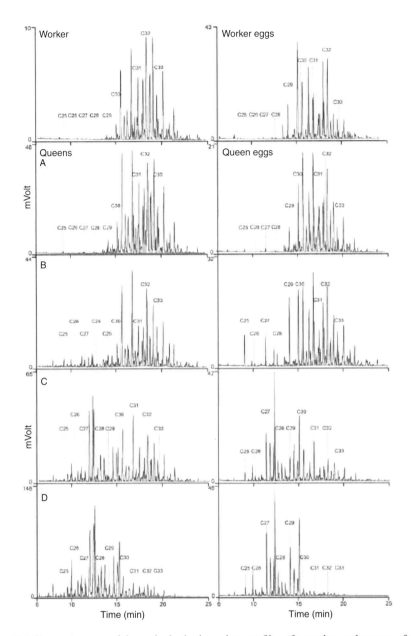

Figure 13.1 Chromatograms of the cuticular hydrocarbon profiles of a worker and queens of the ant species *Camponotus floridanus*, together with the respective hydrocarbon profiles of the surfaces of their eggs. Queens originated from colonies of four different size classes: (A) Founding stage with fewer than 10 workers, (B) 50 to 80 workers, (C) 200 to 300 workers, (D) more than 1000 workers. Only compounds with peaks between 6 and 25 min are shown, which represent the hydrocarbon profile. Minor, not reproducible peaks between 0 and 6 min are not shown. For better comparison, the elution times for n-alkanes with chain length from 25 to 33 are indicated. Increasing queen fertility along with increasing colony size is accompanied by the expression of short-chain hydrocarbons in both the cuticular and egg profiles. The low-fertility workers show profiles similar to those of founding queens (Endler *et al.*, 2006).

hydrocarbon profile is very similar to that of workers. Their profile showed only 0.6% of the compounds typical of the highly fertile queens. When they gradually increased their egg-laying to an average daily rate of 28 eggs in colonies with more than 1000 workers, they finally developed the cuticular profile typical of highly fertile queens (Endler *et al.*, 2006).

Even though the increase in reproductive-specific compounds strongly correlates with egg-laying rate in queens, workers do not show this pattern. The cuticular profiles of egg-laying and non-fertile workers in this species do not differ, indicating that ovarian activity with very low egg-laying rates are not associated with changes in their profile (Endler *et al.*, 2007). This is also supported by the profiles of foundresses that were very similar to those of workers. The lack of changes in the hydrocarbon profile in weakly fertile individuals, however, seems to be specific to this species with strong queen–worker dimorphism. It is not found in species that have workers with higher reproductive potential (see below).

In *L. humile* old reproductive queens also show profiles that differ by almost 50% in the relative abundance of queen-specific compounds from those of workers, which are permanently sterile in this species (de Biseau *et al.*, 2004). In addition, hydrocarbon profiles of virgin non-egg-laying queens are similar to those of workers. However, initiation of egg-laying makes their profiles more like those of mated reproductive queens.

In one case, strong differences in the cuticular profile lead to a visible change. Long-term reproductive mated queens of *Ectatomma tuberculatum* have a matte appearance while virgin weakly fertile queens show a shiny cuticle (Hora *et al.*, 2008). Almost 50% of the profile of the established queens consists of a single alkane (*n*-heptacosane). In virgin queens with weakly active ovaries, this compound represents 13%, but drops to 4% in virgin queens without active ovaries, again showing a strong effect of fertility on the hydrocarbon profile.

Other species like *Cataglyphis iberica* show less pronounced differences between queens and workers (Dahbi and Lenoir, 1998). Here, the concentrations of four compounds (*n*-heptacosane, *n*-octacosane, *n*-nonacosane, *x,y*-dimethylnonacosane) are significantly higher in queens than in workers while a mixture of two compounds (11- and 13-methylheptacosane) is lower in queens. However, they represent less than a third of the total amount of queen hydrocarbons. Even smaller differences in cuticular profiles of queens and workers can still be tracked by using multivariate statistical methods as has been done in two *Leptothorax* ants (Tentschert *et al.*, 2002). These methods were also useful when small differences between queens of different fertility were identified in *Formica fusca* (Hannonen *et al.*, 2002). This species is polygynous and differences among queens may allow workers to preferentially feed the more fertile one. In fact the more fertile queens were preferred in attendance over less fertile ones.

Not only queens express these profiles. Intermediate forms between queen and worker castes, so-called intermorphs, specialize in trophic egg-laying in *Crematogaster smithii* (Heinze *et al.*, 1999). Interestingly, they produce a queen-like profile, while workers produce a profile that differs from those of reproductive queens and intermorphs (Oettler *et al.*, 2008).

Species with queens and workers having a strong reproductive potential

When the reproductive potential of workers is relatively high, they may also produce a queen-like profile when fully fertile. This occurs in some species where workers are confined to male production only. Workers of *Myrmecia gulosa* regularly produce trophic eggs. When they shift to the production of viable eggs, their cuticular profile also shifts towards that of the fertile queen (Dietemann *et al.*, 2003, 2005). Although the profiles of reproductive queens consist of several hydrocarbons that are absent in trophic egg-layers, fertile workers may eventually produce those in similar quantities. Some of the workers exhibiting the intermediate profiles are then the target of physical worker policing (Dietemann *et al.*, 2005).

A similar pattern is present in the ant *Aphaenogaster cockerelli* where workers also lay trophic eggs. Queen profiles differ qualitatively from those of workers that do not lay fertile eggs (Smith *et al.*, 2008, 2009). However, when these workers shift to viable egg production in queenless worker groups, they start producing the hydrocarbons that are typical of reproductive queens (Figure 13.2). Interestingly, such workers have been attacked after reintroduction into their colony (Hölldobler and Carlin, 1989; Smith *et al.*, 2009) which is compatible with the idea that changes in their hydrocarbon profile triggered these attacks. When workers remain isolated for a longer time in queenless worker groups some of them produce the same hydrocarbons at similar levels as reproductive queens (Smith *et al.*, 2008, 2009).

Reproductive potential of workers is generally high in ponerine ant species (Peeters, 1993, 1997). Thus it is not surprising that workers can produce a queen-like cuticular hydrocarbon profile when they are reproductively active for a sufficiently long time. Although workers of *Pachycondyla inversa* do not mate and thus are confined to male production, they compete in a reproductive hierarchy for male production in queenless colonies (Heinze *et al.*, 2002). Here, workers show a relationship between the change in hydrocarbon profile and change in fertility. Non-reproductive workers differ from the queen in their cuticular hydrocarbon profile (D'Ettorre *et al.*, 2004a). Once the workers are reproductive, they switch to the queen-like profile.

The ant *Harpegnathos saltator* differs from the previous species in having gamergates besides queens (Peeters *et al.*, 2000). These gamergates succeed queens when they die (Peeters and Hölldobler, 1995; Liebig, 1998). Gamergates establish themselves during periods of intense fights. Clear changes in the cuticular profile develop in putative gamergates within two weeks in their race for queen succession in groups of workers (Figure 13.3) (Liebig *et al.*, 2000). However, such workers are recognized and attacked when returned to their colony, indicating that the change in their reproductive status is perceived (Liebig *et al.*, 1999). Workers that succeed in becoming established gamergates develop a distinct profile that is queen-like but that does not fully match it (Liebig *et al.*, 2000).

Aggression is not always involved in reproductive regulation. In some species, like *Gnamptogenys striatula*, reproductive status is determined by mating. Nevertheless,

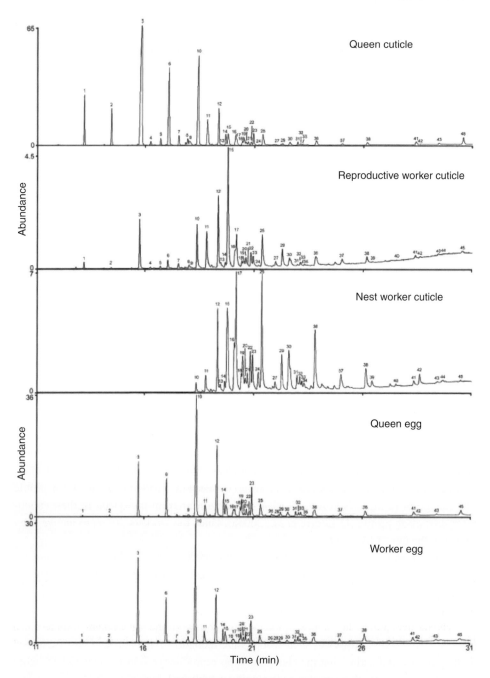

Figure 13.2 Chromatograms of the cuticular hydrocarbon profiles of a reproductive queen, and a reproductive and a non-reproductive worker of the ant species *Aphaenogaster cockerelli*, together with the respective hydrocarbon profiles of the surfaces of their eggs. Major differences relating to reproductive status are found in compounds 1 through 9. The major compounds of reproductives are n-alkanes (1, 2, 3, and 6). In contrast to *Camponotus floridanus* in Figure 13.1, workers express compounds similar to those of the queen. In addition, worker-laid eggs are similar to eggs laid by highly fertile queens that express the compounds typical of reproductives (Smith *et al.*, 2008).

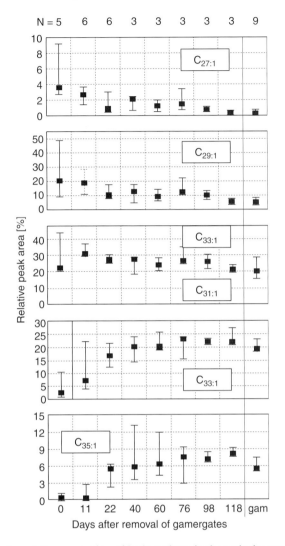

Figure 13.3 Shift in the relative proportion of hydrocarbons in the cuticular profiles of establishing gamergates in the ponerine ant *Harpegnathos saltator*. When workers were isolated from their previous reproductives, some of them developed the profile typical for established gamergates and reproductive queens. In contrast to *Camponotus floridanus* and *Aphaenogaster cockerelli* in Figures 13.1 and 13.2, the fertility-related change in the profile is reflected by a shift to longer-chain hydrocarbons. This example shows the relative proportions of short-chain alkenes, that decrease with increasing fertility, while long-chain alkenes increase in their relative proportion (Liebig *et al.*, 2000).

gamergates still produce hydrocarbon profiles that differ from non-reproductive workers, indicating the presence of reproductives in the colony (Lommelen *et al.*, 2006, 2008).

Pure gamergate species

In ten ponerine genera queens have been permanently lost (Peeters, 1997; Peeters and Ito, 2001). One or several gamergates reproduce instead. One of these species is *Dinoponera quadriceps* where a social hierarchy exists with only the top-ranking individual reproducing (Monnin and Peeters, 1998). Other high-ranking workers are prevented from replacing the top-ranking individual by subordinates biting and holding them (Monnin and Peeters, 1999). This indicates that a mechanism for their recognition should exist. In fact, the relative concentrations of several compounds differ between the top-ranking individuals and subordinates (Monnin *et al.*, 1998; Peeters *et al.*, 1999). The relative concentration of one alkene (9-hentriacontene) positively correlates with the rank of the individual, potentially enabling workers to discriminate between top- and other high-ranking individuals.

A close relative of *Dinoponera*, *Streblognathus peetersi*, shows a similar pattern. The species is permanently queenless and only workers reproduce. The top-ranking individuals show a distinct cuticular hydrocarbon profile (Cuvillier-Hot *et al.*, 2004a, b). In the absence of the highest ranking individual another individual will rise in the hierarchy and replace her. The accompanying changes can be detected by nestmate workers as early as 48 hours after the original reproductive was removed (Cuvillier-Hot *et al.*, 2005). Changes in the cuticular hydrocarbon profiles are associated with several physiological changes. Similar to another queenless ponerine ant, *Diacamma* sp. (Sommer *et al.*, 1993), the JH titer of workers is negatively correlated with social status in *Streblognathus* (Brent *et al.*, 2006). Artifically increasing the effect of JH by applying a JH-analog to top-ranking workers led to the loss of their status despite displaying increased aggression (Cuvillier-Hot *et al.*, 2004b). This loss of status was accompanied by a decrease in vitellogenin titers in the hemolymph (Cuvillier-Hot *et al.*, 2004a). Top-ranking workers have the highest titers of vitellogenin, which becomes a storage protein in the eggs. The drop in vitellogenin titers correlated with a change of their cuticular hydrocarbons towards the profile of low-ranking workers. The change in the profile may have informed nestmate workers about the status change, which finally let the low-ranking workers attack and immobilize the treated top-ranking workers (Cuvillier-Hot *et al.*, 2004a).

When gamergates establish themselves as new reproductives in *Diacamma ceylonense*, they are initially very aggressive (Cuvillier-Hot *et al.*, 2002). However, along with changes in their cuticular hydrocarbon profiles, aggression ceases. This suggests that the distinct profile gamergates exhibit is sufficient to reduce intracolonial conflict, since it allows other nestmates to recognize their reproductive status (Cuvillier-Hot *et al.*, 2001, 2002). Confirmation of the gamergates' status by physical aggression towards nestmate workers is no longer necessary. In fact, callows easily differentiate between nascent gamergates and well established gamergates (Baratte *et al.*, 2006). In another *Diacamma* species, workers do not activate their ovaries when they have direct contact with a gamergate in the colony (Tsuji *et al.*, 1999). Moreover, workers that develop ovaries in the presence of the gamergate are policed in this species (Kikuta and Tsuji, 1999; Kawabata and Tsuji, 2005).

Wasps

Stenogastrine and paper wasps are similar to ponerines with gamergates in the way that all females are capable of becoming the major reproductive in the colony by producing male and female eggs. In contrast to the findings of most studies in ponerine ants, differences in the cuticular hydrocarbon profiles in relation to reproductive status are relatively small in *Polistes dominulus* and four stenogastrine wasps (Bonavita-Cougourdan *et al.*, 1991; Sledge *et al.*, 2001, 2004; Turillazzi *et al.*, 2004). It has also been suggested that the profiles represent social rather then reproductive status in *Polistes* (Dapporto *et al.*, 2004, 2007a). *Polistes metricus* showed similarly weak differences between foundresses and her workers (Layton *et al.*, 1994).

In the monogynous species *Vespa crabro*, differences in the cuticular hydrocarbon profiles between the queen and her workers are more pronounced (Butts *et al.*, 1995). Another study by Butts *et al.* suggests stronger differences in hydrocarbon profiles as well in *Dolichovespula maculata*, *Vespula squamosa* and *Vespula maculifrons* (Butts *et al.*, 1991). Unfortunately, it is not completely clear whether the queens are reproductive or virgin in the latter study although a comparison of the terminology used suggests that reproductive queens had been used.

Especially in small wasp societies, like the *Polistes* paper wasps, living on a comb allows to directly assess the fertility of the foundress. The long-lasting presence of empty cells in the middle of the nest is an indicator of low foundress fertility, since these areas should be quickly filled with eggs. Workers responded to experimental brood removal by ovarian activation and egg-laying in *Polistes dominulus* (Liebig *et al.*, 2005). This may be the reason why differences in hydrocarbon profiles do not correlate with fertility in *Polistes gallicus* (Dapporto *et al.*, 2007b).

Bees

Differences between queens and workers are also present in meliponine bees. Physogastric queens differ from virgin queens and workers in *Melipona bicolor* (Abdalla *et al.*, 2003). Strong differences between virgin queens and workers are present in *Melipona scutellaris*, but unfortunately no mated physogastric queen was included in the analysis (Kerr *et al.*, 2004).

Hydrocarbon profiles and egg discrimination

Reproductive conflicts are sometimes expressed through egg destruction. Either dominant individuals destroy eggs from subordinates or workers mutually eat each others' eggs in the presence of a reproductive–behavior that is known in many ant species, in the honeybee and in several wasp species (Ratnieks *et al.*, 2006). Successful selective egg destruction requires ways of discriminating between eggs of different origin. In ants egg-discrimination is often associated with differences in the hydrocarbon profiles of eggs originating from different sources.

Ants

Camponotus floridanus

Workers of the ant *Camponotus floridanus* differentiate between eggs that are laid by highly fertile queens and other workers (Endler *et al.*, 2004, 2006). When eggs were added to freshly orphaned worker groups that originated from colonies with a highly fertile queen, a median of 52% of the eggs laid by sister workers and a median of 100% eggs laid by their mother were accepted (Endler *et al.*, 2004). This strongly indicates a queen-specific trait of queen-laid eggs that is based on differences in fertility. To exclude that this differentiation is due to physical differences between worker- and queen-laid eggs, workers originating from nests with a highly fertile queen were given the choice between foreign eggs originating either from highly fertile queens, weakly fertile queens, or workers (Endler *et al.*, 2006). In this case, eggs laid by highly fertile queens were still highly preferred with an average acceptance rate of more than 90%. On the other hand, worker-laid eggs were again rejected on average by 100% (median). This was the same average rejection of eggs laid by weakly fertile queens. Since all eggs originated from a foreign source, the only two parameters that varied were the type of egg-layer and the fertility of the egg-layers. Only differences in fertility explain this result, since rejection rates of eggs laid by weakly fertile individuals – represented by queens and workers – were the same, but were much higher than that for eggs of highly fertile queens.

The pattern of egg rejection very closely matched the variations in the pattern of the hydrocarbon profiles present on the egg surfaces. Eggs of highly fertile queens showed profiles that were qualitatively different to those of eggs laid by workers and weakly fertile queens (Endler *et al.*, 2004, 2006; Figure 13.1). On the other hand, profiles of eggs laid by weakly fertile queens and workers did not show qualitative differences. Nevertheless, weak quantitative differences exist between the profiles of eggs laid by weakly fertile queens and workers that, however, do not relate to type of egg-layer or fertility according to the classification pattern of a discriminant analysis (Endler *et al.*, 2006).

Dinoponera quadriceps

Gamergates or a virgin top-ranking reproductive individual (alpha individual) of this species eat the eggs of any other egg-layer in the colony (Monnin and Peeters, 1997). When the beta individual, who follows next in the linear hierarchy, laid eggs, 86% of these eggs were eaten by the alpha. In contrast, alpha did not eat her own eggs when she encountered them shortly after egg-laying, indicating that the fresh status of the egg is not the trigger for egg eating but differences in the eggs relating to rank. In fact, the cuticular hydrocarbon profiles of eggs correlate with ovarian activity, providing the alpha with information needed to identify the eggs of subordinates.

Pachycondyla inversa

When workers in queenright colonies received eggs laid by foreign queens or foreign workers they ate on average 46% of the worker-laid eggs but only 15% of the queen-laid eggs (D'Ettorre *et al.*, 2004b). Differences in egg-eating corresponded to differences in the hydrocarbon profiles of the egg surfaces. The chemicals that mark egg origin are not transferable in this species. Even when worker-laid eggs remained mixed with queen-laid eggs for 45 min, they were readily discriminated afterwards (D'Ettorre *et al.*, 2006).

Myrmecia gulosa and Aphaenogaster cockerelli

In these two species, differential eating of viable eggs was not observed (Dietemann *et al.*, 2005; Smith *et al.*, 2008). This is especially interesting, since workers of both species produce trophic eggs that are non-viable and are only used as an extension of nutrient flow through the colony. However, this result also supports the hypothesis that egg discrimination is mediated by hydrocarbon profiles of the egg surfaces. The lack of egg discrimination suggests that egg profiles are not distinct, preventing accurate egg discrimination between queen- and worker-laid groups. In fact, in both species the hydrocarbon profiles of worker-laid and queen-laid viable eggs were very similar, preventing an accurate classification of these eggs in both cases.

Other ant species

In addition, egg discrimination is known from many other ant species where chemical differences between eggs are unknown. In *Formica fusca*, for example, workers destroy worker-laid eggs (Helanterä and Sundström, 2005, 2007). In addition, workers also discriminate between eggs from nestmates and foreign eggs (Helanterä *et al.*, 2007; Helanterä and Sundström, 2007), behavior which is affected by previous exposure to foreign eggs (Helanterä *et al.*, 2007). Discrimination of viable eggs in some species has not yet been demonstrated, but is likely to be the case. In *Gnamptogenus striatula*, gamergates differ in their profile from other workers (Lommelen *et al.*, 2006). Since egg profiles correlate with cuticular profiles in this species too (Lommelen *et al.*, 2008), differences found among adults may also be present in eggs, which would allow egg discrimination. This is further supported by an abdominal stroking behavior towards laid eggs, that is supposed to be a behavior for marking eggs with cuticle-derived hydrocarbons (Lommelen *et al.*, 2008).

Wasps

Polistes dominulus

Small paper wasp colonies, as in *Polistes dominulus*, have the advantage that egg replacement can be easily tracked due to their location in a cell of the comb (Liebig *et al.*, 2005). Replacement of eggs can be identified when an individual lays an egg in a cell in which another individual has already laid an egg. When these cells were controlled later, usually only a single egg was found.

Conflict over egg replacement occurs between foundresses and between foundresses and workers resulting in the destruction of many eggs (Tsuchida *et al.*, 2003; Liebig *et al.*,

2005; Tsuchida and Suzuki, 2006; Dapporto *et al.*, 2007a). Identification of eggs may be enabled by differences in the hydrocarbon profiles of the egg surface. These are correlated with the rank of the foundress (Dapporto *et al.*, 2007a), which suggests that similar differences exist between eggs laid by foundresses and those laid by workers.

Bees:other potential egg-discrimination mechanisms

Egg policing occurs in a variety of other social Hymenoptera (Ratnieks *et al.*, 2006). The most intensely investigated other Hymenoptera is the honeybee, *Apis mellifera*. It is well documented that worker-laid eggs are regularly policed in queenright colonies in this species (Ratnieks and Visscher, 1989) as well as in other honeybees (Halling *et al.*, 2001; Wattanachaiyingcharoen *et al.*, 2002; Nanork *et al.*, 2007). Despite an extensive study of egg-policing and egg surface chemistry, it is still unclear what allows discrimination of worker- and queen-laid eggs in the honeybee. Hydrocarbons do not seem to play a role in this system (Martin *et al.*, 2004b) although hydrocarbon compounds occur in larger numbers on queen-laid eggs (Katzav-Gozansky *et al.*, 2003). In addition, ultrastructural differences between the two types of egg have not been found (Katzav-Gozansky *et al.*, 2003). Differential viability of queen- and worker-laid eggs has been suggested for *Apis mellifera* (Pirk *et al.*, 2004), but another study could not confirm this result (Beekman and Oldroyd, 2005). Differential viability has also been excluded in the wasp *Vespula vulgaris* as a potential cue for workers to discriminate against worker-laid eggs (Helanterä *et al.*, 2006). Nevertheless, differential viability is a factor that needs to be controlled for in egg discrimination studies (Khila and Abouheif, 2008).

Several observations suggest that esters allow egg discrimination in the honeybee. Mated queens produce much higher levels of esters in their Dufour gland than normal workers (Katzav-Gozansky *et al.*, 1997). Treatment of worker-laid eggs with Dufour gland extracts reduced policing rates, suggesting a role for compounds originating from this gland as a marking pheromone (Ratnieks, 1995). Further indirect support for this idea comes from several studies. Virgin mated queens do not differ in their ester levels from mated queens and workers do not discriminate between their eggs (Beekman *et al.*, 2004). When workers lay eggs for a longer time, these esters become more abundant and egg discrimination decreases with an increased amount of queen-like esters on worker-laid eggs (Martin *et al.*, 2004a, 2005). Despite these differences, all experiments to directly show the function of the esters as egg-discrimination pheromone have failed. Artificial applications of queen-like esters either did not protect the worker-laid eggs at all (Katzav-Gozansky *et al.*, 2001) or delayed egg-eating (Martin *et al.*, 2002).

Hydrocarbon profiles and social hierarchy

Linear social hierarchies that are correlated with reproduction are present in many ants, wasps and bees. In such hierarchies, higher-ranking individuals attack subordinates at varying intensity and frequency while lower-ranking individuals rarely if at all attack higher-ranking individuals. Such types of hierarchy require individual recognition and memory of their own relative rank. Recently, individual recognition of hierarchy members has been demonstrated experimentally in a ponerine ant (D'Ettorre and Heinze, 2005; Dreier *et al.*, 2007). Individual recognition is most likely mediated by cuticular hydrocarbons (D'Ettorre and Heinze, 2005). This may explain rank recognition in other ant species where linear hierarchies are established as well. In paper wasps like *Polistes dominulus*, foundresses form linear hierarchies in foundress associations. Differences in their cuticular hydrocarbon profiles are associated with rank and could provide rank information in addition to information about fertility (Sledge *et al.*, 2001, 2004; Dapporto *et al.*, 2004, 2007a).

Besides individual profiles, gradual differences in certain compounds that correlate with fertility could also provide the necessary information about rank. In *Dinoponera* the second in the hierarchy expresses hydrocarbons typical of reproductive individuals at levels that are between those of top-ranking individuals and workers outside of the hierarchy (Peeters *et al.*, 1999). Similar graded information encoded in the cuticular profile is also present in queens of the ponerine ant *Pachycondyla inversa* (Tentschert *et al.*, 2001). Such a system has however limitations compared to individual recognition as it is hard to imagine how it would work in longer hierarchies. In fact, the fertility profile of *Pachycondyla inversa* workers does not match their specific rank in the linear dominance hierarchy (Heinze *et al.*, 2002).

Other forms of status recognition, like visual cues, are possible (Tibbetts, 2002; Tibbetts and Dale, 2004). Here too, the memory of each others' hierarchy rank is important (Sheehan and Tibbetts, 2008). Nevertheless, this mechanism seems to be unique to paper wasps and cannot be used in the dark of an ant or bee nest. Chemical cues or signals and tactile communication are important in the latter situation.

Profile discrimination and worker reproduction

Recognition of hydrocarbon profiles relating to fertility

Despite the many studies that demonstrate a correlation between cuticular hydrocarbon profiles, egg hydrocarbon profiles, and fertility, only a few studies provide more robust evidence for their importance in the regulation of reproduction in social insects. A study in the ant *Myrmecia gulosa* took advantage of the generally observed high attractiveness of fertile queens to workers (Dietemann *et al.*, 2003). If attractiveness is induced by the specific cuticular profile of these queens, we would expect a higher attractiveness of hydrocarbon extracts from the queen cuticle. In fact, fractionated hydrocarbon extracts of queens were more attractive to workers than fractionated worker extracts. Attractiveness was measured by higher antennation frequency towards queen extracts that were applied on glass slides (Dietemann *et al.*, 2003). This result indicates that workers can discriminate between the different cuticular hydrocarbon profiles of queens and workers, and the higher attractiveness of the queen profile indicates that they may indeed recognize it as queen-like.

Nevertheless, direct proof of a function in the regulation of reproduction would be a bio-assay that indicates either a behavioral response of workers to the queen profile that is seen in a natural context or a physiological response leading to the inhibition of ovarian activity in workers.

Another study in the ant *Pachycondyla inversa* used a more indirect approach (D'Ettorre *et al.*, 2004a). Here, electro-antennograms coupled with gas-chromatography were used to demonstrate that worker antennae detected the main compound (3,11-dimethylhepta-cosane) of the profile of fertile queens. However, it remains unclear why in the same setting other cuticular hydrocarbons did not trigger an antennae response despite their higher con-centrations. This may indicate a specialization of the antennae to perceive this compound. Further work is necessary to understand antennal sensitivity to different compounds of the cuticle. Furthermore, it is also necessary to investigate the function of the queen-specific compound in a bioassay, where again the natural response of workers to this compound can be assessed.

Egg policing based on differences in their surface hydrocarbon profiles was shown in the ant *Camponotus floridanus* (Endler *et al.*, 2004). Workers normally destroy worker-laid eggs in the presence of a fertile queen (Endler *et al.*, 2004, 2006), but the transfer of fractionated queen hydrocarbons onto worker-laid eggs significantly increased the survival of these eggs. A similar test in honeybees suggested that the origin of worker-laid eggs had been covered by the application of queen compounds, since the protective effect of the queen compounds disappeared after 24 h (Martin *et al.*, 2002). Although the protective effect of the queen compounds in *C. floridanus* was still retained after 24 h, it cannot be ruled out that the effect would have been lost later. However, the compounds were applied artificially, so maybe such effects should be expected due to the loss of the compounds by grooming in the worker groups.

The most direct proof of the involvement of cuticular hydrocarbon profiles, so far, has been provided in a worker policing experiment in the ant *Aphaenogaster cockerelli* (Smith *et al.*, 2009). Workers that activate their ovaries get attacked by their nestmate workers (Hölldobler and Carlin, 1989). Since workers that activate their ovaries develop queen-like alkanes, it was suspected that these alkanes may induce worker policing. *n*-Pentacosane was the compound that was most elevated in queens and reproductive workers (Figure 13.2). A transfer onto the cuticle of non-reproductive workers in naturally present amounts induced biting by workers in queenright colonies (Smith *et al.*, 2009). However, a transfer of *n*-nonacosane, which is not associated with fertility, did not elicit aggression, indicating that it is the specific alkane and not the manipulation that induces aggression. Furthermore, pentacosane-treated workers were not attacked in reproductive worker groups. In this case aggression was not expected, since reproductive workers with similar profiles are present in such groups. Workers in these groups are still aggressive, since they attacked foreign workers in a similar fashion to workers from queenright colonies. The experiment directly showed the induction of aggression following pentacosane treatment. Pentacosane is an indicator of fertility and is most likely used by workers to identify and police nestmates that activate their ovaries in the presence of the queen.

Effects on worker reproduction

So far, no study has been able to show that cuticular hydrocarbons are causing the lack of egg-laying by workers in the presence of a reproductive in insect colonies without the involvement of aggression, as occurs in *A. cockerelli*. There is only indirect evidence that hydrocarbon profiles are causally involved. In the ant *Camponotus floridanus*, workers do not start to lay eggs when they are exposed to queen-laid eggs (Endler *et al.*, 2004). However, many workers start laying eggs in the absence of queen-laid eggs, which suggests that workers perceive the presence of queens through the presence of queen-laid eggs. Since workers discriminate queen-laid eggs from worker-laid eggs on the basis of their hydrocarbon profiles, it is most likely that the presence of queen-specific hydrocarbon profiles on queen-laid eggs causes the lack of egg-laying in workers.

Patterns of profile variation

We see differences in the cuticular hydrocarbon profiles of reproductive and non-reproductive individuals across species and insect families. The questions are how the patterns differ and whether there are any general patterns at all. This section demonstrates the huge variety in the kind of differences we see. No compound class exists that could generally reflect differences in fertility, nor does a specific pattern in the composition of a profile separate reproductive from non-reproductive individuals. The only pattern that may emerge relates to the reproductive potential of individuals, colony size and queen–worker dimorphism.

Quantitative versus qualitative differences

Differences between reproductive and non-reproductive individuals can be qualitative, quantitative or both. Two types of quantitative differences exist. Proportional differences refer to compounds that differ in their concentration relative to another compound in the profile. We can also have quantitative differences that relate to the total composition of the profile. Profiles can potentially differ in concentration even though the proportional relations of the individual compounds within the profile are the same. In the case of cuticular hydrocarbons we run into a problem with the latter comparison. Since long-chain hydrocarbons form a solid layer on the cuticle, concentration differences largely indicate a difference in the thickness of the layer. It is not clear whether insects are capable of differentiating between layers of different thickness and how they would perceive this, since the exposure and thus perception of the cuticular hydrocarbons are independent of layer thickness. Therefore, I refer only to proportional differences in this review.

Ants

We often see that reproductives display additional compounds not present in non-reproductives in addition to differences in the relative proportions of compounds both groups share.

This however seems to be restricted to species with relatively large colony size and/or large queen–worker dimorphism.

In *Camponotus floridanus* for example, a species with colony sizes up to 10 000 individuals or more, strong qualitative differences in the profile between queens and workers occur. The strongest differences between queens and workers are expressed in the short-chain part of the spectrum (Endler *et al.*, 2004, 2006; Figure 13.1). Eighteen compounds including *n*-alkanes and mono- and dimethyl-branched alkanes are specific to highly fertile queens. Nevertheless, queens produce a larger proportion of compounds that both parties share leading to quantitative differences (Endler *et al.*, 2006). The long-chain part however is largely invariant. This pattern seems to be due to an up-regulated synthesis of short-chain compounds that does not affect the production of long-chain hydrocarbons.

Similarly to *C. floridanus*, additional short-chain compounds are produced in reproductive queens and workers of *Myrmecia gulosa* (Dietemann *et al.*, 2003) and *Aphaenogaster cockerelli* (Smith *et al.*, 2008). While in *A. cockerelli* nine compounds are produced in reproductives in addition to the normal worker profile, only two compounds (9-pentacosene and 3-methylpentacosane) occur additionally in the short-chain part of the cuticular hydrocarbon profile in *M. gulosa*. In total the differences are less pronounced in *M. gulosa*.

Linepithema humile reveals an additional pattern of qualitative differences. Reproductive queens up-regulate the synthesis of hydrocarbons of intermediate chain length (de Biseau *et al.*, 2004). In workers, additional components in the short- and long-chain areas (13-methylpentacosane and di- and trimethyl-branched alkanes with chain lengths of 33 to 37) are present as well. *Linepithema* also produces large colonies, and workers are sterile.

Yet another pattern leading to qualitative differences is present in the ponerine ant *Harpegnathos saltator*. Reproductive queens and gamergates exhibit a profile that is shifted to longer-chain hydrocarbons (Figure 13.3) (Liebig *et al.*, 2000). When workers become reproductively active, it seems that during biosynthesis compounds will receive one more acetate unit (see *biosynthesis,* Chapter 3, this book), leading to a shift to longer-chain hydrocarbons as was demonstrated for unsaturated hydrocarbons. In the end, this shift leads to weak qualitative differences, since gamergates and queens produce a long-chain compound (13, 23-dimethylheptatriacontane) that is not present in the profile of non-reproductive workers. In contrast to the previous four examples, the basis of the differences is not the production of additional compounds but the elongation of hydrocarbons during biosynthesis. This also results in less pronounced differences, which fit their smaller colony size. *H. saltator* produces colonies of up to 500 individuals when multiple gamergates produce, or up to 250 individuals when a single queen lays eggs (Peeters and Hölldobler, 1995; Peeters *et al.*, 2000).

H. saltator produces less pronounced qualitative differences. Other ponerine ants that are generally characterized by smaller colony sizes and limited queen–worker dimorphism (Peeters, 1993) only show quantitative differences in their cuticular hydrocarbon profiles between reproductives and non-reproductives. Gamergates in *Gnamptogenys striatula* differ from non-reproductive workers by having increased relative amounts of ten cuticular hydrocarbons and by a decrease in six compounds (Lommelen *et al.*, 2006). In *Diacamma ceylonense* four compounds are elevated in reproductive workers and one is strongly increased in foragers (Cuvillier-Hot *et al.*, 2001).

Such a relative increase of compounds in reproductives, together with a relative decrease in non-reproductive workers, is also found in other ant subfamilies. In the myrmecine *Aphaenogaster senilis* five compounds are elevated in reproductive queens while another five are increased in workers (Boulay *et al.*, 2007). In another myrmecine, *Crematogaster smithii*, a mixed pattern of increased and decreased compounds separates workers and virgin queens from fertile intermorphs and fertile queens (Oettler *et al.*, 2008). Here, several compounds are increased in reproductive queens compared to workers but only one separates them from virgin queens, suggesting the presence of some caste-specific compounds that are not related to fertility. However, differences between reproductive and non-reproductive queens can be strong as is evident in the ponerine ant *Ectatomma tuberculatum* (Hora *et al.*, 2008). Here old mated queens differ from virgin non-reproductive queens by the increased as well as decreased proportions of several different compounds. Exclusively related to fertility are the increases of some compounds in *Dinoponera quadriceps*, *Cataglyphis iberica* and *Pachycondyla inversa* (Monnin and Peeters, 1997; Dahbi and Lenoir, 1998; Heinze *et al.*, 2002).

Leptothorax gredleri, *Leptothorax acervorum*, and *Platythyrea punctata* typically produce colonies of fewer than 100 workers (Schilder *et al.*, 1999). All of them show only little differences between reproductive queens or reproductive workers, respectively, and non-reproductive workers. Either only one hydrocarbon plus a polar compound separate the two groups, or the total variation in the profile leads to a less than perfect separation in a discriminant analysis (Tentschert *et al.*, 2002; Hartmann *et al.*, 2005).

Wasps

In wasps, the strongest differences between reproductives and non-reproductives are found in the temperate paper wasps with larger colonies. *Vespa crabro*, *Dolichovespula maculata*, *Vespula squamosa*, and *Vespula maculifrons*, all show either strong quantitative differences in at least one compound or additional qualitative differences (Butts *et al.*, 1991, 1995). On the other hand, *Polistes dominulus*, *Polistes metricus* and several stenogastrine wasps, with smaller colony sizes and either no or low reproductive specialization, show smaller differences (Bonavita-Cougourdan *et al.*, 1991; Sledge *et al.*, 2001, 2004; Dapporto *et al.*, 2004; Turillazzi *et al.*, 2004). If these differences are all related to fertility, then this may indicate that a trend similar to that in ants is present, where differences in hydrocarbon profiles increase with increased colony size and queen–worker dimorphism.

Single versus multiple compounds and compound classes

Qualitative differences are normally associated with multiple compounds involved. The exception here is the ant *Harpegnathos saltator*, since the difference is more a shift in chain length rather than the synthesis of new compounds (Liebig *et al.*, 2000).

Quantitative differences are also usually represented by several compounds. Exceptions are the ants *Pachycondyla inversa*, and *Dinoponera quadriceps*, where one compound largely determines the clear differences between reproductive and non-reproductive individuals (Monnin and Peeters, 1997; Monnin *et al.*, 1998; Peeters *et al.*, 1999; Heinze *et al.*, 2002; D'Ettorre *et al.*, 2004a). Small differences between reproductive queens and workers are largely determined by a single hydrocarbon and one polar compound in *Leptothorax acervorum* and *Leptothorax gredleri* (Tentschert *et al.*, 2002). In the ponerine *Ectatomma tuberculatum*, a single alkane (*n*-heptacosane) largely determines the differences between mated reproductive queens and young virgin non-reproductive queens (Hora *et al.*, 2008).

Compound classes vary greatly among species without a clear pattern. Basically, all hydrocarbon classes that normally occur on the cuticle may be involved in the profile differences between reproductives and non-reproductives in various species. Often this results in differences represented by several compound classes including *n*-alkanes, monomethylalkanes and dimethylalkanes. Occasionally, unsaturated hydrocarbons are involved, too. For example, (Z)-9-hentriacontene (Z9-$C_{31:1}$) is the major compound determining the differences between gamergates and non-reproductive workers in *Dinoponera quadriceps* (Monnin and Peeters, 1997; Monnin *et al.*, 1998; Peeters *et al.*, 1999). Alkadienes are less frequent but are involved in the differences in *Streblognathos peetersi* and *Crematogaster smithii* (Cuvillier-Hot *et al.*, 2004a; Oettler *et al.*, 2008).

Branched alkanes occur in the profile differences of many species. In *Pachycondyla inversa* for example, 3,11-dimethylheptacosane is the major determinant of the differences between reproductives and non-reproductives (Heinze *et al.*, 2002; D'Ettorre *et al.*, 2004a). Queens with different fertility vary primarily in the amount of 9,13-dimethylpentacosane and 5,13-dimethylpentacosane in *Formica fusca* (Hannonen *et al.*, 2002).

Other possibilities for differences are simple *n*-alkanes. *Aphaenogaster cockerelli* queens and reproductive workers produce *n*-tricosane, *n*-pentacosane, *n*-heptacosane and *n*-nonacosane (Smith *et al.*, 2008; Figure 13.2). These compounds do not occur in workers but represent about 33% of the profile of the reproductive queens. The quantitative difference between mated reproductive queens and virgin, non-reproductive queens of *Ectatomma tuberculatum* is represented by *n*-heptacosane, which contributes up to 50% of the old queen profile versus 4% in virgin queens (Hora *et al.*, 2008). This suggests that the structural simplicity of *n*-alkanes does not prevent discrimination of these compounds (see also Chapter 10).

Cuticular profiles versus egg profiles

The hydrocarbon profiles of the cuticles of egg-layers closely match the profiles of their eggs. This is most likely owing to a shuttle mechanism that transports hydrocarbons from the site of synthesis to the cuticle and the ovaries as has been shown in cockroaches (see Chapter 5). Differences in the cuticular profile between differently fertile individuals may consequently also be found on the eggs. Extracts from the hemolymph of egg-layers in the ant *Gnamptogenys striatula* are almost identical to those found on the cuticle (Lommelen *et al.*, 2008). However, some longer-chain hydrocarbons were not found in the ovaries and the eggs.

A better match between cuticular and egg profiles is found in most other ants. The strong differences between the cuticular profiles of queens and workers in *Camponotus floridanus* are also present in eggs (Endler *et al.*, 2004, 2006). The profiles closely match each other, although egg profiles are characterized by a higher amount of alkanes. Since the profiles of reproductive queens and egg-laying workers are very different, egg profiles differ as well.

The cuticular profiles shown in reproductive individuals in *Aphaenogaster cockerelli* closely match their egg profiles (Smith *et al.*, 2008). Only some proportional differences are present. This match is present even though workers that do not lay viable eggs lack the additional alkanes that are present in the reproductive profiles.

When only single compounds on the cuticle relate to differences in fertility, the same differences can be found in eggs. In *Dinoponera quadriceps* the concentration of (Z)9-$C_{31:1}$ present on the cuticle of the egg-layers is represented at equal proportions in the profile of their eggs (Monnin and Peeters, 1997). A similarly close match of hydrocarbon profiles of the cuticle and the eggs is present in queens and workers of the ant *Pachycondyla inversa* (D'Ettorre *et al.*, 2004b).

These similarities are not restricted to ants. So far only one study has looked at the similarities between cuticular and egg profile in wasps. *Polistes dominulus* eggs contain the same compounds as the cuticle of the adults with only slight proportional differences (Dapporto *et al.*, 2007a). Egg profiles of foundresses ranking at the top of the hierarchy were clearly separated from the egg profiles of lower-ranked individuals.

Conclusions

Cuticular hydrocarbons and the hydrocarbon profiles of eggs appear to have a major function in the regulation of reproduction in social insects, adding another important function of hydrocarbon profiles to those described in other chapters of this book. Many species across major groups of social insects show correlations between hydrocarbon profiles and fertility. The broad presence of these intracolonial hydrocarbon patterns suggests a general importance in the regulation of reproduction. It was long assumed that pheromones regulate reproduction in colonies of social insects. Hydrocarbon profiles seem to fulfill this function. Most studies provide only correlative evidence, but a few studies support the presumed function of cuticular hydrocarbons and the hydrocarbon profiles of egg surfaces more strongly. Given that cuticular hydrocarbons differ also in other contexts, reflecting other communicative functions, it seems to be justified to assign these hydrocarbons a major role in the regulation of reproduction.

Nevertheless, studies that directly show a primer or releaser effect of these hydrocarbon profiles are needed to finally assess their importance. So far only one study, in the ant *Aphaenogaster cockerelli* (Smith *et al.*, 2009), directly showed the involvement of cuticular hydrocarbons in the regulation of reproduction of ants. However, this study only showed a releaser function of the hydrocarbons but not a primer effect. Studies on the primer effect are particularly needed. We may however want to consider that such studies seem to be especially difficult. In honeybees, it took four decades to finally verify the inhibitory function of the queen mandibular pheromone (Butler and Fairey, 1963; Hoover *et al.*, 2003).

It is noteworthy that in the study of *A. cockerelli* (Smith *et al.*, 2009), synthetic *n*-alkanes were used, while in other studies usually branched or unsaturated alkanes were synthesized. However, standard synthesis of such compounds leads to a mixture of stereoisomers. If social insects can differentiate between these stereoisomers, bioassays with such synthetic compounds may lead to artifacts. An important future goal is to assess the importance of such stereoisomers and test the respective natural form in the bioassays.

It is also important to consider other potential sources that may contribute to the recognition of reproductive individuals. So far, the vast majority of studies indicate a major role for long-chain hydrocarbons, but polar compounds may occasionally be involved as well (Tentschert *et al.*, 2002; Sramkova *et al.*, 2008). These polar compounds may also be of proteinaceous nature. They are found to differ between foundresses in *Polistes dominulus* (Dapporto *et al.*, 2008). Although polar compounds are involved in the inhibition of ovarian activity in honeybee workers (Hoover *et al.*, 2003) and prevent intracolonial reproduction of new queens in the ant *Solenopsis invicta* (Vargo, 1997, 1998), these findings seem to be exceptions.

We also need to understand how social insects are capable of extracting different kinds of information from the cuticular profile (Peeters and Liebig, 2009). Besides information about fertility, many other types of information are encoded. The interference in nestmate recognition based on cuticular hydrocarbons with information about fertility is particularly interesting. Different thresholds for recognition of different information may be involved (Le Conte and Hefetz, 2008). Further challenges to the understanding of profile discrimination are obvious when we consider task-specific cues (see Chapter 12).

We are just beginning to evaluate why cuticular hydrocarbons are especially suited to regulate reproduction. It could be that the specificity of biosynthesis and transport to ovaries, eggs, and cuticle is important here (Smith *et al.*, 2009). Alterations in the profile may increase water permeability, making hydrocarbon profiles a signal that is reliable because of the associated costs of potentially lower desiccation resistance (Hefetz, 2007). On the other hand reproductive individuals usually stay in nest areas with high humidity.

The many examples of the involvement of cuticular hydrocarbons in sex attraction in solitary insects (see various chapters in part II) suggest a common ground from which the role in the regulation of reproduction in social insects may have been derived. A better understanding of the physiological basis of hydrocarbon production and its linkage to ovarian activity will help to understand how the information about fertility becomes reliable and resistant to cheating (Smith *et al.*, 2009). The combination of these various fields in hydrocarbon research provides the ground for a better understanding of the major property of social insects, reproductive division of labor.

References

Abdalla, F. C., Jones, G. R., Morgan, E. D. and Da Cruz-Landim, C. (2003). Comparative study of the cuticular hydrocarbon composition of *Melipona bicolor* Lepeletier, 1836 (Hymenoptera, Meliponini) workers and queens. *Genet. Mol. Res.*, **2**, 191–199.

Baratte, S., Cobb, M. and Peeters, C. (2006). Reproductive conflicts and mutilation in queenless *Diacamma* ants. *Anim. Behav.*, **72**, 305–311.

Beekman, M., Martin, C. G. and Oldroyd, B. P. (2004). Similar policing rates of eggs laid by virgin and mated honey-bee queens. *Naturwissenschaften*, **91**, 598–601.

Beekman, M. and Oldroyd, B. P. (2005). Honeybee workers use cues other than egg viability for policing. *Biol. Lett.*, **1**, 129–132.

Bonavita-Cougourdan, A., Theraulaz, G., Bagnères, A. G., Roux, M., Pratte, M., Provost, E. and Clement, J. L. (1991). Cuticular hydrocarbons, social organization and ovarian development in a polistine wasp: *Polistes dominulus* Christ. *Comp. Biochem. Physiol. B*, **100**, 667–680.

Boulay, R., Hefetz, A., Cerda, X., Devers, S., Francke, W., Twele, R. and Lenoir, A. (2007). Production of sexuals in a fission-performing ant: dual effects of queen pheromones and colony size. *Behav. Ecol. Sociobiol.*, **61**, 1531–1541.

Bourke, A. F. G. (1988). Worker reproduction in the higher eusocial Hymenoptera. *Quart. Rev. Biol.*, **63**, 291–311.

Bourke, A. F. G. and Franks, N. R. (1995). *Social Evolution in Ants*. Princeton, NJ: Princeton University Press.

Brent, C., Peeters, C., Dietemann, V., Crewe, R. and Vargo, E. (2006). Hormonal correlates of reproductive status in the queenless ponerine ant, *Streblognathus peetersi*. *J. Comp. Physiol. A*, **192**, 315–320.

Butler, C. G. and Fairey, E. M. (1963). The role of the queen in preventing oogenesis in worker honeybees. *J. Apicult. Res.*, **2**, 14–18.

Butts, D. P., Camann, M. A. and Espelie, K. E. (1995). Workers and queens of the European hornet *Vespa crabro* L. have colony-specific cuticular hydrocarbon profiles (Hymenoptera: Vespidae). *Insect. Soc.*, **42**, 45–55.

Butts, D. P., Espelie, K. E. and Hermann, H. R. (1991). Cuticular hydrocarbons of four species of social wasps in the subfamily Vespinae: *Vespa crabro* L, *Dolichovespula maculata* (L), *Vespula squamosa* (Drury), and *Vespula maculifrons* (Buysson). *Comp. Biochem. Physiol. B*, **99**, 87–91.

Choe, J. C. (1988). Worker reproduction and social evolution in ants (Hymenoptera: Formicidae). In *Advances in Myrmecology*, ed. J. C. Trager. Leiden: Britt, pp. 163–187.

Cuvillier-Hot, V., Cobb, M., Malosse, C. and Peeters, C. (2001). Sex, age and ovarian activity affect cuticular hydrocarbons in *Diacamma ceylonense*, a queenless ant. *J. Insect Physiol.*, **47**, 485–493.

Cuvillier-Hot, V., Gadagkar, R., Peeters, C. and Cobb, M. (2002). Regulation of reproduction in a queenless ant: aggression, pheromones and reduction in conflict. *Proc. R. Soc. Lond. B*, **269**, 1295–1300.

Cuvillier-Hot, V., Lenoir, A., Crewe, R., Malosse, C. and Peeters, C. (2004a). Fertility signalling and reproductive skew in queenless ants. *Anim. Behav.*, **68**, 1209–1219.

Cuvillier-Hot, V., Lenoir, A. and Peeters, C. (2004b). Reproductive monopoly enforced by sterile police workers in a queenless ant. *Behav. Ecol.*, **15**, 970–975.

Cuvillier-Hot, V., Renault, V. and Peeters, C. (2005). Rapid modification in the olfactory signal of ants following a change in reproductive status. *Naturwissenschaften*, **92**, 73–77.

D'Ettorre, P. and Heinze, J. (2005). Individual recognition in ant queens. *Curr. Biol.*, **15**, 2170–2174.

D'Ettorre, P., Heinze, E., Schulz, C., Francke, W. and Ayasse, M. (2004a). Does she smell like a queen? Chemoreception of a cuticular hydrocarbon signal in the ant *Pachycondyla inversa. J. Exp. Biol.*, **207**, 1085–1091.

D'Ettorre, P., Heinze, J. and Ratnieks, F. L. W. (2004b). Worker policing by egg eating in the ponerine ant *Pachycondyla inversa. Proc. R. Soc. Lond. B*, **271**, 1427–1434.

D'Ettorre, P., Tofilski, A., Heinze, J. and Ratnieks, F. L. W. (2006). Non-transferable signals on ant queen eggs. *Naturwissenschaften*, **93**, 136–140.

Dahbi, A. and Lenoir, A. (1998). Queen and colony odour in the multiple nest ant species, *Cataglyphis iberica* (Hymenoptera, Formicidae). *Insect Soc.*, **45**, 301–313.

Dapporto, L., Dani, F. R. and Turillazzi, S. (2007a). Social dominance molds cuticular and egg chemical blends in a paper wasp. *Curr. Biol.*, **17**, R504–R505.

Dapporto, L., Lambardi, D. and Turillazzi, S. (2008). Not only cuticular lipids: First evidence of differences between foundresses and their daughters in polar substances in the paper wasp *Polistes dominulus. J. Insect Physiol.*, **54**, 89–95.

Dapporto, L., Santini, A., Dani, F. R. and Turillazzi, S. (2007b). Workers of a *Polistes* paper wasp detect the presence of their queen by chemical cues. *Chem. Sens.*, **32**, 795–802.

Dapporto, L., Theodora, P., Spacchini, C., Pieraccini, G. and Turillazzi, S. (2004). Rank and epicuticular hydrocarbons in different populations of the paper wasp *Polistes dominulus* (Christ) (Hymenoptera, Vespidae). *Insect. Soc.*, **51**, 279–286.

de Biseau, J. C., Passera, L., Daloze, D. and Aron, S. (2004). Ovarian activity correlates with extreme changes in cuticular hydrocarbon profile in the highly polygynous ant, *Linepithema humile. J. Insect Physiol.*, **50**, 585–593.

Dietemann, V., Liebig, J., Hölldobler, B. and Peeters, C. (2005). Changes in the cuticular hydrocarbons of incipient reproductives correlate with triggering of worker policing in the bulldog ant *Myrmecia gulosa. Behav. Ecol. Sociobiol.*, **58**, 486–496.

Dietemann, V., Peeters, C., Liebig, J., Thivet, V. and Hölldobler, B. (2003). Cuticular hydrocarbons mediate recognition of queens and reproductive workers in the ant *Myrmecia gulosa. Proc. Natl. Acad. Sci. USA*, **100**, 10341–10346.

Dreier, S., van Zweden, J. S. and D'Ettorre, P. (2007). Long-term memory of individual identity in ant queens. *Biol. Lett.*, **3**, 459–462.

Endler, A., Hölldobler B. and Liebig, J. (2007). Lack of physical policing and fertility cues in egg-laying workers of the ant *Camponotus floridanus. Anim. Behav.*, **74**, 1171–1180.

Endler, A., Liebig, J. and Hölldobler, B. (2006). Queen fertility, egg marking and colony size in the ant *Camponotus floridanus. Behav. Ecol. Sociobiol.*, **59**, 490–499.

Endler, A., Liebig, J., Schmitt, T., Parker, J. E., Jones, G. R., Schreier, P. and Hölldobler, B. (2004). Surface hydrocarbons of queen eggs regulate worker reproduction in a social insect. *Proc. Natl. Acad. Sci. USA*, **101**, 2945–2950.

Fletcher, D. J. C. and Ross, K. G. (1985). Regulation of reproduction in eusocial Hymenoptera. *Annu. Rev. Entomol.*, **30**, 319–343.

Halling, L. A., Oldroyd, B. P., Wattanachaiyingcharoen, W., Barron, A. B., Nanork, P. and Wongsiri, S. (2001). Worker policing in the bee *Apis florea. Behav. Ecol. Sociobiol.*, **49**, 509–513.

Hannonen, M., Sledge, M. F., Turillazzi, S. and Sundström, L. (2002). Queen reproduction, chemical signalling and worker behaviour in polygyne colonies of the ant *Formica fusca. Anim. Behav.*, **64**, 477–485.

Hartmann, A., D'Ettorre, P., Jones, G. R. and Heinze, J. (2005). Fertility signaling – the proximate mechanism of worker policing in a clonal ant. *Naturwissenschaften*, **92**, 282–286.

Hefetz, A. (2007). The evolution of hydrocarbon pheromone parsimony in ants (Hymenoptera: Formicidae) – interplay of colony odor uniformity and odor idiosyncrasy. *Myrmecol. News*, **10**, 59–68.

Heinze, J. (2004). Reproductive conflict in insect societies. *Adv. Study Behav.*, **34**, 1–57.

Heinze, J., Foitzik, S., Oberstadt, B., Rueppell, O. and Hölldobler, B. (1999). A female caste specialized for the production of unfertilized eggs in the ant *Crematogaster smithi*. *Naturwissenschaften*, **86**, 93–95.

Heinze, J., Stengl, B. and Sledge, M. F. (2002). Worker rank, reproductive status and cuticular hydrocarbon signature in the ant, *Pachycondyla* cf. *inversa*. *Behav. Ecol. Sociobiol.*, **52**, 59–65.

Helanterä, H., Martin, S. J. and Ratnieks, F. L. W. (2007). Prior experience with eggs laid by non-nestmate queens induces egg acceptance errors in ant workers. *Behav. Ecol. Sociobiol.*, **62**, 223–228.

Helanterä, H. and Sundström, L. (2005). Worker reproduction in the ant *Formica fusca*. *J. Evol. Biol.*, **18**, 162–171.

Helanterä, H., and Sundström, L. (2007). Worker policing and nest mate recognition in the ant *Formica fusca*. *Behav. Ecol. Sociobiol.*, **61**, 1143–1149.

Helanterä, H., Tofilski, A., Wenseleers, T. and Ratnieks, F. L. W. (2006). Worker policing in the common wasp *Vespula vulgaris* is not aimed at improving colony hygiene. *Insect Soc.*, **53**, 399–402.

Hölldobler, B. and Bartz, S. H. (1985). Sociobiology of reproduction in ants. In *Experimental Behavioral Ecology and Sociobiology*, ed. B. Hölldobler and M. Lindauer. Stuttgart: Gustav Fischer, pp. 237–257.

Hölldobler, B. and Carlin, N. F. (1989). Colony founding, queen control and worker reproduction in the ant *Aphaenogaster* (= *Novomessor*) *cockerelli* (Hymenoptera: Formicidae). *Psyche*, **96**, 131–151.

Hoover, S. E. R., Keeling, C. I., Winston, M. L. and Slessor K. N. (2003). The effect of queen pheromones on worker honey bee ovary development. *Naturwissenschaften*, **90**, 477–480.

Hora, R. R., Ionescu-Hirsh, A., Simon, T., Delabie, J., Robert, J., Fresneau, D. and Hefetz, A. (2008). Postmating changes in cuticular chemistry and visual appearance in *Ectatomma tuberculatum* queens (Formicidae: Ectatomminae). *Naturwissenschaften*, **95**, 55–60.

Howard, R. W. and Blomquist, G. J. (1982). Chemical ecology and biochemistry of insect hydrocarbons. *Annu. Rev. Entomol.*, **27**, 149–172.

Howard, R. W. and Blomquist, G. J. (2005). Ecological, behavioral, and biochemical aspects of insect hydrocarbons. *Annu. Rev. Entomol.*, **50**, 371–393.

Katzav-Gozansky, T., Soroker, V. and Hefetz, A. (1997). Plasticity of caste-specific Dufour's gland secretion in the honey bee (*Apis mellifera* L.). *Naturwissenschaften*, **84**, 238–241.

Katzav-Gozansky, T., Soroker, V., Ibarra, F., Francke, W. and Hefetz, A. (2001). Dufour's gland secretion of the queen honeybee (*Apis mellifera*): an egg discriminator pheromone or a queen signal? *Behav. Ecol. Sociobiol.*, **51**, 76–86.

Katzav-Gozansky, T., Soroker, V., Kamer, J., Schulz, C. M., Francke, W. and Hefetz, A. (2003). Ultrastructural and chemical characterization of egg surface of honeybee worker and queen-laid eggs. *Chemoecology*, **13**, 129–134.

Kawabata, S. and Tsuji, K. (2005). The policing behavior 'immobilization' towards ovary-developed workers in the ant, *Diacamma* sp. from Japan. *Insect. Soc.,* **52**, 89–95.

Keller, L. and Nonacs, P. (1993). The role of queen pheromones in social insects: queen control or queen signal? *Anim. Behav.*, **45**, 787–794.

Kerr, W. E., Jungnickel, H. and Morgan, E. D. (2004). Workers of the stingless bee *Melipona scutellaris* are more similar to males than to queens in their cuticular compounds. *Apidologie*, **35**, 611–618.

Khila, A. and Abouheif, E. (2008). Reproductive constraint is a developmental mechanism that maintains social harmony in advanced ant societies. *Proc. Natl. Acad. Sci. USA*, **105**, 17884–17889.

Kikuta, N. and Tsuji, K. (1999). Queen and worker policing in the monogynous and monandrous ant, *Diacamma* sp. *Behav. Ecol. Sociobiol.*, **46**, 180–189.

Layton, J. M., Camann, M. A. and Espelie, K. E. (1994). Cuticular lipid profiles of queens, workers, and males of social wasp *Polistes metricus* Say are colony-specific. *J. Chem. Ecol.*, **20**, 2307–2321.

Le Conte, Y. and Hefetz, A. (2008). Primer pheromones in social Hymenoptera. *Annu. Rev. Entomol.*, **53**, 523–542.

Liebig, J. (1998). *Eusociality, female caste dimorphism, and regulation of reproduction in the ponerine Ant Harpegnathos saltator* Jerdon. Berlin: Wissenschaft und Technik Verlag and doctoral thesis, University of Würzburg.

Liebig, J., Hölldobler, B. and Peeters, C. (1998). Are ant workers capable of colony foundation? *Naturwissenschaften*, **85**, 133–135.

Liebig, J., Monnin, T. and Turillazzi, S. (2005). Direct assessment of queen quality and lack of worker suppression in a paper wasp. *Proc. R. Soc. Lond. B*, **272**, 1339–1344.

Liebig, J., Peeters, C. and Hölldobler, B. (1999). Worker policing limits the number of reproductives in a ponerine ant. *Proc. R. Soc. Lond. B*, **266**, 1865–1870.

Liebig, J., Peeters, C., Oldham, N. J., Markstädter, C., and Hölldobler B. (2000). Are variations in cuticular hydrocarbons of queens and workers a reliable signal of fertility in the ant *Harpegnathos saltator*? *Proc. Natl. Acad. Sci. USA*, **97**, 4124–4131.

Lockey, K. H. (1988). Lipids of the insect cuticle: origin, composition and function. *Comp. Biochem. Physiol. B*, **89**, 595–645.

Lommelen, E., Johnson, C. A., Drijfhout, F. P., Billen J. and Gobin, B. (2008). Egg marking in the facultatively queenless ant *Gnamptogenys striatula*: The source and mechanism. *J. Insect Physiol.*, **54**, 727–736.

Lommelen, E., Johnson, C. A., Drijfhout, F. P., Billen, J., Wenseleers, T. and Gobin, B. (2006). Cuticular hydrocarbons provide reliable cues of fertility in the ant *Gnamptogenys striatula*. *J. Chem. Ecol.*, **32**, 2023–2034.

Martin, S., Chaline, N., Drijfhout, F. and Jones, G. (2005). Role of esters in egg removal behaviour in honeybee (*Apis mellifera*) colonies. *Behav. Ecol. Sociobiol.*, **59**, 24–29.

Martin, S. J., Chaline, N., Oldroyd, B. P., Jones, G. R. and Ratnieks, F. L. W. (2004a). Egg marking pheromones of anarchistic worker honeybees (*Apis mellifera*). *Behav. Ecol.*, **15**, 839–844.

Martin, S. J., Jones, G. R., Chaline, N., Middleton, H. and Ratnieks, F. L. W. (2002). Reassessing the role of the honeybee (*Apis mellifera*) Dufour's gland in egg marking. *Naturwissenschaften*, **89**, 528–532.

Martin, S. J., Jones, G. R., Chaline, N., and Ratnieks, F. L. W. (2004b). Role of hydrocarbons in egg recognition in the honeybee. *Physiol. Entomol.*, **29**, 395–399.

Monnin, T. (2006). Chemical recognition of reproductive status in social insects. *Ann. Zool. Fennici*, **43**, 515–530.

Monnin, T., Malosse, C. and Peeters, C. (1998). Solid-phase microextraction and cuticular hydrocarbon differences related to reproductive activity in the queenless ant *Dinoponera quadriceps*. *J. Chem. Ecol.*, **24**, 473–490.

Monnin, T. and Peeters, C. (1997). Cannibalism of subordinates' eggs in the monogynous queenless ant *Dinoponera quadriceps*. *Naturwissenschaften*, **84**, 499–502.

Monnin, T. and Peeters, C. (1998). Monogyny and regulation of worker mating in the queenless ant *Dinoponera quadriceps*. *Anim. Behav.*, **55**, 299–306.

Monnin, T. and Peeters, C. (1999). Dominance hierarchy and reproductive conflicts among subordinates in a monogynous queenless ant. *Behav. Ecol.*, **10**, 323–332.

Nanork, P., Wongsiri, S. and Oldroyd, B. P. (2007). Preservation and loss of the honey bee (*Apis*) egg-marking signal across evolutionary time. *Behav. Ecol. Sociobiol.*, **61**, 1509–1514.

Oettler, J., Schmitt, T., Herzner, G. and Heinze, J. (2008). Chemical profiles of mated and virgin queens, egg-laying intermorphs and workers of the ant *Crematogaster smithi*. *J. Ins. Physiol.*, **54**, 672–679.

Peeters, C. (1993). Monogyny and polygyny in ponerine ants with or without queens. In *Queen Number and Sociality in Insects*, ed. L. Keller. New York: Oxford University Press, pp. 234–261.

Peeters, C. (1997). Morphologically "primitive" ants: comparative review of social characters, and the importance of queen–worker dimorphism. In *The Evolution of Social Behavior in Insects and Arachnids*, ed. J. C. Choe and B. J. Crespi. Cambridge: Cambridge University Press, pp. 372–391.

Peeters, C. and Hölldobler, B. (1995). Reproductive cooperation between queens and their mated workers: The complex life-history of an ant with a valuable nest. *Proc. Natl. Acad. Sci. USA*, **92**, 10977–10979.

Peeters, C. and Ito, F. (2001). Colony dispersal and the evolution of queen morphology in social hymenoptera. *Annu. Rev. Entomol.*, **46**, 601–630.

Peeters, C. and Liebig, J. (2009). Fertility signaling as a general mechanism of regulating reproductive division of labor in ants. In *Organization of Insect Societies: From Genome to Socio-Complexity*, ed. J. Gadau and J. Fewell. Cambridge, MA: Harvard University Press, pp. 220–242.

Peeters, C., Liebig, J. and Hölldobler, B. (2000). Sexual reproduction by both queens and workers in the ponerine ant *Harpegnathos saltator*. *Insect. Soc.*, **47**, 325–332.

Peeters, C., Monnin, T. and Malosse, C. (1999). Cuticular hydrocarbons correlated with reproductive status in a queenless ant. *Proc. R. Soc. Lond. B*, **266**, 1323–1327.

Pirk, C. W. W., Neumann, P., Hepburn, R., Moritz, R. F. A. and Tautz, J. (2004). Egg viability and worker policing in honey bees. *Proc. Natl. Acad. Sci. USA*, **101**, 8649–8651.

Ratnieks, F. L. W. (1988). Reproductive harmony via mutual policing by workers in eusocial Hymenoptera. *Am. Nat.*, **132**, 217–236.

Ratnieks F. L. W. (1995). Evidence for a queen-produced egg-marking pheromone and its use in worker policing in the honey bee. *J. Apicult. Res.*, **34**, 31–37.

Ratnieks, F. L. W., Foster, K. R. and Wenseleers, T. (2006). Conflict resolution in insect societies. *Annu. Rev. Entomol.*, **51**, 581–608.

Ratnieks, F. L. W. and Visscher, P. K. (1989). Worker policing in the honeybee. *Nature*, **342**, 796–797.

Röseler, P.F., Röseler, I. and Vanhonk, C.G.J. (1981). Evidence for inhibition of corpora allata activity in workers of *Bombus terrestris* by a pheromone from the queen's mandibular glands. *Experientia*, **37**, 348–351.

Ross, K.G. and Matthews, R.W., eds. (1991). *The Social Biology of Wasps*. Ithaca, New York: Cornell University Press.

Schilder, K., Heinze, J. and Hölldobler, B. (1999). Colony structure and reproduction in the thelytokous parthenogenetic ant *Platythyrea punctata* (F. Smith) (Hymenoptera, Formicidae). *Insect Soc.*, **46**, 150–158.

Seeley, T.D. (1985). *Honeybee Ecology: a Study of Adaptation in Social Life*. Princeton, NJ: Princeton University Press.

Sheehan, M.J. and Tibbetts, E.A. (2008). Robust long-term social memories in a paper wasp. *Curr. Biol.*, **18**, R851–R852.

Sledge, M.F., Boscaro, F. and Turillazzi, S. (2001). Cuticular hydrocarbons and reproductive status in the social wasp *Polistes dominulus*. *Behav. Ecol. Sociobiol.*, **49**, 401–409.

Sledge, M.F., Trinca, I., Massolo, A., Boscaro, F. and Turillazzi, S. (2004). Variation in cuticular hydrocarbon signatures, hormonal correlates and establishment of reproductive dominance in a polistine wasp. *J. Ins. Physiol.*, **50**, 73–83.

Slessor, K.N., Foster, L.J. and Winston, M.L. (1998). Royal flavours: honey bee queen pheromones. In *Pheromone Communication in Social Insects – Ants, Wasps, Bees, and Termites*, ed. R.K. Vander Meer, M.D. Breed, K.E. Espelie and M.L. Winston. Boulder, CO: Westview Press, pp. 331–343.

Smith, A., Hölldobler, B. and Liebig, J. (2008). Hydrocarbon signals explain the pattern of worker and egg policing in the ant *Aphaenogaster cockerelli*. *J. Chem. Ecol.*, **34**, 1275–1282.

Smith, A., Hölldobler, B. and Liebig, J. (2009). Cuticular hydrocarbons reliably identify cheaters and allow enforcement of altruism in a social insect. *Curr. Biol.*, **19**, 78–81.

Sommer, K., Hölldobler, B. and Rembold, H. (1993). Behavioral and physiological aspects of reproductive control in a *Diacamma* species from Malaysia (Formicidae, Ponerinae). *Ethology*, **94**, 162–170.

Sramkova, A., Schulz, C., Twele, R., Francke, W. and Ayasse, M. (2008). Fertility signals in the bumblebee *Bombus terrestris* (Hymenoptera: Apidae). *Naturwissenschaften*, **95**, 515–522.

Tentschert, J., Bestmann, H.J. and Heinze, J. (2002). Cuticular compounds of workers and queens in two *Leptothorax* ant species – a comparison of results obtained by solvent extraction, solid sampling, and SPME. *Chemoecology*, **12**, 15–21.

Tentschert, J., Kolmer, K., Hölldobler, B., Bestmann, H.J., Delabie, J.H.C. and Heinze, J. (2001). Chemical profiles, division of labor and social status in *Pachycondyla* queens (Hymenoptera : Formicidae). *Naturwissenschaften*, **88**, 175–178.

Tibbetts, E.A. (2002). Visual signals of individual identity in the wasp *Polistes fuscatus*. *Proc. R. Soc. Lond. B*, **269**, 1423–1428.

Tibbetts, E.A. and Dale, J. (2004). A socially enforced signal of quality in a paper wasp. *Nature,* **432**, 218–222.

Tsuchida, K., Saigo, T., Nagata, N., Tsujita, S., Takeuchi, K. and Miyano, S. (2003). Queen-worker conflicts over male production and sex allocation in a primitively eusocial wasp. *Evolution,* **57**, 2365–2373.

Tsuchida, K. and Suzuki, T. (2006). Conflict over sex ratio and male production in paper wasps. *Ann. Zool. Fennici*, **43**, 468–480.

Tsuji, K., Egashira, K. and Hölldobler, B. (1999). Regulation of worker reproduction by direct physical contact in the ant *Diacamma* sp. from Japan. *Anim. Behav.*, **58**, 337–343.

Turillazzi, S., Sledge, M. F., Dapporto, L., Landi, M., Fanelli, D., Fondelli, L., Zanetti, P. and Dani, F. R. (2004). Epicuticular lipids and fertility in primitively social wasps (Hymenoptera Stenogastrinae). *Physiol. Entomol.*, **29**, 464–471.

Vargo, E. L. (1997). Poison gland of queen fire ants (*Solenopsis invicta*) is the source of a primer pheromone. *Naturwissenschaften*, **84**, 507–510.

Vargo, E. L. (1998). Primer pheromones in ants. In *Pheromone Communication in Social Insects – Ants, Wasps, Bees, and Termites*, ed. R. K. Vander Meer, M. D. Breed, K. E. Espelie and M. L. Winston. Boulder, CO: Westview Press, pp. 293–313.

Vargo, E. L. and Hulsey, C. D. (2000). Multiple glandular origins of queen pheromones in the fire ant *Solenopsis invicta*. *J. Insect Physiol.*, **46**, 1151–1159.

Wattanachaiyingcharoen, W., Oldroyd, B. P., Good, G., Halling, L., Ratnieks, F. L. W. and Wongsiri, S. (2002). Lack of worker reproduction in the giant honey bee *Apis dorsata* Fabricius. *Insect Soc.*, **49**, 80–85.

14

Chemical deception/mimicry
using cuticular hydrocarbons

Anne-Geneviève Bagnères and M. Cristina Lorenzi

Chemical deception/mimicry is one of the most amazing phenomena in the field of chemical ecology. It involves a mimicry complex in which multiple living protagonists, i.e., the model, the mime and the dupe (signal receiver), interact at several biochemical and physiological levels. Mimicry complexes have evolved, sometime over thousands of years, throughout the animal and plant kingdoms, and encompass a wide range of organisms from mosses to vertebrates. Chemical deception/mimicry can take many different forms: camouflage, mimesis, crypsis, transparency, and usurpation.

In behavioral ecology, various theories have been proposed to explain inter-individual signaling communication activities, e.g. Dawkins versus Tinbergen (Dawkins and Krebs, 1978; see also Hughes, 2008). Chemical deception/mimicry is generally considered to be a form of co-evolution, i.e., different from convergent evolution, which can be defined in terms of either an arms race when it is used aggressively, or as "reciprocal altruism" or mutualism when it is cobeneficial (e.g. between the bumblebee and orchid). Mimicry processes are usually classified according to function with a basic distinction being made between defensive and aggressive mimicry, while others distinguish between Batesian mimicry, i.e. a deceptive strategy in which a harmless mimic poses as harmful, and Müllerian mimicry, i.e. a nondeceptive strategy in which two harmful species share similar perceived characteristics. In his review, Pasteur (1982) classified mimicry processes into several categories including camouflage in order to escape predators or surprise preys (homochromy), imitation of natural elements or plant organs (homotypy), and temporary or permanent mimicry. In his review, Malcolm (1990) proposed a mimicry paradigm with two main categories: interpecific interactions for defense, foraging and parasitism, on the one hand, and intraspecific interactions for reproduction, on the other. However, Malcolm's discussion focused on aposematic functions as well as general mimicry systems without mentioning chemical deception. Malcolm's call for further study of mimicry seems to have been answered by numerous articles analyzing and testing the phenomenon with evolutionary approaches.

A number of excellent overviews are available on chemical mimicry. Several reviews about pollination mimicry and insect mimicry can be found on the internet (e.g. Legoza, 2005). A more conventional presentation of chemistry, coevolution, and mimicry can be found in two chapters by Stowe (Chapter 17) and Spencer (Chapter 18) in K. C. Spencer's excellent book (1988). The information in the present chapter calls for revision of Spencer's statement that "evidence for coevolution [is] generally weak, but strongest in plant–herbivore association".

Since the main subject of this book is hydrocarbons, we will be focusing on the enormous body of evidence comprising over 150 reports in the literature on mimicry systems using hydrocarbons. Several examples of non-hydrocarbon chemical mimicry however deserve brief mention. Some of them are spectacular examples, e.g. carnivorous plants (Joel, 1988) and bolas spiders (Stowe *et al.*, 1987; Yeargan, 1994). In fact bolas spiders have been included in this chapter since mimicry depends on attracting the male moth prey by emitting female moth sex pheromones that include a few hydrocarbons and derivatives. In a comment made in *Nature*, Gibbs described an interesting mimicry complex in which a virus carried by white flies avoids degradation in the insect hemolymph using a proteinous chaperonin released by protective bacteria (Gibbs, 1999). Another interesting example of non-hydrocarbon chemical deception is phytochemical mimicry used by plants that use phytoestrogens to modulate herbivore fertility (Hughes, 1988). A rare form of non-hydrocarbon mimicry that has spawned an alternative method of pest ant control involves the use of long-chain alcohols and aldehydes by Ichneumonidae wasps to provoke ant warfare (Thomas *et al.*, 2002). In another non-hydrocarbon mimicry complex, the death's head hawkmoth *Acherontia atropos* is able to enter the *Apis mellifera* hive and feed on nectar and honey without being attacked by workers (Moritz *et al.*, 1991). This is especially interesting since, as the authors said, large *Apis* parasites including *Vespa crabro, V. orientalis* and *V. velutina* are unable to enter the hive when guard bees are present at the entrance. The compounds identified in this mimicry system, i.e. four classical fatty acids (C_{16} and C_{18} saturated and unsaturated acids) commonly found in similar concentrations in honeybees, are produced naturally by the moths rather than acquired from the bees. After the authors dissect the moths and perform solvent extracts on body and gland pieces, including the fat body (and certainly the acids from the fat), they problematically compare their extracts with honeybee cuticular washes. They also speculate that, in addition to chemical camouflage, the moth might be able to acoustically mimic the queen honeybee sound. Another example of non-hydrocarbon mimicry by contact is provided by egg-mimicking fungi called "termite balls" that use both morphological and chemical camouflage to induce tending by subterranean termites. Bioassays on methanolic extracts indicated that this mimicry system involves a polar compound recently identified as a protein (Matsuura, 2006; Matsuura *et al.*, 2007). Another recent paper describing three species of fungus-growing ants showed that the chemical profiles of workers, brood and mutualistic fungi comprised a congruent blend of chemical compounds including not only different hydrocarbon classes but also aldehydes, alcohols, acetates, acids, and esters (Viana *et al.,* 2001; Richard *et al.*, 2007). Interestingly, even basic organisms such as fungi seem able to also perform mimicry with flowers (Kaiser, 2006).

In an overview on mating behavior and chemical communication in Hymenoptera, Ayasse *et al.* (2001) described sex pheromones, including some that were hydrocarbons. Some chapters and reviews on the various roles of cuticular hydrocarbons (Howard, 1993;

Singer, 1998; Howard and Blomquist, 2005; Geiselhardt *et al.*, 2007; Provost *et al.*, 2008) have dedicated paragraphs to mimicry systems. Two other papers that thoroughly explore chemical mimicry involving cuticular hydrocarbons are those by Dettner and Liepert (1994) and Lenoir *et al.* (2001). In a recent review, Akino (2008) focused mainly on cuticular hydrocarbons as recognition components involved in chemical strategies, such as mimicry, camouflage, propaganda and phytomimesis used by ants and other arthropods. Nash and Boomsma (2008) also recently reviewed communication between hosts and social parasites.

Finally, before beginning, it is important to point out two conventions that will be used in this chapter. In many papers dealing with chemical strategies employed by parasites, emphasis has been placed on identifying the mechanisms used to achieve resemblance with the host. In this regard, a clear-cut distinction has been made between chemical mimicry, in which the parasite actively biosynthesizes host cues, and camouflage, in which the parasite acquires cues by passive and or active transfer from the host. In practice, the functional difference between these two mechanisms is slight, and both mechanisms are used. Throughout this chapter, which focuses more on adaptive function than on casual factors, we will resolve this difficulty by using the term "mimicry" indiscriminately regardless of the underlying mechanisms.

The second point that should be made concerns the organization of the chapter. Most of the mimicry complexes presented below are disjunct associations involving interaction between different species (as opposed to conjunct associations). For the purpose of chemical communication we have subdivided this chapter into six parts. In the first four parts, examples are grouped according to the type of organism involved in the association: plant–insect, non-social arthropod only, non-social arthropod–social insect and intraspecific mimicry in non-social insects. The last two parts of this chapter present only social insect associations as artificial mixed colonies, and the diverse interactions among social insects, i.e. belonging to different or the same orders, genera or species.

Plant–insect interactions

Plants and insects have developed extensive ecological relationships. Interaction involving mimicry relationships can be classified into two main categories: pollination mimicry and protection mimicry. One of the best-known examples of pollination mimicry involving cuticular hydrocarbons occurs between bees and orchids. This relationship was the topic of what was, to our knowledge, the first description of chemical mimicry, by the Swedish group Bergström, Tengö, and Borg-Karlson (Tengö and Bergström, 1977; Bergström, 1978 in Schiestl *et al.*, 1999; Borg-Karlson and Tengö, 1986; Bergström, 2008). Since the initial 1977 report, a number of chemical similarities have been identified between cuticular extracts from *Andrena* bees (*A. fuscipes* and *A. nigroaenea*) and labelum extracts of *Ophrys* orchids (*O. lutea* and *O. sphegodes*). Taking advantage of the fact that the bee's mating behavior is odor-guided, the plant deceives male bees by releasing volatile compounds for long-distance attraction and hydrocarbon patterns mimicking virgin female pheromones (mainly a blend of C_{25} to C_{29} *n*-alkanes and *n*-alkenes) to elicit male pseudocopulation (Schiestl *et al.*, 1999, 2000). Recently, a similar phenomenon in which (Z)-7-alkenes appeared to play a major role was reported in another solitary bee, *Colletes cunicularius* (Mant *et al.*, 2005a, 2005b) with linalool appearing to function as a long-range attractant and hydrocarbons inducing short-range mating behavior (Figure 14.1). Véla *et al.* (2007) showed that floral chemical signatures provide a good chemotaxonomy tool in the genus *Ophrys*, but did not perform chemical determination. Data showed great variation among the chemical blend of individual flowers, but a good flower species-specific chemical pattern, which might be useful for understanding this pollinator–flower discrimination. The divergent selection among orchid populations may be driven by local pollinator preferences for critical compounds such as alkenes (Mant *et al.*, 2005b). Recent work on a deceptive Asian orchid (*Dendrobium sinense*) shows that flowers of the orchid are able to mimic the alarm pheromone of honey bees, (Z)-11-eicosen-l-ol, in order to attract prey-hunting hornets for pollination (Brodmann *et al.*, 2009).

Several studies have used multitrophic level evaluation to measure biochemical convergence between the surface lipids of host plants and cuticular compounds of insect herbivores and their predators or parasitoids (Espelie and Brown, 1990; Espelie *et al.*, 1991). Espelie and Hermann (1988) explored the mutualistic ant–acacia association, in which the plant produces waxes that mimic ant cuticular waxes. Congruency of cuticular hydrocarbons is the key, allowing some wasps to enter the acacia, otherwise kept intruder-free (Espelie and Hermann, 1988). Nevertheless, as noted by Howard, the data of Espelie and colleagues are "far from definitive and must be interpreted cautiously". Recently, Henrique *et al.* (2005) showed convergence between the lipid profiles of a nymphalidae caterpillar and its host plant (*Solanum tabacifolium*). By switching caterpillars to non-host plants, they showed that camouflage protecting caterpillars from *Camponotus* ant predation depends on the host plant. In another study using multitrophic level evaluation, comparison of the cuticular lipids of a cabbage seedpod weevil with compounds found at the oviposition site on the host plant also showed strong congruency (Richter and Krain, 1980). Caterpillars of the giant geometer *Biston robustum* change their appearance according to their host plant species by varying colors and marks by photo- and chemomimesis. The caterpillars adjust their cuticular chemicals to match the host plant and ants cannot discern them from the tree (Akino *et al.*, 2004). These first examples of chemical congruency among phylogenetically distant organisms such as plants and animals are an indication that relatedness is not a prerequisite in the chemical mimicry process.

Non-social arthropod interactions

Judging from the paucity of references about non-social insects, it is tempting to think that chemical deception/mimicry did not play a major role in their evolution. However there have been too few studies done to draw any definite conclusion, since there are several reports presenting evidence of mimicry in non-social arthropods. Also, another type of mimicry can occur when species such as orthoptera, particularly good in exploiting acoustic signals,

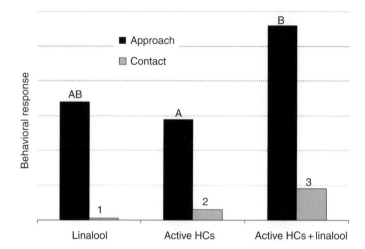

Figure 14.1a Bioessays using synthetic compounds: (1) Linalool alone attracts patrolling *Colletes* bee males without eliciting corresponding contacts; (2) blend of 12 EAD (electro antenno graphic detection)-active alkanes and alkenes (HCs) attracts a similar number of males but elicits more contact; (3) Linalool combined with EAD-active HCs attracts a similar number of males but the number of contacts is significantly higher (all tests are using Mann–Whitney U test with $P < 0.01$, different letters indicate significant differences between groups) (from Mant *et al.*, 2005a; with approximate number of responses).

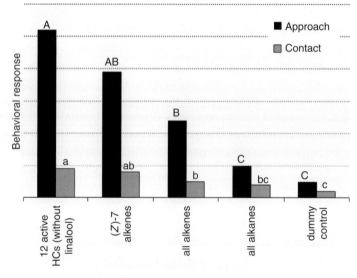

Figure 14.1b Bioessays using various synthetic blends show = mixture of all 12 EAD-active hydrocarbons (HCs) and mixture of (Z)-7 alkenes induced similar number of attractions and contacts (Mann–Whitney U test with $P < 0.005$, different letters indicate significant differences between groups). Behavioral responses to the all-alkene mixture are not different for the (Z)-7 alkene mixture; all alkanes with dummy control show similar response with low approach and contact behaviors (from Mant *et al.*, 2005a with approximate number of responses).

imitate the prey species-specific wing-flick replies of sexually receptive female cicada to attract males (Marshall and Hill, 2009). Howard and Liang (1993) showed that cuticular hydrocarbon profiles were nearly identical in a population of the ectoparasitoid *Choetospila elegans* reared on their bostrichidae host, the grain borer *Rhyzopertha dominica*, and the rice weevil larvae (*Sitophilus oryzae*) cultivated on whole wheat. Major beetle components, mainly *n*-alkanes, perfectly matched those found on adult parasites, but minor components showed some differences. Later in this chapter we will see that a similar association involving a parasitic wasp of a stored grain larvae beetle provides an example of intraspecific mimicry (Steiner *et al.*, 2005). Geiselhardt *et al.* (2006) described chemical mimicry between two closely related coleopterans, i.e. a cleptoparasite and its host, using a series of recognition bioassays, to demonstrate that this mimicry prevents the host from attacking the parasite.

Another interesting example of mimicry in non-social arthropods involves the hymenopteran crabronidae *Philanthus triangulum* (European beewolf) and a highly specialized cuckoo wasp *Hedychrum rutilans* (Hymenoptera, Chrysilidae). Female beewolves are parasites that capture, paralyze and carry honeybees to their nest. Brood cells containing the paralyzed bees are threatened by female cuckoo wasps that enter beewolf nests to oviposit on the paralyzed bees, so that their larvae can feed on beewolf larvae. To investigate whether the cuckoo wasp employs a chemical mimicry strategy, Strohm *et al.* (2008) recently compared host and parasite cuticular signatures. They concluded that *H. rutilans* females closely mimic the composition of cuticular compounds of their host species *P. triangulum*. The presence of isomeric forms of alkenes on the cuticle of the cuckoo wasp, and their absence on female beewolf host cuticle, suggest that cuckoo wasps synthesize the cuticular compounds rather than sequester them from their host. Thus, behavioral data and chemical analysis provide evidence that a specialized cuckoo wasp enters its host nest by odor mimicry using "wolf's hydrocarbon clothing".

In hymenopterans, the Apidae family contains the solitary digger bee *Habropoda pallida*, which is parasitized by the blister beetle *Meloe franciscanus* (Coleoptera, Meloidae). This association provides a remarkable example of aggressive chemical mimicry by the phoretic first-instar beetle known as triungulins (Saul-Gershenz and Millar, 2006). Triungulins cluster together in large groups that visually mimic the shape and color of a female solitary digger bee perching at the tip of a twig (Hafernick and Saul-Gershenz, 2000). To increase their attractiveness to the *H. pallida* male, clustered triungulins also release a subset of the female *H. pallida* hydrocarbon blend that is made up mainly of alkenes (Figure 14.2). When a *H. pallida* male attempts to mate, the triungulins attach to the male and hitch a ride to the next female that the male contacts. They then attach to the female which, in turn, carries them to its nest where triungulins feed on pollen and honey stores. A similar strategy is used by European beetle *Stenoria analis* triungulins, which feed on food stores in the nest of the solitary bee *Colletes hederae*. There is also evidence that *S. analis* triungulins release compounds mimicking the sex pheromones that trigger mating by *C. hederae* males (Vereecken and Mahé, 2007).

One of the best documented examples of aggressive non-social arthropod mimicry is that of bolas spiders that use the female sex pheromones of their lepidopteran hosts, mostly noctuids, to attract moth males. In their study on the moth *Tetanolita mynesalis*, i.e., the

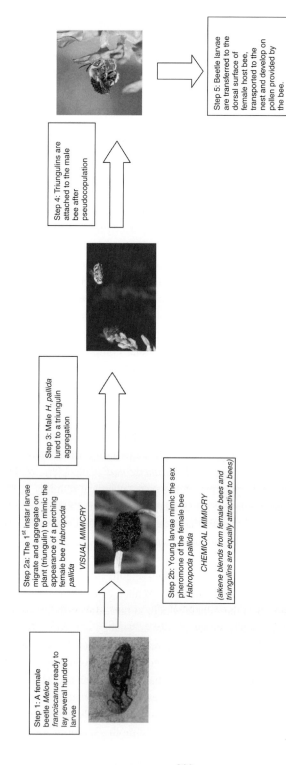

Step 1: A female beetle *Meloe franciscanus* ready to lay several hundred larvae

Step 2a: The 1st instar larvae migrate and aggregate on plant (triungulin) to mimic the appearance of a perching female bee *Habropoda pallida*

VISUAL MIMICRY

Step 2b: Young larvae mimic the sex pheromone of the female bee *Habropoda pallida*

CHEMICAL MIMICRY

(alkene blends from female bees and triungulins are equally attractive to bees)

Step 3: Male *H. pallida* lured to a triungulin aggregation

Step 4: Triungulins are attached to the male bee after pseudocopulation

Step 5: Beetle larvae are transferred to the dorsal surface of female host bee, transported to the nest and develop on pollen provided by the bee.

Figure 14.2 Sexual deception by a phoretic nest parasite (the blister beetle *M. franciscanus*) to resolve the problem of locating a scarce resource in its natural environment (from Hafernik and Saul-Gershenz, 2000 and Saul-Gershenz and Millar, 2006)(Credit photos: L. Saul-Gershenz).

prey of the most widely described bolas spider *Mastophora hutchinsoni*, Haynes *et al.* (1996) identified epoxyheneicosatrienes, thus providing the first description of long-chain oxygenated hydrocarbons used as sex pheromones. Other papers have shown that most of the female sex pheromones in other prey species consist of short-chain acetates and aldehydes (e.g. Gemeno *et al.*, 2000). In J. Millar's chapter (Chapter 18) of this book, we will see that other noctuids can have long-chain hydrocarbons and oxygenated derivatives such as heneicosadienes as well as their epoxy and diepoxy derivatives (e.g. Wakamura *et al.*, 2002).

Non-social arthropod–social insect interactions (Table 14.1)

The excellent book of Schmidt-Hempel (1998) describes the large variety of parasites that infest social insects. Various orders of insect (Lepidoptera, Diptera, Coleoptera, non-social Hymenoptera) and arachnid (Chelicerate Arachnids: Aranea, Acari) parasites use chemical mimicry to manipulate social insect hosts, i.e., ants, bees and termites. Strangely, our perusal of the literature found no references describing this type of chemical interaction in social wasps, even though they are often parasitized by various macroparasites as well as other social insects from the same family, e.g. polistine wasps (see later). However, we found many references describing chemical manipulation by parasites of social hymenoptera and isoptera.

Honeybee parasites

Honeybees have a wide variety of parasites, some of which use mimicry such as the *Acherontia atropos* moths described in the introduction. The honeybee parasites that make the most extensive use of hydrocarbon mimicry are mites. Phelan *et al.* (1991) used a two-choice bioassay to clearly demonstrate the role of hydrocarbon chemical mimicry in the tracheal mite *Acarapis woodi*. In that study, hydrocarbon blend fractionation showed that both the saturated and unsaturated fractions were active, but that the saturated fraction had a stronger effect. One of the most studied honeybee parasites is the acarian *Varroa destructor* (first described as *V. jacobsoni*). In a 1992 study using a servosphere, Rickli *et al.* showed that palmitic acid (one of the four compounds involved in *Acherontia* mimicry) was attractive to mites. Two years later, the same authors showed that the cuticular alkanes of honeybee larvae mediated development arrestment of the parasite (Rickli *et al.*, 1994). Y. LeConte and co-workers have published three reports describing the chemical interaction between *V. destructor* and *Apis mellifera*. The first report, by Salvy *et al.* (2001), described quantitative modifications of honeybee cuticular profiles at various developmental stages in the presence of ectoparasite mites. The basic qualitative make-up of the profiles did not change. The second report, by Martin *et al.* (2001), showed that the degree of chemical mimicry by the mite varied according to honeybee developmental stages. Mimicry was more extensive during the larval and pupal stages of the bee than during the emerging adult stage. The third report, by Martin *et al.* (2002), described a mechanism that honeybees may use to detect mites inside the brood

Table 14.1 *Social insect hosts with their various non-social parasites.*

Social insect	Non-social parasite	Study	References
Honey bee	Acari, tracheal mite (*Acarapi woodi*)	Usaturated alkanes involved in mite locomotion response	Phelan et al., 1991; Gary and Page, 1987
	Ectoparasite (*Varroa destructor*, initially described as *jacobsoni*)	Identical hydrocarbons	Nation *et al.*, 1992
		Larval acids attract *V.d.*	Rickli *et al.*, 1992
		Larval esters attract *V.d.*	LeConte *et al.*, 1989
		Larval alkanes attract *V.d.*	Rickli *et al.*, 1994
		Larval bee cuticular profiles differ in presence of *V.d.*	Salvy *et al.*, 2001
		Degree of mimicry depends of dvlpt stages	Martin *et al.*, 2001
		Bee detects Varroa's hydrocarbons inside brood cell	Martin *et al.*, 2002
	Solitary Hymenoptera: bee wolf (*Philanthus triangulum*)	Use of PPG alkanes to embalm bee	Herzner *et al.*, 2007; Strohm and Lisenmair, 2001
		Use of PPG alkanes as scent marking territory	Kroiss *et al.*, 2006; Strohm *et al.*, 2007
	Cukoo wasp (*Hedychrum rutilans*)	Mimicry of bee wolf's "clothing"	Strohm *et al.*, 2008
Termite	Coleoptera staphylidins	*De novo* biosynthesis of hydrocarbons by parasites	Howard *et al.*, 1980, 1982
Ant	Salticid spider (*Cosmophasis bitaeniata*)	Mimicry of prey mono- and dimethyl-alkanes	Allan and Elgar, 2001; Allan et al., 2002; Elgar and Allan, 2004, 2006
	Hymenoptera parasitic wasps *Orasema sp.*	CHC small passive transfers	Van der Meer *et al*, 1989

Kapala sp.	Dynamic mimicry	Howard *et al.*, 2001
Diverse wasps, aphids and ant associations	Various camouflage-based mimicry systems	Liepert and Dettner, 1996, 1996; Volkl and Machauer, 1993; Völk *et al.*, 1996; Akino and Yamaoka, 1998
Coleoptera	Mimicry by passive transfer	VanderMeer and Wojcik, 1982
Myrmecaphodius	Mimicry of ant alarm pheromones	Stoeffler *et al.*, 2007
Pella	Chemical camouflage	Akino, 2002
Zyras	Review	Geiselhardt *et al.*, 2007
Carabids	Aggressive invasion then acceptance by chemical congruency (feeding)	Dinter *et al.*, 2002; Liang and Silverman, 2000
Diptera: Syrphid fly (*Microdon mutabilis*)	Cocoon mimicry; biosynthesis and chemical mimicry; extreme host specificity; host recognition	Garnett *et al.*, 1985; Howard *et al.*, 1990; Elmes *et al.*, 1999; Schönrogge *et al.*, 2008
Guardian ants and Homoptera (aphid) association with Diptera and Neuroptera predators	Chemical convergence between aphid and predators	Lohman *et al.*, 2006
Lepidoptera: Maculinea sp. Caterpillar; *Feniseca* caterpillar; *Niphanda fusca*	Host adaptation and specificity	Akino *et al.*, 1999; Elmes *et al.*, 2002; Nash *et al.*, 2008
	Chemical camouflage	Youngsteadt and Devries, 2005
	Intracolonial chemical mimicry	Hojo et al., 2007 in Akino 2008
Cricket Myrmecophilus sp.	Chemical camouflage	Akino *et al.*, 1995

cell. Findings showed that cuticular substances allow discrimination between bees and other cell contents. A study in preparation by LeConte, Bagnères and Huang will present hydrocarbon variations between two *Varroa* haplotypes in the presence of *Apis mellifera* versus *A. cerana* (Bagnères *et al.*, 2006, congress presentation abstract).

As described above, female European beewolves (*Philanthus triangulum*) hunt foraging honeybee workers as the exclusive food for their progeny. After capture, the beewolf carries the paralyzed honeybee back to its subterranean nest where the prey is placed inside a brood cell. Before ovipositing and closing the brood cell, the beewolf embalms the prey copiously in a large amount of postpharyngeal gland (PPG) secretion that slows microbial growth and preserves the bee prey. This secretion is composed mainly of linear alkanes and alkenes similar to those in the beewolf's cuticular signature (Herzner *et al.*, 2007). In male beewolves, the PPG serves as a reservoir for the pheromone used for scent-marking mating territory (Kroiss *et al.*, 2006).

Termitophiles

Two reports by Howard *et al.* have described chemical mimicry between various Coleoptera staphylinids and their *Reticulitermes* termite hosts (Howard *et al.*, 1980, 1982). These papers triggered the still-ongoing controversy over the possibility of de novo synthesis of mimicking hydrocarbons by parasites. There is some evidence that de novo synthesis may occur in some cases (see above). Technical limits at that time, and the low available number of termitophiles, hindered further progress. "None have offered any cogent explanation of how we could achieve the results" according to R. W. Howard (personal communication). Other staphylinid termitophile references unfortunately did not review the chemical mimicry phenomenon (Kistner, 2000).

Myrmecophiles

Numerous papers have been published on mimicry by various ant predators including Coleoptera, Hymenoptera, Diptera, Orthoptera, Lepidoptera, and spiders. Allan, Elgar *et al.* (Allan *et al.*, 2002; Elgar and Allan, 2004, 2006) described qualitative mimicry of ant chemical profiles enabling the salticid spider *Cosmophasis bitaeniata* to enter *Oecophylla smaragdina* ant colonies and to feed on ant larvae without being detected by workers. Experiments indicate that the spiders acquire the ant cuticular hydrocarbons necessary to produce a colony-specific chemical signature by eating ant larvae. Behavioral experiments revealed that the spiders can distinguish between nestmate and non-nestmate major ant workers and are less inclined to escape when confined with ant nestmates (Elgar and Allan, 2006; Figure 14.3).

One of the earliest reports published on insect chemical mimicry (Vander Meer and Wojcik, 1982) involved *Myrmecaphodius excavaticolli* beetles (Coleoptera) and *Solenopsis* ants. Both are imported species, almost certainly introduced into North America at the same time. The authors stated that the myrmecophagous beetle acquired the chemicals used for

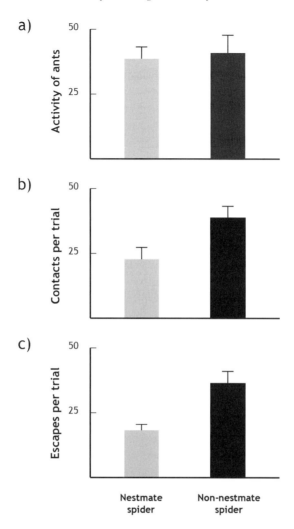

Figure 14.3 Behavior of *Cosmophasis bitaeniata* spiders confined with *Oecophylla smaragdina* major worker groups. Activity of ant groups was not different with nestmate or non-nestmate spider (a); spiders interacted less when ants were nestmate (b); spiders escaped less when contact was with nestmate ants (c) (from Elgar and Allan, 2006).

mimicry by passive transfer from the ants. Other ant-symbiotic beetles, e.g. Staphylinidae and Pselaphidae, are able to enter ant nests by matching host workers' hydrocarbon profiles. A recent report described chemical mimicry of ant alarm pheromones by two staphylinid *Pella* sp. (Stoeffler *et al.*, 2007). Similarly, a recent overview (Geiselhardt *et al.*, 2007) described the use of chemical mimicry by ant guest beetles (Paussinae) to avoid detection by host species belonging mainly to the Formicinae and Myrmicinae ant families. Further study will be needed to confirm congruency with ant hydrocarbon profiles.

The Carabids *Thermophilum* and *Graphipterus* prey on ants in the Sahara desert (Dinter *et al.*, 2002). *Thermophilum* larvae follow ant trails to the nests, which they aggressively invade by attacking workers. Shortly after starting to feed on ants inside the nest, they are accepted and can then move around freely inside the nest. Gas chromatograms performed after acceptance showed congruency between the hydrocarbon profiles of the carabids and some ant hosts (*Tapinoma simrothi* and *Lasius niger*). Liang and Silverman (2000) suggested that carabid larvae obtain ant hydrocarbons by feeding on the ants. Unlike *Thermophilum, Graphipterus* carabid larvae are not tolerated by ants, but there is no available data on surface profiles (Dinter *et al.*, 2002).

One *Myrmecophilus* cricket can invade the nests of several ant species. However, the only paper describing the chemical profiles of the participants involved in this mimicry system was published by Akino *et al.* (1996).

In Hymenopteran parasitic wasps, various camouflage-based mimicry systems have been described, including cuticular lipid mimicry between aphid parasitoid wasps (Aphidiidae), and aphids and their guardian ants (Liepert and Dettner, 1993, 1996; Völkl and Mackauer, 1993; Völkl *et al.,* 1996; Akino and Yamaoka, 1998). Other parasitoid wasps such as *Orasema sp.* (Eucharitidae) mimic the chemical profile of their fire ant hosts (Vander Meer *et al.*, 1989), and *Kapala* sp. (Eucharitidae) wasps are able to mimic the profiles of their ponerine ant hosts (Howard *et al.*, 2001). Lohman *et al.* (2006) compared cuticular hydrocarbons of the mimicry system comprising the North American wooly alder aphid *Prociphilus tessellatus* (Homoptera, Aphididae), their guardian ants, and aphid predators from three other insect orders, i.e., *Feniseca tarquinius* (Lepidoptera, Lycaenidae), *Chrysopa slossonae* (Neuroptera, Chrysopidae) and *Syrphus ribesii* (Diptera, Syrphidae). Results showed convergence of chemical mimicry, with each predatory species being more similar to their aphid prey than to the guardian ants (*Formica glacialis* and *Myrmecia incompleta*). These findings have two important implications about the use of cuticular hydrocarbons in this system. The first is that the guardian ants use hydrocarbons to discriminate trophobionts from potential prey. The second is that mimicking the aphid cuticular signature allows the trophobiont predators to avoid detection by both aphids and guardian ants. The same authors also stated that, although several features of the aphid CHC profile are shared among the chemically mimetic taxa, "the variation in the precision of mimicry among the members of this predatory guild demonstrates that a chemical mimic need not replicate every feature of its model."

By mimicking the ant cocoons that they prey on, some first and second larval instars of Diptera are able to trick worker ants into transporting them into the brood chambers of the nests. All myrmecophilus Diptera with this capability are syrphid flies from the genus *Microdon,* and are obligate parasites of *Camponotus* and *Formica* ants (Garnett *et al.*, 1985; Elmes *et al.*, 1999). Radio-labeling experiments using 14C-precursor indicated that the mimicking hydrocarbons are biosynthesized by the fly rather than acquired from the host, and that the biosynthetic process (especially biosynthesis of rare methyl-branched alkenes) is highly efficient (Howard *et al.*, 1990). In a recent behavioral bioassay, Schönrogge *et al.* (2008) showed that the extreme specialist social parasite *Microdon* also reacts to host volatiles by extending its ovipositors, a behavior which takes place in the complex multiphase strategy of host location.

Youngsteadt and DeVries (2005) described the putative role of chemical mimicry in a lycaenid caterpillar (Lepidoptera: Lycaenidae), i.e., the harvester *Feniseca tarquinius*, that preys on some ant-tended Homoptera (see Aphid section). Chemical analyses and behavioral bioassays indicated that aphid-like camouflage allowed the caterpillar to evade detection by the ants (genera *Camponotus* or *Formica*) without other concealment behavior. The authors stated that mimicry was not an active process in this case.

One of the most sophisticated examples of larval chemical mimicry in ant hosts involves *Maculinea* butterfly caterpillars (Lepidoptera, Lycaenidae) and *Myrmica* ants. The first report on hydrocarbon mimicry of *Myrmica schencki* by *Maculinea rebeli* larvae (Akino *et al.*, 1999) showed that ants tolerate butterfly caterpillar larvae in their brood chambers and continue to feed and tend butterfly caterpillars for months, sometimes in preference to their own offspring, which are then eaten by the caterpillar. In a subsequent report, Elmes *et al.* (2002) compared the cuticular profiles of 49 colonies of five *Myrmica* species (see Chapter 7 on chemotaxonomy) with those of *M. rebeli* before and after exposure to the ant species. Their results indicate that chemical mimicry is host-specific and raises the need for further investigation, since European *Myrmica* ants can also be parasitized by *Maculinea alcon, M. arion, M. teleius* and *M. nausithous*, which have not been studied (Elmes *et al.*, 2002). Hojo *et al.* (2008) showed that the chemical disguise of the caterpillar of the lyceanid butterfly *Niphanda fusca* is host caste specific (host species: the *Camponotus japonicus* ant); *N. fusca* caterpillars mimic larvae and males of the host to be accepted and cared for by the host workers. Another recent study, by Nash *et al.* (2008) (see also Nash and Boomsma, 2008), demonstrated that the speed at which the ants adopt butterfly larvae is directly correlated with the degree to which the butterfly larvae mimic the cuticular profiles of their *Myrmica rubra* and *M. ruginodis* hosts. On the basis of the geographic mosaic theory of coevolution (Thompson, 1994, 2005), it is found, as expected, that caterpillar parasites mimic the local chemical variations of their *Myrmica* hosts. Accordingly, convergence is stronger with the *M. rubra* host that exhibits stronger genetic differentiation among populations (with strong chemical divergence among populations) than with the *M. ruginodis* host that exhibits panmictic populations. As a result, the strategy used by *M. alcon* for *M. ruginodis* is less competitive. The authors have compared the *M. rubra* strategy to a continuing arms race with geographic mosaic hot spots. Thanks to the phylogenetic proximity of the two hosts, *M. alcon* is able to switch from *M. rubra* hotspots to *M. ruginodis* coldspots over time. This study provided clear evidence that chemical mimicry is a dynamic coevolutionary adaptive process in which outcome may differ according to the local strength of the reciprocal selection among the interacting organisms and, to a lesser degree, on the level of gene flow in their populations (Nash *et al*, 2008). To illustrate how complex this coevolutive model can be, it has been shown recently that sounds produced by the parasitic butterfly *Maculinea rebeli* mimic those of queen ants, enabling butterfly larvae to achieve high status within ant societies (Barbero *et al.*, 2009). The geographic mosaic theory of coevolution, correlating trait remixing and rapid chemical changes, can

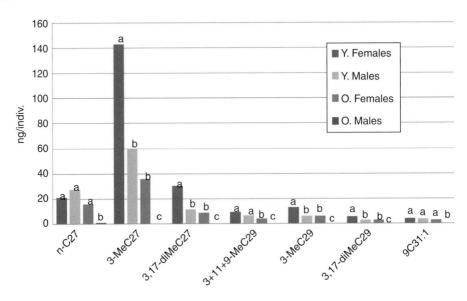

Figure 14.4 Comparative chemical analysis of selected hydrocarbon compounds correlated with biological activity shows similarity between freshly emerged (Y) males and females and 72-h-old (O) females *Lariophagus distinguendus* but not with 72-h-old (O) males (means with different lowercase letters are significantly different within each histogram at $P < 0.05$) (modified from Steiner *et al.*, 2005).

also explain the evolution of non-endogenous vs. native plant-herbivore selective response (Zangerl *et al.*, 2008).

Intraspecific mimicry in insects

Few data have been published on intraspecific mimicry, which is generally related to reproductive strategy. Two reports on the topic involve *Drosophila melanogaster* fruit flies (Scott, 1986) and *Lariophagus distinguendus* parasitic wasps (Steiner *et al.*, 2005). The paper by Steiner *et al.*, presented a novel case of a male mating strategy based on pheromone mimicry in this non-social wasp. Late newly emerged males use female pheromone mimicry to attract and thus distract early emerged males from mating with the females. Chemical analysis demonstrated that the active fraction was comprised mostly of cuticular hydrocarbons (Figure 14.4). This was also the first case of chemically mediated pre-emergent female mimicry in an insect to be described. A more recent paper (Ruther and Steiner, 2008) explains why *L. distinguendus* males get rid of the female odor a few hours after emergence: they are escaping from the aggression (or homosexual mounting) by their congenerous males, which would result in decreased mating chances. "This study provides evidence on how the pheromone function might have evolved secondarily from a primary function relevant for both genders" (J. Ruther, personal communication). A similar strategy involving chemical distraction of conspecific male rivals has been reported in vertebrates, e.g. the red-sided garter snake *(Thamnophis sirtalis parietalis)* (Mason and Crews, 1985). Chemical analysis showed that the mixtures produced by male snakes were composed of long- to very long-chain saturated and unsaturated ketones (LeMaster and Mason, 2002). Reproductive strategies using intraspecific mimicry have also been observed in social insects (see part 6 of the present chapter).

Artificially mixed social insect colonies

Various authors have tried to reproduce the mimicry/camouflage process by founding artificially mixed social insect colonies. Indeed, this easy-to-implement approach provides an excellent tool for elucidating the chemical mechanisms underlying odor recognition not only in ants (Bonavita-Cougourdan *et al.*, 1989; Vienne *et al.*, 1995; Hefetz, 2007), and in termites (Vauchot *et al.*, 1996, 1998), but also in cockroaches (Everaerts *et al.*, 1997). After the pioneering reports by Adele Fielde (1903) and Carlin and Hölldobler (1983), most studies on artificially mixed ant colonies were performed by C. Errard and colleagues, who demonstrated the importance of early conditioning in learning colony odor (Errard, 1986). Contrary to initial reports (Corbara and Errard, 1991), subsequent study showed that ants in mixed colonies of *Manica rubida* and *Formica selysi* perform frequent social behavior, such as trophallaxis and allogrooming (Vienne *et al.*, 1995). There is also evidence that even a low level of social interaction may be sufficient to modify the group odor (Millor *et al.*, 2006). This may explain why qualitative and quantitative blending of the chemical signatures of the two ant species can lead to good congruency in a relatively short time (Bagnères *et al.*, 1991). *Manica* ants acquired unsaturated compounds from *Formica* ants while *Formica* ants acquired the methyl-branched hydrocarbons from *Manica* ants. Surprisingly, the quantity of the non-specific hydrocarbons acquired or synthesized by the other species was higher than that of the normal species (Bagnères *et al.*, 1991). Development of congruent hydrocarbon patterns has also been observed in *Manica rubida* and *Myrmica rubra* heterospecific colonies raised with the queen of the opposite species. The new signature pattern depends on each mixed artificial nest type (Vienne *et al.*, 1990) as shown in termite mixed colonies (Vauchot *et al.*, 1998). This could be also explained by a saturation of the internalization turn-over (see Chapter 5) because of the non-specific nature of the compounds. Chemical analyses were carried out to determine the mechanism underlying cuticular lipid acquisition in mixed *F. selysi/M. rubida* ant colonies. The results demonstrated that transfer was greater than biosynthesis and that the relative proportion of each process differed between species (Vienne *et al.* 1995). Hefetz *et al.* (1992) showed that the post-pharyngeal gland acts as the reservoir of hydrocarbons transferred between ants. The same paper also provided the first evidence that the post-pharyngeal gland served as a Gestalt organ in the Formicidae. This finding was confirmed in subsequent experiments (Soroker *et al.*, 1994; Meskali *et al.*, 1995) (see Chapter 5 for review).

Overall, the experiments with artificial mixed colonies document that free-living species can adjust their chemical signature to social environment. Another interesting finding of studies on mimicry in artificially mixed colonies is the importance of evolutionary pressure.

Although passive transport and active regulation processes appear to play a major role in the acquisition of artificial signatures, experimental evidence of unaggressive behavior toward natural enemies nicely illustrates the complexity of chemical deception systems.

Interactions among social insects

Social parasites exploit the resources of other social species, namely their nests, food stores and workforce to rear their own brood, mainly or exclusively composed of fertile individuals. Social parasites do so by challenging the host recognition ability. Social insect colonies are often large social units and they may be composed of tens to thousands of individuals. Therefore, social parasites face the risk of being discriminated against at any encounter with any legitimate resident member of their host colonies. They incur the risk of being unmasked for days, weeks or even years after first infiltrating host colonies or after laying their eggs, which are then cared for by host workers. The selective pressures on social parasites to fool hosts are consequently extremely intense since they rely completely on their host colonies to reproduce. Lenoir *et al.* (2001) review the different mechanisms employed by ants to invade host nests. Eluding host recognition systems involves a variety of mechanisms depending primarily on the nestmate recognition systems of the hosts, but also on the ecology and life history of the species involved. Moreover, the mechanisms selected to elude host recognition vary among parasite castes as well, because the life histories of parasite queens, their sexuals or, when present, their workers, differ. Yet, the mechanisms used to overcome host detection may change with the physiological status of the parasite. Chemical mimicry of the cuticular hydrocarbons is possibly the most common chemical adaptation that has evolved in social parasites as a response to the hosts' ability to defend their colonies from intruders.

Interactions among social insects belonging to different orders

Chemical mimicry may occur across taxa of social insects as separated as ants and termites. Some species of ants prey on termites. *Anochetus emarginatus* ants patrol the outside of *Nasutitermes* nests and prey on termite individuals without triggering any attack by termite soldiers (Dejean *et al.*, 2007). Similarly, *Hypoponera eduardi* ants prey on termites of the European species *Reticulitermes santonensis* and *R. grassei* without stimulating aggressive behavior in the termites (Lemaire *et al.*, 1986). In both cases, chemical similarity of the cuticular profile may play a role in the lack of reaction against the predatory ants. Indeed, the cuticular hydrocarbon profiles of *A. emarginatus* ants partially mirror those of *Nasutitermes* soldiers (Dejean, 1988), and *H. eduardi* ants share cuticular hydrocarbons with the *Reticulitermes* termites they prey upon (Lemaire *et al.*, 1986). The chemical ecology of these predator/prey systems needs more detailed analyses (see also Dettner and Liepert, 1994), but both examples suggest that among social insects neither parasite/host relationships nor close phylogenetic distances are necessary for the evolution of chemical mimicry of cuticular profiles.

Interactions among social insects belonging to different genera

The life history of slave-making ants is strictly associated with their slaves. Furthermore, large variations in life history exist among castes of slave-making ants (i.e., among queens, workers and sexuals). As we will document in the next paragraphs, the differences in life history may cause a divergence among castes in the chemical adaptation for fooling hosts (Table 14.2).

Queens of the slave-making ants depend on host species for nest foundation. Newly mated queens usurp host colonies and replace the resident queens. Parasitic queens may obtain immediate care by resident host workers. In the species where this occurs, we might expect intense selective pressures in favor of fast acquisition of recognition cues that mimic their hosts.

The slave-making *Polyergus* ants represent a well studied example. When they enter their *Formica* host colonies, *Polyergus rufescens* queens are first attacked and later tolerated by host workers. Chemical analyses of cuticular lipids explain host behavior. Indeed, *P. rufescens* queens have almost no detectable amount of cuticular hydrocarbons when they enter host colonies. This mechanism is named "chemical insignificance" after Lenoir *et al.* (2001) first noticed that the lack of recognition cues was common among social parasites; and dose-dependent responses have been recently reported in a social wasp species (Cini *et al.*, 2009). However, although chemical neutrality possibly helps the slave-making queens to enter host nests unnoticed and quickly acquire the host profile (e.g. by physical contact during the killing of the resident queen), chemical neutrality may not be sufficient to entirely elude the host recognition barrier. Indeed, laboratory observations show that host workers commonly attack invading queens. Parasite queens limit host attacks by secreting chemicals from their Dufour glands, and indeed, after the first qualitative analyses by Bergström and Löfqvist (1968), Dufour gland secretions are viewed as appeasement substances (Topoff *et al.*, 1988; Mori *et al.*, 1995, 2000) or repellents (D'Ettorre *et al.*, 2000). However, immediately after the killing of the host queen, the cuticular hydrocarbon patterns of *P. rufescens* queens change abruptly and mimic those of *Formica cunicularia* host queens (D'Ettorre and Errard, 1998). Similar results, supporting chemical mimicry, were obtained for *P. breviceps* queens and its two *Formica* hosts: newly mated parasite queen profiles transformed dramatically soon after the fatal attack on the host queens, suggesting an effective chemical transfer from the queens of the species killed (with a marked enrichment in branched hydrocarbons, Johnson *et al.*, 2001).

Many more investigations involved *P. rufescens* workers and, interestingly, the results are not coherent. *Polyergus* queens produce workers that raid neighboring host colonies and enslave ants of the host species by stealing host brood from unparasitized colonies. As for recognition cues, we expect a milder selection for chemical mimicry of host recognitions cues on *Polyergus* workers than on *Polyergus* queens. Indeed, slave-making queens are selected to gain acceptance and care by non-enslaved, resident-host workers soon after they infiltrate nests of the host species. In contrast, slave-making workers (as well as their immature stages) are cared for by enslaved host workers. Typically, in ants, workers learn colony odor at eclosion and accept or reject individuals depending upon whether or not their recognition cues match the learned template (e.g., Isingrini *et al.*, 1985; Crozier and

Table 14.2 *The degree of matching of cuticular hydrocarbons with their hosts in social parasitic queens and their offspring.*

Type of parasitic relationship	Parasite/predator species	Epicuticular hydrocarbons in invading parasite queen	Epicuticular hydrocarbons in sterile parasite offspring (adult offspring, if not otherwise stated)	Epicuticular hydrocarbons in fertile parasite offspring (adult offspring, if not otherwise stated)	References
Slave-making ants	*Polyergus rufescens* P. rufescens and P. breviceps	At invasion: chemical neutrality but no chemical mimicry with hosts Soon after host queen killing: chemical mimicry with hosts			Topoff *et al.*, 1988; Mori *et al.*, 1995, 2000; D'Ettorre *et al.*, 2000; D'Ettorre and Errard, 1998; Johnson *et al.*, 2001; Bonavita-Cougourdan, *et al.* 1996
	Polyergus rufescens		Eggs and pupae: no chemical mimicry with rearing hosts Adults: chemical mimicry with rearing hosts		Johnson *et al.*, 2005; D'Ettorre *et al.*, 2002 (but limited matching if mature workers are switched, see text, Bonavita-Cougourdan *et al.*, 1997)
	Harpagoxenus sublaevis		Pupae: no chemical mimicry with rearing hosts Adults: chemical mimicry with rearing hosts		Kaib *et al.*, 1993
	Rossomyrmex minuchae		Larger matching with sympatric than allopatric hosts, but never perfect		Errard *et al.*, 2006b
	Protomagnathus americanus		Larger matching with sympatric than allopatric hosts and within single rather than multiple-host populations		Brandt *et al.*, 2005

Category	Species	Description	Reference
	Chalepoxenus muellerianus	Incomplete matching with rearing species	Beibl et al., 2007
Xenobiotic ants	Formicoxenus nitidulus	No chemical with hosts (host specificity: low)	Lenoir et al., 2001; Martin et al., 2007
	F. quebescensis and F. provancheri	Chemical mimicry with hosts (host specificity: high)	Lenoir et al., 1997
Parabiotic ant associations	Camponotus morosus-Solenopsis gayi;	No chemical mimicry with hosts	Errard et al., 2003
Inquiline bees	Bombus silvestris	Chemical mimicry with non-host species (lab experiments)	Dronnet et al., 2005
Temporary parasitic association	Melipona rufiventris	No chemical mimicry with hosts (biology of the association is poorly known)	Pianaro et al., 2007
Cuckoo ants	Leptothorax kutteri	At invasion: no chemical mimicry with hosts? Soon after host colony invasion: perfect matching with hosts	Allies et al., 1986; Franks et al., 1990
Temporary ant association	Lasius spp.	No chemical mimicry with hosts	Liu et al., 2000
Inquiline ants	Acromyrmex insinuator	No chemical mimicry with hosts	Lambardi et al., 2007
Inquiline wasps	Polistes semenowi	Perfect matching with hosts after host colony invasion	Lorenzi et al., 2004a

(continued)

Table 14.2 (cont.)

Type of parasitic relationship	Parasite/predator species	Epicuticular hydrocarbons in invading parasite queen	Epicuticular hydrocarbons in sterile parasite offspring (adult offspring, if not otherwise stated)	Epicuticular hydrocarbons in fertile parasite offspring (adult offspring, if not otherwise stated)	References
Inquiline wasps	*Polistes sulcifer*	Perfect matching with hosts after host colony invasion	Larvae: no chemical mimicry with rearing species		Turillazzi et al., 2000; Sledge et al., 2001; Cervo et al., 2008
Inquiline wasps	*Polistes atrimandibularis*	Perfect matching with hosts after host colony invasion		No chemical mimicry	Bagnères et al., 1996
Inquiline wasps	*Vespa dibowskii*			No chemical mimicry (both of eggs and adults)	Martin et al., 2008
Facultative parasitic ants	*Linepithema humile*	Matching with adopting colonies (intraspecific usurpation)			Vásquez et al., 2008
Cleptobiotic ants	*Ectatomma ruidum*		Thief ants matching with target colonies		Jeral et al., 1997
Facultative parasitic wasps	*Polistes nimphus*	Matching with adopting colonies in intraspecific usurpation; incomplete matching in interspecific usurpation			Lorenzi et al., in prep.
Facultative parasitic wasps	*Polistes biglumis*	Incomplete matching with adopting colonies in intraspecific usurpation			Lorenzi et al., in prep.

Pamilo, 1996; although template and acceptance thresholds adjust with time, see D'Ettorre *et al.*, 2004a; Brandt *et al.*, 2005; Errard *et al.*, 2006a). Enslaved host workers eclose in parasite colonies, and thus their template resembles the parasite odor, a condition which may let them accept parasites. Indeed, *P. breviceps* immature stages (eggs and pupae) possess distinct hydrocarbon profiles from their host species, but are cared for by *Formica* workers (Johnson *et al.*, 2005) and, similarly, pupal hydrocarbons of *Harpagoxenus sublaevis* do not depend on their host species (Kaib *et al.*, 1993).

However, mature parasite workers also face non-enslaved host workers when, during summer, they capture larvae and pupae from colonies of another species. During raids, resident workers attack parasites in an attempt to defend their colonies. At this stage, parasite workers are selected for avoiding attacks. Such selection may result in chemical similarity with the host species, if similarity in recognition cues facilitates parasite performance during raids by diminishing attacks by the target colonies. Indeed, examples exist where the hydrocarbon profiles of parasite workers of slave-making ants match those of host workers. *Polyergus rufescens* parasite workers emerging in laboratory host colonies possess cuticular hydrocarbon profiles that perfectly matched that of the rearing species, regardless of the rearing species being a host or non-host *Formica* species (D'Ettorre *et al.*, 2002). It is worth noting that in the latter experiment cocoons, not workers, were switched between hosts. These details may help to understand why the degree of matching between host and parasite chemical profiles in the experiment by D'Ettorre *et al.* (2002) was larger than that found by Bonavita-Cougourdan *et al.* (1997) in the same species. In their cross-infectivity lab experiments, Bonavita-Cougourdan and co-workers found that the chemical profiles of *P. rufescens* mature workers showed some adjustments, but "no qualitative homogenization" when the workers were adopted by the alternative host species coming from a different area (allopatric). In social insects, the plasticity of the cuticle hydrocarbon blend declines with age (Lorenzi *et al.*, 2004b). Accordingly, D'Ettorre *et al.* (2002) switched newly emerged parasite workers and found a larger degree of hydrocarbon profile matching with the social environment than was found by Bonavita-Cougourdan *et al.* (1997), who switched mature workers and found limited adjustments. Other factors, however, may be responsible for the different conclusions between the two studies. Indeed, the chemical profile of *P. rufescens* workers collected in the field was distinct from that of both *F. rufibarbis* and *F. cunicularia* hosts (Habersetzer and Bonavita-Cougourdan, 1993; Bonavita-Cougourdan *et al.*, 1996). A subsequent paper by Bonavita-Cougourdan *et al.* (2004) confirmed the existence of qualitative differences between the cuticular blend of *P. rufescens* workers (which contained alkenes) and that of its host *F. rufibarbis* (which was free of alkenes). Contrasting results were also obtained in two chemical ecology studies on another *Polyergus* species, the slave-maker ant *P. samurai* and its host *F. japonica*. Liu *et al.* (2003) found that parasite worker cuticular profiles do not match the host, whereas preliminary results by Yamaoka (1990) on the same interacting species gave evidence of chemical mimicry.

Identifying the reasons for these contrasting results goes beyond the scope of this review. We can only address some possible causes. Beyond the obvious effects of differences related to the analytical method (for example see Liu *et al.*, 2003), we have to take into account that chemical mimicry is the result of a coadaptation process that varies geographically and over time (Nash *et al.*, 2008). The reviewed studies on the *Polyergus/Formica* associations involved parasite populations which could differ in the degree of matching of their host hydrocarbon profiles because of the ongoing evolutionary arms races between social parasites and their hosts. During this process, parasites are expected to evolve better mimicry solely in response to their hosts' enhanced abilities to detect them. Thus, chemical mimicry, as a mutual coadaptation, is restricted to sites of intense and lasting interactions, whereas parasites and hosts may evolve independently in other populations (Nash *et al.*, 2008 for an example involving cuticular hydrocarbons; Thompson, 2005, for the general, theoretical framework).

Hydrocarbon profile match and mismatch have been reported in other host/social parasite systems. The workers of the slave-making ants *Harpagoxenus sublaevis* show the same profile as their hosts, e.g., that of *L. acervorum* if *L. acervorum* was serving as host, and that of *L. muscorum* if *L. muscorum* was parasitized (Kaib *et al.*, 1993). In contrast, the resemblance of the cuticular hydrocarbon profiles between the workers of the slave-making ant *Rossomyrmex minuchae* and its host *Proformica longiseta* is not perfect, and although the parasite worker hydrocarbon patterns qualitatively resemble those of their hosts, quantitative differences are large (Errard *et al.*, 2006b). However, as expected under the geographic mosaic models of coevolution (Thompson, 2005), Errard and co-workers also showed that the parasite chemical signature is more similar in coadapted (sympatric) populations than in allopatric populations. The matching between the chemical signature of the slave-making workers of *Protomagnathus americanus* and its two *Temnothorax* hosts is imperfect as well. Brandt *et al.* (2005) provide evidence that parasite workers exhibit chemical mimicry of their host species' cuticular profile. Moreover, those profiles are locally adapted to different degrees in particular local hosts. Parasite hydrocarbon profiles from populations where the parasite specializes on a single host match their hosts' profile more closely than those from multiple-host populations (and, as expected, are more successful in raiding, Brandt and Foitzik, 2004). Moreover, chemical profiles of parasite workers belonging to different populations, but still using the same host species, are more similar to each other than those of sympatric parasite workers using different host species. As an aid during raids, *P. americanus* worker ants spray Dufour's gland secretions (which also contain light hydrocarbons) as propaganda substances, provoking confusion and agitation among their hosts (Brandt *et al.*, 2006). Overall, the interaction between this parasite species and its hosts appears to be evolving as a geographic mosaic that depends upon the local combination of available host species.

Selection pressures on cuticular chemical resemblance between hosts and parasites are expected to be even weaker when parasite sexuals are taken into account. Indeed, sexuals are cared for by enslaved host workers and live inside the natal colonies until the mating flight. As expected, in *Chalepoxenus muellerianus* (a slave making ant which enslaves multiple hosts of the related *Temnothorax* species), the complex hydrocarbon profiles of sexuals depend on the host-rearing species, with a bias towards the host species used by each parasite population, but with differences among the chemical signatures of parasite sexuals and workers of the host species (Beibl *et al.*, 2007). It is obvious, from the example reported above, that the degree of hydrocarbon matching largely depends on the life history, not only of the parasites and their hosts, but also of their castes.

Evolution seems to have taken different courses among xenobiotic ants, who live independently inside the nests of the host ants. The colonies of the European ant *Formicoxenus nitidulus* may be found in the nests of 9 different *Formica* host species, as well as in *Polyergus rufescens* and in *Tetramorium caespitum* nests (Martin *et al.*, 2007 and references therein). As expected under the hypothesis that mutual coadaptation is restricted to populations where intense and lasting interactions exist among interacting species, *F. nitidulus* cuticular hydrocarbons do not mimic those of their *Formica* host (Lenoir *et al.*, 2001); rather, their species-specific hydrocarbon blend acts as a repellent on hosts (Martin *et al.*, 2007). Because of the multiple potential parasite/host combinations allowing parasites to switch hosts and hosts to be under weak selective pressures to avoid parasitism, the European *Formicoxenus* ants may not be under strong selection to evolve chemical mimicry. In contrast, in the congeneric north-American xenobiotic ants *F. quebescensis* and *F. provancheri*, or shampoo ants, the chemical profiles of parasite workers perfectly match those of their *Myrmica* hosts (Lenoir *et al.*, 1997), possibly via passive acquisition of chemicals by relentless licking of host cuticle. In these systems, the higher degree of host-specificity of *F. quebescensis* and *F. provancheri* probably favors the mutual coadaptation between hosts and parasites.

The lack of intense interactions may also explain the reciprocal tolerance and the lack of chemical mimicry of the cuticular profiles in the occasional parabiotic associations between *Camponotus morosus* and *Solenopsis gayi* or *Crematogaster–Camponotus* (Errard *et al.*, 2003; Menzel *et al.*, 2008).

Interactions among social insects within the same genus

Assuming that the closer the phylogenetical relationships the closer the biochemical pathways will be, one might expect host–parasite phylogenetical relatedness to correlate with the precision of cuticular profile matching. In the following lines we take into account the results of investigations on the epicuticular hydrocarbon patterns among systems where host and parasite belong to the same genus. Contrary to the expectations that a recent common ancestry between host and parasite may grant the sharing of biochemical pathways and therefore of chemical recognition cues, the currently available data in the literature do not consistently confirm this. In fact, examples are reported where parasite–host systems with close, phylogenetical relationships intragenra between close and phylogenetical may or may not show chemical mimicry (see below). These observations suggest that both gain and loss of chemical mimicry may be under strong natural selection.

In bumblebees, social parasites and hosts belong to the same genus, although inquilines constitute "a distinct and monophyletic clade within the group of *Bombus*, only partly comprising the corresponding host species" (Dronnet *et al.*, 2005). The chemical ecology of parasite–host relationships in the bumblebee is largely unknown, making generalizations difficult. *B. norvegicus* parasites may enter host nests spraying repellents (Zimma *et al.*, 2003), but social integration in host colonies may be achieved by chemical mimicry of the host cuticular chemical profile. Indeed, the cuticular profile of *B. silvestris* parasite queens chemically mimicked that of a non-host species (*B. terrestris*) in laboratory mixed colonies (Dronnet *et al.*, 2005). The tested parasitic queens attained chemical mimicry of the host chemical profile in a time period as short as one day, possibly via passive acquisition of host hydrocarbons, facilitated by the initial paucity of recognition cues in the parasite queen's chemical signature. Zimma (2002, in Zimma *et al.*, 2003) suggests that the cuticular signature of *B. norvegicus*, an obligate parasite of *B. hypnorum*, may mimic host colony odor as well. In stingless bees, where the chemical ecology of social parasite/host systems has been rarely studied, Pianaro *et al.* (2007) document that the invasion of a nest of *Melipona scutellaris* by workers of *M. rufiventris* is followed by a larger divergence among the wax chemical profiles of the two bee species rather than by homogenization.

The cuckoo ant *Leptothorax kutteri* enters the nests of the congeneric *L. acervorum* host, ameliorates the severe resident attack by secreting propaganda substances from the Dufour's gland (Allies *et al.*, 1986), and comes to be accepted by resident hosts. Chemical analysis of the cuticular compounds of post-invasion *L. kutteri* queens showed that they match their host workers perfectly, possibly acquiring host compounds by extensive grooming of the host queens (Franks *et al.*, 1990). Quite the opposite results are reported by Liu *et al.* (2000), who describe preliminary data on striking chemical differences between a parasitic *Lasius* ant and its *L. fuliginosus* host occuring during a temporary association.

Similar results were recently obtained by Lambardi *et al.* (2007), who failed to find that the cuticular surface chemistry of *Acromyrmex insinuator* workers mimics that of the host *Acromyrmex echinatior*. They discovered that social parasite workers have a reduced quantity of chemicals on their cuticle, a fact which supports the chemical insignificance hypothesis. Furthermore they found that the parasite hydrocarbon profile largely differs from that of the host in being richer than that of the host in long-chain unsaturated compounds (C_{43}–C_{45}). Such a result is particularly surprising because *A. insinuator* is a recent, incipient parasite, and a sister species of its host (Schultz *et al.*, 1998; Sumner *et al.*, 2004). It will be useful to understand which selective pressures have favored divergence in the chemical signature of the workers of the two species.

Polistes social parasites are not their host's closest relatives. They form a monophyletic clade (Choudhary *et al.*, 1994; Carpenter, 1997) and have different degrees of host specificity. Multiple host species have been reported in *P. atrimandibularis,* for which no host-specific genetic lineages exist (Fanelli *et al.*, 2005), and also in *P. semenowi*, whereas only a single host species has been identified in *P. sulcifer* (Cervo and Dani, 1996; Carpenter, 1997). The recognition cues of free-living species of paper wasps are complex blends of more than 60 compounds, mainly straight-chain and branched hydrocarbons, generally ranging from 23 to about 40 carbon atoms. Notwithstanding the complexity of the chemical signature of their hosts, all three workerless parasitic species of *Polistes* wasps have evolved complex chemical strategies to overcome host recognition barriers (reviewed in Lorenzi, 2006). These strategies include mimicry of the host colony chemical signatures, but differ in many respects among species. In the following paragraphs we will use the term "queens" for brevity to mean the social parasite females which actually invade the host nest, or to mean the host-nest foundresses. We do this in order to easily distinguish them from their female sexual offspring. Usually the word "queen" is avoided in the literature dealing with *Polistes* wasps because queens are not morphologically distinct from workers (see Reeve, 1991).

In the *P. semenowi* queen, the blend of cuticular hydrocarbons is qualitatively poor in compounds during the pre-invasion hibernation period and lacks any parasite-specific compounds. As a result, the chemical profile of this social parasite is not too dissimilar from that of its *P. dominulus* host even before host-nest invasion, because it shares most of its host's compounds (including large proportions of branched alkanes). Two weeks after host-nest invasion, the parasite-queen chemical profile is enriched in quantity and in quality of hydrocarbons (with a relevant increase in long-chain and in straight-chain hydrocarbons) so that it perfectly matches the host profile (Lorenzi *et al.*, 2004a).

P. dominulus is the host species of *P. sulcifer* as well. Although they share a host species, the chemical signature of the *P. sulcifer* parasite queen undergoes larger adjustments than that of the *P. semenowi* queen during integration in the host colonies. Indeed, the pre-invasion chemical profiles of the *P. sulcifer* parasite queen are distinct from those of their *P. dominulus* hosts, but change dramatically soon after invasion. Some variations in the hydrocarbon blend occur within 90 min after entering the host nests and perfect mimicry of the host chemical signature is reached as soon as three days after invasion (Turillazzi *et al.*, 2000).

The host–parasite system *P. atrimandibularis–P. biglumis* is a special and instructive example of chemical mimicry of cuticular hydrocarbons (Bagnères *et al.*, 1996). When they abandon hibernation sites, parasite queens possess chemical signatures which are dramatically different from their hosts. The pre-invasion signature is extremely diluted (Lorenzi and Bagnères, 2002), but extremely rich in alkenes, a property that makes the chemical profile of *P. atrimandibularis* diverge not only from its host but also from any other *Polistes* species, including the other two parasitic species. Notwithstanding the chemical insignificance of the recognition signature of the parasite queen, the single, resident host foundress initially attacks the invading parasite fiercely and then submits to her (Cervo *et al.*, 1990). After the invasion of the host nest, a slow but effective transformation drives the parasite queen's cuticular signature to mimic chemically its *P. biglumis* host (Bagnères *et al.*, 1996). In approximately a month, the parasite-specific alkenes disappear from the parasite queen's cuticular profile and the host-specific compounds appear. However, the extent and the timing of variations differ among the parasite-specific compounds. For example, the heptacosene isomers are dominant compounds in the parasite-queen pre-invasion signature (30% of the whole blend). They completely disappear from the cuticular blend in less than one week after the parasite queen invades the host nest, while other unsaturated species-specific compounds still remain.

When host offspring begin to emerge, the whole profile of the parasite queen is chemically indistinguishable from that of her hosts, and all parasite-specific compounds have disappeared. The parasite-specific blend of the heptacosene isomers appears again in the chemical signature of the parasite queen at the end of the colony activity period, when the queen gradually ceases egg laying and the annual colony approaches its abortion period.

Whereas the cuticular chemical profile of the parasite queen undergoes all these changes during the colony cycle, that of her *P. biglumis* hosts does not change at all. In this respect, the *P. atrimandibularis–P. biglumis* system exemplifies a difference which is common among the properties of the epicuticular chemistry of hosts and their social parasites. The chemical signature of the parasite queen is plastic and adjusts to changing social and physiological conditions: entering the host nest, facing the emergence of host workers or later, of its own offspring, starting or stopping egg laying. That plasticity does not exist in the *P. biglumis* host. Hosts keep their own chemical signature throughout the colony cycle, irrespective of parasite invasion. Indeed, the extent of plasticity in hydrocarbon profiles of mature wasps in free-living species is generally limited, with major variations occurring only soon after emergence (Lorenzi *et al.*, 2004a, but see Dapporto *et al.*, 2005).

Host recognition abilities are impaired in parasitized colonies (e.g., Lorenzi, 2003 and references therein), but parasite eggs, larvae, pupae and adult offspring may face the challenge of avoiding host recognition. In contrast, investigations failed to find any chemical resemblance between parasite offspring (whatever their developmental stage) and host hydrocarbon profiles. The larvae of the obligate social parasite *P. sulcifer* are cared for by *P. dominulus* host workers (Cervo *et al.*, 2004) and do not mimic host larvae, nor exhibit colony specific profiles or reduced recognition cues (Cervo *et al.*, 2008). The cuticular chemical profile of the adult parasite offspring (which spend a few weeks in their natal host colonies) do not match the hosts, and parasite-specific unsaturated compounds are present on both female and male sexual *P. atrimandibularis* epicuticular layers (Bagnères *et al.*, 1996; Lorenzi *et al.*, 1996). At first view, these results suggest that only the parasitic queens have attained chemical mimicry as an adaptation to elude host recognition. However, in social wasps, recognition processes are mediated by the odor of the nest paper, which in turn is covered by hydrocarbons produced by colony members. The odor of the nest is used as a reference by wasps when they learn their colony odor and form a recognition template at emergence (Gamboa, 1996). In mature colonies parasitized by *P. atrimandibularis*, nest paper incorporates parasite-specific unsaturated compounds (Lorenzi *et al.*, 1996). Such a process possibly makes hosts learn the "parasite signature" as their own colony signature. As a result, when the immature parasite brood develops inside the colony, "colony odor" comes to be the parasite odor, not the host odor. By mimicking nest-paper odor, wasp social parasites can socially integrate. By doing that, *Polistes* social parasite queens initially change their chemical signature to a perfect "host signature", but later revert to a "parasite signature", tracking the changes in nest odor (Bagnères *et al.*, 1996).

Comparing the strategies of the three *Polistes* social parasites, it appears that, depending on the species, the pre-invasion chemical profiles of the invading queens may be largely similar to, or quite different from, that of the hosts, but each parasite queen chemically mimics her host at some point during social integration (and here again, the timing of chemical integration largely differs among species). However, in all three social parasite species, queens mimic small details of the host colony signature so that they are indistinguishable from their respective colony, their hydrocarbon profiles perfectly matching those of their host nestmates (as chemical and/or behavioral analyses document, Sledge *et al.*, 2001; Lorenzi, 2003; Lorenzi *et al.*, 2004a).

Although nestmate recognition processes are poorly studied in the genus *Vespa*, it is reasonable to assume that similar mechanisms regulate recognition processes. Indeed, neither the eggs nor the adult female offspring (gynes) of the parasitic *Vespa dibowskii* are chemically mimetic, nor are they insignificant, with either their rare (*Vespa crabro*) or common (*Vespa simillima*) hosts (Martin *et al.*, 2008). Interestingly, branched hydrocarbons constitute only traces of *V. dibowskii* epicuticular layer.

Interaction among social insects within the same species

Chemical strategies are also important in interactions among members of the same species. The adoption of *Linepithema humile* queens in alien conspecific colonies is facilitated by cuticular chemical similarity with adopting colonies (Vásquez *et al.*, 2008). In the tropical ponerine ant *Ectatomma ruidum,* ants can infiltrate conspecific nests where they steal food that they carry to their own colonies (a kind of parasitism called cleptobiosis). Breed *et al.* (1992) documented that individual thief ants were each specialized in robbing specific colonies, and Jeral *et al.* (1997) showed that thief ants possess a specialized cuticular profile with reduced amounts of cuticular hydrocarbons and mimicry of the cuticular pattern of the target colony. This thus suggests that sneaking into target colonies depends both on inhibition of the thief's own colony cues, and on acquisition of the target colony cues. As a rare example of chemical mimicry in contexts different from the parasite/host adaptations, Cremer *et al.* (2002) reported that the hydrocarbon cuticular blend of newly emerged males of the tropical ant *Cardiocondyla obscurior* matches that of virgin queens. By mimicking virgin females, males avoid attacks by a different and aggressive male morphs and increase their mating success.

To our knowledge there has been no investigation of chemical strategies thoroughly focusing on intraspecific parasitism, although the latter is likely ubiquitous among social insect colonies (e.g., Cervo and Dani, 1996). However, recent papers illustrate that chemical mimicry may play a role in conspecific nest usurpation. *Apis mellifera capensis* workers invade colonies of *A. m. scutellata* and take over as queens (Dietemann *et al.*, 2006, 2007). Sole *et al.* (2002) analyzed the Dufour gland secretions in the two species, and documented that in both lineages the glandular secretion (mainly composed of hydrocarbons) of laying workers mimicked the queen's profile. It would be interesting to understand whether this is the result of an arms race leading to a manipulation of recognition cues, or a correlate of the reproductive physiology of the species (i.e., a fertility signal; see for example D'Ettorre *et al.*, 2004b, in ponerine ants).

In the free-living *Polistes nimphus* wasps, behavioral bioassays on manipulated colonies have documented that queens that behave as social parasites and usurp conspecific colonies effectively mimic the host-nest odor and become accepted by resident workers (Lorenzi *et al.*, 2007). The process of concealing identity to host residents by *P. nimphus* usurpers occurs in only 4 days and might involve the absorption of chemicals from the host colonies. The process is so effective that host workers recognize and accept the conspecific usurpers as nestmates the first time they meet them. Conversely, in experiments simulating interspecific usurpations, *P. nimphus* females were unable to mimic the odor of the host colonies, but rather, they deposited their own cuticular blend on the host combs (Lorenzi *et al.*, 2007, for behavioral evidence; Lorenzi *et al.*, in preparation, for chemical evidence). Similar experiments have been performed in the free-living species *P. biglumis*, coupling behavioral tests and chemical analyses. Compared to *P. nimphus*, *P. biglumis* usurpers are less successful in triggering tolerance by host workers, and their chemical profiles are not a consistent match to the host colony odor (Lorenzi *et al.*, in preparation).

Concluding remarks

The examples reported in this chapter point to the independent evolution of chemical mimicry among a broad range of taxa (Tables 14.1 and 14.2). Although it is difficult to test whether chemical mimicry supplies an effective tool to escape recognition systems, in a vast majority of cases in which two unrelated species are locally involved in intense and lasting interactions, a degree of chemical mimicry of the cuticular hydrocarbons is found. Nash's group (Nash *et al.*, 2008) showed that the greater the resemblance in surface chemistry of parasitic butterflies and their host ants, the faster the ants adopt the parasitic caterpillars. The correlation between a strict biological interaction among organisms and chemical mimicry corroborates the hypothesis that, in many insect species, cuticular hydrocarbons do inform organisms about the identity of the encountered individuals, and that chemical mimicry is an effective strategy for manipulating recognition cues and escaping host rejection or attracting hosts. Whereas parasites may use a combination of strategies to manipulate host behavior, they often rely on chemical mimicry.

Our overview also indicates that phylogenetic relationships do not constitute strict barriers for attaining similarity in cuticular chemicals (see Chapter 7 for the use of hybrocarbon profiles as chemotaxonomic tools). Chemicals can be transferred among individuals, even across high-level taxa, through social contacts or contacts with nest walls, with subsequent adsorption of allochemicals (camouflage *sensu* Dettner and Liepert, 1994) (e.g., Vander Meer and Wojcik, 1982; Yamaoka, 1990; Kaib *et al.*, 1993), or may be produced after biosynthetic changes (mimicry *sensu* Dettner and Liepert, 1994) (e.g., Howard *et al.*, 1980, 1982, 1990; Bagnères *et al.*, 1991; Bonavita-Cougourdan *et al.*, 2004), or both (Bagnères *et al.*, 1996; D'Ettorre *et al.*, 2002) (see also Chapter 5). The adsorption of hydrocarbons may contribute to the acquisition of host colony compounds by the social parasite: indeed, experimental evidence suggests that naturally occurring hydrocarbons may be acquired from the environment (in laboratory tests, Lorenzi *et al.*, 2004b). Parasites often exhibit behaviors which involve physical contacts with their hosts, which may in turn give rise to the transfer of cuticular chemicals (by parasite licking, grooming or stroking, e.g., Cervo and Dani, 1996 in wasps; Franks *et al.*, 1990; Errard *et al.*, 1997 in ants, with for example the spectacular shampooing behavior of *Formicoxenus* ants). However, variations in the biochemical pathway responsible for hydrocarbon production, such as inhibition or production of certain compounds and/ or families of compounds, have been invoked as well (Bagnères *et al.*, 1996; Bonavita-Cougourdan *et al.*, 2004). Recently, the study of the chemical integration strategies in parasites which have multiple hosts has shown that parasite chemical profiles are both locally adapted to the primary host and adjust to the rearing species (D'Ettorre *et al.*, 2002). At a larger geographical scale, recent analyses have revealed a coevolutionary arms race in the degree of hydrocarbon-profile matching among populations, and have shown that different parasite populations may undertake different trajectories to a change in surface chemistry (Nash *et al.*, 2008).

Other general conclusions are difficult to draw because of the paucity of data. Although the literature reported here may appear ample, detailed analyses of the chemical strategies employed by social parasites are available for only a limited number of species. For example, in ants, approximately 150 species are social parasites of other ant species (Hölldobler and Wilson, 1990), but chemical strategies of integration have been investigated in less than 10%.

Among variables driving the evolution of chemical mimicry, the degree of dependence of parasites on hosts (how "intimate" the relation is, Combes, 2001), the local prevalence of the parasite, and the impact that the parasite produces on host fitness, are going to play key roles, as well as the properties of the host recognition system mechanism. These variables, rather than phylogenetic relatedness, affect the evolution of chemical mimicry. Indeed, phylogenetic relatedness generally does not correlate with the probability of evolving chemical mimicry (the example of the *Acromyrmex* host–parasite system is instructive), while knowledge of the recognition systems of the species involved, and analysis of the life history of parasites, hosts and their castes, may allow testable hypotheses. Overall, the general conclusion at the end of our screening of the literature on chemical mimicry is that, whenever the life history of parasites or social parasites requires a durable, although limited, interaction among the parasite and its host, chemical mimicry evolves. We have seen examples of chemical mimicry in parasites that spend part of their life on the host-insect body (e.g., mites and bees, Phelan *et al.*, 1991; Martin *et al.*, 2001), in parasitoids that briefly sneak into their host nests (e.g., cuckoo wasps and their hosts, Strohm *et al.*, 2008), in predators that spend time among their prey groups (e.g., predatory ants among their termite preys, Lemaire *et al.*, 1986; Dejean, 1988; aphid predators among their aphid preys and aphid-defending ants, Lohman *et al.*, 2006) and social parasites that deeply integrate within the social structure of the host colonies for most of their lifespan. The common emerging property in such a large and diverse array of interactions among organisms seems to be that, in all cases, there exists a period of time when the two organisms meet each other, and when avoiding host (or prey) detection is advantageous for the parasite (or predator). However, the matching of cuticular hydrocarbon profiles between interacting organisms varies widely. We have seen that the variation between species in the precision of mimicry depends on many different variables, mainly the following: (a) the local degree of the ongoing coadaptation among the organisms involved (e.g., Nash *et al.*, 2008) (Table 14.3), (b) the properties of the recognition system of the

Table 14.3 *Reports on differences in the degree of cuticular hydrocarbon matching among parasite populations of the same species.*

Social parasites	Studies documenting chemical mimicry	Studies documenting mismatching or differences among populations in the degree of matching	Notes
Polyergus rufescens	D'Ettorre *et al.*, 2002	Habersetzer and Bonavita-Cougourdan, 1993; Bonavita-Cougourdan *et al.*, 1996; Bonavita-Cougourdan *et al.*, 2004	Discrepancy may depend on the degree of coadaptation among populations
P. samurai	Liu *et al.*, 2003	Yamaoka, 1990	Discrepancy may depend on GC–MS conditions (Liu *et al.*, 2003) or on the degree of coadaptation among populations
Rossomyrmex minuchae		Errard *et al.*, 2006b	Mimicry depends on coadaptation (higher mimicry in sympatric than allopatric populations) (Errard *et al.*, 2006b)
Protomagnathus americanus		Brandt *et al.*, 2005	Mimicry depends on coadaptation (higher mimicry in single than in multiple host populations) (Brandt *et al.*, 2005)

host or prey species (e.g., mimics may replicate either host comb odor or host queen odor, e.g., in wasps, Lorenzi *et al.*, 1996, or ants, Johnson *et al.*, 2001) and (c) structural chemical properties of the cuticular blends of the mime which may lead to replication of only some features of the model (e.g., Lohman *et al.,* 2006). Within a species, the precision of mimicry may vary among populations, depending on the degree of coadaptation among interacting species (e.g., Nash *et al.*, 2008) (Table 14.3) and with the life stage and/or caste of the parasite (or predator) (Table 14.2).

In theory, chemical insignificance is an alternative adaptive option to chemical mimicry. However, chemical mimicry implies a change in the cuticular hydrocarbon composition, whereas chemical insignificance implies a decrease in the amount of cuticular hydrocarbons. The primary function of cuticular hydrocarbons, i.e., limiting dehydration (see Chapter 6) constrains the degree of chemical insignificance, while such a constraint does not exist for changing the composition of the cuticular blend. Indeed, chemical insignificance is usually reported in combination with chemical mimicry and is often limited to few life stages during the life history of a parasite (or predator) (or even to fractions of one life stage, e.g., Lorenzi and Bagnères, 2002; Lorenzi *et al.*, 2004a, b). Moreover, in social parasites that not only infiltrate host colonies but also actively interact with their hosts, inhibiting their reproduction or manipulating their behavior (e.g., Cervo and Lorenzi, 1996; Nash and Boomsma, 2008; Fucini and Lorenzi, in preparation), chemical insignificance cannot be entertained as a feasible adaptation.

As a final comment, when female social parasites operate as inquilines and enter host colonies mimicking the chemical signature of their hosts, the mechanism of matching has been reported to be incredibly sophisticated in certain cases: each parasite female is able to mimic the chemical signature of exactly her own host colony (e.g., Sledge *et al.*, 2001), occasionally to the point that parasite queens and host workers are more similar to one another than conspecific queens and workers are in unparasitized control colonies (Franks *et al.*, 1990). In this respect, chemical mimicry is a wonderful example of the power of natural selection in shaping the precision of adaptation.

Acknowledgments

We are grateful to Ralph W. Howard, Alain Lenoir and John Thompson for their reading and for their helpful comments.

References

Akino, T. (2002). Chemical camouflage by myrmecophilous beetles *Zyras comes* (Coleoptera: Staphylinidae) and *Diaritiger fossulatus* (Coleoptera: Pselaphidae) to be integrated into the nest of *Lasius fuliginosus* (Hymenoptera: Formicidae). *Chemoecology*, **12**, 83–89.

Akino, T. (2008). Chemical strategies to deal with ants: a review of mimicry, camouflage, propaganda, and phytomimesis by ants (Hymenoptera: Formicidae) and other arthropods. *Myrmecol. News*, **11**, 173–181.

Akino, T. and Yamaoka, R. (1998). Chemical mimicry in the root aphid parasitoid *Paralipsis eikoae* Yasumatsu (Hymenoptera: Aphidiidae) of the aphid-attending ant *Lasius sakagamii* Yamauchi and Hayashida (Hymenoptera: Formicidae). *Chemoecology*, **8**, 153–161.

Akino, T., Knapp, J. J., Thomas, J. A. and Elmes, G. W. (1999). Chemical mimicry and host specificity in the butterfly *Maculinea rebeli*, a social parasite of *Myrmica* ant colonies. *Proc. R. Soc. Lond. B*, **266**, 1419–1426.

Akino, T., Mochizuki, R., Morimoto, M. and Yamaoka, R. (1996). Chemical camouflage of myrmecophilous cricket *Myrmecophilus* sp. to be integrated with several ant species. *Jpn. J. Appl. Entomol. Zool.*, **40**, 39–46. [published in Japanese]

Akino, T., Nakamura, K. and Wakamura, S. (2004). Diet-induced chemical phytomimesis by twig-like caterpillars of *Biston robustum* Butler (Lepidoptera: Geometridae). *Chemoecology*, **14**, 165–174.

Allan, R. A. and Elgar, M. A. (2001). Exploitation of the green tree ant *Oecophylla smaragdina* by the salticid spider *Cosmophasis bitaeniata*. *Aust. J. Zool.*, **49**, 129–137.

Allan, R.A., Capon, R.J., Brown, W.V. and Elgar, M.A. (2002). Mimicry of host cuticular hydrocarbons by salticid spider *Cosmophasis bitaeniata* that preys on larvae of tree ants *Oecophylla smaragdina. J. Chem. Ecol.*, **28**, 835–848.

Allies, A.B., Bourke, A.F.G. and Franks, N.R. (1986). Propaganda substances in the cuckoo ant *Leptothorax kutteri* and the slave-maker *Harpagoxenus sublaevis. J. Chem. Ecol.*, **12**, 1285–1293.

Ayasse, M., Paxton, R. and Tengo, J. (2001). Mating behavior and chemical communication in the order Hymenoptera. *Annu. Rev. Entomol.*, **46**, 31–78.

Bagnères, A.-G., Errard C., Mulheim C., Joulie, C and Lange, C. (1991). Induced mimicry of colony odors in ants. *J. Chem. Ecol.*, **17**, 1641–1664.

Bagnères, A.-G., Huang, Z., Navajas, M., Salvy, M., Christides, J.-P., Zeng, Z. and Le Conte, Y. (2006). Chemical mimicry by the ectoparasitic mite *Varroa destructor* infesting *Apis cerana* and *A. mellifera* broods. In *Proc. ISCE 22nd Meeting, Barcelona*, p. 76.

Bagnères, A.-G., Lorenzi, M.-C., Dusticier, G., Turillazzi, S. and Clément, J.-L. (1996). Chemical usurpation of a nest by paper wasp parasites. *Science*, **272**, 889–892.

Barbero, F., Thomas, J.A., Bonelli, S., Balletto, E. and Schönrogge, K. (2009). Queen ants make distinctive sounds that are mimicked by a butterfly social parasite. *Science*, **323**, 782–785.

Beibl, J., D'Ettorre, P., and Heinze, J. (2007). Cuticular profiles and mating preference in a slave-making ant. *Insectes Soc.*, **54**, 174–182.

Bergström, G. (2008). Chemical communication by behaviour-guiding olfactory signals. *Chem. Comm.*, **34**, 3959–3979.

Bergström, G. and Löfqvist, J. (1968). Odour similarities between the slave-keeping ants *Formica sanguinea* and *Polyergus rufescens* and their slaves *Formica fusca* and *Formica rufibarbis. J. Insect Physiol.*, **14**, 995–1011.

Bonavita-Cougourdan, A., Bagnères, A.-G., Provost E., Dusticier, G. and Clément, J.-L. (1997). Plasticity of the cuticular hydrocarbon profile of the slave-making ant *Polyergus rufescens* depending on the social environment. *Comp. Biochem. Physiol.*, **116B**, 287–302.

Bonavita-Cougourdan, A., Clément, J.-L. and Lange, C. (1989). The role of cuticular hydrocarbons in recognition of larvae by workers of the ant *Camponotus vagus*: changes in chemical signature in response to social environment (Hymenoptera: Formicidae). *Sociobiology*, **16**, 49–74.

Bonavita-Cougourdan, A., Provost, E., Rivière, G., Bagnères, A.-G. and Dusticier, G. (2004). Regulation of cuticular and postpharyngeal hydrocarbons in the slave-making ant *Polyergus rufescens*: effect of *Formica rufibarbis* slaves. *J. Insect Physiol.*, **50**, 285–293.

Bonavita-Cougourdan, A., Rivière, G., Provost, E., Bagnères, A.-G., Roux, M., Dusticier, G. and Clément, J.-L. (1996). Selective adaptation of the cuticular hydrocarbon profiles of the slave making ants *Polyergus rufescens* Latr. and their *Formica rufibarbis* and *F. cunicularia* Latr. slaves. *Comp. Biochem. Physiol.*, **113B**, 313–329.

Borg-Karlson, A.-K. and Tengö, J. (1986). Odor mimetism? Key substances in *Ophrys lutea-Andrena* pollination relationship (Orchidaceae: Andrenidae). *J. Chem. Ecol.*, **12**, 1927–1941.

Brandt, M. and Foitzik, S. (2004). Community context and specialization influence the coevolutionary interactions in a slavemaking ant. *Ecology*, **85**, 2997–3009.

Brandt, M., Heinze, J., Schmitt, T. and Foitzik, S. (2005). A chemical level in the coevolutionary arms race between an ant social parasite and its hosts. *J. Evol. Biol.*, **18**, 576–586.

Brandt, M., Heinze, J., Schmitt, T. and Foitzik, S. (2006). Convergent evolution of the Dufour's gland secretion as a propaganda substance in the slave-making ant genera *Protomognathus* and *Harpagoxenus*. *Insectes Soc.*, **53**, 291–299.

Breed, M. D., Snyder, L. E., Lynn, T. L. and Morhart, J. A. (1992). Acquired chemical camouflage in a tropical ant. *Anim. Behav.*, **44**, 519–523.

Brodmann, J., Twele, R., Francke, W., Yi-bo, L., Xi-qiang, S. and Ayasse, M. (2009). Report orchid mimics honey bee alarm pheromone in order to attract hornets for pollination. *Current Biol.*, **19**, 1368–1372.

Carlin, N. F. and Hölldobler, B. (1983). Nestmate and kin recognition in interspecific mixed colonies of ants. *Science*, **222**, 1027–1029.

Carpenter, J. M. (1997). Phylogenetic relationships among European *Polistes* and the evolution of social parasitism (Hymenoptera: Vespidae; Polistinae). In *The Origin of Biodiversity in Insects: Phylogenetic Tests of Evolutionary scenarios,* ed. P. Grandcolas. Mém. Mus. Natn. Hist. Nat. 173, pp. 135–161.

Cervo, R. and Dani, F. R. (1996). Social parasitism and its evolution in *Polistes*. In *Natural History and Evolution of Paper-Wasps*, ed. S. Turillazzi and M. J. West-Eberhard. Oxford: Oxford University Press, pp. 104–112.

Cervo, R., Dani, F. R., Cotoneschi, C., Scala, C., Lotti, I., Strassmann, J. E., Queller, D. C. and Turillazzi, S. (2008). Why are larvae of the social parasite wasp *Polistes sulcifer* not removed from the host nest? *Behav. Ecol. Sociobiol.*, **62**, 1319–1331.

Cervo, R. and Lorenzi, M. C. (1996). Inhibition of host queen reproductive capacity by the obligate social parasite *Polistes atrimandibularis* (Hymenoptera Vespidae). *Ethology*, **102**, 1042–1047.

Cervo, R., Lorenzi, M. C. and Turillazzi, S. (1990). Nonaggressive usurpation of the nest of *Polistes biglumis bimaculatus* by the social parasite *Sulcopolistes atrimandibularis* (Hymenoptera Vespidae). *Insectes Soc.*, **37**, 333–347.

Cervo, R., Macinai, V., Dechigi, F. and Turillazzi, S. (2004). Fast growth of immature brood in a social parasite wasp: A convergent evolution between avian and insect cuckoos. *Am. Nat.*, **164**, 814–820.

Choudhary, M., Strassmann, J. E., Queller, D. C., Turillazzi, S. and Cervo, R. (1994). Social parasites in polistine wasps are monophyletic: implications for sympatric speciation. *Proc. R. Soc. London B Bio.*, **257**, 31–35.

Cini, A., Gioli, L. and Cervo, R. (2009). A quantitative threshold for nest-mate recognition in a paper social wasp. *Biol. Lett.* **5**, 459–461.

Combes, C. (2001). *Parasitism: The Ecology and Evolution of Intimate Interactions*. Chicago: Chicago University Press.

Corbara, B. and Errard, C. (1991). The organization of artificial heterospecific ant colonies. The case of the *Manica rubida/Formica selysi* association: mixed colony or parallel colonies? *Behav. Proc.*, **23**, 75–87.

Cremer, S., Sledge, M. F. and Heinze, J. (2002). Chemical mimicry: Male ants disguised by the queen's bouquet. *Nature*, **419**, 897.

Crozier, R. H. and Pamilo, P. (1996). *Evolution of Social Insect Colonies*. Oxford: Oxford University Press.

Dapporto, L., Sledge, F. W. and Turillazzi, S. (2005). Dynamics of cuticular chemical profiles of *Polistes dominulus* workers in orphaned nests (Hymenoptera, Vespidae). *J. Insect Physiol.*, **51**, 969–973.

Dawkins, R. and Krebs, J. R. (1978). Animal signals: information or manipulation? In *Behavioural Ecology: an evolutionary approach*, ed. J. R. Krebs and N. B. Davies, 1st edn. Oxford: Blackwell, pp. 282–309.

Dejean, A. (1988). Les écomones impliquées dans la prédation chez les fourmis. *Ann. Soc. Entomol. Fr.*, **24**, 456.

Dejean, A., Kenne, M. and Moreau, C. S. (2007). Predatory abilities favour the success of the invasive ant *Pheidole megacephala* in an introduced area. *J. Appl. Entomol.*, **131**, 625–629.

Dettner, K. and Liepert, C. (1994). Chemical mimicry and camouflage. *Annu. Rev. Entomol.*, **39**, 129–154.

D'Ettorre, P. and Errard, C. (1998). Chemical disguise during colony founding in the dulotic ant *Polyergus rufescens* Latr. (Hymenoptera, Formicidae). *Insect Social Life*, **2**, 71–77.

D'Ettorre, P., Brunner, E., Wenseleers, T. and Heinze, J. (2004a). Knowing your enemies: seasonal dynamics of host-social parasite recognition. *Naturwissenschaften*, **91**, 594–597.

D'Ettorre, P., Errard, C., Ibarra, F., Francke, W. and Hefetz, A. (2000). Sneak in or repel your enemy: Dufour's gland repellent as a strategy for successful usurpation in the slave-maker *Polyergus rufescens*. *Chemoecology*, **10**, 135–142.

D'Ettorre, P., Heinze, J., Schulz, C., Francke, W. and Ayasse, M. (2004b). Does she smell like a queen? Chemoreception of a cuticular hydrocarbon signal in the ant *Pachycondyla inversa*. *J. Experim. Biol.*, **207**, 1085–1091.

D'Ettorre, P., Mondy, N., Lenoir, A. and Errard, C. (2002). Blending in with the crowd: social parasites integrate into their host colonies using a flexible chemical signature. *Proc. R. Soc. Lond. B*, **269**, 1911–1918.

Dietemann, V., Neumann, P., Härtel, S., Pirk, C. W. W. and Crewe, R. M. (2007). Pheromonal dominance and the selection of a socially parasitic honeybee worker lineage (*Apis mellifera capensis* Esch.). *J. Evol. Biol.*, **20**, 997–1007.

Dietemann, V., Pflugfelder, J., Härtel, S., Neumann, P. and Crewe, R. M. (2006). Social parasitism by honeybee workers (*Apis mellifera capensis* Esch.): evidence for pheromonal resistance to host queen's signals. *Behav. Ecol. Sociobiol.*, **60**, 785–793.

Dinter, K., Paarmann, W., Peschke, K. and Arndt, E. (2002). Ecological, behavioural and chemical adaptations to ant predation in species of *Thermophilum* and *Graphipterus* (Coleoptera: Carabidae) in the Sahara Desert. *J. Arid Envir.*, **50**, 267–286.

Dronnet, S., Simon, X., Verhaeghe, J.-C., Rasmont, P. and Errard, C. (2005). Bumblebee inquilinism in *Bombus (Fernaldaepsithyrus) sylvestris* (Hymenoptera, Apidae): behavioural and chemical analyses of host-parasite interactions. *Apidologie*, **36**, 59–70.

Elgar, M. A. and Allan, R. A. (2004). Predatory spider mimics acquire colony-specific cuticular hydrocarbons from their ant model prey. *Naturwissenschaften*, **91**, 143–147.

Elgar, M. A. and Allan, R. A. (2006). Chemical mimicry of the ant *Oecophylla smaragdina* by the myrmecophilous spider *Cosmophasis bitaeniata*: Is it colony-specific? *J. Ethol.*, **24**, 239–246.

Elmes, G. W., Akino, T., Thomas, J. A., Clarke, R. T. and Knapp, J. J. (2002). Interspecific differences in cuticular hydrocarbon profiles of *Myrmica* ants are sufficiently consistent to explain host specificity by *Maculinea* (large blue) butterflies. *Oecologia*, **130**, 525–535.

Elmes, G. W., Barr, B., Thomas, J. A. and Clarke, R. T. (1999). Extreme host specificity by *Microdon mutabilis* (Diptera: Syrphidae), a social parasite of ants. *Proc. R. Soc. Lond. B*, **266**, 447–453.

Errard, C. (1986). Role of early experience in mixed-colony: odor recognition in the ants *Manica rubida* and *Formica selysi*. *Ethology*, **72**, 243–249.

Errard, C., Hefetz, A. and Jaisson, P. (2006a). Social discrimination tuning in ants: template formation and chemical similarity. *Behav. Ecol. Sociobiol.*, **59**, 353–363.

Errard, C., Fresneau, D., Heinze, J., Francoeur, A. and Lenoir, A. (1997). Social organization in the guest-ant *Formicoxenus provancheri*. *Ethology*, **103**, 149–159.

Errard, C., Ipinza Regla, J. and Hefetz, A. (2003). Interspecific recognition in Chilean parabiotic ant species. *Insectes Soc.*, **50**, 268–273.

Errard, C., Ruano F, Richard, F.-J., Lenoir, A., Tinaut, A. and Hefetz, A. (2006b). Co-evolution-driven cuticular hydrocarbon variation between the slave-making ant *Rossomyrmex minuchae* and its host *Proformica longiseta* (Hymenoptera: Formicidae). *Chemoecology*, **16**, 235–240.

Espelie, K. E., Bernay, E. A. and Brown, J. J. (1991). Plant and insect cuticular lipids serve as behavioral cues for insects. *Arch. Insect Biochem. Physiol.*, **17**, 223–233.

Espelie, K. E. and Brown, J. J. (1990). Cuticular hydrocarbons of species which interact on four trophic levels: apple, *Malus pumila* Mill.; codling moth, *Cydia pomonella* L.; a hymenopteran parasitoid, *Ascogaster quadridentata* Wesmael; and a hyperparasite, *Perilampus fulvicornis* Ashmead. *Comp. Biochem. Physiol. B*, **95**, 131–136.

Espelie, K. E. and Hermann, H. R. (1988). Congruent cuticular hydrocarbons: biochemical convergence of a social wasp, an ant and a host plant. *Biochem. Systemat. Ecol.*, **16**, 505–508.

Everaerts, C., Farine, J.-P. and Brossut, R. (1997). Changes of species specific cuticular hydrocarbon profiles in the cockroaches *Nauphoeta cinerea* and *Leucophaea maderae* reared in heterospecific groups. *Entomol. Exp. Appl.*, **85**, 145–150.

Fanelli, D., Henshaw, M., Cervo, R., Turillazzi, S., Queller, D. C. and Strassmann, J. E. (2005). The social parasite wasp *Polistes atrimandibularis* does not form host races. *J. Evol. Biol.*, **18**, 1362–1367.

Fielde, A. (1903). Artificial mixed nests of ants. *Biol. Bull.*, **5**, 320–325.

Franks, N., Blum, M. S., Smith, R.-K. and Allies, A. B. (1990). Behavior and chemical disguise of cuckoo ant *Leptothorax kutteri* in relation to its host *Leptothorax acervorum*. *J. Chem. Ecol.*, **16**, 1431–1444.

Gamboa, G. J. (1996). Kin recognition in social wasps. In *Natural history and evolution of paperwasps*, ed. S. Turillazzi, and M. J. West-Eberhard, Oxford: Oxford University Press, pp. 161–177.

Garnett, W. B., Akre, R. D. and Sehlke, G. (1985). Cocoon mimicry and predation by myrmecophilous diptera (Diptera: Syrphidae). *Florida Entomol.*, **68**, 615–621.

Gary, N. E., and Page, R. E. (1988). Phenotypic variation in susceptibility of honey bees, *Apis mellifera*, to infestation by tracheal mites, *Acarapis woodi*. *Exp. Appl. Acarol.*, **3**, 291–305.

Geiselhardt, S. F., Geiselhardt, S. and Peschke, K. (2006). Chemical mimicry of cuticular hydrocarbons – how does *Eremostibes opacus* gain access to breeding burrows of its host *Parastizopus armaticeps* (Coleoptera, Tenebrionidae)? *Chemoecology*, **16**, 59–68.

Geiselhardt, S. F., Peschke, K. and Nagel, P. (2007). A review of myrmecophily in ant nest beetles (Coleoptera: Carabidae: Paussinae): linking early observations with recent findings. *Naturwissenschaften*, **94**, 871–894.

Gemeno, C., Yeargan, K. and Haynes, K. F. (2000). Aggressive chemical mimicry by the bolas spider *Mastophora hutchinsoni*: identification and quantification of a major prey's sex pheromone components in the spider's volatile emissions. *J. Chem. Ecol.*, **26**, 1235–1243.

Gibbs, M. (1999). Parasitology: Chaperonin camouflage. *Nature*, **399**, p. 415.

Habersetzer, C. and Bonavita-Cougourdan, A. (1993). Cuticular spectra in the slave-making ant *Polyergus rufescens* and the slave species *Formica rufibarbis*. *Physiol. Entomol.*, **18**, 160–166.

Hafernik, J. and Saul-Gershenz, L. (2000). Beetle larvae cooperate to mimic bees. *Nature*, **405**, 35–36.

Haynes, K. F., Yeargan, K. V., Millar, J. G. and Chastain, B. B. (1996). Identification of sex pheromone of *Tetanolita mynesalis* (Lepidoptera: Noctuidae), a prey species of bolas spider, *Mastophora hutchinsoni. J. Chem. Ecol.*, **22**, 75–89.

Hefetz, A. (2007). The evolution of hydrocarbon pheromone parsimony in ants (Hymenoptera: Formicidae) – interplay of colony odor uniformity and odor idiosyncrasy. A review. *Myrmecol. News*, **10**, 59–68.

Hefetz, A., Errard, C. and Cojocaru, M. (1992). Heterospecific substances in the postpharyngeal gland secretion of ants reared in mixed groups. *Naturwissenschaften*, **79**, 417–420.

Henrique, A., Portugal, A. and Trigo, J. R. (2005). Similarity of cuticular lipids between a caterpillar and its host plant: a way to make prey undetectable for predatory ants? *J. Chem. Ecol.*, **31**, 2551–2561.

Herzner, G., Schmitt, T., Peschke, K., Hilpert, A. and Strohm, E. (2007). Food wrapping with the postpharyngeal gland secretion by females of the European beewolf *Philanthus triangulum. J. Chem. Ecol.*, **33**, 849–859.

Hölldobler, B. and Wilson E. O. (1990). *The Ants*. Cambridge, MA: Harvard University Press.

Hojo, M. K., Wada-Katsumata, A., Akino, T., Yamaguchi, S., Ozaki, M. and Yamaoka, R. (2009). Chemical disguise as particular caste host ants in the ant inquiline parasite *Niphanda fusca* (Lepidoptera: Lycaenidae). *Proc. R. Soc. B.*, **276**, 551–558.

Howard, R. W. (1993). Cuticular hydrocarbons and chemical communication. In *Insect Lipids: Chemistry, Biochemistry and Biology*, ed. D. W. Stanley-Samuelson and D. R. Nelson. Lincoln, NE: University of Nebraska Press, pp. 179–226.

Howard, R. W., Akre, R. D. and Garnett, W. B. (1990). Chemical mimicry in an obligate predator of carpenter ants (Hymenoptera: Formicidae). *Ann. Entomol. Soc. Am.*, **83**, 607–616.

Howard, R. W. and Blomquist, G. J. (2005). Ecological, behavioral, and biochemical aspects of insect hydrocarbons. *Annu. Rev. Entomol.*, **50**, 371–392.

Howard, R. W. and Infante, F. (1996). Cuticular hydrocarbons of the host-specific ectoparasitoid *Cephalonomia stephanoderis* (Hymenoptera: Bethylidae) and its host the coffee berry borer (Coleoptera: Scolytidae). *Ann. Entom. Soc. Am.*, **89**, 700–709.

Howard, R. W. and Liang, Y. (1993). Cuticular hydrocarbons of winged and wingless morphs of the ectoparasitoid *Choetospila elegans* Westwood (Hymenoptera: Pteromalidae) and its host, larval lesser grain borer (*Rhyzopertha dominica*) (Coleoptera: Bostrichidae). *Comp. Biochem. Physiol. B*, **106**, 407–414.

Howard, R. W., McDaniel, C. A. and Blomquist, G. J. (1980). Chemical mimicry as an integrating mechanism: cuticular hydrocarbons of a termitophile and its host. *Science*, **210**, 431–433.

Howard, R. W., McDaniel, C. A. and Blomquist, G. J. (1982). Chemical mimicry as an integrating mechanism for three termitophiles associated with *Reticulitermes virginicus. Psyche*, **89**, 157–168.

Howard, R. W., Pérez-Lachaud, G. and Lachaud, J.-P. (2001). Cuticular hydrocarbons of *Kapala sulcifacies* (Hymenoptera: Eucharitidae) and its host, the ponerine ant *Ectatomma ruidum* (Hymenoptera: Formicidae). *Ann. Entomol. Soc. Am.*, **94**, 707–716.

Howard, R. W., Stanley-Samuelson, D. W. and Akre, R. D. (1990). Biosynthesis and chemical mimicry of cuticular hydrocarbons from the obligate predator, *Microdon albicomatus* Novak (Diptera: Syrphidae) and its ant prey, *Myrmica incompleta* Provancher (Hymenoptera: Formicidae). *J. Kansas Entomol. Soc.*, **63**, 437–443.

Hughes, C. L. (1988). Phytochemical mimicry of reproductive hormones and modulation of herbivore fertility by phytoestrogens. *Environ. Health Perspectives*, **78**, 171–174.

Hughes, D. P. (2008). The extended phenotype within the colony and how it obscures social communication. In *Sociobiology of Communication*, ed. P. D'Ettore and D. P. Hughes. Oxford: Oxford University Press, pp. 171–190.

Isingrini, M., Lenoir, A. and Jaisson, P. (1985). Pre-imaginal learning for colony-brood recognition in the ant *Cataglyphis cursor* Fonsc. *Proc. Natl. Acad. Sci. USA*, **82**, 8545–8547.

Jeral, J. M., Breed, M. D. and Hibbard, B. E. (1997). Thief ants have reduced quantities of cuticular compounds in a ponerines ant *Ectatomma ruidum*. *Physiol. Entomol.*, **22**, 207–211.

Joel, D. M. (1988). Mimicry and mutualism in carnivorous pitcher plants (Sarraceniaceae, Nepenthaceae, Cephalotaceae, Bromeliaceae). *Biol. J. Linn. Soc.*, **35**, 185–197.

Johnson, C., Topoff, H., Vander Meer, R. K. and Lavine, B. K. (2005). Do these eggs smell funny to you?: An experimental study of egg discrimination by hosts of the social parasite *Polyergus breviceps* (Hymenoptera: Formicidae). *Behav. Ecol. Sociobiol.*, **57**, 245–255.

Johnson, C., Vander Meer, R. K. and Lavine, B. K. (2001). Changes in the cuticular hydrocarbon profile of the slave-maker ant queen, *Polyergus breviceps*, after killing a *Formica* queen. *J. Chem. Ecol.*, **27**, 1787–1804.

Kaib, M., Heinze, J. and Ortius, D. (1993). Cuticular hydrocarbon profiles in the slave-making ant *Harpagoxenus sublaevis* and its hosts. *Naturwissenschaften*, **80**, 281–285.

Kaiser, R. (2006). Flowers and fungi use scents to mimic each other. *Science*, **311**, 806–807.

Kistner, D. H. (2000). A new genus and species of a queen mimicking termitophile from Brazil (Coleoptera: Staphylinidae). *Sociobiology*, **35**, 191–195.

Kroiss, J., Schmitt, T., Schreier, P., Strohm, E. and Herzner, G. (2006). A selfish function of a "social" gland? A postpharyngeal gland functions as a sex pheromone reservoir in males of the solitary wasp *Philanthus triangulum*. *J. Chem. Ecol.*, **32**, 2763–2776.

Lambardi, D., Dani, F. R., Turillazzi, S. and Boomsma, J. J. (2007). Chemical mimicry in an incipient leaf-cutting ant social parasite. *Behav. Ecol. Sociobiol.*, **61**, 843–851.

Legoza, J. (2005). Chemical mimicry in the insect world. http://www.colostate.edu/Depts/Entomology/courses/en507/papers_2005/legoza.pdf

Lemaire, M., Lange, C., Lefebvre, J. and Clément, J.-L. (1986). Stratégie de camouflage du prédateur *Hypoponera eduardi* dans les sociétés de *Reticulitermes* européens. *Actes Coll. Insect. Soc.*, **3**, 97–101.

LeMaster, M. P. and Mason, R. T. (2002). Variation in a female sexual attractiveness pheromone controls male mate choice in garter snakes. *J. Chem. Ecol.*, **28**, 1269–1285.

Lenoir, A., D'Ettorre, P., Errard, C. and Hefetz, A. (2001). Chemical ecology and social parasitism. *Annu. Rev. Entomol.*, **46**, 573–599.

Lenoir, A., Malosse, C. and Yamaoka, R. (1997). Chemical mimicry between parasitic ants of the genus *Formicoxenus* and their host *Myrmica* (Hymenoptera, Formicidae). *Biochem. Syst. Ecol.*, **25**, 379–389.

Liang, D. and Silverman, J. (2000). You are what you eat: Diet modifies cuticular hydrocarbons and nestmate recognition in the Argentine ant, *Linepithema humile*. *Naturwissenschaften*, **87**, 412–416.

Liepert, C. and Dettner, K. (1993). Recognition of aphid parasitoids by honeydew-collecting ants: The role of cuticular lipids in a chemical mimicry system. *J. Chem. Ecol.*, **19**, 2143–2153.

Liepert, C. and Dettner, K. (1996). Role of cuticular hydrocarbons of aphid parasitoids in their relationship to aphid-attending ants. *J. Chem. Ecol.*, **22**, 695–707.

Liu, Z., Bagnères, A.-G., Yamane, S., Wang, Q. and Kojima, J. (2003). Cuticular hydrocarbons in workers of the slave-making ant *Polyergus samurai* and its slave, *Formica japonica* (Hymenoptera Formicidae). *Entomol. Sci.*, **6**, 125–133.

Liu, Z., Yamane, S., Yamamoto, H. and Wang, Q. (2000). Nestmate discrimination and cuticular profiles of a temporary parasitic ant *Lasius* sp. and its host *L. fuliginosus* (Hymenoptera, Formicidae). *J. Ethol.*, **18**, 69–74.

Le Conte. Y., Arnold, G., Trouiller, J., Massan, C., Chappe B. and Ourisson, G. (1989). Attraction of the parasitic mite *Varroa* to the drone larvae of honey bees by simple aliphatic esters. *Science*, **245**, 638–639.

Lohman, D. J., Liao, Q. and Pierce, N. E. (2006). Convergence of chemical mimicry in a guild of aphid predators. *Ecol. Entomol.*, **31**, 41–51.

Lorenzi, M. C. (2003). Social wasp parasites affect the nestmate recognition abilities of their hosts (*Polistes atrimandibularis* and *P. biglumis*, Hymenoptera: Vespidae). *Insectes Soc.*, **50**, 82–87.

Lorenzi, M. C. (2006). The result of an arms race: the chemical strategies of *Polistes* social parasites. *Ann. Zool. Fennici*, **43**, 550–563.

Lorenzi, M. C. and Bagnères, A.-G. (2002). Concealing identity and mimicking hosts: a dual chemical strategy for a single social parasite? (*Polistes atrimandibularis*, Hymenoptera: Vespidae). *Parasitology*, **125**, 507–512.

Lorenzi, M. C., Bagnères, A.-G. and Clément, J.-L. (1996). The role of cuticular hydrocarbons in social insects: is it the same in paper-wasps? In *Natural History and Evolution of Paper-Wasps*, ed. S. Turillazzi and M. J. West-Eberhard. Oxford: Oxford University Press, pp. 178–189.

Lorenzi, M. C., Caldi, M. and Cervo, R. (2007). The chemical strategies used by *Polistes nimphus* social wasp usurpers (Hymenoptera Vespidae). *Biol. J. Linnean Soc.*, **91**, 505–512.

Lorenzi, M. C., Cervo, R., Zacchi, F., Turillazzi, S. and Bagnères, A.-G. (2004a). Dynamics of chemical mimicry in the social parasite wasp *Polistes semenovi* (Hymenoptera Vespidae). *Parasitology*, **129**, 643–651.

Lorenzi, M. C., Sledge, M. F., Laiolo, P. and Turillazzi, S. (2004b). Cuticular hydrocarbon dynamics in young adult *Polistes dominulus* (Hymenoptera Vespidae) and the role of linear hydrocarbons in nestmate recognition systems. *J. Insect Physiol.*, **50**, 935–941.

Malcolm, S. B. (1990). Mimicry: status of a classical evolutionary paradigm. *Trends Ecol. Evol.*, **5**, 57–62.

Mant, J., Brändli, C., Vereecken, N. J., Schulz, C. M., Francke, W. and Schiestl, F. P. (2005a). Cuticular hydrocarbons as sex pheromone of the bee *Colletes cunicularius* and the key to its mimicry by the sexually deceptive orchid, *Ophrys exaltata*. *J. Chem. Ecol.*, **31**, 1765–1787.

Mant, J., Peakall, R. and Schiestl, F. P. (2005b). Does selection on floral odor promote differentiation among populations and species of the sexually deceptive orchid genus *Ophrys? Evolution*, **59**, 1449–1463.

Marshall, D. C. and Hill, K. B. R. (2009). Versatile aggressive mimicry of cicadas by an Australian predatory katydid. *PloS One*, **4**, e4185.

Martin, C., Provost, E., Bagnères, A.-G., Roux, M., Clément, J.-L. and LeConte, Y. (2002). Potential mechanism for detection of the parasitic mite *Varroa jacobsoni* inside sealed brood cells by *Apis mellifera*. *Physiol. Entomol.*, **27**, 175–189.

Martin, C., Salvy M., Provost, E., Bagnères, A.-G., Roux, M., Crauser, D., Clément, J.-L. and LeConte, Y. (2001). Variations in chemical mimicry by the ectoparasititic mite

Varroa jacobsoni according to the developmental stage of the host honey bee *Apis mellifera. Insect Biochem. Mol. Biol.*, **31**, 365–379.

Martin, S. J., Jenner, E. J. and Drijfhout, F. P. (2007). Chemical deterrent enables a socially parasitic ant to invade multiple hosts. *Proc. R. Soc. B*, **274**, 2717–2722.

Martin, S. J., Takahashi, J.-I., Ono, M. and Drijfhout, F. P. (2008). Is the social parasite *Vespa dybowskii* using chemical transparency to get her eggs accepted? *J. Insect Physiol.*, **54**, 700–707.

Mason, R. T. and Crews, D. (1985). Female mimicry in garter snakes. *Nature*, **316**, 59–60.

Matsuura, K. (2006). Termite-egg mimicry by a sclerotium-forming fungus. *Proc. R. Soc. B*, **273**, 1203–1209.

Matsuura, K., Tamura, T., Kobayashi, N., Yashiro, T. and Tatsumi, S. (2007). The antibacterial protein lysozyme identified as the termite egg recognition pheromone. *PLoS One*, **2**(8), e813.

Menzel, F., Linsenmair, K. E. and Blüthgen, N. (2008). Selective interspecific tolerance in tropical *Crematogaster–Camponotus* associations. *Ann. Behav.*, **75**, 837–846.

Meskali, M., Bonavita-Cougourdan, A., Provost, E., Bagnères, A.-G., Dusticier, G. and Clément, J.-L. (1995). Mechanism underlying cuticular hydrocarbon homogeneity in the ant *Camponotus vagus* (Scop) (Hymenoptera: Formicidae): role of post-pharyngeal glands. *J. Chem. Ecol.*, **21**, 1127–1148.

Millor, J., Halloy, J., Amé, J.-M., Deneubourg, J.-L. (2006). Individual discrimination capability and collective choice in social insects. In *Lecture Notes in Computer Science Serie: Ant Colony Optimization and Swarm Intelligence*. Berlin Heidelberg: Springer Verlag, pp. 167–178.

Mori A., D'Ettorre P., and Le Moli, F. (1995). Host nest usurpation and colony foundation in the European amazon ant, *Polyergus rufescens* Latr (Hymenoptera, Formicidae). *Insectes Soc.*, **42**, 279–286.

Mori, A., Grasso, D. A., Visicchio, R. and Le Moli, F. (2000). Colony founding in *Polyergus rufescens*: the role of the Dufour's gland. *Insectes Soc.*, **47**, 7–10.

Moritz, R. F. A., Kirchner, W. H. and Crewe, R. M. (1991). Chemical camouflage of the death's head hawkmoth (*Acherontia atropos* L.) in honeybee colonies. *Naturwissenschaften*, **78**, 179–182.

Nash, D. R., Als, T. D., Maile, R., Jones, G. R. and Boomsma, J. J. (2008). A mosaic of chemical coevolution in a large blue butterfly. *Science*, **319**, 88–90.

Nash, D. R., and Boomsma, J. J. (2008). Communication between hosts and social parasites. In *Sociobiology of Communication*, ed. P. D'Ettore and D. P. Hugues. Oxford: Oxford University Press, pp. 55–79.

Nation, J. L., Sanford, S. T. and Milne, K. (1992). Cuticular hydrocarbons from *Varroa jacobsoni. Exp. App. Acarol.*, **16**, 331–344.

Pasteur, G. (1982). A classification review of mimicry systems. *Annu. Rev. Ecol. Syst.*, **13**, 169–199.

Phelan, P. L., Smith, A. W. and Needham, G. R. (1991). Mediation of host selection by cuticular hydrocarbons in the honeybee tracheal mite *Acarapis woodi* (Rennie). *J. Chem. Ecol.*, **17**, 463–473.

Pianaro, A., Flach, A., Patricio, E. F. L. R. A., Nogueira-Neto, P. and Marsaioli, A. J. (2007). Chemical changes associated with the invasion of a *Melipona scutellaris* colony by *Melipona rufiventris* workers. *J. Chem. Ecol.*, **33**, 971–984.

Provost, E., Blight, O., Tirard, A. and Renucci, M. (2008). Hydrocarbons and Insects' Social Physiology. In *Insect Physiology: New Research*, ed. P. Rayan. Maes, Hauppauge, NY: Nova Science.

Reeve, H. K. (1991). *Polistes*. In *The Social Biology of Wasp*, ed. Ross, K. G. and Matthews, R. W. Ithaca: Cornell University Press, pp. 99–148.

Richard, F.-J., Poulsen, M., Drijfhout, F., Jones, G. and Boosma, J. J. (2007). Specificity in chemical profiles of workers, brood and mutualistic fungi in *Atta, Acromyrmex,* and *Sericomyrmex* fungus-growing ants. *J. Chem. Ecol.*, **33**, 2281–2292.

Richter, I. and Krain, H. (1980). Cuticular lipid constituents of cabbage seedpod weevils and host plant oviposition sites as potential pheromones. *Lipids*, **15**, 580–586.

Rickli, M., Guerin, P. M. and Diehl, P. A. (1992). Palmitic acid released from honeybee worker larvae attracts the parasitic mite *Varroa jacobsoni* on a servosphere. *Naturwissenschaften*, **79**, 320–322.

Rickli, M., Guerin, P. M. and Diehl, P. A. (1994). Cuticle alkanes of honeybee larvae mediate arrestment of bee parasite *Varroa jacobsoni*. *J. Chem. Ecol.*, **20**, 2437–2453.

Ruther, J. and Steiner, S. (2008). Costs of female odour in males of the parasitic wasp *Lariophagus distinguendus* (Hymenoptera: Pteromalidae). *Naturwissenschaften*, **95**, 547–552.

Salvy, M., Martin, C., Bagnères, A.-G., Provost, E., Roux, M., LeConte, Y. and Clément, J.-L. (2001). Modifications of the cuticular hydrocarbons profile of *Apis mellifera* worker bee in the presence of the ectoparasitic mite *Varroa jacobsoni* in brood cells. *Parasitology*, **122**, 145–159.

Saul-Gershenz, L. S. and Millar, J. G. (2006). Phoretic nest parasites use sexual deception to obtain transport to their host's nest. *Proc. Natl. Acad. Sci. USA*, **103**, 14039–14044.

Schiestl, F. P., Ayasse, M., Paulus, H. F., Löfstedt, C., Hansson, B. S., Ibarra, F. and Francke, W. (1999). Orchid pollination by sexual swindle. *Nature*, **399**, 421.

Schiestl, F. P., Ayasse, M., Paulus, H. F., Löfstedt, C., Hansson, B. S., Ibarra, F. and Francke, W. (2000). Sex pheromone mimicry in the early spider orchid (*Ophrys sphegodes*): patterns of hydrocarbons as the key mechanism for pollination by sexual deception. *J. Comp. Physiol. A,* **186**, 567–574.

Schönrogge, K., Napper, E. K. V., Birkett, M. A., Woodcock, C. M., Pickett, J. A., Wadhams, L. J. and Thomas, J. A. (2008). Host recognition by a specialist hoverfly *Microdon mutabilis*, a social parasite of the ant *Formica lemani*. *J. Chem. Ecol.*, **34**, 168–178.

Schmidt-Hempel, P. (1998). *Parasites in social insects*. Princeton, NJ: Princeton University Press.

Schultz, T. R., Bekkevold, D. and Boomsma, J. J. (1998). *Acromyrmex insinuator* new species: an incipient social parasite of fungus-growing ants. *Insectes Soc.*, **45**, 457–471.

Scott, D. (1986). Sexual mimicry regulates the attractiveness of mated *Drosophila melanogaster* females. *Proc. Natl. Acad. Sci. USA*, **83**, 8429–8433.

Singer, T. L. (1998). Roles of hydrocarbons in the recognition systems of insects. *Am. Zool.*, **38**, 394–405.

Sledge, M. F., Dani, F. R., Cervo, R., Dapporto, L. and Turillazzi, S. (2001). Recognition of social parasites as nest-mates: Adoption of colony-specific host cuticular odours by the paper wasp parasite *Polistes sulcifer*. *Proc. R. Soc. Lond. B*, **268**, 2253–2260.

Sole, C. L., Kryger, P., Hefetz, A., Katzav-Gozansky, T. and Crewe, R. M. (2002). Mimicry of queen Dufour's gland secretions by workers of *Apis mellifera scutellata* and *A. m. capensis*. *Naturwissenschaften*, **89**, 561–564.

Soroker, V., Vienne, C., Hefetz, A. and Nowbahari, E. (1994). The postpharyngeal gland as a "gestalt" organ for nestmate recognition in the ant *Cataglyphis niger*. *Naturwissenschaften*, **81**, 510–513.

Spencer, K.C. (1988). The chemistry of coevolution. Chapter 18. In *Chemical mediation of coevolution*, ed. K.C. Spencer. San Diego, CA: Academic Press Inc., pp. 581–587.

Steiner, S., Steidle, J.L.M. and Ruther, J. (2005). Female sex pheromone in immature insect males – a case of pre-emergence chemical mimicry? *Behav. Ecol. Sociobiol.*, **58**, 111–120.

Stoeffler, M., Maier, T.S., Tolasch, T. and Steidle, J.L.M. (2007). Foreign-language skills in rove-beetles? Evidence for chemical mimicry of ant alarm pheromones in myrmecophilous *Pella* beetles (Coleoptera: Staphylinidae). *J. Chem. Ecol.*, **33**, 1382–1392.

Stowe, M.K. (1988). Chemical mimicry. Chapter 17. In *Chemical mediation of coevolution*, ed. K.C. Spencer. San Diego, CA: Academic Press Inc., pp. 513–580.

Stowe, M.K., Tumlinson, J.H. and Heath, R.R. (1987). Chemical mimicry: bolas spiders emit components of moth prey species sex pheromones. *Science*, **236**, 964–967.

Strohm, E. and Linsenmair, K.E. (2001). Females of the European beewolf preserve their honeybee prey against competing fungi. *Ecol. Entomol.*, **26**, 198–203.

Strohm, E., Kroiss, J., Herzner, G., Laurien-Kehnen, C., Boland, W., Schreier, P. and Schmitt, T. (2008). A cuckoo in wolves' clothing? Chemical mimicry in a specialized cuckoo wasp of the European beewolf (Hymenoptera, Chrysididae and Crabronidae). *Front. Zool.*, **5**, 2.

Sumner, S., Aanen, D.K., Delabie, J. and Boomsma, J.J. (2004). The evolution of social parasitism in *Acromyrmex* leaf-cutting ants: a test of Emery's rule. *Insectes Soc.*, **51**, 37–42.

Tengö, J. and Bergström, G. (1977). Cleptoparasitism and odor mimetism in bees: Do *Nomada* males imitate the odor of *Andrena* females? *Science*, **196**, 1117–1119.

Thomas, J.A., Knapp, J.J., Akino, T., Gerty, S., Wakamura, S., Simcox, D.J., Wardlaw, J.C. and Elmes, G.W. (2002). Parasitoid secretions provoke ant warfare. *Nature*, **417**, 505–506.

Thompson, J.N. (1994). *The coevolutionary process*. Chicago, IL: Univ. of Chicago Press.

Thompson, J.N. (2005). *The geographic mosaic of coevolution*. Chicago, IL: University of Chicago Press.

Topoff, H., Cover, S., Greenberg, L., Goodloe, L. and Sherman, P. (1988). Colony founding by queens of the obligatory slave-making ant, *Polyergus breviceps*: the role of the Dufour's gland. *Ethology*, **78**, 209–218.

Turillazzi, S., Sledge, M.F., Dani, F.R., Cervo, R., Massolo, A. and Fondelli, R. (2000). Social hackers: Integration in the host chemical recognition system by a paper wasp social parasite. *Naturwissenschaften*, **87**, 172–176.

Vánder Meer, R.K., Jouvenaz, D.P. and Wojcik, D.P. (1989). Chemical mimicry in a parasitoid (Hymenoptera: Eucharitidae) of fire ants (Hymenoptera: Formicidae). *J. Chem. Ecol.*, **15**, 2247–2261.

Vander Meer, R.K. and Wojcik, D.P. (1982). Chemical mimicry in the myrmecophilous beetle *Myrmecaphodius excavaticollis*. *Science*, **218**, 806–808.

Vásquez, G.M., Schal, C. and Silverman, J. (2008). Cuticular hydrocarbons as queen adoption cues in the invasive Argentine ant. *J. Exp. Biol.*, **211**, 1249–1256.

Vauchot, B., Provost, E., Bagnères, A.-G. and Clément, J.-L. (1996). Regulation of the chemical signatures of two termite species, *Reticulitermes santonensis* and *R. (l.) grassei*, living in mixed colonies. *J. Insect Physiol.*, **42**, 309–321.

Vauchot, B., Provost, E., Bagnères, A.-G., Rivière, G., Roux, M. and Clément, J.-L. (1998). Differential adsorption of allospecific hydrocarbons by the cuticles of two termite species, *Reticulitermes santonensis* and *R. lucifugus grassei*, living in a mixed colony. *J. Insect Physiol*, **44**, 59–66.

Véla, E., Tirard, A., Renucci, M., Suehs, C. M. and Provost, E. (2007). Floral chemical signatures in the genus *Ophrys* L. *(Orchidaceae)*: A preliminary test of a new tool for taxonomy and evolution. *Plant Molec. Biol. Report*, **25**, 83–97.

Vereecken, N. J. and Mahé, G. (2007). Larval aggregations of the blister beetle *Stenoria analis* (Schaum) (Coleoptera: Meloidae) sexually deceive patrolling males of their host, the solitary bee *Colletes hederae* Schmidt and Westrich (Hymenoptera: Colletidae). *Ann. Soc. Entomol. Fr.,* **43**, 493–496.

Viana, A. M. M., Frézard, A., Malosse, C., Della Luccia, T. M. C., Errard, C. and Lenoir, A. (2001). Colonial recognition in the fungus-growing ant *Acromyrmex subterraneus subterraneus* (Hymenoptera: Formicidae). *Chemoecology*, **11**, 29–36.

Vienne, C., Bagnères, A.-G., Errard, C. and Lange, C. (1990). Etude chimique de la reconnaissance interindividuelle chez *Myrmica rubra* et *Manica rubida* (Formicidae, Myrmicinae) élevées en colonies mixtes artificielles. *Act. Coll. Insectes Soc.*, **6**, 261–265.

Vienne, C., Soroker, V. and Hefetz, A. (1995). Congruency of hydrocarbon patterns in heterospecific groups of ants: tranfer and/or biosynthesis? *Insectes Soc.*, **42**, 267–277.

Völkl, W. and Mackauer, M. (1993). Interactions between ants attending *Aphis fabae* ssp. *cirsiiacanthoidis* on thistles and foraging parasitoid wasps. *J. Insect Behav.*, **6**, 301–312.

Völkl, W., Liepert, C., Birnbach, C. R., Hübner, G. and Dettner, K. (1996). Chemical and tactile communication between the root aphid parasitoid *Paralipsis enervis* and trophobiotic ants: consequences for parasitoid survival. *Cell. Mol. Life Sci.*, **52**, 731–738.

Wakamura, S., Arakaki, N., Yamazawa, H., Nakajima, N., Yamamoto, M. and Ando, T. (2002). Identification of epoxyhenicosadiene and novel diepoxy derivatives as sex pheromone components of the clear-winged tussock moth *Perina nuda*. *J. Chem. Ecol.*, **28**, 449–467.

Yamaoka, R. (1990). Chemical approach to understanding interactions among organisms. *Physiol. Ecol. Jpn.*, **27**, 31–52.

Yeargan, K. V. (1994). Biology of bolas spiders. *Annu. Rev. Entomol.*, **39**, 81–99.

Youngsteadt, E. and DeVries, P. J. (2005). The effects of ants on the entomophagous butterfly caterpillar *Feniseca tarquinius*, and the putative role of chemical camouflage in the *Feniseca*–ant interaction. *J. Chem. Ecol.*, **31**, 2091–2109.

Zangerl, A. R., Stanley, M. C. and Berenbaum, M. R. (2008). Selection for chemical trait remixing in an invasive weed after reassociation with a coevolved specialist. *Proc. Natl. Acad. Sci. USA*, **105**, 4547–4552.

Zimma, B. O., Ayasse, M., Tengö, J., Ibarra, F., Schulz, C. and Francke, W. (2003). Do social parasitic bumblebees use chemical weapons? (Hymenoptera, Apidae). *J. Comp. Physiol. A*, **189**, 769–775.

15

Behavioral and evolutionary roles of cuticular hydrocarbons in Diptera

Jean-François Ferveur and Matthew Cobb

The behavioral role of dipteran cuticular hydrocarbons (CHCs) was first established nearly 40 years ago, in four articles on the behavior and chemistry of the housefly, *Musca domestica*, that came out of the USDA laboratory in Gainesville (Florida). Following Rogoff *et al.* (1964), Mayer and James (1971) showed that the non-polar lipid fraction of housefly feces was attractive to male houseflies; the active components – also found on the bodies of flies – were then identified as hydrocarbons (Silhacek *et al.*, 1972b). Extracts from virgin females were attractive to males, whereas virgin male extracts were not (Silhacek *et al.*, 1972a), indicating that these substances act as sex pheromones. Finally, Carlson *et al.* (1971) used both fractionation and synthetic hydrocarbons to demonstrate that (Z)-9-tricosene, the main compound found on the female cuticle, is a sex pheromone for male houseflies.

This model has come to dominate our understanding of chemical communication in Diptera, in particular those signals involved in courtship and mating. In fact, most dipteran species do not fit the *Musca* model. Although some species, like houseflies, show a marked qualitative sexual dimorphism, many species express only quantitative variation for compounds shared by both sexes (Bartelt *et al.*, 1986; Jallon and David, 1987; Toolson and Kuper-Simbron, 1989; Byrne *et al.*, 1995), while in other species adults of both sexes appear to be virtually identical (Stoffolano *et al.*, 1997; Howard *et al.*, 2003). Furthermore, there are relatively few examples of the kind of rigorous demonstration of a pheromonal role for CHCs that was established in *Musca* (e.g. Carlson *et al.*, 1984; Adams and Holt, 1987; Ferveur and Sureau, 1996; Lacaille *et al.*, 2007).

In this chapter, we will examine the multiple roles that CHCs play in a wide variety of behaviors (courtship, mate discrimination, learning, aggregation, dominance), and in factors related to reproduction (fecundity, sex ratio). We will focus on the variety of effects that have been observed, rather than providing an exhaustive account of all hydrocarbons that have been detected in every species of dipteran. A word of warning may be necessary, however: most studies have focused on the main compound in a given species, and have tended to neglect the role of quantitatively minor compounds or of ratios. The very success of the model established in the early 1970s, coupled with evidence from our own work in *Drosophila*, indicates that other aspects of dipteran pheromones may be overlooked.

Courtship and mating

CHCs have been shown to be involved in courtship and mating in several groups of the sub-order Brachycera: Drosophilidae, Muscidae, Glossinidae, Calliphoridae and Ostridae (for reviews see Stoffolano *et al.*, 1997; Ferveur, 2005; Wicker-Thomas, 2007; Gomes and Trigo, 2008). Given their relatively high molecular weight, these compounds probably act either at a very short distance (a few centimeters) or by contact. The limited evidence we have suggests they are detected by gustatory hairs on the tarsae (Lacaille *et al.*, 2007), the labial palps, the ovipositor and perhaps the wings (Stocker, 1994; see Chapter 10; see also below).

Drosophilidae

The most far-reaching exploration of the role of cuticular hydrocarbons, their biosynthesis, their production and detection, has taken place in the model organism *Drosophila mela-nogaster* and its closely-related species. The first study of *D. melanogaster* cuticular hydro-carbons (Jackson *et al.*, 1981) reported that there was no qualitative difference between males and females, casting doubt on their potential role as sex pheromones. There is in fact a CHC sexual dimorphism in this species, and this initial finding turned out to be an artifact – flies had been taken directly from rearing bottles where both sexes were present, and their CHCs had become blended and their sex-specific profiles were blurred. Similar effects had been noted in *Musca* by Sihalcek *et al.* (1972b) and have biological implica-tions, which will be discussed further below.

From the earliest days of genetics the Morgan laboratory studied *Drosophila* courtship, and described it in visual terms – the male is extremely active, chasing after the female and vibrating his wings, before going on to lick her genitalia and finally copulate (Sturtevant, 1915). This male-centered version of *Drosophila* courtship was effectively inverted by the discovery of female-specific CHCs in *D. melanogaster* and their role in inducing male wing vibration. From this point of view, male behavior is largely a response to chemical stimuli produced by the female – far from the male "persuading" the female by his active courtship, she may be measuring his fitness in terms of the amount of wing vibration he performs in response to her stimuli, and his ability to chase after her. Testing this hypoth-esis will require a degree of experimental ingenuity that has thus far evaded *Drosophila* researchers. However, the existence of multiple signals exchanged between the two sexes suggests that this female-centered view is also too simplistic (see Figure 15.1).

Drosophila species that are not closely related to *D. melanogaster* have been intensively studied and show a range of effects that may be seen in other species of fly. In many cases, behavioral responses are mainly determined by CHC ratios: flies of the three sib-ling species of the cactophilic *D. mojavensis* cluster (*D. repleta* group), *D. mojavensis*, *D. arizonae*, and *D. navojoa* produce a mixture of long-chain components with chain lengths ranging from C_{28} to C_{40}, including *n*-alkanes, methyl-branched alkanes, *n*-alkenes, methyl-branched alkenes, and alkadienes (Etges and Jackson, 2001). In *D. mojavensis*, flies of both sexes show a reciprocal response towards the ratio of the two principal alkadienes

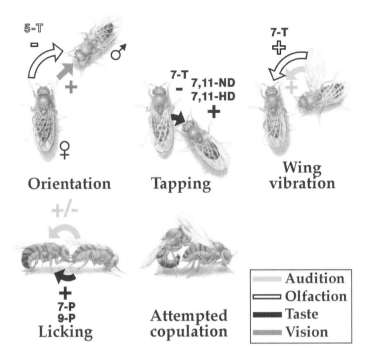

Figure 15.1 The sequence of behaviors in heterosexual courtship of *Drosophila melanogaster*. The arrows represent the known sensory modalities by which flies communicate: (+) for stimulatory and (−) for inhibitory signals. Olfactory and gustatory cues are represented respectively by white and black arrows; auditory and visual cues by light and dark gray arrows. Several cuticular hydrocarbons exchanged during each sequence can affect precise behaviors: 5-and 7-tricosene (respectively 5-T and 7-T), 7,11-hepta- and nonacosadiene (7,11-HD; 7,11-ND) and 7- and 9-pentacosene (7-P; 9-P; see text). (Adapted from Ferveur and Sureau, 1996; Greenspan and Ferveur, 2000.)

($C_{35:2}$ / $C_{37:2}$) – males prefer females with a low ratio whereas females prefer high-ratio males (Markow and Toolson, 1990). Quantitative CHC variation observed between and within these species may be related to the variety of plant the flies live on. CHCs can be affected by interactions between species: in *D. serrata* and *D. birchii* (*D. montium* subgroup), two species that are normally isolated by their niche, but which can interbreed, females prefer males with higher relative abundances of the 2-methyl-C_{28} alkane, but lower relative abundances of (Z,Z)-5,9-tetracosadiene ($C_{24:2}$) and (Z)-9-pentacosene ($C_{25:1}$; Howard *et al.*, 2003). Where the two species are sympatric, their CHC profiles tend to diverge rapidly (Higgie *et al.*, 2000; see below).

Many species, such as *D. pallidosa* (from the *D. ananassae* subgroup), show no differences in the CHC profiles of the two sexes. In this species, natural or synthetic (Z,Z)-5,27-tritriacontadiene ($C_{33:2}$) induces robust courtship by male flies (Nemoto *et al.*, 1994). Some species, however, show quantitative differences between the sexes: *D. virilis* females produce more (Z)-11-pentacosene (11-P) than their males. The application of low doses of

11-P on a dummy tended to increase male courtship responses whereas higher doses tended to inhibit male courtship, indicating that the behavioral role of a given CHC can vary depending on context (Oguma *et al.*, 1992). Finally, although hydrocarbons on the cuticle have been the focus of most attention, there may be other sources of these compounds: in some Hawaiian drosophilids, courting males secrete an anal droplet containing a mixture of hydrocarbons which stimulate female receptivity (Spieth, 1974; Tompkins *et al.*, 1993).

D. melanogaster and *D. simulans* – major CHC components

The nine species of the *D. melanogaster* subgroup of *Drosophila* have been the focus of a great deal of research, due to the wealth of genetic and ecological evidence that is available. Here, we will concentrate on data from the sibling species *D. simulans* and *D. melanogaster* – they are sympatric for much of their range and can produce sterile male hybrids in the laboratory (Lachaise *et al.*, 1988). In *D. simulans*, male and female flies produce qualitatively similar CHC profiles (Jallon and David, 1987; females produce about 10–15% CHCs more than males; Ferveur and Jallon, 1993). *D. simulans* male and female flies (and males and females of four other species in the subgroup) and *D. melanogaster* males show very similar HC profiles, with predominant (Z)-7-monoenes. *D. melanogaster* females, in contrast, produce large amounts of (Z,Z)-7,11-hepta- and nonacosadiene ($C_{27:2}$ and $C_{29:2}$; Antony and Jallon, 1982; Jallon, 1984).

The response of *D. melanogaster* flies of both sexes to (Z)-7-tricosene (7-T) is similar to that observed in *D. mojavensis* flies to the principal CHC in that species. 7-T is abundant on the cuticle of most *D. melanogaster* males and found only at low levels on females (Jallon, 1984; Sureau and Ferveur, 1999). 7-T induces a strong dose–response courtship inhibition in male *D. melanogaster* (see below) whereas females show an increased sexual receptivity to males carrying more 7-T (Ferveur and Sureau, 1996; Grillet *et al.*, 2006). The two sexes show different response thresholds: males respond to hundreds of ng of 7-T whereas females can detect tens of ng. Males detect 7-T with taste receptors on the legs and labial palps, whereas females appear to detect this relatively volatile substance with antennal olfactory sensilla (Grillet *et al.*, 2006; Lacaille *et al.*, 2007; see below).

Courtship inhibition induced by male-specific 7-T is probably related to physico-chemical effects that occur during mating. *D. melanogaster* females will generally mate only once, and males are more attracted to virgin females than to mated females. Some of the most spectacular aspects of this effect are due to the transfer of substances in the male semen, including the non-CHC *cis*-vaccenyl acetate and a series of sex peptides, both of which have been the focus of a great deal of attention because they alter the behavior of the female (e.g. Yapici *et al.*, 2008; Peng *et al.*, 2005). However, during copulation males and females exchange CHCs, thereby "blurring" their respective chemical signatures. After copulation, therefore, a female's attractive CHC blend is at least accompanied by, and may be masked by, the male's inhibitory CHCs. Using direct analysis in real-time (DART) coupled with mass spectrometry, Yew *et al.* (2008) revealed minute variations in several compounds, including some CHCs, during courtship and mating. Similar findings have been found using Solid Phase Micro-Extraction (SPME – Grillet *et al.*, unpublished data). Physical contact is not necessary for these effects – the signals may be exchanged through the substrate, or through close contact with a cloud of CHC particles that may exist around the fly.

This simple physical effect presumably occurs in all insects that possess CHCs (there is evidence for this effect in the Glossinidae – see below); however, it may be of biological significance only in those species that have sexually dimorphic CHCs. This passive transfer effect can also be used experimentally – by housing flies with different CHC profiles in close proximity, their chemical signatures can be modified without damaging the subjects (Coyne *et al.*, 1994). Mating may also have an indirect effect on CHC levels: a slight quantitative post-mating variation (10–15%), probably due to a physiological effect, was found in *D. simulans* females after copulation (Ferveur and Jallon, 1993).

The role of the main *D. melanogaster* female CHCs (7,11-dienes) is more complex than initially suspected (Jallon, 1984; Ferveur *et al.*, 1996). It was at first thought that the amount of male wing vibration was directly related to the amount of 7,11 dienes, and that this would affect mating levels. A survey of more than 70 worldwide geographical strains revealed substantial differences in the levels of 7,11-dienes shown by females of these strains, but despite an initial report by ourselves (Ferveur *et al.*, 1996), these variations in fact have no effect on male behavior (Coyne *et al.*, 1999). By studying the responses of males to genetically-manipulated females that have no 7,11 dienes, and are then "perfumed" with different 7,11 dienes by the passive transfer method described above, subtle intraspecific effects can be detected before, during and after mating. Although genetically identical females with either no 7,11-dienes or covered in these substances induced similar levels of courtship when tested separately, in a "choice" experiment a control male would prefer a 7,11-diene-rich female over a female without these substances. Furthermore, 7,11-diene-rich females induced longer-lasting copulation and yielded a progeny with fewer daughters than females with no 7,11 dienes (Marcillac and Ferveur, 2004).

These same 7,11-dienes exert a strong interspecific effect and inhibit courtship and mating by *D. simulans* males (Savarit *et al.*, 1999). *D. simulans* males will strongly court *D. melanogaster* females with no 7,11 dienes, but when these substances are transferred onto *D. melanogaster* females, *D. simulans* males do not court them, just as with wild-type *D. melanogaster* females. Conversely, the transfer of 7-T onto the cuticle of deprived *D. melanogaster* females strongly enhances *D. simulans* male courtship and mating and inhibits *D. melanogaster* males (Savarit *et al.*, 1999). This finding, together with the evidence of Higgie *et al.* (2000), suggests that at least some dipteran CHCs involved in courtship and mating also act as reinforcers of isolation between species (see also Chapter 7).

Both *D. simulans* and *D. melanogaster* show geographical variation in the amount of their principal CHCs. Whereas most strains of *D. simulans* produce more (Z)-7-tricosene (7-T) and less 7-pentacosene (7-P), those from the Benin Gulf show an inverse ratio. Similarly, *D. melanogaster* males from subtropical areas show a gradual increase of 7-P and a decline in their levels of 7-T. In a separate cline, 7,11-dienes in females from Afro-equatorial and Caribbean strains of *D. melanogaster* are completely replaced by 5,9 isomers (Ferveur *et al.*, 1996). This geographic variation depends on the *desat2* gene (Dallerac *et al.*, 2000; Takahashi *et al.*, 2001), but the selection pressures that cause it to be maintained in the wild are not understood (Coyne and Elwyn, 2006).

D. melanogaster and *D. simulans* – minor CHC components

The CHC profile of most drosophilids includes from 15 to over 30 peaks, depending upon the species, sex, strain, and age. To investigate the role of the minor peaks in courtship, *D. melanogaster* males were stimulated with a large number of live immobilized target flies that produced various blends of CHCs and their behavior was observed (Ferveur and Sureau, 1996; see Figure 15.1).

First, a correlation was found between the occurrence, latency and duration of a series of male behaviors and the absolute and relative levels of some CHCs. Latency to initiate courtship (orientation and tapping) was inversely related to the amount of 5-T. This relatively volatile CHC is produced mainly – but not exclusively – by males and may increase female receptivity in some strains. Males of some Zimbabwean strains produce more 5-T than cosmopolitan males (Grillet *et al.*, unpublished data). Female flies from Zimbabwe show a strong sexual discrimination against cosmopolitan males (e.g. Wu *et al.*, 1995), but it is not yet clear whether 5-T is involved in this strong intraspecific sexual isolation.

Second, the two principal male and female CHCs (7-T and 7,11-dienes) are apparently perceived during gustatory contact at the onset of courtship, when the male taps the female cuticle with its front legs. These two compounds exert an antagonistic effect (negative for 7-T, positive for 7,11-dienes) on the duration of subsequent male behaviors (wing vibration and licking; Ferveur and Sureau, 1996).

Finally, the two pentacosene isomers, 7-P and 9-P, are apparently tasted during male licking and induce a synergistic effect on the frequency and duration of attempted copulation. The "copulatory" effect of 9-P was supported by the discovery that males can learn to associate this CHC with unconditional aversive stimuli produced by an unreceptive fly – conditioned flies attempted to copulate less frequently and less intensively with flies carrying more 9-P (Siwicki *et al.*, 2005).

Calliphoridae

The CHCs of the blow-fly *Phormia regina* are complex mixtures of saturated *n*-, mono-methyl- and dimethylalkanes with 23–33 carbon atoms. These CHCs do not appear to change with age or diet, and differ only slightly between the sexes (Byrne *et al.*, 1995; Stoffolano *et al.*, 1997). Males show the same strong copulatory response to dummies covered with the extract of either sex but a very reduced response to hexane-washed flies (Stoffolano *et al.*, 1997). In *Calliphora vomitoria*, the situation is less clear. Ablation of the female *corpora allata* or of the ovaries leads to an increase in the proportion of monomethylalkanes. However, these two procedures induce divergent effects on male attraction. Moreover, when hydrocarbon production was unchanged, male attraction was reduced, and when overall CHC production was reduced, male attraction was not altered (Trabalon *et al.*, 1994). These contradictory findings may indicate that other CHCs – or other stimuli – were affected and altered the behavior of the tester males.

In the screw-worm fly (*Cochliomyia hominivorax*), courtship is generally very brief; as a result, bioassays are generally based on male copulation levels (Hammack, 1986). Some CHCs show different levels between the sexes and are thought to be involved in courtship and mating (Pomonis, 1989). Similarly, in *Lucilia cuprina*, CHCs are thought to play a role in sexual behavior, but decisive evidence has yet to be produced (Emmens, 1981).

Ostridae

The American warble fly, *Dermatobia hominis*, is a species that attacks humans. Adults produce mainly *n*-saturated alkanes and various mono- and dimethylalkanes with 21–41 carbon atoms. Females produce more *n*-tricosane and *n*-pentacosane than males, which produce more CHCs with at least 29 carbon atoms. Olfactometer tests show a reciprocal attraction of females and males to the extract of the other sex, and suggest that CHCs are used for sexual discrimination (Gomes and Trigo, 2008).

Muscidae

In *Musca domestica* females, the principal CHC, (Z)-9 tricosene (9-T, or musculure) attracts males at a short distance (Carlson *et al.*, 1971). However, a survey of wild-type and laboratory populations revealed that the amount of 9-T can be reduced or even absent in females from some populations, challenging the idea that this CHC is necessary for mating (Darbro *et al.*, 2005). Several minor compounds that are apparently perceived by contact seem to act either as excitants or inhibitors of mating behavior (Uebel *et al.*, 1976; Adams and Holt, 1987). In particular, *cis*-9,10-epoxytricosane allows mate discrimination whereas the two methylalkanes 4,8-dimethylheptacosane and 13-methylnonacosane, reinforce the arrestant activity of other substances. Other non-chemical signals including acoustic and tactile stimuli produced by the male could also be involved in courtship, although these signals do not appear to be crucial for female choice (Meffert and Bryant, 1991; Goulson *et al.*, 1999).

Glossinidae

In the tse-tse fly (*Glossina tachinoïdes*) courtship is very brief and difficult to characterize. First, the male uses visual cues to recognize a potential conspecific partner. Then, several pheromonal components carried by flies of both sexes seem to reciprocally facilitate sexual receptivity whereas other molecules are required for successful mating. When placed on solvent-washed dummies, biological doses of synthetic 11,23-, and (to a lesser extent) of 13,25-dimethylheptatriacontane elicit most of the behaviors normally induced by a female (Carlson *et al.*, 1978; El Messoussi *et al.*, 1994; Carlson *et al.*, 1998). In *G. austeni*, the male copulatory response can be elicited by a mixture of five alkenes including the two principal compounds, 13,17-dimethyltritriacont-1-ene and 13,17-dimethylpentatriacont-1-ene. Biological doses of both synthetic compounds induce male copulatory responses with washed frozen-killed females (Carlson *et al.*, 2005). In contrast, the principal alkene of *G. morsitans morsitans* males, 19,23-dimethyltritriacont-1-ene, which is partially transferred to females during courtship and/or mating, inhibits male sexual activity. When isolated from flies, this component induces a dose-dependent antiaphrodisiac effect on male flies: extracts corresponding to two and five male-equivalents cause respectively 80% and 100% inhibition of copulation attempts (Carlson and Schlein, 1991). This CHC appears to inhibit a male from courting a previously mated female.

Evolutionary and neurobiological considerations

One explanation for the existence of sex pheromones that are not sex-specific may be that other signals (not necessarily chemical) are involved in mate identification and stimulation in these species (Savarit *et al.*, 1999). These other signals may be important precisely because in the wild, where insects come into contact with each other, chemical signals may not be as pure as in strictly controlled laboratory conditions. The existence of species with no apparent sexual dimorphism in their CHC sex pheromones also raises problems about the processing of these signals. Although a male from one of these species is presumably in permanent contact with a source of excitatory pheromones, either on his own cuticle or in traces of CHC that surround him, there is no evidence that he is in a state of permanent sexual excitation. This may indicate that other signals (not necessary chemical) are necessary before the activation of CHC-detecting neurons is transformed into a behavioral response, and/or that the receptor molecules on the dendritic membrane of these neurons may be continually cleared to prevent sensory adaptation from taking place.

Male courtship of other males and of immature flies

Mature *D. simulans* males, like males from other *D. melanogaster* subgroup species that show no CHC sexual dimorphism, often engage in male–male courtship. However, these interactions rarely extend beyond the earliest phases of courtship, suggesting that other signals (chemical, acoustic or visual) are involved in identifying the sex of a potential partner (Paillette *et al.*, 1991). While increasing doses of 7-T elicit strong *D. simulans* male courtship, this has the opposite effect in *D. melanogaster* males (Jallon, 1984; Ferveur and Sureau, 1996; Savarit *et al.*, 1999; Sureau and Ferveur, 1999). A few hundred nanograms of 7-T can inhibit male–male courtship in *D. melanogaster*, but males show variation in their inhibition thresholds (Ferveur and Sureau, 1996; Sureau and Ferveur, 1999). 7-T is detected by peripheral taste neurons that show dose-dependent electrophysiological responses (Lacaille *et al.*, 2007). The threshold of detection of taste neurons located on the labial palps is between 10^{-10} and 10^{-9} M of pure synthetic 7-T.

Interestingly, bitter molecules (caffeine, quinine, berberine) are detected by the same neurons and when painted on the cuticle of immobilized live males also dose-dependently inhibit male–male courtship, suggesting that 7-T tastes "bitter" to *D. melanogaster* males (Lacaille *et al.*, 2007).

The cline shown by *D. melanogaster* males in the ratio of 7-T:7-P, mentioned above, leads to asymmetric male–male courtship between these strains. 7-P-rich males are strongly courted by 7-T-rich males but the inverse does not occur, indicating that 7-P-rich males are not stimulated by 7-T. The variation in courtship response is linked to genetic factors that are independent of those involved in the variation of the 7-T:7-P ratio (Sureau and Ferveur, 1999).

Such effects have been little studied in other dipteran species, but the principal CHC of *G. morsitans morsitans* males, 19,23-dimethyltritriacont-1-ene, strongly inhibits male copulatory behavior (Carlson and Schlein, 1991), and may play a role in restricting male–male interactions in this species.

Although male *D. melanogaster* do not normally court homotypic males, they will intensively court immature young flies of both sexes (McRobert and Tompkins, 1983), presumably attracted by the long-chain CHCs that characterize immature flies of both sexes in this species (Pechiné *et al.*, 1988; see below). Intriguingly, this situation is also found in the two other species of the *D. melanogaster* subgroup that show sexually dimorphic CHCs – *D. erecta* and *D. sechellia* (Cobb and Jallon, 1990), whereas the four monomorphic species that have been studied (*D. simulans, D. mauritiana, D. yakuba* and *D. teissieri*) all show high levels of courtship of mature males, but virtually no courtship of immature flies. The evolutionary significance of these effects is unclear. McRobert and Tompkins (1988) suggest that males that have been courted as young flies may subsequently have an advantage. We have found that when an immature *D. melanogaster* male is exposed to the main adult male 7-monoene (7-T and/or 7-P) during its first 24 hours of adult life, it shows increased dominance behavior with a naïve sibling male four days later (Svetec *et al.*, 2005).

Intra-individual CHC variability

The CHC profile of an individual insect does not remain constant through life. A number of factors have been studied in various dipterans that contribute to these changing profiles.

Cucilidae

Age is a critical determinant of an adult female mosquito's ability to transmit human pathogens and in many species CHCs are a good marker for determining the insect's age (*Anopheles stephensi*; Brei *et al.*, 2004; *Anopheles gambiae*, Caputo *et al.*, 2005; *Anopheles farauti, Aedes aegypti* and *Ochlerotatus vigilax*, Hugo *et al.*, 2006). However, the CHC level in adult females can also change after mating (Polerstock *et al.*, 2002).

Calliphoridae

CHC levels significantly change with age and sexual maturation in *Cochliomyia hominivo-rax*, *Chrysomya bezziana* and *Calliphora vomitoria* (Pomonis, 1989; Brown *et al.*, 1992; Trabalon *et al.*, 1992). CHC levels can also change if individuals are reared in isolation or in groups (Benziane and Campan, 1993). Most of these changes, which may induce variation in courtship intensity and mating success, are presumed to be under hormonal control (Trabalon *et al.*, 1990, 1994).

Muscidae

In *Musca domestica*, a number of new components appear on the cuticle with adult maturation, between two and six days old, during the same period that the level of (Z)-9-tricosene increases and that of (Z)-9-heptacosene decreases (Mpuru *et al.*, 2001).

Drosophilidae

The long chain CHCs (27–35C) are similar on the cuticle of immature male and female flies of the *D. melanogaster* and *D. virilis* group of species (Pechiné *et al.*, 1985, 1988). In *D. melanogaster*, mature CHCs have completely replaced the immature compounds on the fly cuticle after 36 hours (Antony and Jallon, 1982). The maturation of CHCs corresponds to the maturation of sexual behavior, but each maturation process is under separate genetic control: *desat1* for CH production (Marcillac *et al.*, 2005a) and *prospero* for male courtship maturation (Grosjean *et al.*, 2007).

Fitness effects

In *Musca domestica*, the changes in female CHCs that occur during maturation have been shown to be related to ovarian function (Dillwith *et al.*, 1983), and more specifically to the activation of ecdysteroids, as demonstrated by ovarian transplant experiments (Blomquist *et al.*, 1992; see Chapters 3 and 5). We hypothesize that in all Diptera, female CHCs are related to ovarian activity. In many social insects, female CHC profile varies with ovarian function, and can be considered an "honest signal" of fertility (e.g. Cuvillier-Hot *et al.*, 2002; LeConte and Hefetz, 2008). Male flies may use variation in female CHCs as a way of choosing between females; there is little direct evidence for this, but the relationship between variability in mating and courtship and natural intra-populational variation in female CHCs has not been directly investigated. However, there is indirect evidence of such an effect – in *Drosophila serrata* genetic variance underlying female CHC variability accounts for a substantial proportion of the indirect genetic variation in male CHCs, suggesting that males are assessing variability in female CHCs (Petfield *et al.*, 2005).

Interspecific variation and evolutionary effects

Although CHC profiles tend to be species-specific (reviewed by Howard, 1993 and by Howard and Blomquist, 2005), in many species they cannot always be reliably used as phylogenetic markers – this is particularly the case in many drosophilids (Ferveur, 2005; see also Chapter 7). Examples of qualitative geographic variability in *Drosophila* CHCs have been given above. In *Anopheles* mosquitoes, adult CHCs can be used to reliably discriminate between populations in both inter- and intra-specific studies (Phillips *et al.*, 1990; Anyanwu *et al.*, 1993). However, this variability suggests they may not always be reliable for identifying species.

Because the CHCs involved in mating and courtship are not selectively neutral, reinforcing selection may cause species that are very closely related to have distinct profiles. The inter-fertile species *D. serrata* and *D. birchii* show an overlapping geographical distribution and geographic variability in their CHCs. After only nine generations of experimental sympatry, the CHC profile of *D. serrata* flies from strains that normally do not encounter *D. birchii* came to resemble that of *D. serrata* flies that naturally live in sympatry with *D. birchii* (Higgie *et al.*, 2000). This example of character displacement suggests that reinforcing selection can affect CHCs involved in mate recognition (Blows and Allan, 1998). Indicative, but non-experimental, data can be found in the divergence of the *D. melanogaster/D. simulans* sibling species (Jallon and David, 1987) and the differing behavioral effects of their principal CHCs described above. These findings all underline the fact that the pheromonal roles of CHCs may be both intra- and interspecific (Coyne *et al.*, 1994; Coyne and Oyama, 1995; Savarit *et al.*, 1999; Ferveur, 2005).

Intraspecific variation in CHC production and associated behavioral responses may be a critical step in speciation. It can be assumed that variation in CHCs is produced by evolutionary changes in the biosynthetic enzymes underlying pheromone production, as for CHCs and other classes of chemicals in other insect orders (Roelofs *et al.*, 2002; Sillam-Dussès *et al.*, 2005; Eliyahu *et al.*, 2007; see Chapters 16–19). Desaturases, which are highly conserved across many insect orders, seem to be important enzymes with regard to the evolution of pheromones (Xue *et al.*, 2007). Several *D. melanogaster* populations that show a marked sexual isolation also differ in their male and/or female CHCs (Sureau and Ferveur, 1999; Wu *et al.*, 1995; Grillet *et al.*, unpublished). Similar effects are seen in *D. elegans* (Ishii *et al.*, 2001).

For any speciation process based on pheromonal communication to be effective, variation in CHC production should tightly co-evolve with factors regulating CHC processing. This has been observed for the *desat1* gene in *D. melanogaster* (Marcillac *et al.*, 2005b). Such co-evolution requires that pre-existing sensory structures can detect and respond to "new" CHC molecules. This basic assumption of the theory of sexual selection is supported by the observation that taste neurons normally used to detect bitter molecules also serve to detect an aversive sex pheromone (Lacaille *et al.*, 2007). It is thus possible that taste neurons that were initially used by the fly to detect noxious food molecules (bitterness is often associated with alkaloids and toxic molecules) have been more recently used to detect inhibitory sex pheromones.

It is generally assumed that the initial and fundamental function of insect CHCs is protection against desiccation (Gibbs, 1998; see Chapter 6) and that pheromonal effects have been "grafted" on top of this basic function. This suggestion has been supported by the observation of changes in a laboratory population of *D. pseudoobscura* CHCs over a seven-year period (Toolson and Kuper-Simbron, 1989), thereby underlining one potential source of geographic and temporal variability in the CHC signature of a given species. However, an experimental study of water conservation mechanisms in various *Drosophila* species, including cactophilic desert species, found no relation between CHC composition and water loss (Gibbs *et al.*, 2003). This may indicate that the mere presence of CHCs acts as a barrier against desiccation, and that interspecific variability in CHC profiles is primarily a consequence of selective effects (including sexual selection and reinforcing selection) and random drift. Determining the weight of these various factors – natural selection, sexual selection, reinforcing selection and random variation – in the variability shown by any given pair of species will be a matter for experimental investigation. We would not expect there to be a uniform explanation of such variability.

Future prospects

Because of the important economic and applied implications of many dipteran species, and the predominant role of *Drosophila melanogaster* as a model system, the role of CHCs in behavior, and their varying functions and interactions with other modes of communication – chemical and otherwise – will continue to be a focus of intense interest. However, we see three related areas of research on CHCs that have barely been touched in Diptera and that could usefully be pursued.

Social insects may use larval CHCs to recognize their brood (e.g. Bonavita-Cougourdan *et al.*, 1989), although it is not known whether larvae are able to detect the CHCs on their nurses (e.g. in ants, in response to offers of trophallaxis). There is evidence from another social insect, the bumblebee, *Bombus terrestris*, and from the burying beetle, *Nicrophorus vespilloides*, that begging larvae use CHCs to detect the presence of their parents and elicit food from them (den Boer and Duchateau, 2006; Smiseth and Brown, personal communication). No known dipteran species show such altruistic behavior – parental investment in individual offspring is low. However, dipteran larvae may interact with each other behaviorally (Kaiser and Cobb, 2008), while female flies can respond to the presence of eggs and larvae (e.g. *D. melanogaster* and *Musca domestica* – Chess *et al.*, 1990 and Lam *et al.*, 2007, respectively). CHCs perceived during larval development can influence sexual discrimination in *D. paulistorum* flies of different semi-species (Kim *et al.*, 2004) – flies raised communally show a greater ability to discriminate between semi-species (Kim *et al.*, 1996a, b; Kim and Ehrman, 1998). Larval detection of CHCs may require the gustatory receptor gene *Gr66a*, which is involved in the detection of 7-T, and is expressed in the larval terminal organ (Scott *et al.*, 2001).

Most studies of dipteran CHCs have been carried out on brachyceran species, which generally have a terrestrial larval stage. The economic and health implications of some mosquito species have led to a number of investigations of adult Nematocera, which have an aquatic larval stage, beginning with the *Anopheles gambiae* complex (Carlson and Service, 1979). Here, too, the study of larval CHCs and potential larval responses to CHCs may be rewarding. Dipteran embryos are covered with species-specific hydrocarbons that are assumed to be related to ecology and desiccation (Nelson and Leopold, 2003), and *Drosophila melanogaster* larvae carry complex cuticular hydrocarbons (Ferveur, unpublished data). A study of the CHCs in the larvae of the Nematocera could provide new insights into both the identification of these species, and potential new methods of pest control. In turn, the existence of larval CHCs would provide further material for investigating both the assumed selective advantage of CHCs in controlling water balance (either through protection against desiccation or, in the case of aquatic species, the regulation of homeostasis), and also for the investigation of possible CHC-based interactions between larvae or between adults and larvae.

Finally, the detection of CHCs and their processing in the insect brain remain poorly understood (see Chapter 10). It is assumed that because of their relatively low volatility, CHCs are generally perceived by contact, using gustatory organs that are also involved in food detection (Dethier, 1976). In *D. melanogaster*, the male inhibitory pheromone 7-T and several bitter molecules induce a repulsive effect and are processed by the same gustatory neurons housed in taste hairs of the labial palps (Lacaille *et al.*, 2007). All these taste neurons express the same taste receptor molecule (GR66a) that responds to caffeine (Moon *et al.*, 2006). Moreover, males genetically feminized for this specific group of gustatory neurons show increased inhibition when stimulated with 7-T and with caffeine (Lacaille *et al.*, in press). This indicates that these taste neurons perceive CHCs with a lower threshold of sensitivity in females than in males. Another *Drosophila* taste receptor molecule, GR68a, found in male-specific neurons in a small number of tarsal hairs, has also been implicated in the detection of female-specific pheromone(s), although the chemical identity of these compounds remains unknown (Bray and Amrein, 2003). Identifying the precise location of CHC detection, and the way in which receptors identify and neurons process these signals and can potentially discriminate between them, will provide further insight into the function and evolution of these essential components of chemical communication in the Diptera.

References

Adams, T. S. and Holt, G. G. (1987). Effect of pheromone components when applied to different models on male behavior in the housefly, *Musca domestica. J. Insect Physiol.*, **33**, 9–18.

Antony, C. and Jallon, J.-M. (1982). The chemical basis for sex recognition in *Drosophila melanogaster. J. Insect Physiol.*, **28**, 873–880.

Anyanwu, G. I., Davies, D. H., Molyneux, D. H., Phillips, A. and Milligan, P. J. (1993). Cuticular hydrocarbon discrimination variation among strains of the mosquito *Anopheles (Cellia) stephensi* Liston. *Ann. Trop. Med. Parasitol.*, **87**, 269–275.

Bartelt, R. J., Arnold, M. T., Schaner, A. M. and Jackson, L. L. (1986). Comparative analysis of cuticular hydrocarbons in the *Drosophila virilis* species group. *Comp. Biochem. Physiol. B*, **83**, 731–742.

Benziane, T. and Campan, M. (1993). Effects of isolated rearing on gonadotropic development, cuticular hydrocarbons production and mating behavior of *Calliphora vomitoria* (Diptera, Calliphoridae). *Can. J. Zool.*, **71**, 1175–1181.

Blomquist, G., Adams, T.S., Halarnkar, P., Gu, P., MacKay, M. and Brown, L. (1992). Ecdysteroid induction of sex pheromone biosynthesis in the housefly, *Musca domestica*. Are other factors involved? *J. Insect Physiol.*, **38**, 309–318.

Blows, M.W., and Allan, R.A. (1998). Levels of mate recognition within and between two *Drosophila* species and their hybrids. *Am. Nat.*, **152**, 826–837.

Bonavita-Cougourdan, A., Clément, J.-L. and Lange, C. (1989). The role of cuticular hydrocarbons in recognition of larvae by workers of the ant *Camponotus vagus*: Changes in the chemical signature in response to social environment (Hymenoptera: Formicidae). *Sociobiology*, **16**, 49–74.

Bray, S. and Amrein, H. (2003). A putative *Drosophila* pheromone receptor expressed in male-specific taste neurons is required for efficient courtship. *Neuron*, **39**, 1019–1029.

Brei, B., Edman, J.D., Gerade, B. and Clark, J.M. (2004). Relative abundance of two cuticular hydrocarbons indicates whether a mosquito is old enough to transmit malaria parasites. *J. Med. Entomol.*, **41**, 807–809.

Brown, W.V., Morton, R. and Spradbery, J.P. (1992). Cuticular hydrocarbons of the old-world screwworm fly, *Chrysomya bezziana* Villeneuve (Diptera, Calliphoridae) – Chemical characterization and quantification by age and sex. *Comp. Biochem. Physiol. B*, **101**, 665–671.

Byrne, A.L., Camann, M.A., Cyr, T.L., Catts, E.P. and Espelie, K.E. (1995). Forensic implications of biochemical differences among geographic populations of the black blow fly, *Phormia regina* (Meigen). *J. Forensic Sci.*, **40**, 372–377.

Caputo, B., Dani, F.R., Horne, G.L., Petrarca, V., Turillazzi, S., Coluzzi, M., Priestman, A.A. and della Torre, A. (2005). Identification and composition of cuticular hydrocarbons of the major Afrotropical malaria vector *Anopheles gambiae* s.s. (Diptera: Culicidae): analysis of sexual dimorphism and age-related changes. *J. Mass Spectrom.*, **40**, 1595–1604.

Carlson, D.A., Langley P.A. and Huyton, P. (1978). Sex pheromone of the tsetse fly: isolation, identification, and synthesis of contact aphrodisiacs. *Science*, **201**, 750–753.

Carlson, D.A., Mayer, M.S., Silhacek, D.L., James, J.D., Beroza, M. and Bierl, B.A. (1971). Sex attractant pheromone of the house fly: isolation, identification and synthesis. *Science*, **174**, 76–78.

Carlson, D.A., Mramba, F., Sutton, B.D., Bernier, U.R., Geden, C.J. and Mori, K. (2005). Sex pheromone of the tsetse species, *Glossina austeni*: isolation and identification of natural hydrocarbons, and bioassay of synthesized compounds. *Med. Vet. Entomol.*, **19**, 470–479.

Carlson, D.A., Nelson, D.R., Langley, P.A., Coates, T.W., Davis, T.L. and Leegwater-van der Linden, M.E. (1984). Contact sex pheromone in the tsetse fly *Glossina pallidipes* (Austen). Identification and synthesis. *J. Chem. Ecol.*, **10**, 429–450.

Carlson, D.A., Offor, I.I., El Messoussi, S., Matsuyama, K., Mori, K. and Jallon, J.-M. (1998). Sex pheromone of *Glossina tachinoides*: isolation, identification, and synthesis. *J. Chem. Ecol.*, **24**, 1563–1575.

Carlson, D.A. and Schlein, Y. (1991). Unusual polymethyl alkenes in tsetse-flies acting as abstinon in *Glossina morsitans*. *J. Chem. Ecol.*, **17**, 267–284.

Carlson, D.A. and Service, M.W. (1979). Differentiation between species of the *Anopheles gambiae* Giles complex (Diptera: Culicidae) by analysis of cuticular hydrocarbons. *Ann. Trop. Med. Parasitol.*, **73**, 589–592.

Chess, K. F., Ringo, J. M. and Dowse, H. B. (1990). Oviposition by two species of *Drosophila*: behavioral responses to resource distribution and competition. *Ann. Entomol. Soc. Am.*, **83**, 717–724.

Cobb, M. and Jallon, J.-M. (1990). Pheromones, mate recognition and courtship stimulation in the *Drosophila melanogaster* species sub-group. *Anim. Behav.*, **39**, 1058–1067.

Coyne, J. A. and Elwyn, S. (2006). Does the desaturase-2 locus in *Drosophila melanogaster* cause adaptation and sexual isolation? *Evolution, 60*, 279–291.

Coyne, J. A., Mah, K. and Crittenden, A. P. (1994). Pheromonal hydrocarbons and sexual isolation in *Drosophila. Science*, **265**, 1461–1464.

Coyne, J. A. and Oyama, R. (1995). Localization of pheromonal sexual dimorphism in *Drosophila melanogaster* and its effect on sexual isolation. *Proc. Natl. Acad. Sci. USA, 92*, 9505–9509.

Coyne, J. A., Wicker-Thomas, C. and Jallon, J.-M. (1999). A gene responsible for a cuticular hydrocarbon polymorphism in *Drosophila melanogaster. Genet. Res., 73*, 189–203.

Cuvillier-Hot, V., Gadagkar, R., Peeters, C. and Cobb, M. (2002). Regulation of reproduction in a queenless ant: aggression, pheromones and reduction in conflict. *Proc. R. Soc. Lond. B,* **269**, 1295–1300.

Dallerac, R., Labeur, C., Jallon, J.-M., Knipple, D. C., Roelofs, W. L. and Wicker-Thomas, C. (2000). A delta 9 desaturase gene with a different substrate specificity is responsible for the cuticular diene hydrocarbon polymorphism in *Drosophila melanogaster. Proc. Natl. Acad. Sci. USA,* **98**, 3920–3925.

Darbro, J. M., Millar, J. G., McElfresh, J. S. and Mullens, B. A. (2005). Survey of muscalure [(Z)-9-tricosene] on house flies (Diptera: Muscidae) from field populations in California. *Environ. Entomol.*, **34**, 1418–1425.

den Boer, S. P. A. and Duchateau, M. J. H. M. (2006). A larval hunger signal in the bumblebee *Bombus terrestris. Insectes Soc.,* **53**, 369–373.

Dethier, V. G. (1976). *The Hungry Fly: a Physiological Study of the Behavior Associated with Feeding*. Cambridge, MA: Harvard University Press.

Dillwith, J. W., Adams, T. S. and Blomquist, G. J. (1983). Correlation of housefly sex pheromone production with ovarian development. *J. Insect Physiol.*, **29**, 377–386.

Eliyahu, D., Nojima, S., Mori, K. and Schal, C. (2007). New contact sex pheromone components of the German cockroach, *Blattella germanica*, predicted from the proposed biosynthetic pathway. *J. Chem. Ecol.*, **34**, 229–237.

El Messoussi, S., Wicker, C., Arienti, M., Carlson, D. A. and Jallon, J.-M. (1994). Hydrocarbons in species recognition in insects. In *The Identification and Characterization of Insect Pests*, ed. D. L. Hawksworth. Wallingford: CAB International, pp. 277–287.

Emmens, R. L. (1981). Evidence for an attractant in cuticular lipids of female *Lucilia cuprina* (Wied), Australian sheep blowfly. *J. Chem. Ecol.,* **7**, 529–541.

Etges, W. J. and Jackson, L. (2001). Epicuticular hydrocarbon variation in *Drosophila mojavensis* cluster species. *J. Chem. Ecol.,* **27**, 2125–2149.

Ferveur, J.-F. (2005). Cuticular hydrocarbons: their evolution and roles in *Drosophila* pheromonal communication. *Behav. Genet.*, **35**, 279–295.

Ferveur, J.-F., Cobb, M., Boukella, H. and Jallon, J.-M. (1996). World-wide variation in *Drosophila melanogaster* sex pheromone: behavioral effects, genetic bases and potential evolutionary consequences. *Genetica, 97*, 73–80.

Ferveur, J.-F. and Jallon, J.-M. (1993). Genetic control of pheromones in *Drosophila simulans*. II. kété, a locus on the X chromosome. *Genetics,* **133**, 561–567.

Ferveur, J.-F. and Sureau, G. (1996). Simultaneous influence on male courtship of stimulatory and inhibitory pheromones, produced by live sex-mosaic *Drosophila melanogaster*. *Proc. R. Soc. Lond. B*, **263**, 967–973.

Gibbs, A. G. (1998). Water-proofing properties of cuticular lipids. *Am. Zool.*, **38**, 471–482.

Gibbs, A. G., Fukuzato, F. and Matzkin, L. M. (2003). Evolution of water conservation mechanisms in *Drosophila*. *J. Exp. Biol.*, **206**, 1183–1192.

Gomes, C. C. G. and Trigo, J. R. (2008). Sex pheromone of the American warble fly, *Dermatobia hominis*: The role of cuticular hydrocarbons. *J. Chem. Ecol.*, **34**, 636–646.

Goulson, D., Bristow, L., Elderfield, E., Brinklow, K., Parry-Jones, B. and Chapman, J. W. (1999). Size, symmetry, and sexual selection in the housefly, *Musca domestica*. *Evolution*, **53**, 527–534.

Greenspan, R. J. and Ferveur, J.-F. (2000). Courtship in *Drosophila*. *Annu. Rev. Genet.*, **34**, 205–232.

Grillet, M., Dartevelle, L. and Ferveur, J.-F. (2006). A *Drosophila* male pheromone affects female sexual receptivity. *Proc. R. Soc. Lond. B*, **273**, 215–223.

Grosjean, Y., Guenin, L., Bardet, H. M. and Ferveur, J.-F. (2007). Prospero affects precocious sexual behavior in *Drosophila* males. *Behav. Genet.*, **37**, 575–584.

Hammack, L. (1986). Pheromone-mediated copulatory responses of the screwworm fly, *Cochliomyia hominivorax*. *J. Chem. Ecol.*, **12**, 1623–1631.

Higgie, M., Chenoweth, S., and Blows, M. W. (2000). Natural selection and the reinforcement of mate recognition. *Science*, **290**, 519–521.

Howard, R. W. (1993). Cuticular hydrocarbons and chemical communication. In *Insect Lipids: Chemistry, Biochemistry and Biology*, ed. D. W. Stanley-Samuelson and D. D. Nelson. Lincoln, NE: University Nebraska Press, pp. 179–226.

Howard, R. W., Jackson, L. L., Banse, H. and Blows, M. W. (2003). Cuticular hydrocarbons of *Drosophila birchii* and *D. serrata*: identification and role in mate choice in *D. serrata*. *J. Chem. Ecol.*, **29**, 961–976.

Howard, R. W. and Blomquist, G. J. (2005). Ecological, behavioral, and biochemical aspects of insect hydrocarbons. *Annu. Rev. Entomol.*, **50**, 371–393.

Hugo, L. E., Kay, B. H., Eaglesham, G. K., Holling, N. and Ryan, P. A. (2006). Investigation of cuticular hydrocarbons for determining the age and survivorship of Australasian mosquitoes. *Am. J. Trop. Med. Hygiene*, **74**, 462–474.

Ishii, K., Hirai, Y., Katagiri, C. and Kimura, M. T. (2001). Sexual isolation and cuticular hydrocarbons in *Drosophila elegans*. *Heredity*, **87**, 392–399.

Jackson, L. L., Arnold, M. T. and Blomquist, G. J. (1981). Surface lipids of *Drosophila melanogaster*: comparison of the lipids from female and male wild type and sex-linked yellow mutant. *Insect Biochem.*, **11**, 87–91.

Jallon, J.-M. (1984). A few chemical words exchanged by *Drosophila* during courtship and mating. *Behav. Genet.*, **14**, 441–478.

Jallon, J.-M. and David, J. R. (1987). Variations in cuticular hydrocarbons among the eight species of the *Drosophila melanogaster* subgroup. *Evolution*, **41**, 294–302.

Kaiser, M. and Cobb, M. (2008). The behaviour of *Drosophila melanogaster* maggots is affected by social, physiological and temporal factors. *Anim. Behav.*, **75**, 1619–1628.

Kim, Y. K., and Ehrman, L. (1998). Developmental isolation and subsequent adult behavior of *Drosophila paulistorum*. IV. Courtship. *Behav. Genet.*, **28**, 57–65.

Kim, Y. K., Ehrman, L. and Koepfer, H. R. (1996a). Developmental isolation and subsequent adult behavior of *Drosophila paulistorum*. II. Prior experience. *Behav. Genet.*, **26**, 15–25.

Kim, Y. K., Koepfer, H. R. and Ehrman, L. (1996b). Developmental isolation and subsequent adult behavior of *Drosophila paulistorum*. III. Alternative rearing. *Behav. Genet.*, **26**, 27–37.

Kim, Y. K., Phillips, D. R., Chao, T. and Ehrman, L. (2004). Developmental isolation and subsequent adult behavior of *Drosophila paulistorum*. VI. Quantitative variation in cuticular hydrocarbons. *Behav. Genet.*, **34**, 385–394.

Lacaille, F., Hiroi, M., Twele, R., Umemoto, D., Inoshita, T., Manière, G., Marion-Poll, F., Osaki, M., Francke, W., Cobb, M., Everaerts, C., Tanimura, T. and Ferveur, J.-F. (2007). Inhibitory pheromone tastes bitter for *Drosophila* males. *PLoS ONE*, **2**, e661.

Lacaille, F., Everaerts, C. and Ferveur, J.-F. (2009). Feminization and alteration of *Drosophila* taste neurons induce reciprocal effects on male avoidance behavior. *Behav. Genet.*, **39**, 554–563.

Lachaise, D., Cariou, M.-L., David, J. R., Lemeunier, F., Tsacas, L. and Ashburner, M. (1988). Historical biogeography of the *Drosophila melanogaster* species subgroup. *Evol. Biol.*, **22**, 159–225.

Lam, K., Babora, D., Duthiea, D., Babora, E. M., Moorea, M. and Gries, G. (2007). Proliferating bacterial symbionts on house fly eggs affect oviposition behaviour of adult flies. *Anim. Behav.*, **74**, 81–92.

LeConte, Y. and Hefetz, A. (2008). Primer pheromones in social hymenoptera. *Annu. Rev. Entomol.*, **53**, 523–542.

Marcillac, F., Bousquet, F., Alabouvette, J., Savarit, F. and Ferveur, J.-F. (2005a). A mutation with major effects on *Drosophila melanogaster* sex pheromones. *Genetics*, **171**, 1617–1628.

Marcillac, F. and Ferveur, J.-F. (2004). A set of female pheromones affects reproduction before, during and after mating in *Drosophila*. *J. Exp. Biol.*, **207**, 3927–3933.

Marcillac, F., Grosjean, Y. and Ferveur, J.-F. (2005b). A single mutation alters production and discrimination of *Drosophila* sex pheromones. *Proc. R. Soc. Lond. B*, **272**, 303–309.

Markow, T. A. and Toolson, E. C. (1990). Temperature effects on epicuticular hydrocarbons and sexual isolation in *Drosophila mojavensis*. In *Ecological and Evolutionary Genetics of Drosophila*, ed. J. S. F. Barker *et al.* New York: Plenum, pp. 315–331.

Mayer, M. S. and James, J. D. (1971). Response of male *Musca domestica* to a specific olfactory attractant and its initial chemical purification. *J. Insect Physiol.*, **17**, 833–839.

McRobert, S. P. and Tompkins, L. (1983). Courtship of young males is ubiquitous in *Drosophila melanogaster*. *Behav. Genet.*, **13**, 517–523.

McRobert, S. P. and Tompkins, L. (1988). Two consequences of homosexual courtship performed by *Drosophila melanogaster* and *Drosophila affinis* males. *Evolution*, **42**, 1093–1097.

Meffert, L. M. and Bryant, E. H. (1991). Mating propensity and courtship behavior in serially bottlenecked lines of the housefly. *Evolution*, **45**, 293–306.

Moon, S. J., Kottgen, M., Jiao, Y., Xu, H. and Montell, C. (2006). A taste receptor required for the caffeine response in vivo. *Curr. Biol.*, **16**, 1812–1817.

Mpuru, S., Blomquist, G. J., Schal, C., Roux, M., Kuenzli, M., Dusticier, G., Clément, J.-L. and Bagnères, A.-G. (2001). Effect of age and sex on the production of internal and external hydrocarbons and pheromones in the housefly, *Musca domestica*. *Insect Biochem. Mol. Biol.*, **31**, 139–155.

Nelson, D. R. and Leopold, R. A. (2003). Composition of the surface hydrocarbons from the vitelline membranes of dipteran embryos. *Comp. Biochem. Physiol. B*, **136**, 295–308.

Oguma, Y., Nemoto, T. and Kuwahara, Y. (1992). A sex pheromone study of a fruit fly *Drosophila virilis* Sturtevant (Diptera: Drosophilidae): additive effect of cuticular alkadienes to major sex pheromone. *Appl. Entomol. Zool., 27*, 499–505.

Paillette, M., Ikeda, H. and Jallon, J.-M. (1991). A new acoustic message in *Drosophila*: The rejection signal (R.S.) of *Drosophila melanogaster* and *Drosophila simulans. Bioacoustics, 3*, 247–252.

Pechiné, J.-M., Antony, C. and Jallon, J.-M. (1988). Precise characterization of cuticular compounds in young *Drosophila* by mass spectrometry. *J. Chem. Ecol., 14*, 1071–1085.

Pechiné, J.-M., Perez, F., Antony, C. and Jallon, J.-M. (1985). A further characterization of *Drosophila* cuticular monoenes using a mass spectrometry method to localize double bonds in complex mixtures. *Anal. Biochem., 145*, 177–182.

Peng, J., Chen, S., Büsser, S., Liu, H., Honegger, T. and Kubli, E. (2005). Gradual release of sperm bound sex-peptide controls female postmating behavior in *Drosophila. Curr. Biol., 15*, 207–213.

Petfield, D., Chenoweth, S. F., Rundle, H. D. and Blows, M. W. (2005). Genetic variance in female condition predicts indirect genetic variance in male sexual display traits. *Proc. Natl. Acad. Sci. USA, 102*, 6045–6050.

Phillips, A., Sabatini, A., Milligan, P. J. M., Boccolini, D., Broomfield, G. and Molyneux, D. H. (1990). The *Anopheles maculipennis* complex (Diptera, Culicidae) – Comparison of the cuticular hydrocarbon profiles determined in adults of 5 palearctic species. *Bull. Entomol. Res., 80*, 459–464.

Polerstock, A. R., Eigenbrode, S. D. and Klowden, M. J. (2002). Mating alters the cuticular hydrocarbons of female *Anopheles gambiae* sensu stricto and *Aedes aegypti* (Diptera: Culicidae). *J. Med. Entomol., 39*, 545–552.

Pomonis, J. G. (1989). Cuticular hydrocarbons of the screwworm, *Cochliomyia hominivorax* (Diptera, Calliphoridae) – Isolation, identification, and quantification as a function of age, sex, and irradiation. *J. Chem. Ecol., 15*, 2301–2317.

Roelofs, W. L., Liu, W., Hao, G., Jiao, H., Rooney, A. P. and Linn, C. E. (2002). Evolution of moth sex pheromones via ancestral genes. *Proc. Natl. Acad. Sci. USA, 99*, 13621–13626.

Rogoff, W. M., Beltz, A. D., Johnson, J. O. and Plapp, F. W. (1964). A sex pheromone in the housefly, *Musca domestica* L. *J. Insect Physiol., 10*, 239–246.

Savarit, F., Sureau, G., Cobb, M. and Ferveur, J.-F. (1999). Genetic elimination of known pheromones reveals novel chemical bases of mating and isolation in *Drosophila. Proc. Natl. Acad. Sci. USA, 96*, 9015–9020.

Scott, K., Brady, R., Cravchik, A., Morozov, P., Rzhetsky, A., Zuker, C. and Axel, R. (2001). A chemosensory gene family encoding candidate gustatory and olfactory receptors in *Drosophila. Cell, 104*, 661–673.

Silhacek, D. L., Carlson, D. A., Mayer, M. S. and James, J. D. (1972a). Composition and sex attractancy of cuticular hydrocarbons from houseflies: effects of age, sex, and mating. *J. Insect Physiol., 18*, 347–354.

Silhacek, D. L., Mayer, M. S., Carlson, D. A. and James, J. D. (1972b). Chemical classification of a male housefly attractant. *J. Insect Physiol., 18*, 43–51.

Sillam-Dussès, D., Semon, E., Moreau, C., Valterova, I., Sobotnik, J., Robert, A. and Bordereau, C. (2005). Neocembrene A, a major component of the trail-following pheromone in the genus *Prorhinotermes* (Insecta, Isoptera, Rhinotermitidae). *Chemoecology, 15*, 1–6.

Siwicki, K. K., Riccio, P., Ladewski, L., Marcillac, F., Dartevelle, L., Cross, S. A. and Ferveur, J.-F. (2005). The role of cuticular pheromones in courtship conditioning of *Drosophila* males. *Learn. Mem., 12*, 636–645.

Spieth, H. T. (1974). Courtship behavior in *Drosophila. Annu. Rev. Entomol.*, **19**, 385–405.

Stocker, R. F. (1994). The organization of the chemosensory system in *Drosophila melanogaster*: a review. *Cell Tissue Res.*, **275**, 3–26.

Stoffolano, J. G., Schauber, E., Yin, C. M., Tillman, J. A. and Blomquist, G. J. (1997). Cuticular hydrocarbons and their role in copulatory behavior in *Phormia regina* (Meigen). *J. Insect Physiol.*, **43**, 1065–1076.

Sturtevant, A. H. (1915). Experiments on sex recognition and the problem of sexual selection in *Drosophila. J. Anim. Behav.*, **5**, 351–366.

Sureau, G. and Ferveur, J.-F. (1999). Co-adaptation of pheromone production and behavioural responses in *Drosophila melanogaster* males. *Genet. Res.*, **74**, 129–137.

Svetec, N., Cobb, M. and Ferveur, J.-F. (2005). Chemical stimuli induce courtship dominance in *Drosophila. Curr. Biol.*, **15**, R790–792.

Takahashi, A., Tsaur, S. C., Coyne, J. A. and Wu, C. I. (2001). The nucleotide changes governing cuticular hydrocarbon variation and their evolution in *Drosophila melanogaster. Proc. Natl. Acad. Sci. USA*, **97**, 9449–9454.

Tompkins, L., McRobert, S. P. and Kaneshiro, K. Y. (1993). Chemical communication in Hawaiian *Drosophila. Evolution*, **45**, 1407–1419.

Toolson, R. C., and Kuper-Simbron, R. (1989). Laboratory evolution of epicuticular hydrocarbon composition and cuticular permeability in *Drosophila pseudoobscura*: effects on sexual dimorphism and thermal-acclimatation ability. *Evolution*, **43**, 468–473.

Trabalon, M., Campan, M., Clément, J.-L., Lange, C. and Miquel, M. T. (1992). Cuticular hydrocarbons of *Calliphora vomitoria* (Diptera): relation to age and sex. *Gen. Comp. Endocrinol.*, **85**, 208–216.

Trabalon, M., Campan, M., Hartmann, N., Baehr, J.-C., Porcheron, P. and Clément, J.-L. (1994). Effects of allatectomy and ovariectomy on cuticular hydrocarbons in *Calliphora vomitoria* (Diptera). *Arch. Insect Biochem. Physiol.*, **25**, 363–373.

Trabalon, M., Campan, M., Porcheron, P., Clément, J.-L., Baehr, J.-C., Morinière, M. and Joulie, C. (1990). Relationships among hormonal changes, cuticular hydrocarbons, and attractiveness during the first gonadotropic cycle of the female *Calliphora vomitoria* (Diptera). *Gen. Comp. Endocrinol.*, **80**, 216–222.

Uebel, E., Sonnet, P. E. and Miller, R. (1976). House fly sex pheromone: enhancement of mating strike activity by combination of (Z)-9-tricosene with branched saturated hydrocarbons. *Environ. Entomol.*, **5**, 905–908.

Wicker-Thomas, C. (2007). Pheromonal communication involved in courtship behavior in Diptera. *J. Insect Physiol.*, **53**, 1089–1100.

Wu, C. I., Hollocher, H., Begun, D. J., Aquadro, C. F., Xu, Y. and Wu, M. L. (1995). Sexual isolation in *Drosophila melanogaster*: a possible case of incipient speciation. *Proc. Natl. Acad. Sci. USA*, **92**, 2519–2523.

Xue, B. Y., Rooney, A. P., Kajikawa, M., Okada, N., and Roelofs, W. L. (2007). Novel sex pheromone desaturases in the genomes of corn borers generated through gene duplication and retroposon fusion. *Proc. Natl. Acad. Sci. USA*, **104**, 4467–4472.

Yapici, N., Kim, Y. J., Ribeiro, C. and Dickson, B. J. (2008). A receptor that mediates the post-mating switch in *Drosophila* reproductive behaviour. *Nature*, **451**, 33–37.

Yew, J. Y., Cody, R. B., and Kravitz, E. A. (2008). Cuticular hydrocarbon analysis of an awake behaving fly using direct analysis in real-time time-of-flight mass spectrometry. *Proc. Natl. Acad. Sci. USA*, **105**, 7135–7140.

16

Contact recognition pheromones in spiders and scorpions

Marie Trabalon and Anne-Geneviève Bagnères

Most of the present book is dedicated to one class of Arthropoda, the Insecta, because chemical communication research in this class is the most complete and broadly illustrated. This type of research on the chelicerate arthropods of the class Arachnida is, by contrast, poorly developed. We saw for example in Chapter 7, studies of chemical ecology interactions with Acari and particularly mite–insect interactions, and a few examples of chemical interaction with spiders were also shown in the same chapter on chemical mimicry, even though spiders are the most familiar and numerous of the arachnids. We undertook some work and about 15–10 years ago on contact chemical signal description and its relationship with behavior, physiology and reproduction, in different types of Aranea (spiders). We will present here a distillation of this work with a review of studies on the subject by different authors. Most notable here is the poverty of research on contact recognition signals and relative behavioral works on the order Scorpionida, the scorpions. Some of the few chemical data available are published here for the first time.

Contact recognition pheromones in spiders

Spiders are generally known as solitary predators with strong territorial behavior. Territory defense often leads to cannibalism, even when male spiders are displaying courtship behavior to females. However interactions between solitary spiders are not always aggressive and conspecifics may interact peacefully during reproduction. Reproductive behavior refers to the set of behaviors that allows the perpetuation of the species: sexual behavior and parental behavior. Sexual behavior consists of a set of behavior linked to the attraction between the sexes and leads to mating. In most species, the sole purpose of sexual behavior is the perpetuation of the species. Parental behavior includes all activity aimed at satisfying the needs and ensuring the survival of the young (whether one's own or under one's parental care). In sexual reproduction, parental behavior may begin before egg fertilization and may include aspects as diverse as inhibition of aggressive behavior (cannibalism), nest building, food transfer and supplying the birth cell. Reproductive behavior prerequisites for gamete exchanges and maternal behavior are inhibition of cannibalism and physiological maturation.

Spiders emit pheromones from their body cuticle and/or web silk. Spider silk is a well-known vector for vibratory, tactile and chemical signals. Using silk or webs for pheromone transmission may provide spiders with the advantage of continuous pheromone emission without the need to actively emit pheromones from glands (Schulz, 1997). Silk pheromones may be produced in the silk glands (Schulz, 2004) or could be cuticle in origin, and only applied to the silk during web construction (Trabalon *et al.*, 2005). Cuticular pheromones may be synthesized near the integument and transported to the epicuticle for emission (Trabalon *et al.*, 1996), possibly via modified cuticular wax glands or abdominal organs (Pollard *et al.*, 1987). The specific glands that secrete pheromones in spiders are unknown. Studies of spider pheromones generally focus on pheromones emitted by females.

Spider contact chemoreceptors were found on the legs (pretarsus and tarsus) and on the pedipalps and share a common organization pattern with insects, but differ from them in a number of respects. Contact chemoreceptors on the body are elongated and possess a pore at the tip that lets outside molecules diffuse inside the hair and to the dendrites of the sensory neurons. Whereas insect contact chemoreceptors usually include four chemosensitive neurons and one mechanoreceptor (Altner and Prillinger, 1980; Zacharuk, 1980), spider contact chemoreceptors were reported to include up to 19 chemosensitive and two mechanosensitive neurons (Foelix and Chu-Wang, 1973). When spiders move or come into contact with a substrate, these chemoreceptors are likely to play a crucial role in the analysis of the chemical environment. In *Tegenaria atrica* (*Agelenidae*) they are functional immediately after spiders leave the cocoon (Vallet *et al.*, 1998).

Identification of silk and cuticle lipids

The silk and cuticle of spiders are covered by a lipid layer. These lipid compounds protect from external damaging factors such as humidity, which modifies the cuticle's or silk's physical properties. The frequency of water exposure in natural circumstances suggests that wetting with water or similar solvents must have negligible or manageable effects on pheromone function.

The main components of this layer are generally long-chain aliphatic hydrocarbons and fatty acids, as well as smaller amounts of methyl esters, long-chain aliphatic alcohols and aldehydes, glycerides, and cholesterol. Similar compounds are present on the web and spider cuticle. It has been established that these compounds may be responsible for the antifungal and antibacterial properties of silk and also play a role in communication.

For example the total pentane extracts of the web and cuticle from two European *Agelenidae* and *Amaurobiidae* comprised a complex mixture of fatty acids, alcohols and long-chain aliphatic hydrocarbons (Prouvost *et al.*, 1999; Trabalon *et al.*, 1996, 1997; Trabalon and Assi-Bessekon, 2008; Tables 16.1 and 16.2). The qualitative composition of cuticular extract obtained was similar to that reported for most insects and scorpions. Hydrocarbons in spiders form complex and varied mixtures with unsaturated hydrocarbons. In those species, the hydrocarbon fraction consists of *n*-, monomethyl- and

Table 16.1 *Wax components from web in* Agelenidae *(Prouvost* et al., *1999) and in* Amaurobiidae *(Trabalon* and *Assi-Bessekon, 2008).*

Chemical compounds of methanol eluate	Chemical compounds of pentane eluate
Acids	**Hydrocarbons**
tridecanoic acid	*n*-tetradecane
tetradecanoic acid	*n*-pentadecane
pentadecanoic acid	*n*-hexadecane
hexadecanoic acid (palmitic acid)	*n*-heptadecane
heptadecanoic acid	octadecene
octadecadienoic acid	*n*-octadecane
octadecenoic acid	*n*-nonadecane
eicosanoic acid	eicosene
docosanoic acid	*n*-eicosane
hexacosanoic acid	*n*-heneicosane
	n-docosane
Alcohols	*n*-tricosane
1-octadecanol	*n*-tetracosane
1-eicosanol	*n*-pentacosane
1-docosanol	13-+11-methylpentacosane
1-tetracosanol	3-methylpentacosane
1-pentacosanol	*n*-hexacosane
1-hexacosanol	4-methylhexacosane
	2-methylhexacosane
Esters	*n*-heptacosane
methyl tetradecanoate (myristate)	13-+11-+9-methylheptacosane
methyl pentadecanoate	3-methylheptacosane
methyl hexadecenoate	2-methylheptacosane
methyl hexadecanoate (palmitate)	*n*-octacosane
methyl octadecadienoate	4-methyloctacosane
methyl octadecenoate	*n*-nonacosane
methyl octadecanoate (stearate)	15-+13-+11-+9-methylnonacosane
	7-methylnonacosane
Ketones	5-methylnonacosane
2-nonadecanone	3-methylnonacosane
2-henicosanone	*n*-triacontane
1-docosen-3-one	*n*-hentriacontane
2-tricosanone	15-+13-+11-methylhentriacontane
	9,21-+11,17-+11,19-dimethylhentriacontane
	3-methylhentriacontane

Table 16.1 *(cont.)*

Chemical compounds of methanol eluate	Chemical compounds of pentane eluate
	2-methylhentriacontane
	n-docotriacontane
	n-tritriacontane
	17-+15-+13-+11-methyltritriacontane
	11,21-+13,17-+13,19-dimethyltritriacontane
	n-tetratriacontane
	12-methyltetratriacontane
	n-pentatriacontane
	17-+15-+13-+11-methylpentatriacontane

dimethylalkanes, containing a relatively high proportion of even-numbered carbon chain components. Neither species exhibited unsaturated hydrocarbons. The abundance of even-numbered carbon chain alkanes and odd-numbered carbon chain fatty acyl groups, along with abundant methyl branches, suggests that the propionyl-CoA and its carboxylated product, methyl-malonyl-CoA, play important roles in the biosynthesis of these unique waxes.

The predominance of *n*-alkanes in cuticular lipids of spiders is in agreement with previous finding in insects. The *n*-alkane fraction represents 35–51% of total hydrocarbons in three *Tegenaria spp.* (Trabalon *et al.*, 1996, 1997) as in the scorpion *Paruroctonus mesaensis* (Hadley and Jackson, 1977). The predominance of pentacosanes, heptacosanes and nonacosanes is similar to what one finds in insect cuticular lipids. The origin of the alkanes is at present unknown; the possibility that they come from the insect diet has not been investigated. The monomethylalkanes, branched at positions 15, 13, 11, 9, 5, and 3 were the most abundant constituents in both species (i.e. *T. atrica, T. domestica* and *T. pagana*). Low levels of dimethylalkanes were found (9% of the total extract) in both species.

The pentane cuticular extract of the social spider of Guiana, *Anelosimus eximius* (*Theridiidae*), contains hydrocarbons, fatty acids and their methyl esters, and a series of novel propyl esters of long-chain methyl-branched fatty acids (Bagnères *et al.*, 1997; Table 16.3). The 30+ propyl esters quantified comprise almost three-fourths of the extract and consist predominantly of odd-numbered carbon chain components. Mass spectrometric analysis of the propyl esters, their methyl esters and cyanide derivatives of some of those showed that mono-, di- and trimethyl-branched components on even-numbered carbons predominate with linear propyl esters. The major components are propyl 6,20- and 6,30-dimethylhentriacontanoate and propyl 6,20- and 6,30-trimethylhentriacontanoate. The physiological role of these compounds is, at present, completely unknown but a biosynthetic pathway has been proposed to understand the possible biochemical origin of each chain/methylated part of these original esters (Bagnères *et al.*, 1997).

Table 16.2 *Wax components from cuticle of* Tegenaria sp. (Agelenidae) *(Trabalon* et al.*, 1996, 1997; Prouvost* et al.*, 1999).*

Chemical compounds eluate

Chemical compounds of methanol eluate	Chemical compounds of methanol eluate (cont.)
Acids	**Methylalkanes**
tetradecanoic acid (myristic)	13-+11-+9-methylpentacosane
hexadecanoic acid (palmitic)	3-methylpentacosane
octadecadienoic acid (linoleic)	3,11-dimethylpentacosane
octadecanoic acid (stearic)	12-+10-+8-methylhexacosane
	2-methylhexacosane
Esters	3-methylhexacosane
methyl tetradecanoate (myristate)	13-+11-+9-methylheptacosane
methyl pentadecanoate	5-methylheptacosane
methyl hexadecenoate	9,13-+9,15-+9,17-dimethylheptacosane
methyl hexadecanoate (palmitate)	2-methylheptacosane
methyl heptadecanoate	3-methylheptacosane
methyl octadecadienoate linoleate)	5,17-dimethylheptacosane
methyl octadecenoate (oleate)	3,9-+3,11-dimethylheptacosane
methyl octadecanoate (stearate)	12-+10-methyloctacosane
	6-methyloctacosane
Chemical compounds of pentane eluate	2-methyloctacosane
***n*-alkanes**	3-methyloctacosane
n-heptadecane	15-+13-+11-+9-methylnonacosane
n-octadecane	5-methylnonacosane
n-nonadecane	9,13-+7,17-dimethylnonacosane
n-eicosane	5,18-dimethylnonacosane
n-heneicosane	3-methylnonacosane
n-docosane	14-+12-+10-methyltriacontane
n-tricosane	2-methyltriacontane
n-tetracosane	3-methyltriacontane
n-pentacosane	14-+12-+10-methyltriacontane
n-hexacosane	15-+13-+11-+9-methylhentriacontane
n-heptacosane	7-methylhentriacontane
n-octacosane	13,17-+11,17-+9,17-dimethylhentriacontane
n-nonacosane	9,21-+7,19-dimethylhentriacontane
n-triacontane	2-methylhentriacontane
n-hentriacontane	16-+14-+12-+10-methyldocotriacontane
n-dotriacontane	17-+15-+13-+11-methyltritriacontane
n-tritriacontane	12-+14-+16-+18-methyltetratriacontane
n-tetratriacontane	15,19-dimethylpentatriacontane
n-pentatriacontane	12-methylhexatriacotane
n-hexatriacontane	
n-heptatriacontane	
n-octatriacontane	
n-nonatriacontane	

Table 16.3 *Propyl esters from cuticle of the spider* Anelosimus
eximius *(Bagnères* et al., *1997).*

Compound	Formula
Propyl 20-methylheneicosanoate	$C_{25}H_{50}O_2$
Propyl docosanoate	$C_{25}H_{50}O_2$
Propyltricosanoate	$C_{26}H_{52}O_2$
Propyl 22-methyltricosanoate	$C_{27}H_{54}O_2$
Propyl tetracosanoate	$C_{27}H_{54}O_2$
Propyl pentacosanoate	$C_{28}H_{56}O_2$
Propyl 24-methylpentacosanoate	$C_{29}H_{58}O_2$
Propyl hexacosanoate	$C_{29}H_{58}O_2$
Propyl heptacosanoate	$C_{30}H_{60}O_2$
Propyl 16-methylheptacosanoate	$C_{31}H_{62}O_2$
Propyl 26-methylheptacosanoate	$C_{31}H_{62}O_2$
Propyl octacosanoate	$C_{31}H_{62}O_2$
Propyl nonacosanoate	$C_{32}H_{64}O_2$
Propyl 18-methylnonacosanoate	$C_{33}H_{66}O_2$
Propyl 28-methylnonacosanoate	$C_{33}H_{66}O_2$
Propyl 4,18-dimethylnonacosanoate	$C_{34}H_{68}O_2$
Propyl 4,28-dimethylnonacosanoate	$C_{34}H_{68}O_2$
Propyl 20,28-dimethylnonacosanoate	$C_{34}H_{68}O_2$
Propyl 4,20,28-trimethylnonacosanoate	$C_{35}H_{70}O_2$
Propyl 20-methylhentriacontanoate	$C_{35}H_{70}O_2$
Propyl 4,20-dimethylhentriacontanoate	$C_{36}H_{72}O_2$
Propyl 4,30-dimethylhentriacontanoate	$C_{36}H_{72}O_2$
Propyl 6,20-dimethylhentriacontanoate	$C_{36}H_{72}O_2$
Propyl 6,30-dimethylhentriacontanoate	$C_{36}H_{72}O_2$
Propyl 20,30-dimethylhentriacontanoate	$C_{36}H_{72}O_2$
Propyl 4,20,30-trimethylhentriacontanoate	$C_{37}H_{74}O_2$
Propyl 6,20,30-trimethylhentriacontanoate	$C_{37}H_{74}O_2$
Propyl 22-methyltritriacontanoate	$C_{37}H_{74}O_2$
Propyl 6,22-dimethyltritriacontanoate	$C_{38}H_{76}O_2$
Propyl 4,22-dimethyltritriacontanoate	$C_{38}H_{76}O_2$
Propyl 6,x-dimethyltritriacontanoate	$C_{38}H_{76}O_2$
Propyl 4,x-dimethyltritriacontanoate	$C_{38}H_{76}O_2$
Propyl 4,x,x-trimethyltritriacontanoate	$C_{39}H_{78}O_2$
Propyl 6,x,x-trimethyltritriacontanoate	$C_{39}H_{78}O_2$

Contact recognition pheromones and sexual behavior in spiders

In most spider species, females lead sedentary and solitary web-bound lives. It is usually the male that leaves its web in search of a suitable mate. There are two types of chemical communication system that she can use to attract the male: chemical communication by means of airborne sex pheromones that are capable of attracting males (Gaskett, 2007) and tactochemical communication by means of contact pheromones associated with the silk or cuticle. Currently, airborne sex pheromones have been conclusively identified in only three spider species: *Linyphia triangularis* (Schulz and Toft, 1993b), *Cupiennius salei* (Papke *et al.*, 2000) and *Agelenopsis aperta* (Papke *et al.*, 2001). These different signals enable the male to locate and identify a mate, but they can also provide precise information about the female's physiological state. After detecting the signals, the male responds by exhibiting one of the following: orientation behavior, increased exploratory behavior or courtship behavior. Unlike the case regarding insects, there has been little research done to identify sex contact pheromones in spiders. This has been notoriously difficult because many compounds of unknown function are typically present in extracts of spider silk and cuticles.

The presence of contact pheromones associated with the female's web and cuticle has been mentioned by numerous authors (Table 16.4). The silk-bound contact pheromones bring about, in males, an increase in exploratory behavior, and the adoption of orientation behavior, as well as courtship behavior (Hegdekar and Dondale, 1969; Blanke, 1975; Dijkstra, 1976; Tietjen, 1979; Suter and Renkes, 1982; Roland, 1984; Schulz, 1997; Prouvost *et al.*, 1999; Trabalon *et al.*, 2005). For example the male of *Tegenaria atrica* (*Agelenidae*) develops a different behavior whether in the presence of a receptive female and her web or of a non-receptive female and her web (Prouvost *et al.*, 1999). Furthermore, in 90% of cases, the male produces abdominal vibrations when in the presence of a receptive female who is present in her web, as against only 20% when in the presence of a non-receptive female. These actions produce short repeated vibrations. According to Szlep (1964) and Platnick (1971), this behavior inhibits the female's predatory behavior and reduces the risk of the male being cannibalized. In *T. atrica*, these vibrations allow males to test the female's reactivity from a distance and enable male orientation toward the female. Similar results have been observed in numerous species belonging to families far removed from each other. Furthermore, males do not produce abdominal vibrations when in contact with the webs of absent non-receptive females. They do, however, produce vibrations when in contact with the empty web of a receptive female or with a dead receptive female placed outside the web. To date, no study has been able to show a difference in the physical structure of webs belonging to individuals of different sexes, ages and physiological states. We can therefore assume that males change their behavior whether in the presence of the web of a receptive female or that of a non-receptive female according to the absence or presence of contact sex pheromones. These compounds would therefore provoke abdominal vibrations. This could be vital for cannibalistic or aggressive species because males that can correctly identify a female's qualities from her silk have the strong selective advantage of being able to reject an unattractive female before entering her web and risking death (Krafft, 1982; Tietjen and Rovner, 1982; Trabalon *et al.,* 1997; Gaskett *et al.,* 2004). This pattern is apparent in the ground-dwelling lycosid *Schizocosa ocreata*: males respond most quickly to the silk of virgin adult females, which are more sexually receptive and less aggressive than mated females (Roberts and Uetz,

Table 16.4 *Summary of literature investigating contact pheromones in spiders.*

Spider taxa	Substratum	References
Agelenidae		
Tegenaria atrica (Koch)	silk and cuticle	Prouvost *et al.*, 1999; Pourié and Trabalon, 1999a, 1999b, 2001; Pourié *et al.,* 2005; Trabalon *et al.*, 1996, 1998, 2005
Tegenaria domestica (Clerck)	silk and cuticle	Roland, 1984; Trabalon *et al.*, 1997
Tegenaria pagana (Koch)	silk and cuticle	Roland, 1984; Trabalon *et al.*, 1997
Amaurobiidae		
Coelotes terrestris (Wider)	silk and cuticle	Krafft, 1978; Roland, 1984; Trabalon and Assi-Bessekon, 2008
Araneidae		
Araneus sclopetarius (Clerck)	silk	Roland, 1984
Ctenidae		
Cupiennius sp.	silk	Barth and Schmitt, 1991; Barth, 1993; Schuster *et al.*, 1994; Papke *et al.*, 2000, 2001; Tichy *et al.*, 2001; Schulz, 2004
Dysderidae		
Dysdera crocata (Koch)	silk and cuticle	Jackson and Pollard, 1982
Eresidae		
Stegodyphus sarasinorum	silk	Roland, 1984
Linyphiidae		
Frontinella communis (Hentz)	silk and cuticle	Suter and Renkes, 1982; Suter and Hirscheimer, 1986; Suter *et al.*, 1987
Linyphia sp.	silk	Locket and Bristowe, 1926; Schulz and Toft, 1993b
Microlinyphia impigra	silk	Schulz and Toft, 1993a
Neriene sp.	silk	Schulz and Toft, 1993a
Lycosidae		
Lycosa sp.	silk	Tietjen, 1979; Lizotte and Rovner, 1989; Costa *et al.*, 2000
Pardosa sp.	silk	Hegdekar and Dondale, 1969; Richter *et al.*, 1971; Dondale and Hegdekar, 1973; Dumais and Perron, 1973
Schizocosa sp.	silk and cuticle	Hegdekar and Dondale, 1969; Uetz and Denterlein, 1979; Tietjen and Rovner, 1982; Roberts and Uetz, 2005

Table 16.4 (*cont.*)

Spider taxa	Substratum	References
Pisauridae		
Dolomedes fimbriatus	silk	Arnqvist, 1992
Salticidae		
Anasaitis canosa	silk	Jackson, 1987
Bavia aericeps (Simon)	silk	Jackson, 1987
Carrhotus xanthogramma	silk	Yoshida and Suzuki, 1981
Cosmophasis micarioides	silk	Jackson, 1987
Helpis minitabunda (Koch)	silk	Jackson, 1987
Jacksonoides kochi (Simon)	silk	Jackson, 1987
Marpissa marina (Goyen)	silk	Jackson, 1987; Jackson and Cooper, 1990
Menemerus sp.	silk	Jackson, 1987
Mopsus mormon (Karsch)	silk	Jackson, 1987
Mymarachne lupala (Koch)	silk	Jackson, 1987
Phidippus sp.	silk	Jackson, 1981, 1986, 1987
Plexippus paykulli (Andouin)	silk	Jackson, 1987; Jackson and Cooper, 1990
Pseudicius sp.	silk	Jackson, 1987
Simaetha sp.	silk	Jackson, 1987; Jackson and Cooper, 1990
Thiodena sylvana (Hentz)	silk	Jackson, 1987
Trite sp.	silk	Jackson, 1987; Jackson and Cooper, 1990; Taylor, 1998
Tetragnathidae		
Nephila clavipes	silk	Schulz, 2001
Theridiidae		
Anelosimus eximius	cuticle	Bagnères *et al.*, 1997; Pasquet *et al.*, 1997
Argyrodes antipodianus	silk	Whitehouse and Jackson, 1994
Latrodectus mactans	silk	Ross and Smith, 1979; Breene and Sweet, 1985

2005). The silk pheromones of female *Lycosa tristani* and *L. longitarsis* are unaffected by water (Lizotte and Rovner, 1989).

In addition to locating the nest, completion of male courtship behavior in many spider species requires close-range recognition by the mating partners (Kaston, 1936; Suter and Renkes, 1982; Suter and Hirscheimer, 1986). For example, females of some *Linyphiidae* species share a long-range pheromone that attracts both conspecific and heterospecific males (Schulz and Toft, 1993a). Females in this group appear to have additional species-specific pheromones that males use to recognize conspecific females at short range. Males of *Dysdera crocata* (Jackson and Pollard, 1982), *Linyphia triangularis* (Schulz, 2004) and *Dolomedes scriptus* (Kaston, 1936) respond to female cuticle extracts obtained with lipophilic solvents such as diethyl ether or dichloromethane. Other species with sex cuticular pheromones that are soluble in other solvents (methanol, pentane) include *Pardosa lapidicina* (Dondale and Hegdekar, 1973), *Frontinella communis* (Suter and Renkes, 1982), *Tegenaria atrica* (Prouvost *et al.*, 1999), and several species of *Salticidae* (Jackson, 1987).

Quantitative changes in lipid compounds on the silk and cuticle of females correlate significantly with changes in female sexual receptivity in spiders. For example, female *T. atrica* attach a contact sex pheromone to their web (Trabalon *et al.,* 1997, 2005; Prouvost *et al.,* 1999). This pheromone consists of a complex mixture of saturated hydrocarbons, methyl esters (methyl tetradecanoate, methyl pentadecanoate, methyl hexadecanoate, and methyl octadecanoate) and their fatty acids (tetradecanoic, pentadecanoic, hexadecanoic, and cis,cis-9,12-octadecadienoic acids). The female uses cuticular compounds, which are applied to the silk in substantial amounts during web construction. Modification of chemical profiles makes the female attractive to males (Trabalon *et al.*, 2005). Receptive females are different to unreceptive ones with respect to three fatty acids (hexadecanoic, octadecadienoic and octadecenoic acids) and three methyl esters (linoleate, oleate, and stearate) present on both the web and the cuticle. Our combined results from chemical analyses and behavioral assays demonstrate clearly that these contact compounds are quantitatively correlated with the behavior of spiders.

After mating, females of many spider species are no longer attractive to males (Tietjen and Rovner, 1982). Mated females can have a different blend of pheromone compounds, as identified for virgin and mated female *T. atrica* (Trabalon *et al.*, 1998). Alternatively, mated females of *Linyphia triangularis* (Schulz and Toft, 1993b) and *Agelenopsis aperta* (Papke *et al.*, 2001) could stop producing the attractants they emit when virgin. Post-mating changes in female pheromone emission are generally associated with changes web and cuticular chemical compounds and increased cannibalistic behavior (Pollard *et al.*, 1987; Trabalon *et al.*, 1998; Herberstein *et al.*, 2002). For example in webs of post-mating female *Coelotes terrestris* (Trabalon and Assi-Bessekon, 2008) there was significantly less palmitic acid, 1-octadecanol, 13-+11-methylpentacosane and 3-methylheptacosane, and significantly more 1-docosanol, *n*-pentacosane, *n*-hentriacontane and 17-+15-+13-+11-methylpentatriacontane than in the virgin webs. These post-mating webs were attractive for copulated females and repulsive for males. It is unclear whether these changes in female pheromone emission and behavior are controlled autonomously by females, or are due to interference by her mate.

The genetics and regulation of sex pheromone production in spiders remain largely unstudied (Schulz, 2004), although a hormone involved in the synthesis or release of female sex pheromones has been identified for *T. atrica* (Trabalon *et al.*, 2005). Injection of 20-hydroxyecdysone, normally only found in virgin female *T. atrica* during oocyte development, stimulated vitellogenesis (Pourié and Trabalon, 2003), increased the quantity of male attracting compounds (methyl esters and fatty acids) on female silk and cuticle, decreased female cannibalism, and increased her sexual receptivity (Trabalon *et al.*, 2005).

Contact recognition pheromones and social behavior in spiders

After mating, some spider species, like *Theridiidae* (Hirschberg, 1969), *Eresidae* (Krafft, 1971), *Agelenidae* (Trabalon *et al.*, 1996) and *Amaurobiidae* (Krafft, 1978), exhibit parental behavior, often followed by a time-limited gregarious period in the young. This period begins inside the cocoon, persists for several days after emergence, and is accompanied by metabolizing of the vitelline reserves by the young (Krafft *et al.*, 1986; Ramousse, 1986). For example, the young *Tegenaria atrica* gather immediately after leaving the cocoon, form a dense ball near the cocoon and show reduced locomotor activity for one week (Trabalon *et al.*, 1996). During this period, the young simultaneously undergo a molt. They react neither to the weakening of the web produced by the displacements of the mother nor to the vibrations provoked by prey. The mother stays close to the cocoon but no corporal contact is observed between the mother and the young. One week after emergence, the locomotor activity of the young increases. Aggregation of the individuals becomes less dense, and after a further molt, the young progressively overrun the maternal web. The 20-day-old young move quickly and the resulting weakening of the web alerts the mother. The young, not yet tending to disperse out of the maternal web, feed on prey killed and left by the female. During this period, the young frequently make contact with one another, in fortuitous encounters or when feeding. These contacts are accompanied by agonistic behavior which tends to increase with age, whereas contact with the mother produces no agonistic behavior, either in the young or in the adult. At the end of the third week of development, the young start catching small prey and in the course of the fourth week, after a further molt, the young leave the maternal web and settle on individual webs.

Different experiments demonstrated that the transition from the gregarious phase to the solitary phase is coupled with a change in the tolerance behavior of the mother in relation to changes in the composition of the cuticular compounds of the young (Trabalon *et al.*, 1996, 1998). Five compounds characteristic of mother females (*n*-eicosane, 2-methylhexacosane, *n*-octacosane, *n*-nonacosane and *n*-tritriacontane) are present on the cuticle of tolerated gregarious young. When the young are 20 days old they begin to develop avoidance behavior towards each other, and, at the same time, their cuticular chemical profiles begin to change. Consequently, two new chemicals (5-methylhentriacontane and methyloctadecadienoate) emerge and four different compounds are variously synthesized/released (*n*-octadecane, *n*-octacosane, octadecadienoic and octadecenoic acids). Nevertheless, these changes do not appear sufficient to cause tolerance behavior to cease completely, as the young continue to show tolerance towards each other. The compounds that vary during the dispersal phase, as well as the occurrence of cannibalism in young, are linked to four compounds that are characteristic of solitary virgin females (methyloctadecanoate, *n*-tricosane, *n*-pentacosane and *n*-heptacosane). After dispersal, young are no longer tolerated by other conspecifics and exhibit modified cuticular chemical profiles: *n*-heneicosane, 3-methylpentacosane and 14-+12-+10-methyltriacontane emerge in them and changes in the synthesis/release of three compounds (*n*-heptadecane, methyltetradecanoate and *n*-octadecane) are observed. So, tactochemical information plays an important role in modulating agonistic behavior after close contact between female spiders and young. The cannibalism of females appears to correspond with the increase in polar compound levels (methyl esters and fatty acids) and the decrease in apolar compound levels (hydrocarbons) in young of different ages.

The interactions expressed by *Tegenaria atrica* towards conspecific intruders are not always of a predator–prey nature. Individuals of this solitary species are capable of adjusting their behavior according to the intruder, whether it is a conspecific adult or juvenile. This adjustment is linked to the age and physiological state of the observed individuals (Trabalon *et al.*, 1998). Therefore, no matter what their physiological state, adult females of *T. atrica* show tolerance towards gregarious juveniles. Nevertheless, the level of tolerance shown by these different types of female decrease at the same rate as the young develop. This form of tolerance towards very recently hatched young has led females of some species of spider to care for their brood to a greater or lesser extent.

Parental females tolerate not only their own young, but also unrelated gregarious juveniles (conspecific and sympatric). The mother–young grouping in the solitary spider *T. atrica* (Pourié and Trabalon, 1999a), in *Coelotes terrestris* (Assi-Bessekon, 1997) is therefore not a closed family group. Adult females without parental experience show a level of tolerance towards gregarious young comparable with that of parental females. Thus, the females' physiological state or parental experiences are not the only factors responsible for the tolerance behavior. The combination of the two stimuli, vibrations and cuticular chemical compounds, could explain the frequent occurrence of cannibalistic behavior in females towards solitary young. We have been provided with part of the answer through a discrimination test where solitary adult female *T. atrica* are given a choice between a gregarious young, solitary spiderling and a prey (Pourié and Trabalon, 1999a; Pourié *et al.*, 2005). It is observed that no matter what the test intruder, the female exhibits the same type of behavior until physical contact occurs between her and the intruder. It is not until after physical contact has been made that the female adjusts her agonistic behavior according to the nature of the intruder. Thus, it appears that vibratory information is insufficient for the adult female to precisely determine the nature of the intruder and therefore does not allow her to adapt her behavior at a distance. Pourié and Trabalon (1999a) demonstrated that extracts from solitary young lead to high levels of biting comparable to those induced by the extract from a prey. Conversely, the extracts from gregarious young cause only low levels of biting. Females can therefore, upon contact, distinguish not only between the contact chemical signature of a potential prey and that of a conspecific, but also between a gregarious and a solitary young.

Moreover, observations of a gregarious phase where juveniles of spiders show mutual tolerance provide the foundations for further research on the factors capable of inducing these juveniles to remain in groups, and to tolerate each other until they reach the adult stage. One of these factors may be linked to prey density. According to Ruttan (1990), the growing youngs' increasing need for food drives them to disperse more or less rapidly according to the prey availability in that particular area. When young *Coelotes terrestris* (Gundermann *et al.*, 1993) or *T. atrica* (Pourié and Trabalon, 1999b) are prevented from dispersing while being fed *ad-libitum*, a significant decrease in the cannibalism rate can be observed. Thus, the trophic factor has a considerable influence on maintaining tolerance behavior or inhibition of cannibalism in spider groups. According to these observations, the occurrence of cannibalism could be related to the first signs of "hunting" or predatory behavior at the time of the natural dispersal phase of the young. Under natural conditions, when prey density is low, cannibalism would be an alternative to foraging. Pourié and Trabalon (2001) indicate that the decrease in the cannibalism rate in experimental groups of young *T. atrica* coincides with variations in their cuticular chemical signature. In spiders reared in isolation, the development of chemicals during ontogeny remains unchanged regardless of the level of nutrition, and the chemical signature becomes enriched as the individual becomes older, with increasingly long-chain compounds. The level of food intake therefore does not influence the synthesis/release of the chemical compounds found on the cuticle. We observe the same development in the cuticular chemical signature of young that have been reared in groups and fed *ad-libitum*. However, the increasing complexity of their chemical signature does not follow the same development pattern as that of isolated individuals. More precisely, palmitic acid and 13,17- +11,17- +9,17-dimethyl-hentriacontane are not detected in grouped individuals. Furthermore, in grouped individuals, methyloctadecenoate and *n*-heptatriacontane do not emerge until after the fifth molt, in contrast to isolated young, where these compounds emerge just before dispersal. These compounds are perhaps the cause of inhibition of cannibalism in our experimental groups of solitary young (Pourié and Trabalon, 2001). It is therefore reasonable to believe that the absence of competition with regard to access to food may drive spiders not to differentiate themselves by means of their chemical signals and thereby enhance their defenses against being bitten.

However, another hypothesis must be taken into consideration. In experiments on experimental groups of solitary young, the necessary factor for keeping together a group of relatively stable size is the mutual tolerance between growing individuals. This tolerance is prompted by the absence of competition for food, and is perhaps maintained by a change in the chemical communication between individuals. The interactions that progressively lead to mutual tolerance between young could be compared to learning processes. Indeed, throughout the eight months of experiments involving these groups, through repeated interactions in a rich environment, the individuals have progressively learned to restrict their agonistic interactions by assessing the advantages and disadvantages of such a situation. This allows adults from these groups to display a higher level of tolerance behavior and conversely, a very low cannibalism rate towards adult conspecifics. Consequently, it appears very likely that if favorable environmental factors prompt the young not to disperse, then it is possible that the adults preserve the communal lifestyle based on the maintenance of social interactions (i.e. social tolerance).

It is clear, however, that it is not a question of comparing experimental groups of adult *Coelotes terrestris* or *T. atrica* to spider societies. These results simply demonstrate the pre-existence of a certain behavioral plasticity even in solitary species. Normally solitary individuals are capable of showing mutual tolerance according to one essential environmental factor: food resource availability. This social tolerance appears to be linked to a very slight change in their cuticular chemical signature during ontogeny.

Permanent social spiders also exhibit a developed parental behavior and a gregarious period in the young. Therefore, one can suppose that interactions occurring during the "periodic-social" phase in solitary spiders correspond to a pre-adaptation to permanent social life. In 1971, Krafft suggested that tactochemical communication plays a role in the colony cohesion and organization of the social spider *Agelena consociata*. In a behavioral and chemical study of the social spiders *Anelosimus eximius* we have observed quantitative chemical differences between individuals from different colonies (Pasquet *et al.*, 1997). These differences are not linked to the geographical distribution or to the lack of group closure, nor to the competition between colonies. They seem to be linked, however, to both genetic variation and ecological factors. The cuticular chemical compounds we have identified in these social spiders are hydrocarbons, fatty acids and, particularly, novel propyl esters of long-chain methyl-branched fatty acids (Bagnères *et al.*, 1997). We have made the hypothesis that these propyl esters (Table 16.3) must play a part in social interactions (Trabalon, 2000). Indeed, how can one imagine volatile molecules being involved in the communication between two individuals within a huge community, when the "conversation" between these two individuals would be "drowned out" and imperceptible inside an "odorous cloud" emitted by the whole colony? The different interactions within the group would eventually be jumbled up, creating a cacophony. The question that remains, for the time being, is whether the non-volatile cuticular compounds found in spiders are used in social interactions.

Contact recognition pheromones in scorpions

As arachnid arthropods, scorpions provide an important out-group for comparing chemical ecology data obtained in insects and better-studied groups of arachnids such as spiders. Scorpions are easy to keep as laboratory animals but subtle aspects of their behavior require optimized conditions for natural courtship and reproductive behaviors to occur (Brownell and Polis, 2001, introductory chapter). Most scorpions are nocturnal predators by habit with life history patterns that can be strongly influenced by cannibalism. Thus, while many of their behaviors rely on tactile mechano and chemosensory signals (Brownell, 2001) it is still a striking fact that so few chemical studies have been done on scorpions, with the exception of studies on peptidic neurotoxic venoms. These neurotoxics are well described in numerous papers, for example Carbonell *et al.* (1988) describing the synthesis of a gene coding for an insect-specific scorpion neurotoxin. The team of H. Rochat published various precursor works including a description of the three-dimensional structure of the potent toxin II (Fontecilla-Camps *et al.*, 1988) and characterization of antibodies specific for the α-neurotoxin I (Aah I) from the venom of the dangerous *Androctonus australis* hector scorpion (Clot-Faybesse *et al.*, 1999). Of special interest to us are the early works on scorpion cuticular lipids and permeability (Hadley, 1970, 1981; Hadley and Quinlan, 1987), and the very first descriptions of the epicuticular lipid composition of several species (for example in Toolson and Hadley, 1979 on *Centruroides sculpturatus*; Hadley and Jackson, 1977 and Hadley and Hall, 1980 on *Paruroctonus mesaensis*; Hadley and Filshie, 1979 on *Hadrurus arizonensis*). A more recent work on comparative osmoregulative capability of two scorpion families (Scorpionidae and Buthidae) is indirectly related to the cuticular permeability topic (Gefen and Ar, 2004).

Despite these studies the poor state of knowledge on scorpion signaling chemistry is a "fact of life" to be noted again (e.g. P. Brownell, personal communication, 2008). To date we have been able to find chemical ecology works on only two desert scorpions: *Smeringerus* (formerly *Paruroctonus) mesaensis* and *Hadrurus arizonensis*. Accordingly this part of the chapter will review works on these two species with special focus on semiochemicals mediating intraspecific interactions, particularly during courtship and mating behavior.

Hadley and Jackson (1977) analyzed the chemical composition of the epicuticular lipids of *S. mesaensis* using a combination of thin-layer and gas chromatography. Their findings indicated that hydrocarbons (HCs) were the most abundant constituents (28 to 33%) but sterols (mainly cholesterol), primary alcohols (C_{20} to C_{32}), free fatty acids (C_{14} to C_{30}) and triacylglycerols were also detected. HCs were fully saturated, with chain length ranging from 21 to 39 carbons. Normal alkanes accounted for 54 to 65% of the total HC fraction and branched alkanes for 38 to 46%. Hadley and Hall (1980) examined cuticular lipid biosynthesis in *S. mesaensis* using radiolabeled precursors (^{14}C-acetate and ^{14}C-palmitic acid) and found that the rate of biosynthesis of HCs was lower than that of cuticular alcohols and free fatty acids during intermolt periods. The same authors also reported (assumedly for the first time) that cuticular and hemolymph HCs were similar in composition but present in different relative proportions. This finding suggests that extraction of surface lipids must be carefully controlled to avoid contamination by hemolymph. Hadley and Filshie (1979) performed scanning and transmission electron microscopy on surface and transverse sections of the epicuticle of *Hadrurus arizonensis*. Images made after extraction of surface waxes clearly demonstrated the presence of empty wax canals that were probably filled by surface waxes in vivo. It is noteworthy that these observations date back to the 1970s and were among the first descriptions of the arthropod epicuticle.

Further experiment in collaboration with Philip Brownell were performed many years later in an attempt to identify the chemical signals detected by the scorpion's major chemosensory organs during reproductive behavior. Brownell and colleagues (Brownell and Polis, 2001) have shown detection of substrate-deposited semiochemicals by their ventral appendage, called the pectines, in various species of scorpion, such as *Hadrurus arizonensis* (Melville *et al.,* 2003 and end of this chapter) and *Centruroides vittatus* (Steinmetz *et al.,* 2004).

Chemosensory experimentation and cuticular compounds of Smeringerus mesaensis

Farley and Polis studied reproduction and development of *Smeringerus mesaensis* (for review see Farley, 2001). Natural populations of *S. mesaensis* (Scorpionida: Vaejovidae) are found only on sandy substrates, for which they have specially adapted sensory systems

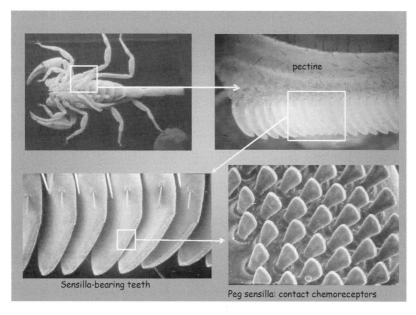

Figure 16.1 Ventral view of male *S. mesaensis* pectines (photos P. Brownell).

(Brownell, 2001). Due to the subtlety and complexity of reproduction in scorpions, field-testing is problematic but laboratory conditions may have a biasing effect (Benton, 2001). Current evidence has shown that *Smeringerus* is a well-documented surface forager showing clear behavioral responses in the field. Discrimination of vibration signals is particularly well developed and allows scorpions to detect the presence of a prey or congenerous of the opposite sex. After processing the orientation signal, behavioral responses are modulated according to the chemical properties of the target: food, water or mate (Brownell, 2001). Gaffin and Brownell (1992) have reported evidence of intraspecific chemical communication between scorpions and in particular in *Smeringerus*. Discrete variations in male behavior were observed in the presence of substrates impregnated with chloroform:methanol (2:1) dried cuticular extracts from females. The same authors performed experiments to evaluate pectine reactivity. They observed that after touching sand previously exposed to a conspecific female with the pectines, which male scorpions initiated juddering, which They observed that after touching sand previously exposed to a conspecific female with the pectines, male scorpions initiated juddering, which was followed by a female-seizing lunge and a "promenade à deux" if a conspecific female was encountered. was followed by a female-seizing lunge and a "promenade à deux" if a conspecific female was encountered. Gaffin and Brownell (1997a, b) studied response properties of chemosensory peg sensilla of scorpion pectines and presented evidence that they are sensitive to odorants and tastants like insect antennae. Though pectines are the largest and most elaborate scorpion sensory appendages, there are also numerous pore-tipped sensilla on the pedipalps and tarsal leg segments that may mediate chemosensitivity (see Figure 16.1). In addition chemosensory hairs in *Smeringerus* extending from cuticular surfaces are concentrated ventrally so that they are in contact with the substrate (Brownell, 2001; Gaffin and Brownell, 2001).

Almost 20 years after the first description of epicuticular lipids by Hadley, we re-examined the cuticular composition of *Smeringerus mesaensis* using GC and GC–MS. Quantitative and qualitative data for about 100 cuticular compounds (71 peaks) was obtained in adult and juvenile males and females (see Table 16.7). Some long-chain amides (from C_{14} to C_{22}) in particular Erucamide (a C_{22} amide) observed in initial analyses and then never detected again, were probably contaminants from polyolefin films (Shuler *et al.*, 2004) and plasticizers. Long-chain alcohols (C_{18} to C_{24}), trace amounts of aldehydes (C_{20} to C_{26}), and a few acids (mainly C_{18}s) were identified and quantified. Hydrocarbons – *n*-alkanes (from C_{20} to C_{34}) and methyl branched alkanes (C_{21} to C_{39}) with mono-methyl to trimethyl-branches – were also identified and quantified. Trace amounts of shorter (C_{14} to C_{19}) and longer (C_{35} to C_{39}) alkanes were also detected. Principal component analysis (PCA) showed a clear differentiation between juveniles and adults on the first axis (35% variation) and differentiation of males and females on the second axis (10% variation) (Figure not shown). Amides and contaminants never played a role in discrimination between groups. These data provide an interesting indication of chemical signal priorities in *Smeringerus*, with juvenile/adult discrimination taking priority over male/female discrimination. The quantities of cuticular compounds extracted per gram of individual body weight were higher in juveniles (133.80 and 224.57μg for females and males, respectively) than in adults (78.04 and 101.47μg for females and males, respectively). Curiously the quantity of HC did not depend on individual size or weight, females generally being larger than males. Some peaks seem to play key roles in discrimination. For example the proportion of central methyl C_{33} (peak 62) was 1.38% in adult females (*N*=10), 2.31% in adult males (*N*=9), 0.75% in juvenile females (*N*=7) and 0.94% in juvenile males (*N*=7). In comparison with the previous description by Hadley and Jackson (1977), our findings showed different proportions of normal and branched alkanes in the total HC fraction (around 30% and 70%, respectively) but the composition was similar in terms of the chemical families of the compounds (Table 16.5).

Table 16.5 *Proportions of the different compound families in males and females of* Smeringerus mesaensis.

	Alkanes	Oxygenated compounds	Esters
Males	40.54%	3.37%	56.09%
Females	66.07%	5.63%	28.30%

Identification and assay of chemical recognition signals *in* Hadrurus arizonensis

Scorpions have never really been used as a model to study the role of chemical signals in intraspecific and sexual communication. Up to now sex pheromones have been described

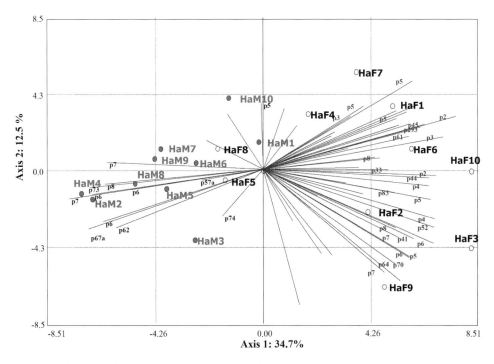

Figure 16.2 Principal component analysis biplot using the total cuticular mixtures of *Smeringerus mesaensis*, summer adult males (10 HaM) and females (10 HaF) with peaks' projection (p1 to 83) involved in sex-differrention.

in only three arachnids, i.e., a tick and two spiders (see review by Melville *et al.*, 2003). We used *Hadrurus arizonensis* (Scorpionida: Iuridae), the largest North American scorpion, as a model to study isolation and identify chemical signals in the species. Recent studies provided the first description of courtship and mating behavior in this scorpion family. According to Tallarovic *et al.* (2000), mating in *H. arizonensis* can be divided into four behavioral phases, i.e., initiation, *promenade à deux*, sperm transfer and termination. During the *promenade* phase, male actively use their pectines and first pair of legs to feel objects. The same authors (Melville *et al.*, 2003) reported another study aimed at determining if males are able to follow receptive females and if mate trailing was mediated by a gender-specific signal. Their findings suggested that *Hadrurus* males follow and respond to substrate-borne signals from conspecific females. Tail wagging probably represents precourtship signaling. Gender-specific signals from female scorpions may serve as both a male attractant and a courtship releaser according to the quality and quantity released. Various observations suggest that pectines are used for pheromone detection. Male scorpions increase pectinal sweeping activity after detection of female releases (Gaffin and Brownell, 2001).

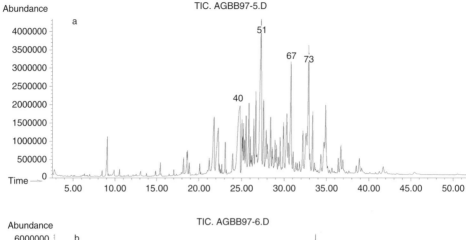

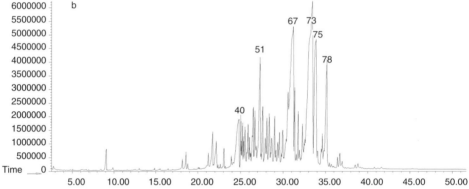

Figure 16.3 (a) Female and (b) male total ion chromatograms (TIC) of *Hadrurus arizonensis*.

We performed GC and GC–MS on whole extracts from male and female *Hadrurus arizonensis* scorpions collected during two seasons (winter and summer). Findings demonstrated a mixture of about 90 compounds including long-chain hydrocarbons (C_{20} to C_{41}), n-alcohols (C_{18} to C_{26}), aldehydes (C_{22} to C_{31}) and uncommon acid esters with long-chain alcohols (C_{26} to C_{30}) esterified to short-chain (C_2 to C_6) acids (Tables 16.6 and 16.7). The apolar nature of these molecules suggests that they are well suited to deposition on a natural substrate (sand) for detection by chemoreceptive organs. Clear sexual dimorphism was noted (Figures 16.2 and 16.3), with males expressing elevated levels of several esters (56.09% versus 28.3% for females) and fewer long-chain alkanes (44.54% versus 66.07% for females). The proportion of oxygenated compounds was low but higher in females than males (5.63% versus 3.37%) (Thevenieau, 1999; Bagnères *et al.*, 2001). As in *Paruroctonus*, amide traces were detected in *Smeringerus* in the same proportions in males and females, supporting the aforesaid contamination hypothesis.

Table 16.6 *Principal esters from* Hadrurus arizonensis.

Ester	R_1	R_2
octacosyl acetate	$C_{28}H_{57}$	CH_3
octacosyl butanoate	$C_{28}H_{57}$	C_3H_7
triacontyl acetate	$C_{30}H_{61}$	CH_3
triacontyl butanoate	$C_{30}H_{61}$	C_3H_7
triacontyl pentanoate	$C_{30}H_{61}$	C_4H_9
triacontyl hexanoate	$C_{30}H_{61}$	C_5H_{11}

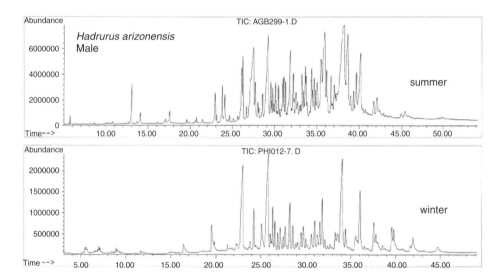

Figure 16.4 Winter (bottom) and summer (top) blends of *Hadrurus arizonensis* males.

Comparison of content in winter and summer blends showed not only fewer compounds but also different proportions in summer, i.e. the active mating season (Figure 16.4). This finding is consistent with sexual dimorphism in desiccation responses shown in another sand scorpion (Gefen, 2008) and also with seasonal changes in cuticular permeability (Toolson and Hadley, 1979). Esters were particularly abundant in tail gland extracts of reproductively active males, suggesting that they may act as aphrodisiacs or attractants. The main esters were synthesized (Brunetti, 1999; Bagnères *et al.*, 2001) and used in behavioral assays. Using the main male compounds, i.e. triacontyl butyrate (or butanoate) (peak 75) (13.5% versus 3.77% in females for the C_4) and acetate (10.2% versus 4.42% for the C_2), tail-rubbing behavior was elicited from males (see spectrum in Figure 16.5). Mature/receptive female blends attracted males and provoked courtship behavior during encounters.

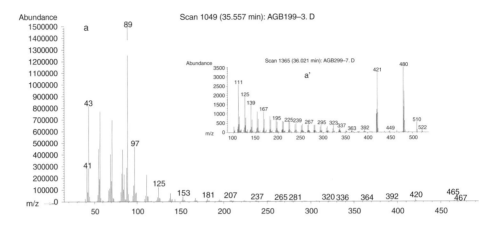

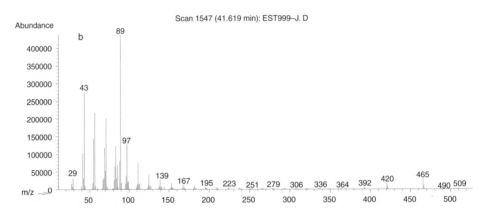

Figure 16.5 EI (a) and CI (a′) mass spectrum of the natural and (b) synthetic triacontyl butyrate where *m/z* 465 represents the M−43 and *m/z* 509 the M+1.

Conclusion on the role of the determined mixtures for the two scorpion species

Comparison of the cuticular blends of *Smeringerus mesaensis* and *Hadrurus arizonensis* (Table 16.7) indicates a high degree of chemospecificity, in agreement with the subtlety of their behavior and the particularities of their habitat. Scorpions are generally sedentary and forage near their retreat. However during the reproductive season males can travel great distances (sexual vagrancy) in a short time to increase encounters with females (Benton, 1992, 2001). In accordance with this duality, cuticular blends in both species may function not only as pheromones for conspecific recognition but also as kairomones. This would be of great utility in the harsh desert environments shared by the two species in the American southwest. For instance, in the Coachella Valley (California, Riverside County, USA), *Smeringerus* is the dominant species (up to 95% according to Polis, 2001) and *Hadrurus* could easily depend on cuticular blends to distinguish between its female or male conspecifics and also identify non-Iuridae scorpions such as *Smeringerus* that is one of its preys. It should also be emphasized that competition between scorpions, even conspecifics, can be particularly hard when the food supply is limited. While *Hadrurus* is rarely (if ever) preyed upon, *Smeringerus* can perform intraguilde and also intraspecific predation (Polis, 2001).

This work needs to be completed by performing more behavioral tests and also electrophysiological assays to test synthetic and natural blends as well as isolated compounds. We know that non-congenerous HCs can be detected by ant antennae (see Chapter 10 for a review) so it should be possible, since the chemicals are available, to perform similar sensory physiology tests on pectines, which seem to have a function in the detection of contact chemicals (G. and B., 2001) or other chemosensory sensitive parts of the scorpion body. In conclusion this chemical ecology work is only preliminary, but it appears to point in the same direction as behavioral studies by our American colleagues.

Conclusion

Arachnids (spiders and scorpions) present two modes of living: solitary most of the time but also communal during the first days of life. Under natural conditions, the young disperse to live isolated from each other. In solitary arachnids, intraspecific interactions often take the form of conflicts or fights. These displays have been described as agonistic behavior, particularly cannibalism. But interactions between solitary arachnids are not always aggressive. Under favorable biotic or abiotic conditions, arachnids can modulate their behavior to interact peacefully with conspecifics. These usually short periods of non-aggressive relations have been described as sexual and maternal tolerance. But there is another surprising kind of tolerance for such exclusively predatory and cannibalistic animals: mutual social tolerance during the gregarious period of young spiders and scorpions. In these gregarious or communal groups, arachnids must be able to distinguish between members of their own group and others. This is a real problem during the capture of prey, because the attacking arachnids have to decide instantly whether they should bite or not. Before engaging in a killing bite, arachnids touch the unknown animal. This contact allows them to identify the animal as a prey or as a conspecific. Contact chemoreceptors seem thus to play a key role in discriminating between prey and conspecifics. In arachnids, contact chemoreceptors are important in sexual recognition, but few data are available on the chemical cues used in social recognition by arachnids. Social behavior shifts linked to their development could be induced by changes in the chemical composition of the cuticle.

The chemical composition of the arachnid cuticle, especially the lipid layer, can be used for information transfer. These substances act as releaser pheromones and are identified by the arachnid after contact with another animal. Behavioral observations on arachnids demonstrate that chemical contact compounds are able to inhibit aggressive behavior between conspecifics (prevent cannibalism) and are used for sex recognition. The production of the cuticular compounds is sex- and age-dependent. Different studies have shown that qualitative and / or quantitative changes in cuticular lipids could play a role in intra- and interspecific relationships in arachnids, as has been demonstrated in insects. Knowledge of the function and mode of action of cuticular compounds is only fragmentary in arachnid groups. In arachnids, the endocrine regulation of pheromone synthesis and reproductive behavior is not known and the role of ecdysteroids is a very neglected field of research.

So further work will be necessary to test the effect of contact compounds by electrophysiological and bioassays methods, and to study the role of hormones in contact pheromone biosynthesis in arachnids.

Table 16.7 *Determination of cuticular waxes of the two scorpion species (t: traces).*

Pk No	Components	Hadrurus arizonensis	Smeringerus mesaensis
	n-nonadecane		t
1	tetradecanamide	+	+
	9-+11-methylnonacosane		t
2	palmitic acid		t
3	n-eicosane	+	+
4	n-octadecanol		+
5	n-heneicosane		+
6	linoleic and oleic acids	t	t
7	hexadecenamide	+	
8	hexadecanamide	+	+
9	n-docosane	+	+
12	heptadecanamide	+	
13	n-eicosanol	+	+
14	n-tricosane	+	+
15+16	octadecenamide	+	+
17a	n-tetracontane	+	+
17b	docosanal	+	
18	n-docosanol	+	+
19	n-pentacosane	+	+
20	11-(+9)-methylpentacosane		+
21	5- methylpentacosane		+
23	5,11-dimethylpentacosane	+	+
24	n-hexacosane	+	+
25	tetracosanal	+	
	+12-methylhexacosane		
26	4/2-methylhexacosane	+	+
27	n-tetracosanol	+	+
28	n-heptacosane	+	+
29	11-methylheptacosane		+
30	7- methylheptacosane		+
31	5-methylheptacosane	+	+

Table 16.7 (*cont.*)

Pk No	Components	*Hadrurus arizonensis*	*Smeringerus mesaensis*
32	4/2-methylheptacosane	+	+
33	3-methylheptacosane	+	+
	amide		
	amide		
36	*n*-octacosane	+	+
37a	hexacosanal	+	+
	+12-methyloctacosane		
37b	6- methyloctacosane		
38	4/2-methylhexacosane	+	+
39	*n*-hexacosanol	+	
40	*n*-nonacosane	+	+
41	15-+13-+11-methylnonacosane	+	+
42	7-methylnonacosane	+	+
43	5-methylnonacosane	+	+
44	4/2-methylnonacosane +11,	+	+
	15-dimethylnonacosane		
45a	3-methylnonacosane	+	+
45b	trimethylnonacosane	t	
46	tetracosyl butyrate	+	
47	*n*-triacontane	+	+
48a	hexacosyl acetate	+	
48b	14-+12-methyltriacontane	+	
48c	octacosanal + tetracosyl valerate	+	+
49	3-methyltriacontane	+	+
50	4/2-methyltriacontane	+	+
	(+octacosanol)		
51	*n*-hentriacontane	+	+
52	15-+13-methylhentriacontane	+	+
53	7-methylhentriacontane	+	+
	(+pentacosyl valerate)		
54a	5-methylhentriacontane	+	
54b	11,15-+11,17-	+	+
	dimethylhentriacontane		
55	3-methylhentriacontane	+	+
56	*n*-dotriacontane + hexacosyl	+	
	butyrate		
57a	octacosyl acetate	+	
57b	triacontenal + 16-+12-	+	+
	methyldotriacontane		
58	triacontanal +	+	+
	3-methyldotriacontane		
59	4/2-methyldotriacontane	+	+
60	heptacosyl butyrate	t	

Table 16.7 (*cont.*)

Pk No	Components	*Hadrurus arizonensis*	*Smeringerus mesaensis*
61	*n*-tritriacontane (+ triacontanol)	+	+
62	nonacosyl acetate	t	
63	17-+15-+13-methyltritriacontane	+	+
64a	7-methyltritriacontane + hentriacontenal	+	
64b	13,19-+11,15- dimethyltritriacontane	+	+
64c	9,19-dimethyltritriacontane	+	+
65	3-methylritriacontane	+	
66	*n*-tetratriacontane + octacosyl butyrate	t	
67a	triacontyl acetate	+	
67b	14-methyltetratriacontane	+	
68	4/2-methyltetratriacontane + octacosyl valerate	+	
69	*n*-pentatriacontane + nonacosyl butyrate	+	
70a	hentriacontyl acetate + triacontyl propionate	+	
70b	17-+15-+13- methylpentatriacontane	+	+
71	nonacosyl valerate	t	
72a	15,19-dimethylpentatriacontane	+	
72b	11,19-+11,21- dimethylpentatria- contane	+	+
73	triacontyl butyrate (+*n*-hexatriacontane)	+	
74	docotriacontyl acetate	t	
75	triacontyl valerate	+	
76a	nonacosyl caproate + *n*-heptatriacontane	⊤	
76b	hentriacontyl butyrate	+	
77	17-+15-+13-+11- methylheptatriacontane	+	+
78	triacontyl caproate + hentriacontyl valerate	+	
79	15,19-+13,17- dimethylheptatriacontane	+	+
80	(*n*-octatriacontane) + triacontyl caproate	t	
81	17-+15-+13- methylnonatriacontane	+	+

Table 16.7 (*cont.*)

Pk No	Components	*Hadrurus arizonensis*	*Smeringerus mesaensis*
82	15,17-+13,17-dimethylnonatriacontane	+	+
83	5,17-dimethylnonatriacontane	+	
	15,19-dimethylhentriacontane	t	

Acknowledgments

Anne-Geneviève Bagnères wishes to thank Phil Brownell for his enthusiastic assistance. Part of this work was performed during periods of sabbatical leave taken by Anne-Geneviève Bagnères in Reno, UNR, USA in 1997 and of PB in Marseille, France in 2001. Finances for both stays were provided by the CNRS.

References

Altner, H. and Prillinger, L. (1980). Ultrastructure of invertebrate chemo-, thermo- and hygroreceptors and its functional significance. *Int. Rev. Cytol.*, **67**, 69–139.

Arnqvist, G. (1992). Courtship behaviour and sexual cannibalism in the semi-aquatic fishing spider, *Dolomedes fimbriatus* (Clerck) (Araneae: Pisauridae). *J. Arachnol.*, **20**, 222–226.

Assi-Bessekon, D. (1997). Intraspecific identification and tolerance in the social maternal behaviour of *Coelotes terrestris* (Araneae, Agelenidae). *Behav. Process.*, **39**, 231–239.

Bagnères, A.-G., Trabalon, M., Blomquist, G. and Schulz, S. (1997). Waxes of the social spider *Anelosimus eximius* (Aranea, Theridiidae): Abundance of novel *n*-propyl esters of long-chain methyl-branched fatty acids. *Arch. Insect Biochem. Physiol.*, **36**, 295–314.

Bagnères, A.-G., Brownell, P. H., Brunetti, G., Clément, J.-L., Blomquist, G. J. and Kuenzli, M. (2001). Identification and assay of chemical recognition signals for scorpions (Scorpionida: *Hadrurus arizonensis*). In *Proc. 18th Annual Meeting of the ISCE (Lake Tahoe, USA)*, p. 101.

Barth, F. G. (1993). Sensory guidance in spider pre-copulatory behaviour. *Comp. Biochem. Physiol.*, **104A**, 717–733.

Barth, F. G. and Schmitt, A. (1991). Species recognition and species isolation in wandering spiders (*Cupiennius* spp., Ctenidae). *Behav. Ecol. Sociobiol.*, **29**, 333–339.

Benton, T. G. (1992). Determinants of male mating success in a scorpion. *Anim. Behav.*, **43**, 125–135.

Benton, T. (2001). Reproductive ecology. In *Scorpion Biology and Research*, ed. P. Brownell and G. Polis. Oxford: Oxford University Press, pp. 278–301.

Blanke, R. (1975). Untersuchungen zum sexualverhalten von *Cyrtophora cicatrosa* (Stoliczka) (Araneae, Araneidae). *Zeitschrift Tierpsychol.*, **37**, 62–74.

Breene, R. and Sweet, M. (1985). Evidence of insemination of multiple females by the male black widow spider, *Latrodectus mactans* (Araneae, Theridiidae). *J. Arachnol.*, **13**, 331–335.

Brownell, P. (2001). Sensory ecology and orientational behaviors. In *Scorpion biology and research*, ed. P. Brownell and G. Polis. Oxford: Oxford University Press, pp. 159–183.

Brownell, P. and Polis, G. (2001). *Scorpion Biology and Research*, ed. P. Brownell and G. Polis. Oxford: Oxford University Press, pp. 159–183.

Brunetti, G. (1999). *Synthèse de phéromones et d'allomones de contact de scorpions.* Unpublished report E.N.S.C.M (Montpellier, France), 34p.

Carbonell, L. F., Hodge, M. R., Tomalski, M. D. and Miller, L. K. (1988). Synthesis of a gene coding for an insect-specific scorpion neurotoxin and attempts to express it using baculovirus vectors. *Gene*, **73**, 409–418.

Clot-Faybesse, O., Juin, M., Rochat, H. and Devaux, C. (1999). Monoclonal antibodies against the *Androctonus australis* hector scorpion neurotoxin I: characterisation and use for venom neutralisation. *FEBS Lett.*, **458**, 313–318.

Costa, F. G., Viera, C. and Francescoli, G. (2000). A comparative study of sexual behaviour in two synmorphic species of the genus *Lycosa* (Araneae, Lycosidae) and their hybrid progeny. *J. Arachnol.*, **28**, 237–240.

Dijkstra, H. (1976). Seaching behaviour and tactochemical orientation in males of the wolf spider *Pardosa amentata* (Cl.) (Araneae Lycosidae). *Proc. K. Ned. Akad. Wet.*, **79**, 235–244.

Dondale, C. D. and Hegdekar, B. M. (1973). The contact sex pheromone of *Pardosa lapidicina* Emerton (Araneideae, Lycosidae). *Can. J. Zool.*, **51**, 400–401.

Dumais, J. and Perron, J. M. (1973). Elements du comportement sexuel chez *Pardosa xerampelina* (Keyserling) (Araneidae: Lycosidae). *Can. J. Zool.*, **51**, 265–271.

Farley, R. (2001). Structure, reproduction and development. In *Scorpion Biology and Research*, ed. P. Brownell and G. Polis. Oxford: Oxford University Press, pp. 13–78.

Foelix, R. F. and Chu-Wang, I. (1973). The morphology of spider sensilla II. Chemoreceptors. *Tissue Cell*, **5**, 461–478.

Fontecilla-Camps, J.-C., Habersetzer-Rochat, C. and Rochat, H. (1988). Orthorhombic crystals and three-dimensional structure of the potent toxin II from the scorpion *Androctonus australis* Hector. *Proc. Natl. Acad. Sci. USA*, **85**, 7443–7447.

Gaffin, D. and Brownell, P. (1992). Evidence of chemical signaling in the sand scorpion, *Paruroctonus mesaensis* (Scorpionida: Vaejovidae). *Ethology*, **91**, 59–69.

Gaffin, D. and Brownell, P. (1997a). Response properties of chemosensory peg sensilla on the pectines of scorpions. *J. Comp. Physiol. A*, **181**, 291–300.

Gaffin, D. and Brownell, P. (1997b). Electrophysiological evidence of synaptic interactions within chemosensory sensilla of scorpion pectines. *J. Comp. Physiol. A*, **181**, 301–307.

Gaffin, D. and Brownell, P. (2001). Chemosensory behavior and physiology. In *Scorpion Biology and Research*, ed. P. Brownell and G. Polis. Oxford: Oxford University Press, pp. 184–203.

Gaskett, A. C. (2007). Spider sex pheromones: emission, reception, structures, and functions. *Biol. Rev.*, **82**, 27–48.

Gaskett, A. C., Herberstein, M. E., Downes, B. and Elgar, M. A. (2004). Changes in male mate choice in a sexually cannibalistic orb-web spider (Araneae: Araneidae). *Behav.*, **141**, 1197–1210.

Gefen, E. (2008). Sexual dimorphism in dessication response of the sand scorpion *Smeringurus mesaensis* (Vaejovidae). *J. Insect Physiol.*, **54**, 798–805.

Gefen, E. and Ar, A. (2004). Comparative water relations of four species of scorpions in Israel: evidence for phylogenetic differences. *J. Exp. Biol.*, **207**, 1017–1025.

Gundermann, J.L., Horel, A. and Krafft, B. (1993). Experimental manipulation of social tendencies in subsocial spider *Coelotes terrestris*. *Insectes Soc.*, **40**, 219–229.

Hadley, N.F. (1970). Water relations of the desert scorpion, *Hadrurus arizonensis*. *J. Exp. Biol.*, **53**, 547–558.

Hadley, N.F. (1981). Cuticular lipids of terrestrial plants and arthropods: a comparison of their structure, composition, and waterproofing function. *Biol. Rev.*, **56**, 23–47.

Hadley, N.F. and Filshie, B.K. (1979). Fine structure of the epicuticle of the desert scorpion, *Hadrurus arizonensis*, with reference to location of lipids. *Tissue Cell*, **11**, 263–275.

Hadley, N.F. and Hall, R.L. (1980). Cuticular lipid biosynthesis in the scorpion, *Paruroctonus mesaensis*. *J. Exp. Zool.*, **212**, 373–379.

Hadley, N.F. and Jackson, L.L. (1977). Chemical composition of the epicuticular lipids of the scorpion, *Paruroctonus mesaensis*. *Insect Biochem.*, **7**, 85–89.

Hadley, N.F. and Quinlan, M.C. (1987). Permeability of arthrodial membrane to water: A first measurement using in vivo techniques. *Cell. Mol. Life Sci.*, **43**, 164–166.

Hegdekar, B.M. and Dondale, C.D. (1969). A contact sex pheromone and some response parameters in lycosid spiders. *Can. J. Zool.*, **47**, 1–4.

Herberstein, M.E., Schneider, J.M. and Elgar, M.A. (2002). Costs of courtship and mating in a sexually cannibalistic orbweb spider: female mating strategies and their consequences for males. *Behav. Ecol. Sociobiol.*, **51**, 440–446.

Hirschberg, D. (1969). Beiträge zur Biologie, insbesondere zur Brutpflege einiger Theridiiden. *Zeit. Wissensch. Zool.*, **179**, 189–252.

Jackson, R.R. (1981). Comparative studies of *Dictyna* and *Mallos* (Araneae, Dictynidae), IV: Silk-mediated interattraction. *Insectes Soc.*, **29**, 15–24.

Jackson, R.R. (1986). Use of pheromones by males of *Phidippus johnsoni* (Araneae, Saticidae) to detect subadult females that are about to molt. *J. Arachnol.*, **14**, 137–139.

Jackson, R.R. (1987). Comparative study of releaser pheromones associated with the silk of jumping spiders (Araneae, Salticidae). *N. Z. J. Zool.*, **14**, 1–10.

Jackson, R.R. and Cooper, K.J. (1990). Variability of the responses of jumping spiders (Araneae, Salticidae) to sex pheromones. *N. Z. J. Zool.*, **17**, 39–42.

Jackson, R.R. and Pollard, S.D. (1982). The biology of *Dysdera crocata* (Araneae; Dysderidae): intraspecific interactions. *J. Zool.*, **198**, 197–214.

Kaston, B.J. (1936). The senses involved in the courtship of some vagabond spiders. *Entomol. Am.*, **16**, 97–166.

Krafft, B. (1971). Les interactions entre les individus chez *Agelena consociata* araignée sociale du Gabon. *Arachno. Congres. Inter.*, **V**, 159–164.

Krafft, B. (1978). The recording of vibratory signals performed by spiders during courtship. *Symp. Zool. Soc. London*, **42**, 59–67.

Krafft, B. (1982). The significance and complexity of communication in spiders. In *Spider communication: mechanisms and ecological significance*, ed. P.N. Witt, and J.S. Rovner, Princeton, NJ: Princeton University Press., pp. 15–66.

Krafft, B., Horel, A. and Julita, J. M. (1986). Influence of food supply on the duration of the gregarious phase of a maternal-social spider, *Coelotes terrestris* (Araneae, Agelenidae). *J. Arachnol.*, **14**, 219–226.

Lizotte, R. and Rovner, J. S. (1989). Water-resistant sex pheromones in Lycosid spiders from a tropical wet forest. *J. Arachnol.*, **17**, 123–124.

Locket, G. H. and Bristowe, W. S. (1926). Observations on the mating habits of some web spinning spiders. *Proc. Zool. Soc. London*, **94**, 1125–1143.

Melville, J. M., Tallarovic, S. K. and Brownell, P. H. (2003). Evidence of mate trailing in the giant hairy desert scorpion *Hadrurus arizonensis* (Scorpionida, Iuridae). *J. Insect Behav.*, **16**, 97–115.

Papke, M., Schulz, S., Tichy, H., Gingi, E. and Ehn, R. (2000). Identification of a new sex pheromone from the silk dragline of the tropical wandering spider *Cupiennius salei*. *Angewandte Chemie Inter. Ed.*, **39**, 4339–4341.

Papke, M. D., Riechert, S. E. and Schulz, S. (2001). An airborne female pheromone associated with male attraction and courtship in a desert spider. *Anim. Behav.*, **61**, 877–886.

Pasquet, A., Trabalon, A., Bagnères, A.-G. and Leborgne, R. (1997). Does group closure exist in the social spider *Anelosimus eximius*? Behavioural and chemical approaches. *Insectes Soc.*, **44**, 159–169.

Platnick, N. (1971). The evolution of courtship behaviour in spiders. *Bull. British Arachnol. Soc.*, **2**, 40–47.

Pollard, S. D., Macnab, A. M. and Jackson, R. R. (1987). Communication with chemicals: pheromones and spiders. In *Ecophysiology of Spiders*, ed. W. Nentwig, Berlin: Springer, pp. 133–141.

Polis, G. A. (2001). Population and community ecology of desert sorpions. In *Scorpion Biology and Research*, ed. P. Brownell and G. Polis. Oxford: Oxford University Press, pp. 302–316.

Pourié, G. and Trabalon, M. (1999a). Agonistic behaviour of female *Tegenaria atrica* in the presence of different aged spiderlings. *Physiol. Entomol.*, **24**, 143–149.

Pourié, G. and Trabalon, M. (1999b). Relationships among food and contact signals in experimental group-living young of *Tegenaria atrica*. *Arch. Insect Biochem. Physiol.*, **42**, 188–197.

Pourié, G. and Trabalon, M. (2001). Plasticity of agonistic behaviour in relation to diet and contact signals in experimentally group-living of *Tegenaria atrica*. *Chemoecology*, **11**, 175–181.

Pourié, G. and Trabalon, M. (2003). The role of 20-hydroxyecdysone on the control of spider vitellogenesis. *Gen. Comp. Endocrinol.*, **131**, 250–257.

Pourié, G., Ibarra, F., Francke, W. and Trabalon, M. (2005). Fatty acids mediate agressive behaviour in the spider *Tegeneria atrica*. *Chemoecology*, **15**, 161–166.

Prouvost, O., Trabalon, M., Papke, M. and Schulz, S. (1999). Contact sex signals on web and cuticle of *Tegenaria atrica* (Araneae, Agelenidae). *Arch. Insect Biochem. Physiol.*, **40**, 194–202.

Ramousse, R. (1986). Oophagie et croissance des stades grégaires chez *Araneus suspicax*. *Mem. Sc. Belge Entomol.*, **33**, 179–186.

Richter, C. J. J., Stoling, H. C. J. and Vlijm, L. (1971). Silk production in adult female of the wolf spider *Pardosa amentata* (Lycosidae, Araneae). *J. Zool.*, **165**, 285–290.

Roberts, J. A. and Uetz, G. W. (2005). Information content of female chemical signals in the wolf spider, *Schizocosa ocreata*: male discrimination of reproductive state and receptivity. *Anim. Behav.*, **70**, 217–223.

Roland, C. (1984). Chemical signals bound to the silk in spider communication (Arachnidae, Araneae). *J. Arachnol.*, **11**, 309–314.

Ross, K. and Smith, R. L. (1979). Aspects of the courtship behaviour of the black widow spider, *Ladrodectus hesperus* (Araneae: Theridiidae), with evidence for the existence of a contact sex pheromone. *J. Arachnol.*, **7**, 69–77.

Ruttan, L. M. (1990). Experimental manipulations of dispersal in the subsocial spider, *Theridion pictum. Behav. Ecol. Sociobiol.*, **27**, 169–173.

Schulz, S. (1997). The chemistry of spider toxins and spider silk. *Angewandte Chemie*, **36**, 314–326.

Schulz, S. (2001). Composition of the silk lipids of the spider *Nephila clavipes. Lipids*, **36**, 637–647.

Schulz, S. (2004). *Semiochemistry of Spiders: Advances in Chemical Ecology. Vol. 1*, ed. J. Millar and J. Acree, New York: Academic, pp. 110–150.

Schulz, S. and Toft, S. (1993a). Branched long chain alkyl methyl ethers: a new class of lipids from spider silk. *Tetrahedron*, **49**, 6805–6820.

Schulz, S. and Toft, S. (1993b). Identification of a sex pheromone from a spider. *Science*, **260**, 1635–1637.

Schuster, M., Baurecht, D., Mitter, E., Schmitt, A. and Barth, F. G. (1994). Field observations on the population structure of three Ctenid spiders (*Cupiennius, Araneae, Ctenidae*). *J. Arachnol.*, **22**, 32–38.

Shuler, C. A., Janorkar, A. V. and Hirt, D. E. (2004). Fate of erucamide in polyolefin films at elevated temperature. *Polym. Engin. Sci.*, **44**, 2247–2253.

Steinmetz, S. B., Bost, K. C. and Gaffin, D. D. (2004). Response of male *Centruroides vittatus* (Scorpiones: Buthidae) to aerial and substrate-borne chemical signals. *Euscorpius*, **12**, 1–6.

Suter, R. B. and Renkes, G. (1982). Linyphiid spider courtship: Releaser and attractant functions of a contact sex pheromone. *Anim. Behav.*, **30**, 714–718.

Suter, R. B. and Hirscheimer, A. J. (1986). Multiple web-borne pheromones in a spider *Frontinella pyramitela* (Araneae, Linyphiidae). *Anim. Behav.*, **34**, 748–753.

Suter, R. B., Shane, C. M. and Hirscheimer, A. J. (1987). Communication by cuticular pheromones in a Linyphiid spider. *J. Arachnol.*, **15**, 157–162.

Szlep, R. (1964). Change in the response of spiders to repeated web vibrations. *Behav.*, **23**, 203–239.

Tallarovic, S. K., Melville, J. M. and Brownell, P. H. (2000). Courtship and mating in the giant hairy desert scorpion, *Hadrurus arizonensis* (Scorpionida, Iuridae). *J. Insect Behav.*, **13**, 827–838.

Taylor, P. W. (1998). Dragline-mediated mate-searching in *Trite planiceps* (Araneae, Salticidae). *J. Arachnol.*, **26**, 330–334.

Thevenieau, L. (1999). Phéromones de contact chez le scorpion *Hadrurus arizonensis*. Unpublished report of Master thesis, Univ. Aix-Marseille II, 14p.

Tichy, H., Gingl, E., Ehn, R., Papke, M. and Schulz, S. (2001). Female sex pheromone of a wandering spider (*Cupiennius salei*): identification and sensory reception. *J. Comp. Physiol. A*, **187**, 75–78.

Tietjen, W. J. (1979). Tests for olfactory communication in four species of wolf spiders (Araneae, Lycosidae). *J. Arachnol.*, **6**, 197–206.

Tietjen, W. J. and Rovner, J. S. (1982). Chemical communication in *Lycosida* and other spiders. In *Spider Communication: Mechanisms and Ecological Significance*, ed. P. N. Witt, and J. S. Rovner, Princeton, NJ: Princeton University Press, p. 249.

Toolson, E. C. and Hadley, N. F. (1979). Seasonal effects on cuticular permeability and epicuticular lipid composition in *Centruroides sculpturatus* Ewing 1928 (Scorpiones: Buthidae). *J. Comp. Physiol. B*, **129**, 319–325.

Trabalon, M. (2000). The forms of communication in spiders. *Act. Coll. Insectes Sociaux*, **13**, 1–11.

Trabalon, M. and Assi-Bessekon, D. (2008). Effects of web chemical signatures on intraspecific recognition in a subsocial spider, *Coelotes terrestris* (Araneae). *Anim. Behav.*, **76**, 1571–1578.

Trabalon, M., Bagnères, A.-G., Hartmann, N. and Vallet, A.-M. (1996). Change in cuticular compounds composition during the gregarious period and after dispersal of the young in *Tegenaria atrica* (Araneae, Agelenidae). *Insect Biochem. Mol. Biol.*, **26**, 77–84.

Trabalon, M., Bagnères, A.-G. and Roland, C. (1997). Contact sex signals in two sympatric spider species, *Tegenaria domestica* and *T. pagana. J. Chem. Ecol.*, **23**, 747–758.

Trabalon, M., Niogret, J. and Legrand-Frossi, C. (2005). Effect of 20-hydroxyecdysone on cannibalism, sexual behavior, and contact sex pheromone in the solitary female spider, *Tegenaria atrica. Gen. Comp. Endocrinol.*, **144**, 60–66.

Trabalon, M., Pourié, G. and Hartmann, N. (1998). Relationships among cannibalism, contact signals, ovarian development and ecdysteroid levels in *Tegenaria atrica* (Araneae, Agelenidae). *Insect Biochem. Mol. Biol.*, **28**, 751–758.

Uetz, G. W. and Denterlein, G. (1979). Courtship behaviour, habitat, and reproductive isolation in *Schizocosa rovneri* Uetz and Dondale (Araneae: Lycosidae). *J. Arachnol.*, **7**, 121–128.

Vallet, A.-M., Marion-Poll, F. and Trabalon, M. (1998). Preliminary electrophysiological study of the contact chemoreceptors in a spider. *C. R. Acad. Sc. Paris, serie III*, **321**, 463–469.

Whitehouse, M. E. A. and Jackson, R. R. (1994). Intraspecific interactions of *Argyrodes antipodiana*, a kleptoparasitic spider from New Zealand. *N. Z. J. Zool.*, **21**, 253–268.

Yoshida, H. and Suzuki, Y. (1981). Silk as a cue for mate location in the jumping spider *Carrhotus xanthogramma* (Latreille) (Araneae: Salticidae). *App. Entomol. Zool.*, **16**, 315–317.

Zacharuk, R. Y. (1980). Ultrastructure and function of insect chemosensilla. *Annu. Rev. Entomol.*, **25**, 27–47.

17

Hydrocarbons as contact pheromones of longhorned beetles (Coleoptera: Cerambycidae)

Matthew D. Ginzel

The cuticular wax layer of insects is comprised of a complex mixture of long-chain fatty acids, alcohols, esters, aldehydes, ketones, and hydrocarbons that protect insects from desiccation (Gibbs, 1998). The components of the wax layer, particularly the hydrocarbons, also act as contact pheromones that mediate mate recognition (Howard and Blomquist, 2005). In fact, there is a growing body of evidence that contact pheromones play important roles in the mating systems of beetles in the family Cerambycidae, the longhorned beetles (e.g., Kim *et al.*, 1993; Fukaya *et al.*, 1996, 1997, 2000; Wang, 1998; Ginzel *et al.*, 2003a, 2003b, 2006; Ginzel and Hanks, 2003, 2005; Lacey *et al.*, 2008). These chemical signals elicit mating responses from males and have been confirmed in a number of cerambycid species, including those in the primitive subfamily Prioninae (Barbour *et al.*, 2007), and more derived subfamilies of Cerambycinae (e.g., Ginzel and Hanks, 2003; Ginzel *et al.*, 2003a, 2003b, 2006; Lacey *et al.*, 2008) and Lamiinae (e.g., Wang, 1998; Fukaya *et al.*, 2000; Ginzel and Hanks, 2003; Yasui *et al.*, 2003; Zhang *et al.*, 2003; see Table 17.1).

Cerambycid beetles are collectively known as longhorned beetles because many species have elongate filiform antennae (Linsley, 1964), and this diverse family includes more than 35,000 species in ~4,000 genera (Lawrence, 1982). Our understanding of chemical communication in cerambycids pales in comparison to what is known about the geographical distribution and taxonomy of this family (Hanks, 1999). Nevertheless, longhorned beetles are among the most important pests of woody plants in natural and managed systems worldwide, and many attack domestic broadleaf trees and shrubs (Solomon, 1995). The larvae are phytophagous, and colonize hosts that vary greatly in quality and may be healthy, moribund, recently killed or even decomposing (Hanks, 1999). Stressed and dying host trees become available sporadically and unpredictably when they are damaged or weakened by environmental stresses (reviewed by Hanks, 1999), and their nutritional quality declines rapidly as subcortical tissues are degraded by xylophagous competitors, including buprestid and scolytid beetles, as well as by other cerambycid species and saprophytes. Thus, early colonizers of moribund trees are afforded better larval nutrition, and selection favors behaviors in adults that expedite mate location and recognition and rapid mating and oviposition. These behaviors include the use of contact pheromones in mate recognition. This chapter first discusses the use of contact pheromones in the mating systems of cerambycids, then details the copulatory behavior of longhorned beetles, and explores the

Table 17.1 *Summary of published research on contact pheromones of longhorned beetles and bioassays used to confirm activity.*

Subfamily (Tribe)	Species	Stimulus	Biological activity	References
Cerambycinae (Callidiini)	*Semanotus japonicus* (Lacordaire)	Ether extracts of ♀	Stimulates mating behavior in ♂	Kim *et al*, 1993
Cerambycinae (Cerambycini)	*Nadezhdiella cantori* (Hope)	Hexane extracts of ♀	Freshly killed ♀ and ♀ extracts are bioactive	Wang *et al.*, 2002
Cerambycinae (Clytini)	*Megacyllene caryae* (Gahan)	Z9-C_{29}	Stimulates mating behavior in ♂	Ginzel and Hanks, 2003 Ginzel *et al.*, 2006
Cerambycinae (Clytini)	*Megacyllene robiniae* (Förster)	Filter paper exposed to ♀ Z9-C_{25}	♂ remain longer with filter paper exposed to females than control Stimulates mating behavior	Galford, 1977 Ginzel and Hanks, 2003 Ginzel *et al.*, 2003
Cerambycinae (Clytini)	*Neoclytus a. acuminatus* (F.)	7-MeC_{27}, 7-MeC_{25}, 9-MeC_{25}	7Me-C27 is the major component of the contact sex pheromone and 7Me-C25 and 9Me-C27 act as synergists.	Lacey *et al.*, 2008
Cerambycinae (Clytini)	*Neoclytus m. mucronatus* (F.)	Hexane extracts of ♀	♂ attempt to mate with reconstituted ♀	Ginzel and Hanks, 2003
Cerambycinae (Clytini)	*Xylotrechus colonus* (F.)	nC_{25}, 9-MeC_{25}, 3-MeC_{25}	♂ attempt to mate with hexane-washed ♀ treated with all three compounds	Ginzel *et al.*, 2003
Laminae (Acanthocinini)	*Dectes texanus* LeConte	Ether extracts of ♀	♂ attempt to mate with reconstituted ♀	Crook *et al.*, 2004
Laminae (Gleneini)	*Paraglenea fortunei* Saunders	Hexane extracts of ♀	♂ attempt to mate with reconstituted ♀	Wang *et al.*, 1991
Laminae (Lamiini)	*Anoplophora chinensis* (Förster)	Dead ♀ and hexane extracts of ♀	~50% of ♂ attempt to mate with recently killed ♀ and dummies treat with extract	Wang, 1998

Subfamily (Tribe)	Species	Compound(s)	Bioassay	Reference
Lamiinae (Lamiini)	*Anoplophora malasiaca* Thompson	Hexane and ether fractions of crude ether extracts of ♀ elytra Solvent extracts of ♀ and ♂	♀ attempt to mate with glass rods treated with both saturated subfraction of hexane fraction and ether fraction. ♂ attempt to mate with glass dummies treated with ♀ extract, addition of ♂ reduces bioactivity	Fukaya *et al.*, 2000 Akino *et al.*, 2001 Yasui *et al.* 2003
		10-heptacosanone 12-heptacosanone (Z)-18-heptacosen-10-one, (18Z, 21Z)-heptacosa-18,21-dien-10-one, (18Z, 21Z, 24Z)-heptacosa- 18,21,24-trien-10-one	♂ attempt to mate with glass rod treated with all five compounds. Synthetic blend of all five compounds except 12-heptacosanone ~ twice as bioactive	
Lamiinae (Lamiini)	*Anoplophora glabripennis* (Motschulsky)	$Z9\text{-}C_{23}$, $Z9\text{-}C_{25}$, $Z7\text{-}C_{25}$, $Z9\text{-}C_{27}$, $Z7\text{-}C_{27}$	Mixture of all five compounds applied to glass dummies elicited abdominal bending in ♂	Zhang *et al.*, 2003
Lamiinae (Lamiini)	*Plectrodera scalator* (F.)	Hexane extracts of ♀	♂ attempt to mate with reconstituted ♀	Ginzel and Hanks, 2003
Lamiinae (Lamiini)	*Psacothea hilaris* (Pascoe)	Ether extracts of ♀ $Z8_{21}Me\text{-}C_{35}$	♂ attempt to mate with treated dummies; ♂ display a dose-dependant response to treated gelatin capsules, but weaker than that elicited by crude extract	Fukaya and Honda, 1995 Fukaya *et al.*, 1996
Lamiinae (Monochamini)	*Acalolepta luxuriosa* Bates	Solvent extracts of ♀ elytra	♂ attempt to mate with dummies treated with extracts	Kubok *et al.* 1985
Lamiinae (Monochamini)	*Monochamus alternatus* Hope	Hexane extracts of ♀	♂ attempt to mate with dummies treat with 3-5 female equivalents of extract	Kim *et al.*, 1992
Prioninae (Prionini)	*Prionus californicus* Motschulsky	Hexane extracts of ♀	56% of ♂ display abdominal bending toward reconstituted ♀ (extracts applied to solvent-washed ♀)	Barbour *et al.* 2007

Adapted from Allison *et al.*, 2004.

use of bioassays to demonstrate the role of hydrocarbons in mate recognition. Two methods of sampling cuticular components – traditional whole-body solvent extraction and solid phase microextraction (SPME) wipe sampling – are detailed. Quantitative and qualitative differences in the hydrocarbon profiles of males and females are then discussed, as well as the biological relevance of cuticular hydrocarbons specific to males. The focus then shifts towards the use of single components and blends of compounds as contact pheromones of cerambycids.

Contact pheromones in the mating systems of longhorned beetles

Volatile male-produced sex and aggregation pheromones have been identified for a number of species, including eleven species in three tribes of the subfamily Cerambycinae alone (see Lacey *et al.*, 2004, 2007; Hanks *et al.*, 2007). Most of these pheromones share a common structural motif: six-, eight-, or ten-carbon chains with hydroxyl or carbonyl groups on C_2 or C_3. Some of these compounds (e.g., (2, 3) hexanediol and (*R*)-3-hydroxy-2-hexanone) may even serve as a widespread aggregation pheromone for this diverse and speciose subfamily. In fact, live traps baited with generic blends of racemic 2-hydroxy-3-hexanone and 3-hydroxy-2-hexanone captured both males and females of three sympatric cerambycine species (Hanks *et al.*, 2007). The structure of these compounds is unique to the Cerambycidae; among the beetles, semiochemicals used by members of even closely-related families are quite different. Mate location and recognition in beetles in the sub-family Cerambycinae appear to involve three sequential behavioral stages: (1) both sexes are independently attracted to volatiles emanating from the weakened larval host plant; (2) males attract females over shorter distances with aggregation pheromones; and (3) males recognize females by contact pheromones (Ginzel and Hanks, 2005).

In cerambycids of diverse subfamilies, males orient to females only after contacting them with their antennae, and it appears that mate recognition is mediated by contact chem-oreception alone (see Hanks, 1999). Males may even come within millimeters of females and not recognize them if antennal contact is not made. The reliance on chemical rather than visual cues for mate recognition may allow males to more readily detect females on the adult food source, larval host or in the dark. In some species there is marked sexual dimorphism in antennal length; the antennae of males are often more than double the length of those of females (e.g., see volumes indexed by Linsley and Chemsak, 1997). These elon-gate antennae may aid in locating mates. For example, male *Phoracantha semipuncatata* walk along the larval host trees with their antennae outstretched before them searching for females. Males recognize females by antennal contact alone and longer antennae contrib-ute to a wider antennal spread, increasing the probability of encountering a mate (Hanks *et al.*, 1996). The antennae of male *P. semipunctata* contain *sensilla trichodea*, which may serve as contact chemoreceptors (Lopes *et al.*, 2005). After initial antennal contact with a female, male cerambycids of different subfamilies display a similar progression of behav-iors that lead to copulation (see Ginzel *et al.*, 2003a). The male stops walking immedi-ately after touching a female with his antennae, and then aligns his body with the female, grasping the pronotum or elytra with his forelegs. The male then bends his abdomen to couple the genitalia and mates with the female.

The use of bioassays to study contact chemoreception of cerambycids

The first empirical evidence that cerambycids use contact pheromones for mate recognition came from Heintz (1925), who reported that males of flower-visiting lepterines recognized females solely by antennal contact (see Linsley, 1959). Interestingly, male beetles whose antennae were removed were unable to recognize conspecific females. Our understanding of contact chemoreception in this economically important family has subsequently increased greatly through the use of bioassays. In fact, most recent studies (see Table 17.1) have relied on a common bioassay to demonstrate that contact pheromones mediate mate recognition in cerambycids. In the assay, a female beetle is freeze-killed, and after being allowed to warm to room temperature, is presented to a male in a Petri dish arena. If the male attempts to mate with the female, it demonstrates that recognition cues on the cuticle are intact and behavior is not involved in mate recognition. The hydrocarbons are then stripped from the cuticle of the female by immersing her in successive aliquots of hexane or some other nonpolar solvent. These washes presumably contain all of the hydrocarbon components of the cuticle. The female is then allowed to air dry before she is reintroduced to the male to test whether he will respond to her in any way. If the male does not respond, it confirms that solvent washing removed chemical cues that mediate mate recognition. To test whether the extract resulting from the solvent washes contains the pheromone, the extract is pipetted back onto the female's body and the reconstituted female is presented again to the same male. The bioactivity of these extracts is often tested at 0.1 female-equivalent (FE) increments, where one FE is the total amount of hydrocarbons extracted from an individual female. This assay is useful for testing not only the bioactivity of the crude extract, but also that of fractions of the crude extract and synthetic compounds.

Interestingly, males of some species respond more readily to freeze-killed females than to solvent-washed females that have been reconstituted with crude cuticular extracts. For example, only 40% of male *Megacyllene caryae* tested display the full progression of mating behavior ending with abdominal bending toward reconstituted females, suggesting that solvent extraction may scramble the profile of hydrocarbons present in the stratified wax layer (Ginzel *et al.*, 2006). Additionally, males of *Neocytus acuminatus acuminatus* do not respond as strongly to freeze-killed females as to live ones. Freeze-killed females elicited abdominal bending from only 35% of males, suggesting that freezing may also alter the chemical cue or perhaps behavioral or physiological cues are necessary for mate recognition (Lacey *et al.*, 2008).

Models or dummies have also been used to demonstrate the use of contact pheromones in the Cerambycidae (see Table 17.1). For example, male *Acalolepta luxuriosa* display abdominal bending toward a model made from solvent-washed elytra of females to which crude hexane and ether extracts of females have been applied. Kim *et al.* (1992) immersed female Japanese pine sawyers, *Monochamus alternatus*, in hexane and tested the bioactivity of the resulting extract in a bioassay using a dummy made of a glass rod. A dummy coated with 3 to 5 FEs of extract elicited male mating behavior. Similarly, male *Psacothea hilaris* attempt to mate with gelatin capsules treated with female extract; the primary component of the pheromone was later identified as (Z)-21-methyl-8-pentatriacontene ((Z)-21MeC$_{35:1}$; Fukaya *et al.*, 1996, 1997). More recently, it was also demonstrated that male *Prionus californicus*, a member of the primitive subfamily Prioninae, attempt to mate with a ground glass stopper coated with one FE of crude cuticular extract (Barbour *et al.*, 2007). These assays support the notion that neither vision nor the behavior of females is important for mate recognition in many longhorned beetles.

Sampling cuticular hydrocarbons

Solvent extraction: Contact pheromones have usually been identified by comparing the hydrocarbon profiles of whole-body solvent extracts of male and female beetles (see references in Table 17.1). As males seldom attempt to mate with conspecific males (Ginzel *et al.*, 2006), the compounds that mediate mate recognition are often unique to the cuticular extracts of females. To identify female-specific compounds that may serve as contact pheromones, solvent extracts of males and females are analyzed by coupled gas chromatography–mass spectrometry (GC-MS). In some cases, such as the contact pheromones of *Xylotrechus colonus,* only a few compounds are unique to the female cuticle. For example, there are three early-eluting compounds (i.e., *n*-pentacosane, 9-methylpentacosane, and 3-methylpentacosane) in the extracts of females that are either absent or present in very small amounts in extracts of males (see Figure 17.1; Table 17.2; Ginzel *et al.*, 2003a). Interestingly, when the bioactivity of these compounds was tested in assays like those described above, all three were necessary to elicit the full sequence of mating behavior in males.

Solid phase microextraction: Solid phase microextraction (SPME) has recently been used as an alternative to solvent extraction for studying the cuticular hydrocarbons of insects, including cerambycids (e.g., Turillazzi *et al.*, 1998; Peeters *et al.*, 1999; Liebig *et al.*, 2000; Sledge *et al.*, 2000; Roux *et al.*, 2002; Ginzel *et al.*, 2003b, 2006; Lacey *et al.*, 2008). SPME is a solvent-less sampling technique and yields samples that are qualitatively and quantitatively similar to those obtained by solvent extraction (Moneti *et al.*, 1997; Monnin *et al.*, 1998; Bland *et al.*, 2001; Tentschert *et al.*, 2002). Moreover, wiping the SPME fiber over the cuticle primarily samples the outer surface of the wax layer, making extracts free from internal body lipids and exocrine gland secretions that may contaminate solvent extracts. An SPME apparatus resembles a syringe and contains a retractable fused-silica fiber, coated with a thin polymer film (Millar and Sims, 1998). The most common SPME coating used in sampling cuticular hydrocarbons of longhorned beetles is 100 μm polydimethylsiloxane (Ginzel *et al.*, 2003b; Ginzel *et al.*, 2006; Lacey *et al.*, 2008). During sampling, the fiber is wiped across the cuticular surface with the polymer essentially acting as a sponge. The fiber is then retracted back into the protective sheath and extended again inside a heated GC inlet where the analytes are thermally desorbed.

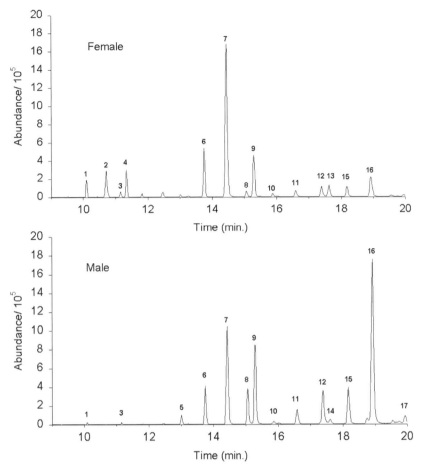

Figure 17.1 Representative gas chromatograms of hexame extracts of a *Xylotrechus colonus* female (top) and male (bottom). Reproduced from Ginzel *et al.*, 2003a with permission of Springer Science and Business Media.

SPME has been used in conjunction with solvent extraction to identify the contact pheromones of *Megacyllene robiniae* (Ginzel *et al.*, 2003b), *M. caryae* (Ginzel *et al.*, 2006), and more recently *N. a. acuminatus* (Lacey *et al.*, 2008). It appears that wipe sampling by SPME may yield a more representative profile of cuticular components than solvent extraction (Ginzel *et al.*, 2003b). For example, the contact pheromone of *M. robiniae*, (Z)-9-pentacosene (Z9-$C_{25:1}$), comprised ~16% of the total hydrocarbons in hexane extracts of females and was co-dominant with two other hydrocarbons that were not biologically active. In contrast, Z9-$C_{25:1}$ was dominant in the SPME wipe samples of female elytra, thoracic tergites and abdominal sternites and represented ~34% to 36% of the sampled hydrocarbons, suggesting that hydrocarbons that cue mate recognition are more abundant on the surface of the wax layer of females where they are readily accessible to the antennae of males (Ginzel *et al.*, 2003b). Interestingly, Z9-$C_{25:1}$ was also present in the hexane extract

Table 17.2 *Cuticular hydrocarbons of female and male* Xylotrechus colonus.[1]

Peak Number	Hydrocarbon	Female	Male	Diagnostic ions
1	n-C$_{25}$	+	+	352 (M+)
2	9-MeC$_{25}$	+	−	140, 252/253, 366 (M+)
2	11-MeC$_{25}$ (trace)	+	−	168/169, 224/225
3	2-MeC$_{25}$	+	+	323, 351, 366 (M+)
4	3-MeC$_{25}$	+	−	309, 337, 366 (M+)
5	2-MeC$_{26}$	−	+	337, 365, 380 (M+)
6	nC$_{27}$	+	+	380 (M+)
7	11,13-MeC$_{27}$	+	+	168/169, 196/197, 224/225, 252/253, 394 (M+)
8	2-MeC$_{27}$	+	+	351, 379, 394 (M+)
9	3-MeC$_{27}$	+	+	337, 365, 394 (M+)
10	n-C$_{28}$	+	+	394 (M+)
11	13-MeC$_{28}$	+	+	196/197, 238/239
11	12, 11-MeC$_{28}$ (trace)	+	+	168/169, 182/183, 252/253, 266/267
12	C$_{29}$:1	+	+	406 (M+)
13	C$_{29}$:1	+	−	406 (M+)
14	3-MeC$_{28}$	−	+	351, 379, 408 (M+)
15	n-C$_{29}$	+	+	408 (M+)
16	11, 13, 15-MeC$_{29}$	+	+	168/169, 196/197, 224/225, 252/253, 280/281, 422 (M+)
17	C$_{31}$:1	−	+	434 (M+)

[1] Peak numbers correspond with Figure 17.1; "+" indicates compound is present and "−" indicates it is absent. 11-MeC$_{25}$ and 12,11-MeC$_{28}$ coeluted in trace amounts with other compounds. Peaks 12 and 13 represent isomers of the same alkene. Reproduced from Ginzel *et al.*, 2003a with permission of Springer Science and Business Media.

of males but represented a negligible proportion of the SPME samples of the male cuticle, suggesting that this compound is not on the surface of the cuticle of males where it would be detected by the antennae, but rather sequestered deeper in the wax layer. Moreover, in the congener *M. caryae* an important component of the contact pheromone, Z9-C$_{29:1}$, was the only compound among the dominant hydrocarbons that was present in higher abundance in SPME than in solvent extracts (Ginzel *et al.*, 2006). SPME was also recently used to identify minor components of the *N. a. acuminatus* contact (Lacey *et al.*, 2008). There are a number of advantages to SPME over solvent extraction. Namely, SPME may provide a clearer representation of the semiochemicals present in the wax layer, and it is a non-destructive sampling technique. Living insects can be repeatedly sampled by SPME and later used in bioassays – an important consideration given the difficulty in rearing the large numbers of insects required for meaningful statistically-based laboratory experiments.

These sampling techniques are not without limitations. For example, high-molecular weight compounds that are beyond the analytical range of GC and electron-ionization MS (EI-MS) may be present in the wax layer and serve as synergists or minor components of contact pheromones. Recently, matrix-assisted laser desorption/ionization (MALDI) time-of-flight (TOF) mass spectrometry was used to demonstrate that the wax layer of 12 insect species of diverse taxa contains high-molecular weight hydrocarbons as long as C_{70} (Cvačka *et al.*, 2006). These long-chain length compounds would likely decompose even in high-temperature GC columns, and identification by EI-MS would be very difficult as the molecular ion would likely not survive.

Qualitative and quantitative differences in hydrocarbon profiles of the sexes

Cuticular hydrocarbon profiles of male and female beetles of some cerambycid species are qualitatively quite similar, with most cuticular components occurring in both sexes. For example, in *X. colonus* there are only four compounds that are unique to the female cuticle and another three that are male specific (Ginzel *et al.*, 2003a). Moreover, cuticular profiles of male and female *Anoplophora glabripennis* share a series of saturated, branched, and unsaturated compounds, but five alkenes are far more prevalent in the extracts of females. Interestingly, all five of these compounds are necessary to elicit abdominal bending in males (Zhang *et al.*, 2003). In *M. caryae,* however, cuticular profiles of females contain a number of aliphatic hydrocarbons that are not present in the wax layer of males. In fact, extracts of males and females were qualitatively very different and sex-specific compounds represented almost half of the hydrocarbons of females and a third of the hydrocarbons of males (Ginzel *et al.*, 2006). In cases where there are many qualitative differences in the profiles of males and females, it is often most expedient to first test the biological activity of these compounds by functional group. For example, the bioactivity of the most abundant synthetic straight-chain alkanes, branched alkanes and monoenes present in the extract of female *M. robiniae* were tested by functional group, and only the synthetic monoenes elicited a mating response in males similar to that of crude extact. Moreover, of the monoenes, males responded most strongly to Z9-C_{29}:1 alone – the contact pheromone. Cuticular extracts of males also contained a greater proportion of longer-chain hydrocarbons, which also appears to be true for *M. caryae* and *X. colonus* (Ginzel *et al.*, 2003a, 2006). On the other hand, solvent extracts of female *Anoplophora malasiaca* contain longer-chained hydrocarbons which are lacking in the hydrocarbon profiles of males (Akino *et al.*, 2001). Sex-based differences in alkyl chain lengths of cuticular hydrocarbons have also been reported in other insects such as the tsetse fly (Nelson and Carlson, 1986) and the bark beetle *Ips lecontei* (Page *et al.*, 1997).

Males of many cerambycid species compete aggressively for mates and after copulation a male often guards a female by grasping her elytra with his forelegs and accompanying her as she seeks out oviposition sites (see Hanks, 1999). Males may even remain paired with the female for long periods of time, repeatedly copulating with her, while defending her from challenging males. Rival males deploy a variety of tactics to displace paired males, including antennal lashing, biting, and head butting. There is some evidence that male-specific cuticular components mediate aggressive behavior and male–male competition. For example, male *Nadezhdiella cantori* lash with their antennae and front legs and also violently bite dead conspecific males and even paper rolls treated with solvent extracts of males (Wang *et al.*, 2002). There is also evidence to suggest these male-specific hydrocarbons may be acting as chemical deterrents to mating or abstinons. For example, hexane extracts of female *Semanotus japonicus* elicit abdominal bending in males but, when combined with extracts of males, their bioactivity is greatly reduced (Kim *et al.*, 1993).

Contact pheromones as single compounds or blends

Mate recognition in cerambycids can be mediated by either a single component or blends of several compounds (Table 17.1). In the subfamily Cerambycinae, for example, contact pheromones of the congeners *M. caryae* and *M. robiniae* are single alkenes and also chain-length analogs (Ginzel *et al.*, 2003b, 2006). Although the contact pheromones of these two closely related species are quite similar, it appears that the composition of cerambycid contact pheromones is not phylogenetically conserved. In fact, the contact pheromone of *M. robiniae*, $Z9$-$C_{25:1}$, is also one of five alkenes that mediate mate recognition in the lamiine *A. glabripennis* (Zhang *et al.*, 2003). Even other members of the tribe containing the *Megacyllene* species use contact pheromones that are blends of saturated *n*- and methyl-branched compounds. The contact pheromone of *X. colonus*, for example, is a mixture of *n*-C_{25} and two branched alkanes (Ginzel *et al.*, 2003a). Moreover, a branched alkane, 7-methylheptacosane (7Me-C_{27}), mediates mate recognition in *N. a. acuminatus*, but two other branched compounds, 9Me-C_{27} and 7Me-C_{25}, act as synergists (Lacey *et al.*, 2008). Interestingly, 9Me-C_{27} is part of the cuticular profile of female *M. caryae* (Ginzel *et al.*, 2006) and apparently a component of the contact pheromone of *A. malasiaca*, a member of the subfamily Lamiinae (Fukaya *et al.*, 2000), further suggesting that some hydrocarbons may be common to cerambycids. In addition to 9Me-C_{27}, the contact pheromone of *A. malasiaca* consists of a blend of seven other hydrocarbons and five ketones, and it also appears that three gomadalactones (oxabicyclo[3.3.0]octane compounds with an aliphatic chain) serve as synergists (Fukaya *et al.*, 2000; Yasui *et al.*, 2003, 2007). The structure of this pheromone is unique among the cerambycids because it is composed of two classes of compound. To date, no other polar compounds have been identified as contact pheromones of longhorned beetles.

A number of cerambycid contact pheromones are methyl-branched alkanes that have chiral carbons. In the case of *N. a. acuminatus*, for example, the three bioactive methyl-branched alkanes are chiral. However, these compounds are available in vanishingly small amounts from each insect, and with current analytical limitations it is nearly impossible to determine their natural enantiomeric ratio in the wax layer. Moreover, there are currently no chiral stationary phase GC or LC columns capable of resolving enantiomers of long-chain methy-branched hydrocarbons (Lacey *et al.*, 2008). Irrespective of the naturally occurring enantiomeric ratios, male *N. a. acuminatus* respond to racemic standards, suggesting that even if females produce only one enantiomer its activity is not influenced by the presence of the other. In *P. hilaris*, the major component of the contact pheromone (Z8–21Me-$C_{35:1}$) is a long-chain methyl-branched alkene, and accounts for approximately 60% of the hydrocarbons extracted from the female elytra. Nevertheless, the bioactivity of the synthetic compound is considerably less than that of the crude solvent extracts of females, suggesting that enantiomeric composition may influence its activity. Although only the (Z)-configuration of 8–21Me-$C_{35:1}$ was found in extracts of females, Fukaya *et al.* (1997) evaluated the bioactivities of synthetic (*R*)- and (*S*)-enantiomers of both (*Z*)- and (*E*)-isomers of the synthetic compound. Males displayed a greater response to the (*Z*,*R*)- and (*Z*,*S*)-isomers but the response to the different optical enantiomers of this compound were similar, suggesting that males can distinguish between the two geometric isomers, but optical rotation has little influence on the bioactivity of the pheromone. It is likely that males respond less readily to the synthetic pheromone because it lacks minor components that are present in the crude extract that may act as synergists.

Conclusions

Contact pheromones play an essential role in the mating systems of many longhorned beetles and may also be important in the divergence of this speciose group by acting as prezygotic mating isolation mechanisms. Our understanding of the chemically-mediated mate location strategies of cerambycids suggests that volatile sex pheromones are highly conserved and share a common structural motif. In fact, even species that share the same host plant may be attracted to one another's volatile aggregation pheromones (see Hanks *et al.*, 2007). Furthermore, the niches of closely related cerambycid species often overlap and species-specific contact pheromones may play a vital role in maintaining reproductive isolation.

To date, research on the chemical ecology and contact chemoreception of cerambycids remains largely descriptive in nature. Although our understanding of mate recognition in this group has improved considerably in recent years, most work has focused on the more advanced subfamilies. Further research on the primitive subfamilies, including the Parandrinae, Prioninae and Aseminae, will shed light on the evolution of mating systems in this economically important family. Finally, the biosynthesis, regulation and transport of these semiochemicals remain virtually unexplored in the longhorned beetles. By applying powerful tools of molecular biology and physiology to understanding these processes, effective pest-management tactics targeting the chemically-mediated mating behavior of the beetles are likely on the horizon.

Acknowledgments

I thank Lawrence M. Hanks and Larry L. Murdock for helpful comments during the preparation of this manuscript.

References

Akino, T., Fukaya, M., Yasui, H. and Wakamura, S. (2001). Sexual dimorphism in cuticular hydrocarbons of the white-spotted longicorn beetle, *Anoplophora malasiaca* (Coleoptera: Cerambycidae). *Entomol. Sci.*, **4**, 271–277.

Allison, J.D., Borden, J.H. and Seybold, S.J. (2004). A review of the chemical ecology of the Cerambycidae (Coleoptera). *Chemoecology*, **14**, 123–150.

Barbour, J.D., Lacey, E.S. and Hanks, L.M. (2007). Cuticular hydrocarbons mediate mate recognition in a species of longhorned beetle (Coleoptera: Cerambycidae) of the primitive subfamily Prioninae. *Ann. Entomol. Soc. Am.*, **100**, 333–338.

Bland, J.M., Osbrink, W.L.A., Cornelius, M.L., Lax, A.R. and Vigo, C.B. (2001). Solid-phase microextraction for the detection of termite cuticular hydrocarbons. *J. Chromatogr. A*, **932**, 119–127.

Crook, D.J., Hopper, J.A., Ramaswamy, S.B. and Higgins, R.A. (2004). Courtship behavior of the soybean stem borer *Dectes texanus texanus* (Coleoptera: Cerambycidae): Evidence for a female contact sex pheromone. *Ann. Entomol. Soc. Am.*, **97**, 600–604.

Cvačka, J., Jiroš, P., Šobotník, J., Hanus, R. and Svatoš, A. (2006). Analysis of insect cuticular hydrocarbons using matrix-assisted laser desorption/ionization mass spectrometry. *J. Chem. Ecol.*, **32**, 409–434.

Fukaya, M., Akino, T., Yasuda, T., Wakamura, S., Satoda, S. and Senda, S. (2000). Hydrocarbon components in contact sex pheromone of the white-spotted longicorn beetle, *Anoplophora malasiaca* (Thomson) (Coleoptera: Cerambycidae) and pheromonal activity of synthetic hydrocarbons. *Entomol. Sci.*, **3**, 211–218.

Fukaya, M. and Honda, H. (1995). Reproductive biology of the yellow-spotted longicorn beetle, *Psacothea hilaris* (Pascoe) (Coleoptera: Cerambycidae). 2. Evidence for two female pheromone components with different functions. *Appl. Entomol. Zool.*, **30**, 467–470.

Fukaya, M., Wakamura, S., Yasuda, T., Senda, S., Omata, T. and Fukusaki, E. (1997). Sex pheromonal activity of geometric and optical isomers of synthetic contact pheromone to males of the yellow-spotted longicorn beetle, *Psacothea hilaris* (Pascoe) (Coleoptera: Cerambycidae). *Appl. Entomol. Zool.*, **32**, 654–656.

Fukaya, M., Yasuda, T., Wakamura, S. and Honda, H. (1996). Reproductive biology of the yellow-spotted longicorn beetle, *Psacothea hilaris* (Pascoe) (Coleoptera: Cerambycidae). 3. Identification of contact sex pheromone on female body surface. *J. Chem. Ecol.*, **22**, 259–270.

Fukaya, M., Yasui, H., Yasuda, T., Akino, T. and Wakamura, S. (2005). Female orientation to the male in the white-spotted longicorn beetle, *Anoplophora malasiaca* (Thomson) (Coleoptera: Cerambycidae) by visual and olfactory cues. *Appl. Entomol. Zool.*, **40**, 63–68.

Galford, J.R. (1977). Evidence for a pheromone in the locust borer. Research Note NE-240. Washington, DC: US Department of Agriculture, Forest Service.

Gibbs, A.G. (1998). Water-proofing properties of cuticular lipids. *Am. Zool.*, **38**, 471–482.

Ginzel, M. D., Blomquist, G. J., Millar, J. G. and Hanks, L. M. (2003a). Role of contact pheromones in mate recognition in *Xylotrechus colonus*. *J. Chem. Ecol.*, **29**, 533–545.

Ginzel, M. D. and Hanks, L. M. (2003). Contact pheromones as mate recognition cues of four species of longhorned beetles (Coleoptera: Cerambycidae). *J. Insect Behav.*, **16**, 181–187.

Ginzel, M. D. and Hanks, L. M. (2005). Role of host plant volatiles in mate location for three species of longhorned beetles. *J. Chem. Ecol.*, **31**, 213–217.

Ginzel, M. D., Millar, J. G. and Hanks, L. M. (2003b). (Z)-9-pentacosene – contact sex pheromone of the locust borer, *Megacyllene robiniae*. *Chemoecology*, **13**, 135–141.

Ginzel, M. D., Moreira, J. A., Ray, A. M., Millar, J. G. and Hanks, L. M. (2006). (Z)-9-Nonacosene – Major component of the contact sex pheromone of the beetle *Megacyllene caryae*. *J. Chem. Ecol.*, **32**, 435–451.

Hanks, L. M. (1999). Influence of the larval host plant on reproduction strategies of cerambycid beetles. *Ann. Rev. Entomol.*, **44**, 483–505.

Hanks, L. M., Millar, J. G., Moreira, J. A., Barbour, J. D., Lacey, E. S., McElfresh, J. S., Reuter, F. R. and Ray, A. M. (2007). Using generic pheromone lures to expedite identification of aggregation pheromones for the cerambycid beetles *Xylotrechus nauticus*, *Phymatodes lecontei*, and *Neoclytus modestus modestus*. *J. Chem. Ecol.*, **33**, 889–907.

Hanks, L. M., Millar, J. G. and Paine, T. D. (1996). Mating behavior of the eucalyptus longhorned borer (Coleoptera: Cerambycidae) and the adaptive significance of long "horns". *J. Insect Behav.*, **9**, 383–393.

Heintz, A. (1925). Lepturinernas blombesök och sekundära könskaraktärer. *Entomol. Tidskr.*, **46**, 21–34.

Howard, R. W. and Blomquist, G. J. (2005). Ecological, behavioral, and biochemical aspects of insect hydrocarbons. *Ann. Rev. Entomol.*, **50**, 371–393.

Ibeas, F., Diez, J. J. and Pajares, J. A. (2008). Olfactory sex attraction and mating behaviour in the pine sawyer *Monochamus galloprovincialis* (Coleoptera: Cerambycidae). *J. Insect Behav.*, **21**, 101–110.

Kim, G. H., Takabayashi, J., Takahashi, S. and Tabata, K. (1992). Function of pheromones in mating behavior of the Japanese pine sawyer beetle, *Monochamus alternatus* Hope. *Appl. Entomol. Zool.*, **27**, 489–497.

Kim, G. H., Takabayashi, J., Takahashi, S. and Tabata, K. (1993). Function of contact pheromone in the mating behavior of the cryptomeria bark borer, *Semanotus japonicus* Lacordaire (Coleoptera: Cerambycidae). *Appl. Entomol. Zool.*, **28**, 525–535.

Kobayashi, H., Yamane, A. and Iwata, R. (2003). Mating behavior of the pine sawyer, *Monochamus saltuarius* (Coleoptera : Cerambycidae). *Appl. Entomol. Zool.*, **38**, 141–148.

Kuboki, M., Akutsu, K., Sakai, A. and Chuman, T. (1985). Bioassay of the sex pheromone of the udo longicorn beetle, *Acalolepta luxuriosa* Bates (Coleoptera: Cerambycidae). *Appl. Entomol. Zool.*, **20**, 88–89.

Lacey, E. S., Ginzel, M. D., Millar, J. G. and Hanks, L. M. (2004). Male-produced aggregation pheromone of the cerambycid beetle *Neoclytus acuminatus acuminatus*. *J. Chem. Ecol.*, **30**, 1493–1507.

Lacey, E. S., Ginzel, M. D., Millar, J. G. and Hanks, L. M. (2008). A major component of the contact sex pheromone of the cerambycid beetle, *Neoclytus acuminatus*

acuminatus is 7-methylheptacosane. *Physiol. Entomol.*, **33**, 209–216.

Lacey, E. S., Moreira, J. A., Millar, J. G., Ray, A. M. and Hanks, L. M. (2007). Male-produced aggregation pheromone of the cerambycid beetle *Neoclytus mucronatus mucronatus*. *Entomol. Exp. Appl.*, **122**, 171–179.

Lawrence, J. F. (1982). Coleoptera. In *Synopsis and Classification of Living Organisms*, Vol. 2., ed. S. P. Parker. New York: McGraw-Hill, pp. 482–553.

Liebig, J., Peeters, C., Oldham, N. J., Markstadter, C. and Hölldobler, B. (2000). Are variations in cuticular hydrocarbons of queens and workers a reliable signal of fertility in the ant *Harpegnathos saltator*? *Proc. Natl. Acad. Sci. USA*, **97**, 4124–4131.

Linsley, E. G. (1959). Ecology of Cerambycidae. *Ann. Rev. Entomol.*, **4**, 99–138.

Linsley, E. G. (1964). The Cerambycidae of North America, Part V. *Univ. Calif. Publ. Entomol.*, **22**, 1–197.

Linsley, E. G. and Chemsak, J. A. (1997). The Cerambycidae of North America, Part VIII. *Univ. Calif. Publ. Entomol.*, **117**, 1-534.

Lopes, O., Marques, P. C. and Araujo, J. (2005). The role of antennae in mate recognition in *Phoracantha semipunctata* (Coleoptera: Cerambycidae). *J. Insect Behav.*, **18**, 243–257.

Lu, W., Wang, Q., Tian, M. Y., He, X. Z., Zeng, X. L. and Zhong, Y. X. (2007). Mate location and recognition in *Glenea cantor* (Fabr.) (Coleoptera: Cerambycidae: Lamiinae): Roles of host plant health, female sex pheromone, and vision. *Environ. Entomol.*, **36**, 864–870.

Millar, J. G. and Sims, J. J. (1998). Preparation, cleanup, and preliminary fractionation of extracts. In *Methods in Chemical Ecology*, Vol. 1, ed. J. G. Millar and K. F. Haynes. New York: Kluwer, pp. 1–37.

Moneti, G., Dani, F. R., Pieraccini, G. and Turillazzi, S. (1997). Solid-phase microextraction of insect epicuticular hydrocarbons for gas chromatographic mass spectrometric analysis. *Rapid Commun. Mass Sp.*, **11**, 857–862.

Monnin, T., Malosse, C. and Peeters, C. (1998). Solid-phase microextraction and cuticular hydrocarbon differences related to reproductive activity in the queenless ant *Dinoponera quadriceps*. *J. Chem. Ecol.*, **24**, 473–490.

Mori, K. (2007). Absolute configuration of gomadalactones A, B and C, the components of the contact sex pheromone of *Anoplophora malasiaca*. *Tetrahedron Lett.*, **48**, 5609–5611.

Nelson, D. R. (1993). Methyl-branched lipids in insects. In *Insect Lipids: Chemistry, Biochemistry, and Biology*, ed. D. W. Stanley-Samuelson and D. R. Nelson. Lincoln, NE: University of Nebraska Press, pp. 271–315.

Nelson, D. R. and Blomquist, G. J. (1995). Insect waxes. In *Waxes: Chemistry, Molecular Biology, and Function*, ed. R. J. Hamilton. Dundee, Scotland: The Oily Press, pp. 1–90.

Nelson, D. R. and Carlson, D. A. (1986). Cuticular hydrocarbons of the tsetse flies *Glossina morsitans morsitans, G. austeni, and G. pallidipes*. *Insect Biochem.*, **16**, 403–416.

Page, M. Nelson, L. J., Blomquist, G. J. and Seybold, S. J. (1997). Cuticular hydrocarbons as chemotaxonomic characters of pine engraver beetles (*Ips* spp.) in the grandicollis subgeneric group. *J. Chem. Ecol.*, **23**, 1053–1099.

Peeters, C., Monnin, T. and Malosse, C. (1999). Cuticular hydrocarbons correlated with reproductive status in a queenless ant. *Proc. R. Soc. B*, **266**, 1323–1327.

Roux, E., Sreng, L., Provost, E., Roux, M. and Clément, J.-L. (2002). Cuticular hydrocarbon profiles of dominant versus subordinate male *Nauphoeta cinerea* cockroaches. *J. Chem. Ecol.*, **28**, 1221–1235.

Sledge, M. J., Moneti, G., Pieraccini, G. and Turillazzi, S. (2000). Use of solid phase microextraction in the investigation of chemical communication in social wasps. *J. Chromatogr. A*, **873**, 73–77.

Solomon, J. D. (1995). Guide to insect borers of North American broadleaf trees and shrubs. *Agricultural Handbook 706*, Washington, DC: US Department of Agriculture, Forest Service.

Tentschert, J., Bestmann, H. J. and Heinze, J. (2002). Cuticular compounds of workers and queens in two *Leptothorax* ant species – a comparison of results obtained by solvent extraction, solid sampling, and SPME. *Chemoecology*, **12**, 15–21.

Turillazzi, S., Sledge, M. F. and Moneti, G. (1998). Use of a simple method for sampling cuticular hydrocarbons from live social wasps. *Ethol. Ecol. Evol.*, **10**, 293–297.

Wang, Q. (1998). Evidence for a contact female sex pheromone in *Anoplophora chinensis* (Forster) (Coleoptera: Cerambycidae: Lamiinae). *Coleopts. Bull.*, **52**, 363–368.

Wang, Q., Li, J. S., Zeng, W. Y. and Yin, X. M. (1991). Sex recognition by males and evidence for a female sex pheromone in *Paraglenea fortunei* (Coleoptera: Cerambycidae). *Ann. Entomol. Soc. Am.*, **84**, 107–110.

Wang, Q., Zeng, W. Y., Chen, L. Y., Li, J. S. and Yin, X. M. (2002). Circadian reproductive rhythms, pair-bonding, and evidence for sex-specific pheromones in *Nadezhdiella cantori* (Coleoptera: Cerambycidae). *J. Insect Behav.*, **15**, 527–539.

Yasui, H., Akino, T., Yasuda, T., Fukaya, M., Ono, H. and Wakamura, S. (2003). Ketone components in the contact sex pheromone of the white-spotted longicorn beetle, *Anoplophora malasiaca*, and pheromonal activity of synthetic ketones. *Entomol. Exp. Appl.*, **107**, 167–176.

Yasui, H., Akino, T., Yasuda, T., Fukaya, M., Wakamura, S. and Ono, H. (2007). Gomadalactones A, B, and C: novel 3-oxabicyclo[3.3.0]octane compounds in the contact sex pheromone of the white-spotted longicorn beetle, *Anoplophora malasiaca*. *Tetrahedron Lett.*, **48**, 2395–2400.

Zhang, A. J., Oliver, J. E., Chauhan, K., Zhao, B. G., Xia, L. Q. and Xu, Z. C. (2003). Evidence for contact sex recognition pheromone of the Asian longhorned beetle, *Anoplophora glabripennis* (Coleoptera: Cerambycidae). *Naturwissenschaften*, **90**, 410–413.

18

Polyene hydrocarbons, epoxides, and related compounds as components of lepidopteran pheromone blends

Jocelyn G. Millar

Moth species rely heavily to exclusively on sex attractant pheromones for long-range location of mates. In most but not all species, the attractant pheromone is produced by females, and males have evolved extremely sensitive antennae, the receptors of which are tuned specifically to detect the scent of conspecific females. It is clearly a very effective mate location system, because male moths have been shown to be attracted to females over distances of a kilometer or more (Mell, 1922; Collins and Potts, 1932), despite the fact that females may only produce micrograms to nanograms of pheromone per hour. To be effective as long-range, species-specific attractants, sex attractant pheromones should fit several criteria. First, they must have sufficient vapor pressure under ambient temperature conditions to provide a detectable signal for males to find and follow. Thus, attractant pheromones are relatively small molecules, with most having molecular weights of less than 300 Daltons. They also tend to have few or even no polar functional groups, because polar functional groups decrease volatility. Second, the individual components of the pheromone, or the blend of components, must be sufficiently distinctive that they form a unique, species-specific blend, so that there is no cross-attraction among sympatric species (see Chapter 7). Whereas there are cases known in which species appear to produce unique compounds, the majority of insects appear to create species-specific signals by blending two or more compounds, each of which may be shared by several species, in specific ratios. Furthermore, many different structures can be created from a simple straight-chain backbone by selective placement and manipulation of the stereochemistry of double bonds and other functional groups. Third, the pheromone components should be sufficiently stable that they persist for long enough to provide a reliable signal, but not so stable that a substrate on which a female has been calling, and consequently which may have become contaminated with pheromone, will continue to attract males for hours or days after the female has left.

Thus, the majority of known lepidopteran pheromones are comprised of compounds that are between 10 and 23 carbons in length, with one or no polar functional groups. Within these broad limits, there are two major structural classes, each of which arises from different biosynthetic pathways. The first class, designated as Type I pheromones (Ando *et al.*, 2004) consists primarily of alcohols, aldehydes, and acetates with unbranched carbon chains of 10–18 carbons, and with 0–3 double bonds. Unusual structures or functional groups, such as triple bonds, nitro groups, or esters other than acetates, are occasionally found within this class, but they are clearly still part of this general class. Furthermore, for those Type I pheromones with two or more double bonds, those bonds are not found in a "skipped-methylene" motif, that is, with a single methylene group between the two double bonds. Type I pheromones were the first to be discovered – in the late 1950s – and they currently comprise the majority of known lepidopteran pheromones (Ando *et al.*, 2004). They are biosynthesized from saturated carboxylic acid intermediates, which by various combinations of chain lengthening, chain shortening, desaturation, and decarboxylation steps to build the basic skeleton, followed by reduction, oxidation, and esterification steps to adjust the functional groups, are converted to the many known pheromone compounds of this general class (Bjostad *et al.*, 1987).

The second major structural class, Type II pheromones, consist of polyene hydrocarbons and related epoxides and ketones with unbranched chains that are 17–25 carbons in length. In all cases, the double bonds that identify them as members of this class are specifically placed at positions 3, 6, 9, 12, or 15 from one end of the chain, and all of these diagnostic double bonds have the Z-configuration in a typical skipped-methylene pattern. The basic polyene structures can be further modified by epoxidation of one or two of the double bonds, by having additional double bonds, or both. Another group of compounds that probably belong in this class are the mono- and dienyl ketones of 21 carbons found in some lymantrid pheromones, and the 18-carbon trienals found in several arctiid pheromones (see tables). In contrast to Type I pheromones, the evidence suggests that Type II pheromones are biosynthesized from diunsaturated linoleic acid (for 6Z,9Z-hydrocarbons and derivatives) or triunsaturated linolenic acid (3Z,6Z,9Z-trienes and derivatives), both of which are not produced by the insect but must be obtained from their diet. By using these substrates as starting materials, the diagnostic all-Z double bonds that characterize members of this class are already in place. The biosynthesis is described in more detail in a subsequent section. The first Type II pheromones to be described were found in the late 1970s, some 20 years after the identification of the first Type I pheromones, but numerous other examples have since been identified (*vide infra*), and it is clear that they constitute a large class of moth pheromones in their own right. Furthermore, many of them are chiral, unlike the Type I pheromones, which provides additional avenues for creating species-specific messages, including antagonism by the incorrect enantiomer, and synergism between enantiomers.

The goal of this chapter is to provide an overview of the occurrence of Type II polyene pheromones and their derivatives, and their chemistry, including their biosynthesis and synthesis. The older literature in this subject area was reviewed in Millar (2000), and summarized more recently in Ando *et al.* (2004). Thus, this chapter will provide a comprehensive summary of all known Type II pheromone structures and their occurrence in Tables 1–4, whereas the text will focus more on work over the past ten years. The interested reader is further directed to three useful online databases, two of which focus on lepidopteran pheromones (www.tuat.ac.jp/~antetsu/review/e-List.pdf; Ando, 2003; www-pherolist.slu. se/pherolist.php; Witzgall *et al.*, 2004) and the third of which covers insect pheromones in general (www.pherobase.com; El-Sayed, 2008).

Types of structure and nomenclature

Almost all known Type II pheromones are listed in Tables 18.1–18.4 (Arctiids, Geometrids, Noctuids, and Lymantriids, respectively). The few exceptions where Type II pheromones have been found in other lepidopteran families (e.g., Crambidae or Pyralidae) are discussed in the section below on taxonomic distribution of these compounds.

Representative examples of the various types of structure that might be included in the Type II pheromone class are shown in Scheme 1, along with abbreviated versions of their names. These shorthand forms are used almost universally by researchers because they can be immediately and unequivocally associated with a structure, whereas the formal names used by IUPAC or Chemical Abstracts are complex, and although unequivocal, they require time to decipher and associate with a particular structure. Furthermore, the epoxides are chiral, and the formal chemical names become even more complicated and less intuitive as a result. For example, in the formal naming system, the oxirane ring of an epoxide forms the base structure for the name, with the oxygen designated position 1, and then the two alkyl chains appended to the oxirane are assigned precedence in naming *in alphabetical order*, so that positions 2 and 3 of the oxirane change depending on the alphabetical precedence of names of the side chains rather than their size. Even from this description, the reader can understand how confusing it can be to work with the formal names. In any event, illustrative examples of the formal chemical names and the shorthand versions are given below for a representative compound of each type.

The simplest Type II compounds are the polyunsaturated hydrocarbons with 17 to ~25 carbon chain lengths. The base compounds from which all others arise include the (6Z,9Z)-alkadienes (shorthand version 6Z,9Z-X:H, where X = the chain length; formal name for the 19 carbon compound would be (6Z,9Z)-nonadeca-6,9-diene) and (3Z,6Z,9Z)-alkatrienes (e.g. 3Z,6Z,9Z-19:H). Documented positions of additional double bonds include the 1 position (1,3Z,6Z,9Z-19:H; the terminal double bond between carbons one and two has only one configuration so no stereochemical designator is necessary), 4 position (4E,6Z,9Z-19:H; Francke *et al.*, 1998, 1999), or the 11 position (3Z,6Z,9Z,11Z-19:H or 3Z,6Z,9Z,11E-19:H) (e.g., Wong *et al.*, 1984a, b; Szöcs *et al.*, 1998a,b). In addition, pheromone components with the skipped-methylene motif extended by two double bonds to give 3Z,6Z,9Z,12Z,15Z-X:H (X = 23 or 25) have recently been identified (Leal *et al.*, 2005; Millar *et al.*, 2005; Grant *et al.*, 2008).

The base hydrocarbon compounds can then be epoxidized in any one of the double-bond positions, giving in the case of a 3Z,6Z,9Z-19:H base structure, the monoepoxides 6Z,9Z-cis-3,4-epo-19:H (formal name (2R*,3S*)-2-ethyl-3-(2Z,5Z)-2,5-pentadecadien-1-yl-oxirane), 3Z,9Z-cis-6,7-epo-19:H ((2R*,3S*)-2-(2Z)-2-dodecen-1-yl-3-(2Z)-2-penten-1-yl-oxirane), or 3Z,6Z-cis-9,10-epo-19:H ((2S*,3R*)-2-nonyl-3-(2Z,5Z)-2,5-octadien-1-yl-oxirane), in which only the relative stereochemistry of the epoxide is indicated. For the chiral compounds, the names are just changed slightly so that the stereochemistry is explicitly stated (e.g., for the 3,4-epoxide, 6Z,9Z-3R,4S-epo-19:H). In addition to *cis*-epoxides, two examples of *trans*-epoxide are known (Francke *et al.*, 1998, 1999; Wakamura *et al.*, 2002), but

Table 18.1 *Known and possible female-produced sex pheromone components for arctiid moths. Compounds in **bold** have been found in pheromone gland extracts or aeration extracts and have been shown to be active in behavioral bioassays or field trials; compounds in normal font have been found in pheromone gland or aeration extracts; and compounds in italics have been shown to attract males in field screening trials.*

Subfamily, genus, species	Polyene hydrocarbons	Polyene epoxides	Aldehydes	Other	References
Lithosiinae					
Cyana spp.		*3Z,6Z-cis-9,10-epo-18:H*			Hai *et al.*, 2002
Schistophleps sp. 1		*3Z,6Z-cis-9,10-epo-19:H*			Hai *et al.*, 2002
Schistophleps sp. 2	3Z,6Z,9Z-19:H				Hai *et al.*, 2002
Schistophleps sp. 3	3Z,6Z,9Z-21:H				Hai *et al.*, 2002
Arctiinae					
Amata (=*Syntomis*) *phegea* L.	3Z,6Z,9Z-21:H				Szöcs *et al.*, 1987
Arachnis picta Packard	6Z,9Z-20:H 6Z,9Z-21:H 3Z,6Z,9Z-21:H				J.G. Millar and T.C. Baker, unpub.
Amsacta albistriga (Walker)	3Z,6Z,9Z-21:H		9Z,12Z-18:Ald 9Z,12Z,15Z-18:Ald, 18:Ald		Persoons *et al.*, 1993
Apantesis phalerata (Harris)		3Z,6Z-cis-9,10-epo-21:H			Meyer, 1984
Arctia caja (L.)	3Z,6Z,9Z-21:H	3Z,6Z-cis-9,10-epo-21:H 3Z,6Z-cis-9,10-epo-20:H			Bestmann *et al.*, 1992

(*continued*)

Table 18.1 (cont.)

Subfamily, genus, species	Polyene hydrocarbons	Polyene epoxides	Aldehydes	Other	References
Arctia villica L.	3Z,6Z,9Z-21:H 1,3Z,6Z,9Z-21:H				Einhorn *et al.*, 1984
Creatonotus gangis L.	3Z,6Z,9Z-21:H 3Z,6Z,9Z-23:H 6Z,9Z-21:H	3Z,6Z-cis-9,10-epo-21:H			Bell and Meinwald, 1986 Wunderer *et al.*, 1986
Creatonotos transiens Walker	3Z,6Z,9Z-21:H 3Z,6Z,9Z-23:H 6Z,9Z-21:H	3Z,6Z-cis-9,10-epo-21:H			Bell and Meinwald, 1986 Wunderer *et al.*, 1986
Diacrisia (= *Spilosoma*) *obliqua* Walker	**3Z,6Z,9Z-21:H**	**3Z,6Z-cis-9,10-epo-21:H 1,3Z,6Z-cis-9,10-epo-21:H**	**9Z,12Z-18:Ald 9Z,12Z,15Z-18:Ald**		Persoons *et al.*, 1993 Yadav *et al.*, 2001
Estigmene acrea (Drury)		**3Z,6Z-cis-9,10-epo-21:H**	**9Z,12Z-18:Ald 9Z,12Z,15Z-18:Ald**		Hill and Roelofs, 1981
Euchaetes egle (Drury)	1,3Z,6Z,9Z-21:H 3Z,6Z,9Z-21:H 3Z,6Z,9Z-20:H 6Z,9Z-21:H			2Me-18:H	Simmons *et al.*, 1998
Grammia blakei (*Grote*)	3Z,6Z,9Z-20:H 3Z,6Z,9Z-21:H	3Z,6Z-cis-9,10-epo-21:H			J. G. Millar and J. R. Byers, unpub. data
Grammia virgo (L.)		3Z,6Z-cis-9,10-epo-21:H			Meyer, 1984
Halysidota leda Druce	3Z,6Z,9Z-21:H, analogs, homologs	3Z,6Z-cis-9,10-epo-21:H			Déscoins *et al.*, 1989

Species	Compounds	Epoxides	Aldehydes	References
Halysidota tessellaris (JE Smith)		**3Z,6Z-9R,10S-epo-21:H**		Meyer, 1984; Landolt *et al.*, 1996
Hyphantrea cunea Drury	1,3Z,6Z,9Z-21:H 3Z,6Z,9Z-21:H	**3Z,6Z-9S,10R-epo-21:H** **1,3Z,6Z-9S,10R-epo-21:H** 3Z,6Z- 9S,10R-epo-20:H **1,3Z,6Z-9S,10R-epo-20:H**	**9Z,12Z-18:Ald** **9Z,12Z,15Z-18:Ald**	Hill *et al.*, 1982 Einhorn *et al.*, 1982 Tóth *et al.*, 1989 Binda *et al.*, 1990 Marcelli *et al.*, 1992 Zhang *et al.*, 1998 El-Sayed *et al.*, 2005a
Panaxia (=*Callimorpha*) *quadripunctaria* Poda	6Z,9Z-21:H 6Z,9Z-20:H Homologs, analogs			Schneider *et al.*, 1992
Pareuchaetes pseudoinsulata Rego Barres	1,3Z,6Z,9Z-21:H 3Z,6Z,9Z-21:H 6Z,9Z-21:H Homologs, analogs			Schneider *et al.*, 1992 Frérot *et al.*, 1993
Phragmatobia fuliginosa (L.)	3Z,6Z,9Z-21:H 6Z,9Z-21:H	**3Z,6Z-cis-9,10-epo-21:H** 6Z-cis-9,10-epo-21:H	*9Z,12Z-18:Ald* *9Z,12Z,15Z-18:Ald*	Déscoins and Frérot, 1984 Kovalev and Nikolaeva, 1986; Rule and Roelofs, 1989
Spilosoma imparilis Butler	1,3Z,6Z,9Z-21:H 3Z,6Z,9Z-21:H 3Z,6Z,9Z-23:H 6Z,9Z-23:H			Wei *et al.*, 2003
Spilosoma lubricipeda L.	3Z,6Z-cis-9,10-epo-21:H 1,3Z,6Z-cis-9,10-epo-21:H		*9Z,12Z,15Z-18:Ald*	Ostrauskas, 2004

(*continued*)

Table 18.1 (cont.)

Subfamily, genus, species	Polyene hydrocarbons	Polyene epoxides	Aldehydes	Other	References
Spilosoma luteum Hufnagel		3Z,6Z-9S,10R-epo-21:H 1,3Z,6Z-9S,10R-epo-21:H	9Z,12Z,15Z-18:Ald		Maini and Belifiori, 1996
Spilosoma virginica (F.)	3Z,6Z,9Z-21:H	3Z,6Z-cis-9,10-epo-21:H			Meyer, 1984
Syntomoides imaon Cramer		**3Z,6Z-9Z,9Z-21:H 1,3Z,6Z, 9Z-21:H**			
Tyria jacobaeae L.	3Z,6Z,9Z-21:H	3Z,6Z-9S,10R-epo-21:H 3Z,6Z-cis-9,10-epo-20:H			Frérot *et al.*, 1988a Bestmann *et al.*, 1994 Clarke *et al.*, 1996
Utetheisa ornatrix (L.)	3Z,6Z,9Z-20:H **3Z,6Z,9Z-21:H** 1,3Z,6Z,9Z-21:H 6Z,9Z-21:H				Conner *et al.*, 1980 Jain *et al.*, 1983, Lim *et al.*, 2007
Callimorphinae					
Callimorpha dominula L.		3Z,6Z-9S,10R-epo-21:H			Clarke *et al.*, 1996
Cymbalophora pudica Esper		3Z,6Z-cis-9,10-epo-21:H			Frérot *et al.*, 1988b
Ctenuchinae					
Antichloris viridis Druce	3Z,6Z,9Z-20:H 3Z,6Z,9Z-21:H	3Z,6Z-cis-9,10-epo-20:H 3Z,6Z-cis-9,10-epo-21:H			Meyer, 1984
Cisseps fulvicolis (Hübner)	3Z,6Z,9Z-21:H	3Z,6Z-cis-9,10-epo-20:H 3Z,6Z-cis-9,10-epo-21:H			Meyer, 1984

Species	Compound	Reference
Ctenucha virginica (Esper)	(3Z,6Z-cis-9,10-epo-21:H)	Meyer, 1984
Dysauxes ancilla L.	1,3Z,6Z,9Z-19:H	Szöcs et al., 1987
Empyreuma affinis Rothschild	3Z,6Z,9Z-21:H	Déscoins et al., 1989
	3Z,6Z-cis-9,10-epo-21:H	Déscoins et al., 1989
Empyreuma mucro Zerny	3Z,6Z,9Z-21:H	Déscoins et al., 1989
	3Z,6Z-cis-9,10-epo-21:H	
Syntomis (=Amata) phegea L.	3Z,6Z,9Z-21:H	Szöcs et al., 1987
Syntomeida epilais (Walker)	3Z,6Z,9Z-21:H	Déscoins et al., 1989
	3Z,6Z-cis-9,10-epo-21:H	
Syntomoides imaon Cramer	3Z,6Z,9Z-20:H	Ando et al., 2008
	3Z,6Z,9Z-21:H	
	1,3Z,6Z,9Z-20:H	
	1,3Z,6Z,9Z-21:H	

Table 18.2 *Polyene hydrocarbons, epoxides, and related compounds known or suspected as sex pheromones for geometrid moths.*
Compounds in **bold** *have been found in pheromone gland extracts or aeration extracts and have been shown to be active in behavioral*
bioassays or field trials, compounds in normal font have been found in pheromone gland or aeration extracts, compounds in italics
have been shown to attract males in field screening trials, and underlined compounds have been shown to be antagonistic.

Subfamily, genus, species	Polyene hydrocarbons	Polyene epoxides	Other compounds	References
Ennominae				
Abraxas grossulariata L.	**3Z,6Z,9Z:17:H**	**6Z,9Z-3S,4R-epo-17:H**		Tóth *et al.*, 1992, 1994
Abraxas niphonibia Wehrli		*6Z,9Z-cis-3,4-epo-18:H*		Ando *et al.*, 1995
Abraxas sylvata Scopoli		*6Z,9Z-3R,4S-epo-17:H*		Tóth *et al.*, 1994
Agriopis aurantiaria (H.)		**6Z,9Z-3S,4R-epo-19:H**		Szöcs *et al.*, 1993
Agriopis (= Erannis) bajaria Denis and Schiffermüller	**3Z,6Z,9Z-18:H** **3Z,6Z,9Z-19:H**			Szöcs *et al.*, 1996 Lecheva, 1999 Goller *et al.*, 2007
Agriopis leucophearia Denis and Schiffermüller		*6Z,9Z-3S,4R-epo-19:H*		Szöcs *et al.*, 1993
Agriopis marginaria F.	*3Z,6Z,9Z-19:H*	*3Z,9Z-6S,7R-epo-19:H*		Hansson *et al.*, 1990; Szöcs *et al.*, 1993
Alcis angulifera Butler		*6Z,9Z-3S,4R-epo-19:H*		Ando *et al.*, 1993, 1997
Alcis (=Boarmia) repandata L.		*6Z,9Z-19:H* *3Z,6Z,9Z-19:H* *1,3Z,6Z,9Z-19:H*		Bogenschütz *et al.*, 1985; Subchev *et al.*, 1986
Anacamptodes humaria Guenée	**6Z,9Z-19:H**	**9Z-6R,7S-epo-19:H**		Millar *et al.*, 1991b
Anavitrinella pampinaria Guenée		*6Z,9Z-cis-3,4-epo-19:H* *3Z,9Z-6S,7R-epo-19:H*		Millar *et al.*, 1990c

Species		Compounds	References
Ascotis (=*Boarmia*) *selenaria* Denis and Schiffermüller	3Z,6Z,9Z-19:H	**6Z,9Z-3S,4R-epo-19:H**	Becker *et al.*, 1983, 1990; Cossé *et al.*, 1992
Ascotis selenaria cretacea Butler	**3Z,6Z,9Z-19:H**	**6Z,9Z-3R,4S-epo-19:H** **6Z,9Z-3S,4R-epo-19:H** 3Z,9Z-cis-6,7-epo-19:H 3Z,6Z-cis-9,10-epo-19:H	Ando *et al.*, 1993, 1997; Witjaksono *et al.*, 1999
Biston robustum Butler	**6Z,9Z-19:H** **3Z,6Z,9Z-19:H**	**9Z-6S,7R-epo-19:H** **3Z,9Z-6S,7R-epo-19:H**	Yamamoto *et al.*, 2000
Bupalus piniarius L.	6Z,9Z-19:H **4E,6Z,9Z-19:H**	**6Z,9Z-4S,5S-epo-19:H** **9Z-6S,7R-epo-19:H** 6Z,9Z-4R,5R-epo-19:H	Bestmann *et al.*, 1982; Francke *et al.*, 1998, 1999
Cabera erythemaria Guenée	3Z,6Z,9Z-19:H	3Z,6Z-9S,10R-epo-19:H 3Z,6Z-9R,10S-epo-19:H	Wong *et al.*, 1985
Campea margaritata L.	6Z,9Z-19:H 3Z,6Z,9Z-19:H 1,3Z,6Z,9Z-19:H		Bogenschütz *et al.*, 1985
Caripeta angustiorata Walker	3Z,6Z,9Z-19:H	3Z,9Z-6R,7S-epo-19:H 6Z-cis-9,10-epo-21:H	Millar *et al.*, 1990b
Caustoloma flavicaria Denis and Schiffermüller			Kovalev, and Nikolaeva, 1986
Colotois pennaria L.	**3Z,6Z,9Z-19:H** 3Z,6Z,9Z-21:H	**3Z,9Z-6R,7S-epo-19:H**	Hansson *et al.*, 1990; Szöcs *et al.*, 1993
Colotois pennaria ussuriensis Bang-Haas		6Z,9Z-cis-3,4-epo-20:H 3Z,9Z-cis-6,7-epo-19:H 3Z,9Z-cis-6,7-epo-20:H	Ando *et al.*, 1993 Ando *et al.*, 1995
Ectropis excellens (Butler)		9Z-cis-6,7-epo-18:H 6Z-cis-9,10-epo-18:H 3Z, 9Z-cis-6,7-epo-18:H	Ando *et al.*, 1995

(*continued*)

Table 18.2 (cont.)

Subfamily, genus, species	Polyene hydrocarbons	Polyene epoxides	Other compounds	References
Ectropis obliqua Prout	3Z,6Z,9Z-18:H 3Z,6Z,9Z-19:H 3Z,6Z,9Z-22:H 3Z,6Z,9Z-24:H	3Z,9Z-cis-6,7-epo-18:H		Yao *et al.*, 1991 Yin *et al.*, 1994
Epelis truncataria Walker		6Z,9Z-cis-3,4-epo-17:H		Millar *et al.*, 1990c
Erannis defoliaria Clerck	**3Z,6Z,9Z-19:H**	**3Z,9Z-6S,7R-epo-19:H** 3Z,9Z-6R,7S-epo-19:H		Hansson *et al.*, 1990 Szöcs *et al.*, 1993, 1998a
Erannis golda Djakonov		3Z,9Z-cis-6,7-epo-18:H		Ando *et al.*, 1995
Erannis tiliaria tiliaria Harris	3Z,6Z,9Z-19:H	3Z,9Z-6S,7R-epo-19:H 3Z,9Z-6R,7S-epo-19:H		Szöcs *et al.*, 1998a, 2001
Erannis tiliaria vancouverensis Hulst	3Z,6Z,9Z-19:H	3Z,9Z-6S,7R-epo-19:H 3Z,9Z-6R,7S-epo-19:H		Szöcs *et al.*, 1998a, 2001
Euchlaena madusaria Walker	**6Z,9Z-19:H**	6Z-9S,10R-epo-19:H 6Z-9R,10S-epo-19:H 3Z,9Z-cis-6,7-epo-19:H		Millar *et al.*, 1991b
Eufidonia convergaria Walker	**3Z,6Z,9Z-19:H**	**3Z,9Z-6S,7R-epo-19:H**		Millar *et al.*, 1990b,c, d
Hemerophila atrilineata Butler		**6Z-9S,10R-epo-18:H** **3Z,6Z-9S,10R-epo-18:H**		Tan *et al.*, 1996 Pu *et al.*, 1999
Itame brunneata Thunberg	3Z,6Z,9Z-17:H	6Z,9Z-3S,4R-epo-17:H 6Z,9Z-3R,4S-epo-17:H		Millar *et al.*, 1990c
Itame occiduaria Packard	3Z,6Z,9Z-17:H	6Z,9Z-3R,4S-epo-17:H 6Z,9Z-3S,4R-epo-17:H		Millar *et al.*, 1990c

Species			Reference
Lomographa semiclarata Walker	**3Z,6Z,9Z-17:H**		Millar *et al.*, 1990a
Lycia ursaria Walker	*3Z,6Z,9Z-19:H*	*6Z,9Z-3S,4R-epo-19:H*	Millar *et al.*, 1990c
Menophra (=Hemerophila) atrilineata Butler		**3Z,6Z-9S,10R-epo-18:H** **6Z-9S,10R-epo-18:H**	Tan *et al.*, 1996 Pu *et al.*, 1999
Menophra senilis (Butler)		*9Z-cis-6,7-epo-19:H*	Ando *et al.*, 1995
Metanema inatomaria Guenée	*3Z,6Z,9Z,11E-19:H* 3Z,6Z,9Z-19:H	*3Z,6Z-9R,10S-epo-19:H*	Wong *et al.*, 1985
Mnesampela privata (Guenée)	**3Z,6Z,9Z-19:H**		Steinbauer *et al.*, 2004
Ourapteryx sambucaria L.	*3Z,6Z,9Z-19:H*		Subchev *et al.*, 1986
Pachyerannis obliquaria Motschulsky		*6Z,9Z-3R,4S-epo-19:H* 6Z,9Z-3S,4R-epo-19:H	Qin *et al.*, 1997
Pachyligia dolosa Butler		*3Z,9Z-cis-6,7-epo-21:H* *3Z, 6Z-cis-9,10-epo-21:H*	Ando *et al.*, 1997
Paleacrita vernata Peck	**6Z,9Z-19:H** **3Z,6Z,9Z-19:H** **3Z,6Z,9Z-20:H**	6Z,9Z-cis-3,4-epo-19:H	Millar *et al.*, 1990d
Peribatodes rhomboidaria Denis and Schiffermüller	**3Z,6Z,9Z-19:H**	**6Z,9Z-3-one-19:H**	Buser *et al.*, 1985 Tóth *et al.*, 1987
Plagodis alcoolaria Guenée	*3Z,6Z,9Z-20:H*	*3Z,9Z-6R,7S-epo-20:H*	Millar *et al.*, 1991a
Plagodis pulveraria L.		*3Z,9Z-cis-6,7-epo-19:H*	Szöcs *et al.*, 2002

(continued)

401

Table 18.2 (*cont.*)

Subfamily, genus, species	Polyene hydrocarbons	Polyene epoxides	Other compounds	References
Probole alienaria Herrich-Schäffer		*3Z,9Z-6R,7S-epo-19:H* *3Z,9Z-6S,7R-epo-19:H*		Landolt *et al.*, 1996
Probole amicaria Herrich-Schäffer	**3Z,6Z,9Z-19:H**	**6Z,9Z-3S,4R-epo-19:H** **3Z,9Z-6R,7S-epo-19:H** <u>*3Z,9Z-6S,7R-epo-19:H*</u>		Millar *et al.*, 1990a,c
Prochoerodes transversata Drury	*3Z,6Z,9Z-19:H*	*3Z,6Z-9S,10R-epo-19:H*		Wong *et al.*, 1985
Pseudocoremia suavis (Butler)		**6Z-cis-9,10-epo-19:H** **3Z,6Z-cis-9,10-epo-19:H**		Gibb *et al.*, 2006
Rhynchobapta cervinaria bilineata (Leech)	*6Z,9Z-17:H*			Ando *et al.*, 1995
Sabulodes caberata Guenée	*9Z-19:H* *6Z,9Z-19:H*			McDonough *et al.*, 1986
Semiothisa bicolorata F.	**3Z,6Z,9Z-17:H**	*3Z,9Z-cis-6,7-epo-17:H* <u>*6Z,9Z-cis-3,4-epo-17:H*</u>		Millar *et al.*, 1987
Semiothisa cineraria Bremer and Grey	*3Z,6Z,9Z-17:H*	*6Z,9Z-cis-3,4-epo-17:H*		Li *et al.*, 1993
Semiothisa (=Chiasmia) clathrata L.	**3Z,6Z,9Z-17:H**	**6Z,9Z-3R,4S-epo-17:H**		Tóth *et al.*, 1991, 1994
Semiothisa delectata Hulst	*3Z,6Z,9Z-17:H*	*3Z,9Z-cis-6,7-epo-17:H*		Millar *et al.*, 1987
Semiothisa marmorata Ferguson	*3Z,6Z,9Z-17:H*	*6Z,9Z-3R,4S-epo-17:H*		Gries *et al.*, 1993

Species	Compound	Compound	Reference
Semiothisa neptaria Guenée	**3Z,6Z,9Z-17:H**	**6Z,9Z-3R,4S-epo-17:H** **6Z,9Z-3S,4R-epo-17:H**	Millar *et al.*, 1987 Gries *et al.*, 1993
Semiothisa sexmaculata Packard	<u>3Z,6Z,9Z-17:H</u>	**6Z,9Z-3R,4S-epo-17:H**	Gries *et al.*, 1993
Semiothisa signaria dispuncta Packard	**3Z,6Z,9Z-17:H** 3Z,6Z,9Z-18:H	**6Z,9Z-3R,4S-epo-17:H** **6Z,9Z-3S,4R-epo-17:H** *6Z,9Z-cis-3,4-epo-18:H*	Millar *et al.*, 1987, 1990c
Semiothisa ulsterata Pearsall	<u>3Z,6Z,9Z-17:H</u>	**3Z,9Z-6S,7R-epo-17:H**	Millar *et al.*, 1987, 1990c
Sicya macularia Harris	3Z,6Z,9Z-19:H	*6Z-3S,4R-epo-19:H* <u>3Z,9Z-6R,7S-epo-19:H</u> <u>3Z,9Z-6S,7R-epo-19:H</u> <u>3Z,6Z-cis-9,10-epo-19:H</u>	Millar *et al.*, 1990a, c
Synaxis jubararia Hulst	3Z,6Z,9Z-19:H	*3Z,6Z-cis-9,10-epo-19:H*	Wong *et al.*, 1985
Synegia esther Butler	*6Z-21:H*		Ando *et al.*, 1995
Tephrina arenacearia Hübner	3Z,6Z,9Z-17:H	**6Z,9Z-3S,4R-epo-17:H** 3Z,9Z-cis-6,7-epo-17:H	Tóth *et al.*, 1991, 1994
Theria rupicapraria (Denis and Schiffermüller)	**6Z,9Z-20:H** **6Z,9Z-21:H** 6Z,9Z-22:H 3Z,6Z,9Z-19:H 3Z,6Z,9Z-20:H 3Z,6Z,9Z-21:H		Szöcs *et al.*, 1996
Xanthotype sospeta Drury		*6Z-9S,10R-epo-19:H* <u>6Z-9R,10S-epo-19:H</u>	Millar *et al.*, 1991b

(*continued*)

403

Table 18.2 *(cont.)*

Subfamily, genus, species	Polyene hydrocarbons	Polyene epoxides	Other compounds	References
Zethenia albonotaria nesiotis Wehrlic		*6Z,9Z-3S,4R-epo-19:H* <u>*6Z,9Z-3R,4S-epo-19:H*</u>		Qin *et al.*, 1997
Oenochominae				
Inurois fumosa Inoue		*6Z,9Z-cis-3,4-epo-21:H*		Ando *et al.*, 1993
Inurois membranaria (Christoph)		*6Z,9Z-cis-3,4-epo-21:H*		Ando *et al.*, 1993
Milionia basalis pryeri Druce		**6Z,9Z-3S,4R-epo-19:H** <u>*6Z,9Z-3S,4R-epo-19:H*</u>		Yasui *et al.*, 2005
Alsophila				
Alsophila aescularia Denis and Schiffermuller	3Z,6Z,9Z-19:H 3Z,6Z,9Z-23:H 3Z,6Z,9Z,11E-21:H 3Z,6Z,9Z,11Z-21:H			Szöcs *et al.*, 1998a
Alsophila japonensis Warren	3Z,6Z,9Z-19:H			Ando *et al.*, 1993
Alsophila pometaria Harris	**3Z,6Z,9Z-19:H** **3Z,6Z,9Z,11E-19:H** **3Z,6Z,9Z,11Z-19:H**			Wong *et al.*, 1984a, b
Alsophila quadripunctata Esper	6Z,9Z-19:H 3Z,6Z,9Z-19:H			Szöcs *et al.*, 1984 Subchev *et al.*, 1986

Geometrinae

Agathia carissima Butler	3Z,6Z,9Z-20:H	Ando *et al.*, 1993
Agathia visenda visenda Prout	6Z,9Z-20:H 6Z,9Z-21:H	Ando *et al.*, 1993
Pachyodes superans Butler	3Z,6Z,9Z-20:H	Ando *et al.*, 1993

Laurentiinae

Anticlea vasiliata Guenée	**3Z,6Z,9Z-21:H**	Millar *et al.*, 1991a
Costaconvexa polygrammata Borkhausen	3Z,6Z,9Z-20:H	Szöcs *et al.*, 1987
Diactinia silaceata Denis and Schiffermüller	6Z,9Z-19:H 3Z,6Z,9Z-19:H 1,3Z,6Z,9Z-19:H	Bogenschütz *et al.*, 1985
Dysstroma brunneata ethela Hulst	3Z,6Z-9S,10R-epo-20:H	Wong *et al.*, 1985
Dysstroma sp. (*citrina?*)	3Z,6Z,9Z-20:H 3Z,6Z,9Z-21:H	Wong *et al.*, 1985
Epirrhoe sperryi Herbulot	3Z,6Z,9Z-18:H **3Z,6Z,9Z-19:H** <u>6Z,9Z-19:H</u>	Wong *et al.*, 1985 Millar *et al.*, 1992
Epirrita autumnata Borkhausen	**1,3Z,6Z,9Z-21:H**	Zhu *et al.*, 1995
Epirrita viridipurpurescens Prout	3Z,6Z,9Z-21:H	Ando *et al.*, 1993

(continued)

Table 18.2 (*cont.*)

Subfamily, genus, species	Polyene hydrocarbons	Polyene epoxides	Other compounds	References
Esakiopteryx volitans Butler	3Z,6Z,9Z-19:H			Ando et al., 1993
Eulithis testata L.	3Z,6Z,9Z-20:H 3Z,6Z,9Z-21:H	C_{20} *epoxydienes* C_{21} *epoxydienes*		Wong et al., 1985
Eupithecia assimilata Doubleday	<u>3Z,6Z,9Z-21:H</u>	**3Z,6Z-9S,10R-epo-21:H** <u>3Z,6Z-9R,10S-epo-21:H</u>		Campbell et al., 2007
Eupithecia annulata Hulst	3Z,6Z,9Z-21:H	C_{19} *epoxydienes* C_{20} *epoxydienes* C_{21} *epoxydienes* 3Z,6Z-9S,10R-epo-20:H **3Z,6Z-9S,10R-epo-21:H**		Wong et al., 1985 Millar et al., 1991c
Eupithecia subnotata Hübner	3Z,6Z,9Z-19:H			Szöcs et al., 1987
Eupithecia vulgata Hw.		3Z,6Z-*cis-9,10-epo-21:H*		Ostrauskas, 2004
Lobophora nivigerata Walker	6Z,9Z-21:H **3Z,6Z,9Z-21:H**			Millar et al., 1992
Mesoleuca ruficillata Guenée	3Z,6Z,9Z-20:H 3Z,6Z,9Z-21:H 3Z,6Z,9Z-22:H			Wong et al., 1985 Millar et al., 1991a
Operophtera bruceata Hulst	**1,3Z,6Z,9Z-19:H**			Roelofs et al., 1982 Underhill et al., 1987
Operophtera brumata L.	**1,3Z,6Z,9Z-19:H** <u>1,3E,6Z,9Z-19:H</u> <u>6Z,9Z-19:H</u>			Roelofs et al., 1982 Bestmann et al., 1982 Albert et al., 1984 Knauf et al., 1984 Underhill et al., 1987 Szöcs et al., 2004

Species	Pheromone components		Reference
Operophtera fagata Scharf.	9Z-19:H **6Z,9Z-19:H** **1,3Z,6Z,9Z-19:H**		Szöcs et al., 1998b, 2004
Operophtera occidentalis (Hulst)	1,3Z,6Z,9Z-19:H		Roelofs et al., 1982
Operophtera relegata Prout	3Z,6Z-9S,10R-epo-19:H 3Z,6Z-9R,10S-epo-19:H		Qin et al., 1997
Sibatania mactata Felder and Rogenhofer	3Z,6Z,9Z-21:H		Ando et al., 1993
Triphosa haesitata affirmata Guenée	3Z,6Z,9Z-20:H 3Z,6Z,9Z-21:H	9Z,12Z,15Z-18:OAc	Wong et al., 1985
Xanthorhoe abrasaria aquilonaria Herrich-Schiffermüller	3Z,6Z,9Z-21:H C21 epoxydienes 3Z,9Z-6S,7R-epo-21:H 3Z,6Z-9S,10R-epo-21:H		Wong et al., 1985 Millar et al., 1991a
Xanthorhoe ferrugata Clerck	3Z,9Z-6S,7R-epo-21:H		Millar et al., 1991a
Xanthorhoe iduata Guenée	3Z,6Z,9Z-21:H C21 epoxydienes		Wong et al., 1985
Xanthorhoe munitata Hübner	3Z,6Z,9Z-21:H C21 epoxydienes 3Z,9Z-6S,7R-epo-21:H		Wong et al., 1985 Millar et al., 1991c

Table 18.3 *Polyene hydrocarbons, epoxides, and related compounds known or suspected as sex pheromones for noctuid moths. Compounds in* **bold** *have been found in pheromone gland extracts or aeration extracts and have been shown to be active in behavioral bioassays or field trials, compounds in normal font have been found in pheromone gland or aeration extracts, compounds in italics have been shown to attract males in field screening trials, and underlined compounds have been shown to be antagonistic.*

Subfamily, genus, species	Polyene hydrocarbons	Polyene epoxides	Other compounds	References
Herminiinae				
Adrapsa notigera (Butler)		3Z,9Z-cis-6,7-epo-22:H		Ando et al., 1995
		3Z,9Z-cis-6,7-epo-23:H		
Bleptina caradrinalis (Guenée)		3Z,9Z-6S,7R-epo-20:H		Wong et al., 1985
		3Z,9Z-6R,7S-epo-20:H		Millar et al., 1991a
		3Z,9Z-6S,7R-epo-21:H		
		3Z,9Z-6R,7S-epo-21:H		
		<u>3Z,9Z-6S,7R-epo-19:H</u>		
		<u>3Z,9Z-6R,7S-epo-19:H</u>		
Idia aemula Hübner		*C₂₁ diene epoxides*		Wong et al., 1985
Idia americalis (Guenée)		3Z,9Z-6S,7R-epo-20:H		Wong et al., 1985
		3Z,9Z-6S,7R-epo-21:H		Millar et al., 1991a
Palthis angulalis Hübner		9Z-6S,7R-epo-19:H		Millar et al., 1991b
		<u>9Z-6R,7S-epo-19:H</u>		
Paracolax pryeri (Butler)		3Z,9Z-cis-6,7-epo-20:H		Ando et al., 1993
Renia salusalis Guenée		3Z,9Z-6S,7R-epo-19:H		Landolt et al., 1996
Tetanolita mynesalis (Walker)	**3Z,6Z,9Z-21:H**	**3Z,9Z-6S,7R-epo-21:H**		Haynes et al., 1996
		<u>3Z,9Z-6R,7S-epo-21:H</u>		
Zanclognatha sp.		3Z,9Z-cis-6,7-epo-19:H		Hai et al., 2002

Ophiderinae

Species			Reference
Aedia leucomelas L.	3Z,6Z,9Z-21:H		Tamaki et al., 1996
Hypenomorpha calamina (Butler)	3Z,6Z,9Z-21:H	6Z-cis-9,10-epo-20:H	Ando et al., 1995
Hypersypnoides submarginata Walker	3Z,6Z,9Z-21:H	6Z,9Z-cis-3,4-epo-21:H	Witjaksono et al., 1999
Hypocala rostrata (F.)		3Z,9Z-6S,7R-epo-21:H	Wakamura et al., 2002
Luceria sp.		3Z,6Z-cis-9,10-epo-18:H	Hai et al., 2002
Neachrostia bipuncta Sugi	3Z,6Z,9Z-18:H		Ando et al., 1995
Panagrapta trimantesalis (Walker)		3Z,6Z-cis-9,10-epo-23:H	Ando et al., 1995
Paragabara flavomacula (Oberthür)		3Z,9Z-cis-6,7-epo-21:H	Ando et al., 1993
Rivula leucanoides (Walker)		3Z,9Z-cis-6,7-epo-18:H	Ando et al., 1995
Rivula propinquinalis Guenée	3Z,6Z,9Z-19:H	3Z,9Z-6S,7R-epo-19:H	Wong et al., 1985 / Millar et al., 1990b
Rivula sasaphila Sugi		3Z,9Z-cis-6,7-epo-19:H	Ando et al., 1993
Rivula sericealis (Scopoli)		3Z,9Z-cis-6,7-epo-19:H	Ando et al., 1993

Hypeninae

Species			Reference
Bomolocha palparia (Walker)		3Z,6Z-9S,10R-epo-21:H	Wong et al., 1985
Hypena sp.		3Z,6Z-cis-9,10-epo-21:H	Hai et al., 2002
Lomonaltes eductalis Walker		3Z,6Z-9S,10R-epo-21:H	Wong et al., 1985

(continued)

Table 18.3 (cont.)

Subfamily, genus, species	Polyene hydrocarbons	Polyene epoxides	Other compounds	References
Spargoloma sexpunctata Grote		3Z,6Z-9S,10R-epo-21:H		Wong et al., 1985
Catocalinae				
Achaea janata L.	**3Z,6Z,9Z-21:H*** **6Z,9Z-21:H**		**9Z,12Z-18: Ald*, 21:H**	Persoons et al., 1993 Krishnakumari et al., 1998 Jyothi et al., 2005
Alabama argillacea Hübner	6Z,9Z-21:H 6Z,9Z-23:H		**(S)-9Me-19:H** (R)-9Me-19:H	Hall et al., 1993
Anomis texana F.	6Z,9Z-21:H 6Z,9Z-23:H		**(S)-7Me-17:H**	Hall et al., 1993
Anticarsia gemmetalis (Hübner)	**3Z,6Z,9Z-20:H** **3Z,6Z,9Z-21:H***			Heath et al., 1983, 1988 McLaughlin and Heath, 1989
Caenurgina distincta Cramer	3Z,6Z,9Z-20:H	3Z,6Z-cis-9,10-epo-20:H		Wong et al., 1985 Millar et al., 1991c
Caenurgina erechtea (Cramer)	**3Z,6Z,9Z-20:H** **3Z,6Z,9Z-21:H**			Underhill et al., 1983
Catocala briseis Walker	3Z,6Z,9Z-21:H	3Z,6Z-cis-9,10-epo-21:H		J. G. Millar and D. Hawks, unpub. data
Catocala grotiana Bailey	3Z,6Z,9Z-21:H	3Z,6Z-cis-9,10-epo-21:H		J. G. Millar and D. Hawks, unpub. data

410

Species			Reference
Catocala hermia H. Edwards	3Z,6Z,9Z-21:H	3Z,6Z-cis-9,10-epo-21:H	J. G. Millar and D. Hawks, unpub. data
Catocala irene Behr	3Z,6Z,9Z-21:H	3Z,6Z-cis-9,10-epo-21:H	J. G. Millar and D. Hawks, unpub. data
Catocala relicta Walker	3Z,6Z,9Z-21:H	3Z,6Z-cis-9,10-epo-21:H	J. G. Millar and D. Hawks, unpub. data
Euclidea cuspidea Hübner	3Z,6Z,9Z-20:H **3Z,6Z,9Z-21:H**	**3Z,6Z-cis-9,10-epo-21:H** 3Z,9Z-cis-6,7-epo-21:H 6Z,9Z-cis-3,4-epo-21:H	Wong et al., 1985; Millar et al., 1991c
Mocis disseverans (Walker)	**3Z,6Z,9Z-20:H** **3Z,6Z,9Z-21:H**		Landolt et al., 1986; McLaughlin and Heath, 1989
Mocis latipes (Guenée)	**6Z,9Z-21:H** 3Z,6Z,9Z-20:H **3Z,6Z,9Z-21:H**		Déscoins et al., 1986; Lalanne-Cassou et al., 1988; Landolt and Heath, 1989; McLaughlin and Heath, 1989
Mocis megas (Guenée)	3Z,6Z,9Z-21:H + homologs, analogs	3Z,6Z-cis-9,10-epo-21:H	Déscoins et al., 1990
Oraesia excavata (Butler)		**6Z-9S,10R-epo-21:H 3Z** 3Z, 6Z-cis-9,10-epo-21:H	Ohmasa et al., 1991 Yamamoto et al., 1999
Zale duplicata Bethune		**3Z,6Z-9R,10S-epo-21:H**	Wong et al., 1985
Zale lunifera Hübner		*3Z,6Z-9R,10S-epo-21:H*	Landolt et al., 1996

*Produced by males as well as females

Table 18.4 *Polyene hydrocarbons, epoxides, ketones, and related compounds known or suspected as sex pheromones for lymantriid moths. Because of the uncertainty of subfamilies in the Lymantriidae, species have simply been listed alphabetically. Compounds in* **bold** *have been found in pheromone gland extracts or aeration extracts and have been shown to be active in behavioral bioassays or field trials, compounds in normal font have been found in pheromone gland or aeration extracts, compounds in italics have been shown to attract males in field screening trials, and underlined compounds have been shown to be antagonistic.*

Genus, species	Unsaturated ketones	Polyenes and epoxides	Other compounds	References
Dasychira grisefacta ella Bryk.	*6Z-11-keto-21:H*			Daterman et al., 1976
Dasychira plageata Walker	*6Z-11-keto-21:H*			Grant, 1977
Dasychira vagans grisea Barnes and McDonough	*6Z-11-keto-21:H*			Daterman et al., 1976
Euproctis chrysorrhoea L.			**7Z,13Z,16Z,19Z-22:isobutyrate**	Leonhardt et al., 1991
Euproctis pulverea Leech			**11Z,14Z,17Z-20: isobutyrate, 11Z,14Z,17Z-20: 3-methylbutyrate**	Wakamura et al., 2001a
Gynaeophora ginghainensis Chou and Ying		**3Z,6Z,9Z-20:H** **3Z,6Z,9Z-21:H**		Chen, 1980

Species		Compound	Reference
Leucoma salicis L.		**3Z-6R,7S,9R,10S-diepo-21:H**	Gries *et al.*, 1997a
		3Z,9Z-cis-6,7-epo-21:H	Szöcs *et al.*, 2005
		3Z,6Z-cis-9,10-epo-21:H	
Lymantria mathura L.	2Me–cis–epo-18	3Z,6Z,9Z-19:H	Odell *et al.*, 1992
		3Z,6Z-9S,10R-epo-19:H	Oliver *et al.*, 1999
		3Z,6Z-9R,10S-epo-19:H	Gries *et al.*, 1999
Orgyia antiqua L.		*6Z-11-keto-21:H*	Daterman *et al.*, 1976
Orgyia cana Edwards		*6Z-11-keto-21:H*	Daterman *et al.*, 1976
Orgyia gonostigma L.		*6Z-11-keto-21:H*	Kovalev *et al.*, 1985
Orgyia leucostigma		**6Z-11-keto-21:H**	Grant, 1997
J.E. Smith		**6Z,9Z-11-keto-21:H**	Grant *et al.*, 2003
Orgyia postica Walker		**6Z,9Z-trans-11,12-epo-21:H**	Wakamura *et al.*,
Taiwan strain		**6Z,9Z-11S,12S-epo-21:H**	2001b, 2005
Okinawa strain		6Z,9Z-21:H	Chow *et al.*, 2001
		6Z-11-keto-21:H	Wakamura *et al.*,
Ishigaki strain		**6Z,9Z-11S,12S-epo-21:H**	2001b, 2005
		6Z,9Z-21:H	
		6Z-11-keto-21:H	
Orgyia pseudotsugata		**1,6Z-11-keto-21:H**	Smith *et al.*, 1975
McDunnough		**6Z-11-keto-21:H**	Smith, 1978
		6Z,8E-11-keto:21:H	Daterman *et al.*, 1976
			Gries *et al.*, 1997b

(*continued*)

413

Table 18.4 (cont.)

Genus, species	Unsaturated ketones	Polyenes and epoxides	Other compounds	References
Orgyia thyellina Butler	**6Z-11-keto-21:H** **6Z-9-keto:21:H**			Gries *et al.*, 1999
Perina nuda F.		**3Z,9Z-6S,7R-21:H** **9Z-3S,4R,6S,7R-diepo-21:H** **9Z-3R,4S,6S,7R-diepo-21:H** 3Z,9Z-cis-6,7–20:H		Yamazawa *et al.*, 2001 Wakamura *et al.*, 2002
Teia anartoides (Walker)	**6Z,9Z-11-keto-21:H** 6Z-11-keto-21:H 7E,9E-6-keto-11-keto-21:H 7E,11E-6-ol-11-keto-21:H 6Z,9Z-11-ol-21:H 6Z,8E-11-keto-21:H	**6Z,9Z-21:H** 6Z,9R,10S-epo-20:H 6Z,9R,10S-epo-21:H		El-Sayed *et al.*, 2005b Gries *et al.*, 2005

414

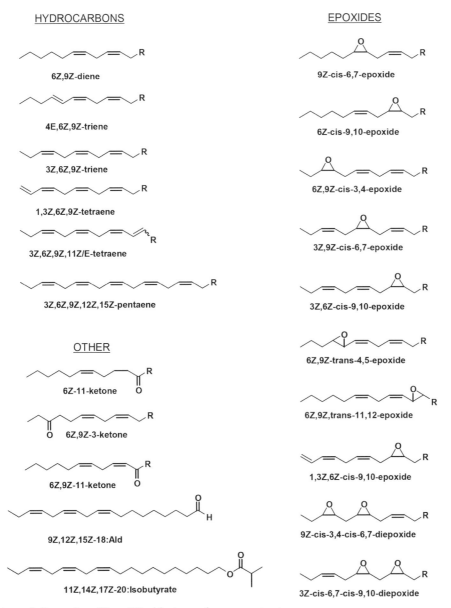

Scheme 1. Examples of Type II lepidopteran pheromone structures

in both cases the *trans*-epoxide is in a position other than the 3,4, 6,7, or 9,10 positions of the original double bonds that define Type II pheromones. Furthermore, there are now two examples known of diepoxides (Gries *et al.*, 1997a; Yamazawa *et al.*, 2001), such as 3Z-6R,7S-epo-9R,10S-epo-21:H (formal name: (2S,3R)-2-[[(2R,3S)-3-(2Z)-2-pentenyloxira-nyl]methyl]-3-undecyl-oxirane).

In addition to the epoxides, which have been unequivocally linked to the polyene base structures, I will include a third group of compounds in this discussion, the mono- and diene ketones that constitute sex pheromone components for some lymantrid species. These include 6Z-11-keto-21:H (formal name (Z)-heneicos-6-en-11-one), 6Z,9Z-11-keto-21:H (formal name: (6Z,9Z)-heneicosa-6,9-dien-11-one), and some structural variants. Even though it has not yet been shown that these compounds do indeed arise from the core Type II pathway, they have been included for the following reasons. First, they bear obvious structural similarities to the other Type II pheromones, including the chain lengths and positioning of the double bonds. Second, in at least two lymantrid species (*Orgyia postica* and *Teia anartoides*), they are found in combination with typical Type II polyenes and epoxides (see Table 18.4 for details and references).

There are also some miscellaneous compounds that probably belong in the Type II group of pheromones. These include 9Z,12Z-18:Ald ((9Z,12Z)-octadeca-9,12-dienal) and 9Z,12Z,15Z-18:Ald, found in a few arctiids (Table 18.1), that bear obvious similarities to linoleic and linolenic acids, respectively, and the polyene alcohol esters found in some lymantrid species (see Table 18.4). For example, the browntailed moth *Euproctis chrysorrhoea* produces 7Z,13Z,16Z,19Z-22:isobutyrate, which is the largest sex attractant pheromone known from any moth species. For all these miscellaneous compounds, although their biosynthesis has not been studied, the skipped-methylene pattern of the diagnostic double bonds, and their position beginning at carbon three or six from the end of the chain, strongly suggest a Type II pathway.

Taxonomic distribution of Type II pheromone compounds

The taxonomic distribution of Type II pheromones has been previously reviewed in some detail (Millar, 2000; Ando *et al.*, 2004), and so I will focus only on the main patterns and trends, along with a discussion of more recent findings. The reader is also cautioned that these patterns and trends are heavily biased by the fact that the identification of lepidopteran pheromones has not been conducted methodically. Rather, species that are of economic importance have been most intensively studied; for example, pheromones or sex attractants are known for hundreds of tortricid species, whereas few or no pheromones at all have been identified for members of other families (Ando *et al.*, 2004).

Generally speaking, Type II pheromones appear to be restricted primarily to four lepidopteran families, the Arctiidae, the Geometridae, the Noctuidae, and the Lymantriidae. Even within these families, Type II pheromones are by no means the only types of pheromone used by family members. Thus, for the Arctiidae, Type II pheromones are known from species in the subfamilies Lithosiinae, Arctiinae, Callimorphinae, and Ctenuchinae, but a large group of arctiine species, primarily in the genus *Virbia*, use saturated methyl-branched hydrocarbons as pheromones instead. Studies of the biosynthesis of these methyl-branched pheromones (Charlton and Roelofs, 1991) have shown that these compounds are unrelated to either the Type I or Type II pheromones. In addition, a lithosiine species has recently been shown to use saturated methyl-branched ketones as pheromone components (Yamamoto *et al.*, 2007).

In the Geometrids, with seven subfamilies according to one of the most current phylogenies (www.tolweb.org/Geometridae) and with the genus *Alsophila* not placed, pheromones or sex attractants are known for representatives of five of the seven families plus *Alsophila spp.*, and Type II pheromones predominate (Table 18.2). However, of the ten species in the subfamily Sterrhinae for which pheromones or attractants are known, none are Type II. In the remaining four subfamilies, at least one species does not use Type II pheromones.

In the Noctuidae, Type II pheromones are restricted to four of the approximately fifteen recognized subfamilies, the Hermiinae, Ophiderinae, Hypeninae, and Catocalinae, which constitute the Quadrifinae group within the family. Within these four subfamilies, almost all produce Type II pheromones, the only exceptions being one species in the Catocalinae (*Tyta luctuosa*; Cao *et al.*, 2003), and three species in the Opheridinae (*Alabama argillacea* and *Anomis texana*, Hall *et al.*, 1993; *Scoliopteryx libatrix*, Francke *et al.*, 2000; Toshova *et al.*, 2003). Interestingly, for the latter two species, pheromone gland extracts of both species contain Type II dienes, but they do not appear to be part of the active pheromone. Another curious anomaly has also turned up in the Noctuidae: males of the species *Anticarsia gemmetalis* (Heath *et al.*, 1988) and *Achaea janata* (Jyothi *et al.*, 2005) have been shown to produce components of the female pheromone in their hair pencil scent, and other males are attracted to the scent. Heath *et al.* (1988) suggested that males may use these compounds as aphrodisiacs to render females receptive to mating, and that nearby males are exploiting this courtship signal as an alternative method of finding mates.

Within the Lymantriidae, patterns of pheromone use are not yet clear, and there appears to be the greatest diversity of pheromone structures. Roughly two-thirds of the approximately thirty species for which pheromones or sex attractants have been reported have polyenes, epoxides, ketones, or polyunsaturated esters that could be classified as belonging to the Type II class, whereas the remainder use branched-chain epoxyalkane pheromones. Remarkably, even within a genus (e.g., *Lymantria* or *Euproctis*), congeners produce pheromones of different classes (Table 18.4).

One of the more significant developments of the past decade in our understanding of lepidopteran pheromone chemistry has been the discovery of moth species that use pheromone components of both Type I and Type II. Furthermore, the examples uncovered to date have all been in lepidopteran families that had not previously been shown to use Type II pheromones at all. The first example discovered was the tomato fruit borer *Neoleucinodes elegantalis*, in the family Crambidae. The pheromone of this moth is a synergistic blend of (*E*)-11-hexadecenol and 3Z,6Z,9Z-23:H (Cabrera *et al.*, 2001). Another crambid moth, *Deanolis sublimbalis*, has a similar pheromone blend, consisting of (*Z*)-11-hexadecenal and 3Z,6Z,9Z-23:H (Gibb *et al.*, 2007). Several other examples also have come to light, all in the family Pyralidae. For example, the navel orangeworm *Amyelois transitella* produces a pheromone blend that consists primarily of (11Z,13Z)-hexadecadienal, but an important synergist is the unprecedented 3Z,6Z,9Z,12Z,15Z-23:H (Leal *et al.*, 2005; Millar *et al.*, 2005; Kuenen *et al.*, 2010), and pheromone gland extracts also contain the homologous 3Z,6Z,9Z,12Z,15Z-25:H, although the role of this compound is not yet clear. 3Z,6Z,9Z,12Z,15Z-23:H was also found in extracts of pheromone glands of *Pyralis farinalis* when using this insect as a model organism to try and uncover minor components in the navel orangeworm blend (Kuenen *et al.*, 2010), because it was known that these two species cross-attract (Landolt and Curtis, 1982). The homolog, 3Z,6Z,9Z,12Z,15Z-25:H, has been shown to be a crucial part of the pheromone blend of the pyralid *Dioryctria abietivorella*, the other component being the Type I compound (9Z,11E)-tetradeca-9,11-dienyl acetate (Millar *et al.*, 2005), and 3Z,6Z,9Z,12Z,15Z-25:H has since been implicated in the pheromone blends of several other *Dioryctria* spp. (e.g., Grant *et al.*, 2008; Miller *et al.*, 2009). Finally, 6Z,9Z-23:H has recently turned up in pheromone glands of *Stenoma catenifer*, family Elachistidae, although its role remains unclear (Millar *et al.*, 2008; Hoddle *et al.*, 2009).

Overall, the importance of these blends of Type I and Type II pheromones to the development of our understanding of lepidopteran pheromone chemistry cannot be underestimated: they proved to be the key in working out previously intractable pheromones that had eluded some of the best groups of pheromone researchers for decades. For example, (11Z,13Z)-hexadeca-11,13-dienal, the major component of the navel orangeworm pheromone, was first reported by Coffelt *et al.* in 1979, more than twenty-five years before the crucial pentaene component was finally discovered. Similarly, the major component of the pheromone of *Dioryctria abietivorella*, (9Z,11E)-tetradeca-9,11-dienyl acetate, had been known for at least two decades before the discovery of 3Z,6Z,9Z,12Z,15Z-25:H, but the acetate was inactive as a single component (Millar *et al.*, 2005). Thus, it is certain that additional examples of further blends of Type I and Type II pheromones will turn up, with all their interesting implications with regard to the insects probably having two independent biosynthetic pathways for production of the two types of pheromone component (see below).

Analysis and identification of Type II pheromones

The techniques that can be applied to the identification of any insect pheromone are limited by the very small amounts of material that are available, typically in the nanogram to microgram range from a reasonable number of insects. Thus, coupled gas chromatography–electroantennogram detection is used routinely to locate potential pheromone components in extracts of pheromone glands or headspace collections from calling female moths, followed by coupled gas chromatography–mass spectrometry to obtain information about the structures of the compounds that elicit antennal responses. The GC retention characteristics of 6Z,9Z-dienes, 3Z,6Z,9Z,-trienes, 1,3Z,6Z,9Z-tetraenes and the corresponding monoepoxides were reviewed in Millar (2000), and the characteristic mass spectral fragmentation patterns and diagnostic fragments were reviewed in both Millar (2000) and Ando *et al.* (2004), along with representative spectra of the various structural types. Furthermore, Ando and his co-workers have published a series of papers containing detailed tables of diagnostic ions and ion ratios for the following classes of compound: 6,9,11-trienes, 3,6,9,11-tetraenes, and 1,3,6,9-tetraenes (Yamamoto *et al.*, 2008); C_{17}–C_{23} monoene epoxides with the epoxide in the 6 or 9 position (Ando *et al.*, 1995); C_{19}–C_{21} monoepoxydienes with the epoxide in the 3, 6, or 9 position (Ando *et al.*, 1993); 3,6-diepoxymonoenes (Yamazawa *et al.*, 2001; Wakamura *et al.*, 2002). Detailed mass spectral data can be found for the less common compounds by consulting the references cited in the tables.

Other methods are also available to aid in identification of nanogram or less amounts of pheromones. For example, the differences in the retention characteristics of compounds on polar and nonpolar GC columns, quantified in the form of Kovats indices under temperature programmed conditions (Van den Dool and Kratz, 1963), can provide valuable information as to the possible number and types of functional groups present in a molecule (e.g., Cabrera *et al.*, 2001; Millar *et al.*, 2005). Gries and co-workers have used this technique particularly effectively during the identification of trace quantities of pheromones from a wide variety of insects (e.g., Gries *et al.*, 1997a, 1999). In fact, this technique may be the only available option for tracking pheromone components during analyses if pheromone components are present in amounts that are too small to be detected by MS or the standard flame ionization detector used with gas chromatography (i.e., quantities < a few picograms), but which can still be detected by GC–EAD.

Microchemical reactions, which with care and suitably sized microscale equipment can be carried out on nanogram amounts of material, can be used to determine the presence or absence of specific functional groups, or determine the numbers, positions, and even geometries of double bonds. The application of microchemical reactions to pheromone identification has been reviewed in detail by Attygalle (1998). Coupled GC–Fourier transform infrared spectroscopy has also found occasional use in pheromone identification (Attygalle *et al.*, 1995; review, Leal, 1998).

One of the thorniest problems in the identification of both mono- and diepoxide pheromones has been the unambiguous determination of their absolute configurations. If the insects only produced a single, pure enantiomer, then in theory, the question could be answered indirectly by simply running bioassays using each of the pure enantiomers. In practice, there are two complications. First, it has been shown that a number of species actually require a specific ratio of both enantiomers, with each pure enantiomer being inactive (i.e., enantiomeric synergism; see tables for examples). Second, it may be difficult to produce compounds that are 100% enantiomerically pure. Because an insect may respond to very tiny amounts of a pheromone, even traces of enantiomeric impurities in synthesized compounds could still elicit behavioral responses, leading to potentially erroneous conclusions as to which enantiomer the insect produces.

The problem has been partially resolved by the development of cyclodextrin based chiral stationary phases for GC. A few of these phases can resolve 3,4-epoxides (Ando *et al.*, 1997; Koenig and Gehrcke, 1993; Szöcs *et al.*, 1993), 6,7-epoxides (Koenig and Gehrcke, 1993; Szöcs *et al.*, 1993), or 9,10-epoxides (Gries *et al.*, 1999), but to my knowledge only one of these columns is commercially available (Chiraldex A-PH, Astec Inc., used by Ando *et al.*, 1997).

The epoxide enantiomers can also be resolved on chiral stationary phase HPLC columns as follows:

1. 6Z,9Z-*cis*-3,4-epoxides: Chiralpak AS, Daicel Chemical Industries, Tokyo (Ando *et al.*, 1997; Qin *et al.*, 1997; Yasui *et al.*, 2005) or Chiralcel OJ-R, Diacel Chemical Industries, Tokyo (Pu *et al.*, 1999).
2. 3Z,9Z-*cis*-6,7-epoxides, Chiralpak AD (Qin *et al.*, 1997; Yamamoto *et al.*, 2000), Chiralcel OJ-R (Qin *et al.*, 1997).
3. 3Z,6Z-*cis*-9,10-epoxides: Chiralpak AD (Qin *et al.*, 1997) or Chiralcel OJ-R for chain lengths <C$_{20}$ (Qin *et al.*, 1997).
4. Monoene epoxides: Chiralpak AD or AS (Yamamoto *et al.*, 1999).
5. 9Z-*cis*-3,4-*cis*-6,7-diepoxides: Chiralpak AD (Wakamura *et al.*, 2002).

The main drawback in the resolution of enantiomers and/or determination of the absolute configuration of insect-produced compounds by HPLC is the limited sensitivity, because the epoxides do not have chromophores and thus are poorly detected by the UV detectors commonly used with HPLC. With such detectors, epoxides have to be injected in comparatively large (~microgram) amounts in order to be detected. Alternatively, fractions can be taken by time and then checked by GC. Until recently, LC–MS methods suffered from a similar lack of sensitivity, but the introduction of more sensitive time-of-flight mass spectrometers, operating with electrospray ionization, may eliminate this limitation and produce sensitivities similar to those achievable with GC–MS (Yamazawa *et al.*, 2003).

Biosynthesis of Type II pheromones

There are several fundamental differences between the biosynthetic pathways used to produce the Type I straight-chain alcohol, acetate, and aldehyde pheromone components of many Lepidoptera, and the Type II unsaturated hydrocarbons, epoxides, aldehydes, and related structures. First, Type I compounds are synthesized de novo in the pheromone gland from saturated fatty acyl precursors by a series of chain shortening and desaturation steps, followed by adjustment of the functional group on the terminal carbon to produce the desired alcohol, acetate, or aldehyde (Roelofs and Bjostad, 1984; Bjostad *et al.*, 1987). In contrast, all evidence to date (*vide infra*) indicates that the polyunsaturated Type II pheromones are derived from linoleic or linolenic acid precursors derived from the diet, with the double-bond positions and geometries of these unsaturated precursors being preserved in the final pheromone products. In the case of epoxides derived from polyunsaturated precursors, the double bond is stereospecifically oxidized to a *cis*-epoxide in which the configuration of the original double bond is echoed.

Second, the decarboxylase enzymes required to produce hydrocarbons apparently are not present in lepidopteran pheromone glands, but these enzymes are present in oenocyte cells where most insect hydrocarbon synthesis takes place (Blomquist *et al.*, 1987). Thus, Lepidoptera that produce Type II pheromones must have a mechanism for transport of hydrocarbon pheromone components from the oenocytes to the pheromone gland, where they are released as is, or after further structural modifications (see Chapter 5). Several studies have now shown that lipophorin proteins in the hemolymph perform this function (Schal *et al.*, 1998; Matsuoka *et al.*, 2006).

Third, whereas Type I pheromone components are both produced in and released from the pheromone gland, production and release of Type II pheromones take place by a mixed mechanism. That is, for Type II pheromones that consist entirely of hydrocarbons, the components probably are produced in oenocyte cells outside the gland and then transported to the gland for release. However, if the pheromone blend consists of a blend of diene or triene hydrocarbons with more highly functionalized compounds such as epoxides, then the hydrocarbon precursors are transported to the gland as before, but the final biosynthetic step of epoxidation occurs in the pheromone gland (see below for detailed case studies). Thus, for these types of pheromone, the pheromone gland may be involved only in the release of the pheromone, or in both the latter stages of biosynthesis and the release of the pheromone.

Fourth, there appear to be differences in the endocrine regulation of the production of Type I and Type II pheromones. With Type I pheromones, pheromone biosynthesis is switched on by pheromone biosynthesis activating neuropeptides (PBANs) generated in the subesophageal ganglion (SOG) (reviewed in Blomquist and Vogt, 2003). In contrast, from the limited data available to date, PBANs are not involved in the production of Type II pheromones directly. Instead, the evidence suggests that PBANs in these insects may regulate the release of the pheromone (Choi *et al.*, 2007), or the uptake of hydrocarbon pheromones into the gland, where they are then acted on by monooxygenases to produce epoxides (Wei *et al.*, 2004).

The first detailed study of the biosyntheses of Type II pheromones was carried out by Rule and Roelofs (1989), working with the two arctiid species *Estigmene acrea* and *Phragmatobia fuliginosa*, which produce polyunsaturated aldehyde and epoxide pheromone components. Thus, labeled linolenic acid (9Z,12Z,15Z-18:COOH) when topically applied to the pheromone gland or injected into pupae was incorporated into 9Z,12Z,15Z-18:Ald, implicating linolenic acid as a direct precursor to the aldehyde. However, application of likely precursors of 3Z,6Z-*cis*-9,10-epo:21:H to the gland did not result in production of the epoxide, suggesting that this compound, or the triene precursor, was not synthesized in the gland. Injection of 9Z,12Z,15Z-18:COOH, 11Z,14Z,17Z-20:COOH, or 13Z,16Z,19Z-22-:COOH, but not 12Z,15Z,18Z-21:COOH, into pupae resulted in production of the epoxide, demonstrating that chain elongation of 9Z,12Z,15Z-18:COOH by two two-carbon units to 13Z,16Z,19Z-22:COOH, followed by decarboxylation and epoxidation were likely steps in the biosynthesis. Saturated 18:COOH was not incorporated, corroborating the idea that pheromone production started with a precursor like linolenic acid in which some or all of the double bonds were already in place.

Recent studies have provided additional details of Type II pheromone biosynthesis in two other arctiid species. For example, injection of labeled 9Z,12Z,15Z-18:COOH into pupae of female *Utetheisa ornatrix* resulted in incorporation of label into 3Z,6Z,9Z-21:H as expected, and also into 1,3Z,6Z,9Z-21:H (Choi *et al.*, 2007), supporting the chain elongation and decarboxylation pathway first proposed by Rule and Roelofs (1989). However, labeled 3Z,6Z,9Z-21:H was not incorporated into 1,3Z,6Z,9Z-21:H, indicating that the terminal double bond in this tetraene must be inserted earlier in the biosynthesis, probably prior to the decarboxylation step. This study also provided some further details. First, decapitated females continued to produce the unsaturated hydrocarbon pheromone components but did not release them, and injection of PBAN did not affect pheromone production. Both these points suggest that PBAN is not directly involved in the production of the hydrocarbon pheromones, although it may regulate their release. Second, both of the pheromone components were detected in late-stage pupae, indicating that pheromone production begins even before eclosion. Third, the hemolymph contained only small amounts of the triene and tetraene, in contrast to analogous results with other Type II species (see below), which would not be expected if these components were indeed transported through the hemolymph to the pheromone gland. As one of several possible explanations, the authors suggested that the pheromone-producing oenocytes in this species may be very close to the pheromone gland, so that transport through the hemolymph would be unnecessary.

In contrast, the hemolymph of females of the arctiid moth *Spilosoma imparilis* were found to contain significant levels of the polyunsaturated hydrocarbons corresponding to the epoxide pheromone components produced by this species (Wei *et al.*, 2003). In a biosynthetic study with the arctiid *Syntomoides imaon*, the pheromone of which consists of a blend of 3Z,6Z,9Z-21:H and 1,3Z,6Z,9Z-21:H (Matsuoka *et al.*, 2008), the lipids extracted from oenocytes and peripheral fat bodies associated with the abdominal integument contained both (11Z,14Z,17Z)-eicosa-11,14,17-trienoic acid and (13Z,16Z,19Z)-docosa-13,16,19-trienoic acid, the intermediates predicted by elongation of linolenic acid by one or two cycles of 2-carbon chain extension (Ando *et al.*, 2008). The latter acid is likely to be the direct biosynthetic precursor to 3Z,6Z,9Z-21:H (Ando *et al.*, 2008).

The most comprehensive study of Type II pheromone biosynthesis is being carried out by Tetsu Ando and co-workers, primarily using the above-mentioned geometrid species *Ascotis selenaria cretacea* as a model species. The pheromone of this species consists primarily of 3Z,6Z,9Z-19:H and both enantiomers of the corresponding 6Z,9Z-*cis*-3,4-epo-19:H (Ando *et al.*, 1993, 1997). Ando and co-workers began their studies by examining the substrate specificity of the epoxidation reaction. Injection or topical application of 3Z,6Z,9Z-19:H or a variety of analogs to decapitated females simultaneously with injection of PBAN resulted in stereospecific epoxidation of any substrate with a 3Z double bond, regardless of chain length or the presence or absence of further unsaturations (Miyamoto *et al.*, 1999). However, (3E)-, (2Z)-, and (4Z)-alkenes were not epoxidized, showing that whereas the monooxygenase may accept a variety of substrates, it will only epoxidize a 3Z double bond.

In a follow-up study, Wei *et al.* (2004) then compared the epoxidation in *A. selenaria cretacea* and a second geometrid species, *Hemerophila artilineata*, which produces a 10:1 mixture of 3Z,6Z-*cis*-9,10-epo-18:H and 6Z-*cis*-9,10-epo-18:H as its pheromone blend (Tan *et al.*, 1996; Pu *et al.*, 1999). Whereas the epoxidation in both species was confined entirely to one position (3Z in *A. selenaria cretacea* and 9Z in *H. artilineata*), both the lipophorin proteins that transported the hydrocarbons through the hemolymph to the phe-romone glands and the monooxygenases in the glands accepted a variety of homologs of chain lengths C_{17} to C_{23}. This lack of selectivity was further demonstrated by injecting two triene substrates simultaneously; the resulting ratio of epoxides reflected the ratio of trienes that had been injected. When females were decapitated, the epoxide titer in the pheromone glands dropped to undetectable levels, but injection of PBAN restored the production of epoxides. Furthermore, when labeled triene precursors were topically applied to glands of these decapitated females with zero titers of epoxides and without application of PBAN, the glands produced only labeled epoxides. In sum, these results led to several key conclu-sions. First, the fact that the lipophorin carrier proteins and the monooxygenases in both species accepted a range of substrates suggested that the pheromone blend was the result of tight control over the production of specific precursors, rather than the result of highly selective carrier proteins or monooxygenases. Second, the fact that topical application of triene precursors to the glands of decapitated females resulted in production of epoxides without the application of PBAN suggested that the epoxidation activity was turned on all the time, and was not under the control of PBANs or analogous peptides. This, together with the fact that application of PBAN restored the production of epoxides in the pherom-one glands of decapitated females, suggested that the role of PBAN was in regulation of the uptake of pheromone precursors by the gland, rather than regulation of epoxidation or release of the pheromone.

In a parallel study, Wei *et al.* (2003) showed that the triene precursors to epoxide pherom-ones were present, along with a variety of other hydrocarbons, in extracts of hemolymph from females of two geometrid (*A. selenaria cretacea* and *H. artilineata*) and one arctiid species (*Spilosoma imparilis*). The epoxides were not found in the hemolymph extracts, nor were the trienes found in extracts of hemolymph from males. When labeled trienes were injected into the abdomens of females, labeled epoxides were later recovered from pheromone glands, demonstrating that the injected compounds had been transported to and taken up by the glands, and that triene hydrocarbons, synthesized somewhere outside the glands (probably in oenocytes), were the direct precursors to the epoxide pheromone components.

The lipophorin proteins in the hemolymph of *A. selenaria cretacea* were subsequently identified, and as predicted from results of the previous work, were shown to accept a variety of hydrocarbon substrates (Matsuoka *et al.*, 2006). Labeled 3Z,6Z,9Z-19:H was transported through the hemolymph and selectively delivered to the pheromone gland for epoxidation. Interestingly, lipophorins from hemolymph of both males and females were shown to bind the 3Z,6Z,9Z-19:H triene precursor to the epoxide pheromones, even though the triene was not present in males.

Kawai *et al.* (2007a,b) then went on to identify a PBAN from *A. selenaria cretacea*, and showed that it was quite different from the 15 PBANs identified from lepidoptera up to that point, with <46% sequence homology with other PBANs. As expected, injection of this PBAN into decapitated females restored the production of epoxides in the pheromone gland, presumably through restoring uptake of the triene precursors from the hemolymph as described above. This is the first and so far the only PBAN that has been identified from a lepidopteran insect producing Type II pheromones.

Fujii *et al.* (2007) pinned down the location of the tissues with mono-oxygenase activity to a cell layer in the intersegmental membrane between the 8th and 9th abdominal segments, and made several further observations that corroborated previous predictions by Ando's group. First, the 3Z,6Z,9Z-19:H precursor to the epoxides did not accumulate in the abdomen tip, and second, intersegmental tissue dissected during either the photophase or the scotophase displayed similar epoxidation activity. These findings verified that epoxidation activity was turned on all the time, and that production of the epoxide pheromones was instead regulated by uptake of the 3Z,6Z,9Z-19:H precursor into the gland.

Watanabe *et al.* (2007) now have isolated and fully sequenced one tissue-specific pheromone binding protein (PBP) from antennae of *A. selenaria cretacea*, with a partial sequence of a second. Whereas the fully sequenced PBP was found in the antennae of both sexes, it was present in much larger quantities in antennae from males. In binding studies, the PBP was found to strongly bind the natural pheromone components (3Z,6Z,9Z-19:H and 6Z,9Z-*cis*-3,4-epo-19:H). Interestingly, the PBP also bound 3Z,9Z-*cis*-6,7-epo-19:H and 3Z,6Z-*cis*-9,10-epo-19:H, which have been shown to antagonize the behavioral response to the pheromone (Witjaksono *et al.*, 1999).

In the final installment of this story to date, Matsuoka *et al.* (2008) determined that female *A. selenaria cretacea* contained (11Z,14Z,17Z)-eicosa-11,14,17-trienoic acid but not the longer-chain (13Z,16Z,19Z)-docosa-13,16,19-trienoic acid, in line with the fact that the pheromone of this species consists of 3Z,6Z,9Z-19:H and a corresponding monoepoxide (Matsuoka *et al.*, 2008). That is, this geometrid only requires a C_{20} fatty acid precursor to decarboxylate to its C_{19} pheromone compounds. In contrast, as mentioned above, the arctiid species *Syntomoides imaon*, which produces C_{21} triene and tetraene pheromone components, was found to contain both (11Z,14Z,17Z)-eicosa-11,14,17-trienoic acid and the longer-chain (13Z,16Z,19Z)-docosa-13,16,19-trienoic acid (Matsuoka *et al.*, 2008). That is, the arctiid species requires the C_{22} precursor in order to produce its C_{21} pheromone components by decarboxylation, whereas the geometrid species only requires the C_{20} precursor, because its pheromone is composed of the shorter-chain C_{19} compounds. These data suggest that the chain elongation of linolenic acid and related precursors is under precise control.

Several other isolated studies by other groups have shed light on various aspects of the biosynthesis of Type II pheromones. For example, there are examples of Type II pheromones with both odd- and even-numbered chain lengths. Whereas the odd-numbered carbon chains can be visualized to arise by decarboxylation of even-numbered precursors built up by 2-carbon chain extensions of linoleic or linolenic acids, the pathway to the even-numbered chains was unclear. Goller *et al.* (2007) have now elucidated some of this pathway in the geometrid species *Erranis bajaria*, which produces both 3Z,6Z,9Z-18:H and 3Z,6Z,9Z-19:H. Females incorporated both 10Z,13Z,16Z-19:COOH and 11Z,14Z,17Z-20:COOH into 3Z,6Z,9Z-18:H, providing support for a pathway consisting of 2-carbon chain extension of linolenic acid, followed by chain shortening by α-oxidation and decarboxylation to produce 10Z,13Z,16Z-19:COOH. This key intermediate is then reduced and decarbo-nylated, or decarboxylated, to produce 3Z,6Z,9Z-18:H.

In the Lymantriidae, biosynthetic studies have focused on production of the gypsy moth pheromone, 2-methyl-7R,8S-epo-18:H. Kasang *et al.* (1974) demonstrated that injection of the alkene precursor resulted in production of the epoxide in the pheromone gland, and Jurenka *et al.* (2003) filled in many of the details some 30 years later. In parallel with the results from the Type II pheromone-producing arctiids and geometrids, the alkene pre-cursor to the pheromone, (Z)-2-methyloctadec-7-ene, probably synthesized in oenocytes, is transported through the hemolymph to the pheromone gland, where it is specifically taken up by the gland and epoxidized (Jurenka *et al.*, 2003). However, in distinct contrast to the Type II pheromones, in which the pathway begins with linoleic or linolenic acids derived from the diet, the alkene precursor was assembled de novo, with valine providing the methyl branch.

To date, the biosynthesis of the monoene and diene ketones found in some lymantrid pheromones, and which have been included in this chapter both because of their structural similarities to other Type II pheromones and because some lymantrid species produce mix-tures of ketones and "normal" Type II pheromones, has not been studied. Such studies may be complicated by the instability of some of the diene ketones. One of the other significant unresolved questions in the biosynthesis of Type II pheromones is the desaturation that results in the production of 1,3Z,6Z,9Z-tetraenes, for which the corresponding 3Z,6Z,9Z-trienes are not the precursors (Choi *et al.*, 2007).

Chemical synthesis of Type II pheromones

The syntheses of Type II dienes, trienes, tetraenes, and the related epoxides were covered in some detail in Millar (2000), and so only the main points, key references, or relatively recent information will be described here. 6Z,9Z-Dienes and 3Z,6Z,9Z-trienes of 18 car-bons or more are most easily synthesized by reduction of linoleic or linolenic acids to the corresponding alcohols, followed by conversion of the alcohol to a leaving group and reduction with LiAlH$_4$ (18 carbons) or chain extension (>18 carbons), as first described by Conner *et al.* (1980) (Scheme 2A). Almost 30 years after its original publication, this remains the method of choice for the synthesis of these compounds, although improve-ments have been claimed. In the first, Wang and Zhang (2007) claimed slightly better yields by using triflate (= trifluoromethanesulfonate) instead of tosylate (= *p*-toluenesulphonate) as the leaving group. In the second, Davies *et al.* (2007) reported that addition of butylated hydroxytoluene (BHT) as an antioxidant to reaction mixtures and crude products substan-tially improved yields by minimizing degradation. Because shorter-chain homologs of linoleic and linolenic acids are not readily available, 6Z,9Z-17:H and 3Z,6Z,9Z-17:H must be made from scratch. A straightforward route involves coupling of propargyl tosylates with alkynes, followed by simultaneous reduction of the resulting two or three triple bonds to Z double bonds (Millar *et al.*, 1987). Pohnert and Boland (2000) more recently published

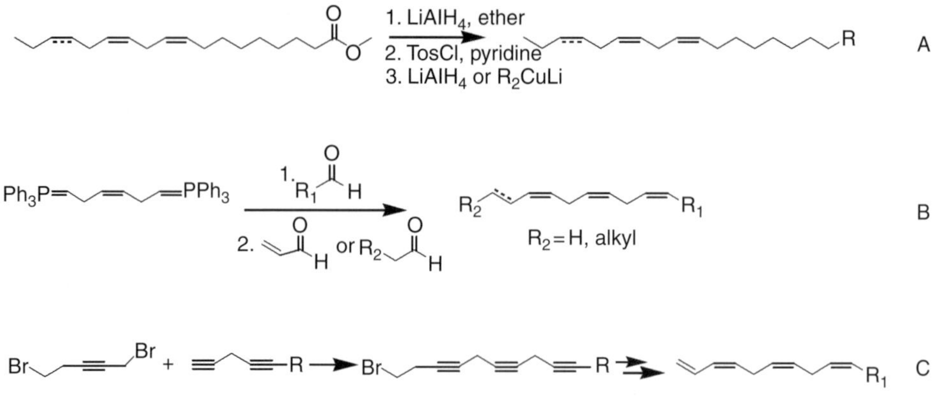

Scheme 2.

a very short route to these trienes and related compounds based on double-ended Wittig reagents prepared from (Z)-1,6-diiodo-3-hexene, which are reacted sequentially with two different aldehydes to place the second and third double bonds, respectively, with <5% of each of the E-isomers of each of the newly introduced double bonds (Scheme 2B).

Similarly, 1,3Z,6Z,9Z-tetraenes must be assembled from scratch because there are no suitable commercial precursors with the tetraene structure in place. Most syntheses have used iterative coupling of propargyl units to produce a triynol, followed by reduction of the triple bonds and elimination of HX (where X = a leaving group) to place the terminal double bond (e.g., Huang et al., 1983). A short and flexible route allowing the synthesis of tetraenes of any desired length required 5 sequential steps (Millar and Underhill, 1986a) (Scheme 2C). Alternatively, sequential addition of propenal and a second aldehyde to double-ended Wittig reagents was used in a short synthesis of these (Pohnert and Boland, 2000) (Scheme 2B) and related tetraenes and trienes (Yamamoto et al., 2008).

The choice of route for the synthesis of unsaturated epoxide pheromones depends on the amount required, and whether or not enantiomerically pure compounds are required. Small quantities of racemic materials for use as GC or HPLC standards can be readily produced by partial epoxidation of the corresponding 6Z,9Z-dienes (Ando et al., 1995), 3Z,6Z,9Z-trienes (Wong et al., 1985; Hansson et al., 1990; Ando et al., 1995), or 1,3Z,6Z,9Z-tetraenes (Tóth et al., 1987, 1989) with metachloroperbenzoic acid, followed by liquid chromatographic separation of the regioisomers. The racemic mixtures can then be resolved in milligram amounts using chiral stationary phase HPLC columns (see above for columns and references). In an extension of this method, Yamazawa et al. (2001) used analogous sequential separations on achiral and chiral HPLC columns to produce pure enantiomers of diepoxides, starting with the mixture of regio- and stereoisomers generated by nonselective diepoxidation of 3Z,6Z,9Z-triene precursors with metachloroperbenzoic acid. Obviously this method is not appropriate for the large-scale production of chiral pheromones, but it is probably the fastest and easiest method of generating small quantities as identification standards and for preliminary field trials.

Several strategies have been used to synthesize the epoxides in enantiomerically enriched or enantiomerically pure forms. The first strategy used Sharpless asymmetric epoxidation of appropriate achiral allylic alcohol precursors to place the key epoxide function (Scheme 3) (e.g., Wong *et al.*, 1985; Millar and Underhill, 1986b; Mori and Brevet, 1991). The alcohol function in the resulting epoxyalcohol intermediates is then converted to a leaving group, followed by alkylation with a chain of any desired length. Whereas either epoxide enantiomer is available from a single precursor by appropriate choice of the ligands used to form the chiral epoxidation catalyst, the method suffers from the drawback that the asymmetric epoxidation reaction on the 1,2-disubstituted Z-alkenol precursors is stereoselective rather than stereospecific (maximum enantiomeric excess, ~90%), and the small amount of the other enantiomer may be inhibitory (see examples in tables). The enantiomeric impurity can be removed by recrystallization of the epoxyalcohol intermediates if they are crystalline (e.g., Ebata and Mori, 1989). If the epoxyalcohols are not crystalline, then they can be recrystallized as their 2,4-dinitrobenzoate derivatives, followed by hydrolysis to yield the enantiomerically pure epoxyalcohols for further elaboration (e.g., Mori and Ebata, 1981, 1986; Mori and Takeuchi, 1989; Brevet and Mori, 1993; Khrimian *et al.*, 2004).

Another general method of making racemic or enantiomerically pure (or enriched) unsaturated epoxides takes advantage of two related reactions. The first of these is the Payne rearrangement of epoxyalcohols, in which 1,2-epoxy-3-ols or 2,3-epoxy-1-ols interconvert upon treatment with base (Scheme 4A), with the position of the equilibrium between the two forms varying with the structure (e.g., Soulié *et al.*, 1995a). In the second reaction, termed alkylative epoxide rearrangement (Scheme 4B), alkylation of the tosylate of a 1,2-epoxy-3-ol occurs selectively at C_1, and the resulting alkoxide displaces the tosylate to form a new epoxide between C_2 and C_3. In both cases, the stereospecificity of the nucleophilic substitution reactions in either the formation or opening of the epoxide rings means that the stereochemistry of the epoxyalcohol starting materials is reliably and predictably translated to the epoxide products. Thus, the key synthons in this strategy are chiral, *syn*-1,2-epoxy-3-ols, which can be generated in a number of ways. For example, in the first application of the method to unsaturated epoxide pheromones, Bell and Ciaccio (1988) generated the racemic but diastereomerically pure tosylate of *syn*-1,2-epoxy-tetradecan-3-ol by nonselective epoxidation of tetradec-1-en-3-ol and chromatographic separation of the resulting diastereomers after tosylation. Reaction of the *syn*-tosylate with 1-lithio-1,4-heptadiyne and BF_3 etherate at –78°C gave a racemic diyne epoxide. Catalytic reduction then furnished the desired 3Z,6Z-*cis*-9,10-epoxide. Bell and Ciaccio (1993) then extended the method to chiral epoxides by generating the key *syn*-1,2-epoxy-3-ol intermediate in chiral form in five steps from protected glyceraldehyde. Yadav *et al.* (1998) devised a variation on this strategy, whereby chiral tetradec-1-en-3-ol (from Sharpless asymmetric epoxidation of (*E*)-tetradec-2-en-1-ol and rearrangement) was nonselectively epoxidized, followed by tosylation and chromatographic separation of the resulting *syn*- and *anti*-1,2-epoxy-tetradecan-3-ols, and alkylative epoxide rearrangement as before.

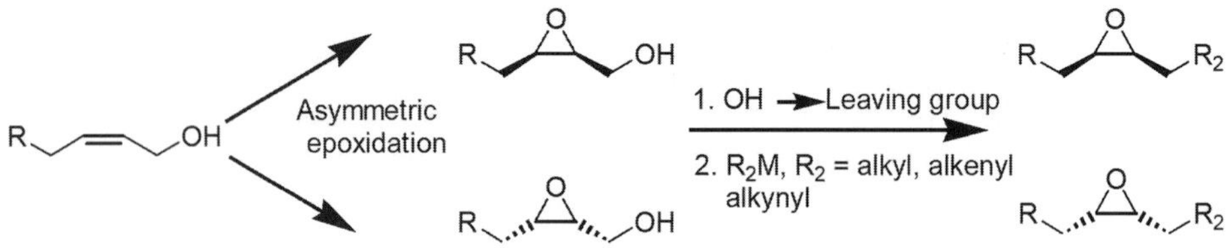

Scheme 3.

It is worth noting that either enantiomer of a desired epoxide product is available from one chiral *syn*-1,2-epoxy-3-ol. That is, tosylation followed by alkylative epoxide rearrangement gives one epoxide enantiomer (Scheme 4B), whereas initial Payne rearrangement to a 2,3-epoxy-1-ol followed by alkylation (after conversion of the alcohol to a leaving group, analogous to Scheme 3) yields the other. The opposite enantiomer is also accessible in another way, by protection of the alcohol function of the initial *syn*-1,2-epoxy-3-ol rather than tosylation (Scheme 4C). Regioselective alkylation of the epoxide at C_1 then gives an alcohol at C_2. Tosylation of this alcohol, followed by deprotection of the alcohol on the vicinal carbon and simultaneous ring closure then provides the other epoxide enantiomer (e.g., Wimalaratne and Slessor, 2004).

Zhang *et al.* (1999, 2000) and Che and Zhang (2005) have devised the shortest and most efficient method of using the alkylative epoxide rearrangement strategy (Scheme 4D). Their route uses Sharpless asymmetric dihydroxylation of (*E*)-2-alken-1-ols or (*E*)-1-chloro-2-alkenes to generate chiral *syn*-1,2,3-triols or *syn*-1-chloro-2,3-diols, respectively, in 95–97% enantiomeric excess. These intermediates are then converted in one or two steps respectively to chiral *syn*-1,2-epoxy-3-ol tosylates, which are then elaborated to the final epoxide products by alkylative epoxide rearrangement and reduction (Scheme 4D). Because of its brevity and efficiency, this may be the method of choice for making most chiral monoene or diene epoxides. Somewhat longer routes may be required for 6,7-epoxydienes where the epoxide is placed between two unsaturations, because the substrate for the asymmetric dihydroxylation can have only a single double bond. Thus, this synthon would need to possess additional functionality for inserting another double bond, after the alkylative epoxide rearrangement step.

A third general method of synthesis of epoxide pheromones uses double alkylation of a synthon consisting of a chiral epoxide flanked on either side by two different groups, which can be sequentially converted to leaving groups and replaced. By reversing the order in which each group is replaced, both enantiomers of an epoxide are accessible from a single synthon.

This strategy was used repeatedly by Mori and co-workers (Brevet and Mori, 1992; Muto and Mori, 2003a, b; Nakanishi and Mori, 2005), first using enzyme-based kinetic resolution of an achiral, *meso*-diacetate to generate an enantiomerically enriched epoxy-alcohol synthon (Scheme 5A), which could be further purified if necessary by dinitrobenzoate derivatization and recrystallization. Conversion of the alcohol to a leaving group,

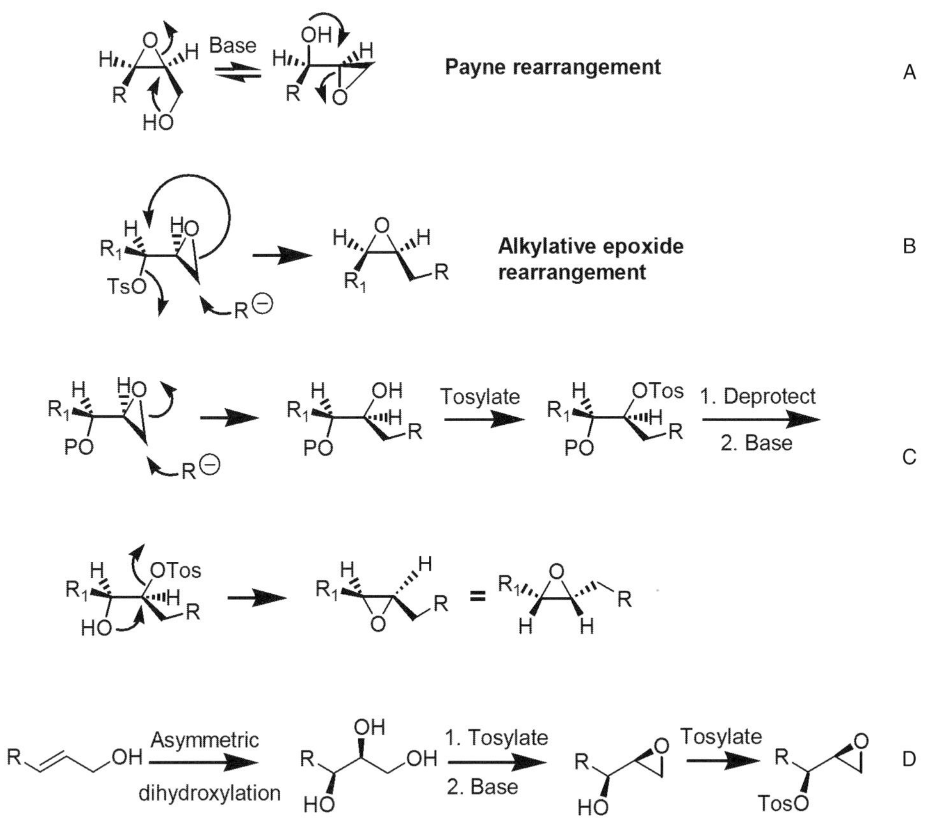

Scheme 4.

and alkylation installed the required chain on one side of the epoxide, and an analogous series of steps on the other side of the epoxide then completed the synthesis. The strategy was improved in later iterations (Muto and Mori, 2003a, b; Nakanishi and Mori, 2005) by lipase-catalyzed acetylation kinetic resolution of *cis*-2,3-epoxy-1,4-butanediol monoprotected as the *t*-butyldiphenylsilyl ether, using vinyl acetate as the source of acetate (Scheme 5B). The kinetic resolution was exceptional, providing the (2*R*,3*S*)-alcohol and the (2*S*,3*R*)-acetate in >98% yield, respectively.

In all of the above reactions of synthons containing both an epoxide and a leaving group on an adjacent carbon, successful regiospecific reaction with alkyl, alkenyl, or alkynyl organometallic reagents, without elimination or other side reactions, is critically dependent on the reaction conditions. Alkylations appear to work best by reacting tosylates with dialkyllithium cuprates in ether (e.g., Brevet and Mori, 1992), whereas alkenylations work well with an iodide leaving group reacted with alkenyl Grignard reagents with cuprous salt catalysis in THF/HMPA solvent mixtures (e.g., Millar and Underhill, 1986b; Brevet and Mori, 1992; Khrimian *et al.*, 2004), and alkynylations require alkynyllithium reagents

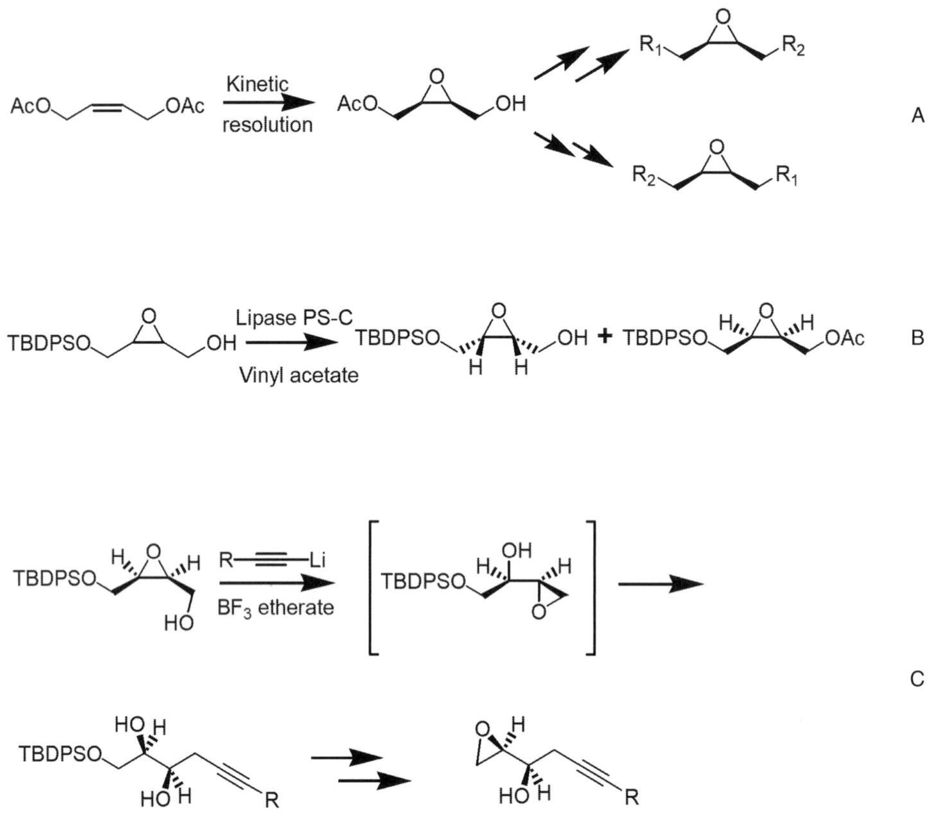

Scheme 5.

reacted with triflates in THF (Nakanishi and Mori, 2005), or less polar ether/hexane mixtures (Khrimian *et al.*, 2004).

Soulié *et al.* (1992, 1995a,b) developed a variation that combines elements of two methods described above, using a similar double-ended chiral epoxide synthon but with the alcohol function unprotected (Scheme 5C). This synthon was reacted directly with an alkynyllithium and BF$_3$ etherate. Under the influence of the strong Lewis acid, the epoxyalcohol first underwent an *in situ* Payne rearrangement to give the 1,2-epoxide, which then reacted with the alkynyllithium to give a diol, with an additional protected primary alcohol on carbon 1. Removal of the protecting group, selective tosylation of the resulting primary alcohol, and ring closure then gave a chiral 1,2-epoxy-3-ol, which was then elaborated further by alkylative epoxide rearrangement as described above.

Preparation of the stereoisomers of the diepoxide pheromones represents more of a synthetic challenge, and it has been approached in several ways. As mentioned above, small quantities can be prepared in several steps from 3Z,6Z,9Z-trienes (Yamamoto *et al.*, 1999; Yamazawa *et al.*, 2001). First, nonselective partial epoxidation of 3Z,6Z,9Z-trienes,

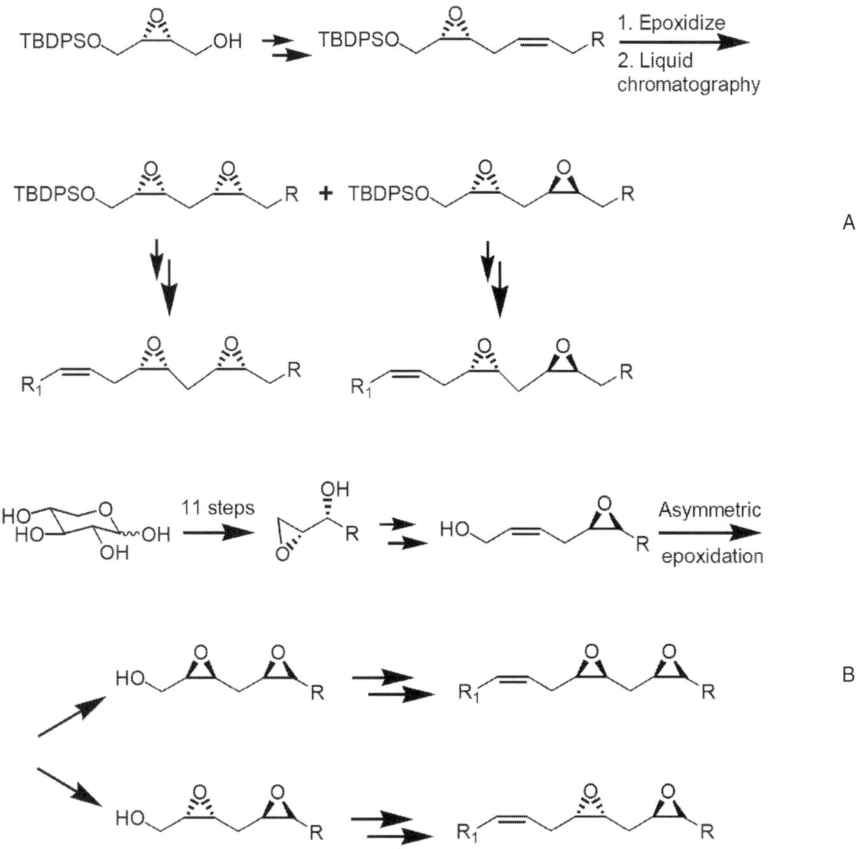

Scheme 6.

separation of regioisomers by HPLC, and resolution of the two enantiomers of a regio-isomer by HPLC on a chiral column gives the pure enantiomer of a monoepoxide. A second cycle of nonselective partial epoxidation then gives pairs of diastereomers of each regio-isomer, which can be separated by HPLC.

Mori and co-workers used an extension of their double-ended epoxide synthon method of making monoepoxides to develop a direct synthetic route to racemic (Lizarraga and Mori, 2001) and chiral (Muto and Mori, 2003b) diepoxide pheromones (Scheme 6A). Thus, alkenylation of the epoxide synthon, epoxidation of the double bond, and liquid chroma-tographic separation of the resulting diastereomers provided diepoxides of defined rela-tive and/or absolute configuration. Straightforward elaboration of these intermediates by methods analogous to those used in the syntheses of monoepoxide pheromones then com-pleted the syntheses. Wimalaratne and Slessor (2004) developed an alternative but longer synthetic route from a chiral 1,2-epoxy-3-alkanol synthon, prepared from D-xylose in 11 steps (Scheme 6B). These were chain extended with alkylative epoxide rearrangements to form the first epoxide ring while simultaneously introducing a propargyl alcohol function, which was reduced to the corresponding Z-allylic alcohol. Sharpless asymmetric epoxidation then introduced the second epoxide with a known configuration. The syntheses were completed by tosylation of the remaining free alcohol, and alkylation with the appropriate dialkyllithium cuprate as described above for monoepoxide syntheses. Using this method, all four stereoisomers of 3Z-*cis*-6,7-epo-*cis*-9,10-epo-21:H were prepared.

Among the miscellaneous Type II compounds, 6Z-11-keto-21:H is commercially available from several sources (Bedoukian Research, www.bedoukian.com, or Wako Pure Chemicals, www.wakousa.com), whereas the syntheses of the other ketone compounds, most of which have only been found in one or a few species, can be found in the papers describing the original identifications of the compounds (see tables). 9Z,12Z-18:Ald and 9Z,12Z,15Z-18:Ald are simply made by textbook methods, by reduction of linoleic or linolenic acids, respectively, in ether solvents with LiAlH$_4$, followed by oxidation of the resulting alcohols to aldehydes using Swern oxidation, pyridinium dichromate, or a variety of other standard methods for converting alcohols to aldehydes. Finally, polyunsaturated alcohol esters such as 7Z,13Z,16Z,19Z-22:Isobutyrate, the remarkable pheromone of the browntailed moth (Leonhardt *et al.*, 1991), have been made by a combination of Wittig and acetylene chemistry (Khrimian *et al.*, 2008). Surprisingly, the synthesis of this pheromone was not reported until some 17 years after it was first identified.

While this manuscript was in press, Ando and coworkers reported the first examples of two new types of polyene hydrocarbon pheromones, from two emerald moth species in the geometrid subfamily Geometrinae from which no pheromones had been previously identified (Yamakawa *et al.*, 2009). These were (6Z,9Z,12Z)-octadeca-6,9,12-triene, from *Hemithea tritonaria*, and the tetraene, (3Z,6Z,9Z,12Z)-icosadeca-3,6,9,12-tetraene, from *Thalassodes immissaria intaminata* females. It is very likely that these structural motifs will be found in pheromones of related species as additional species are studied.

Conclusions

Since their first discovery almost 30 years ago, it has become clear that Type II pheromones form a large class of pheromones, all linked by their common derivation from diet-derived linoleic and linolenic acids. Whereas most research during the first couple of decades after their discovery was directed towards the identification of new pheromones and exploring the taxonomic distribution of these compounds, the past decade has seen a major shift towards developing an understanding of the biosynthesis, biochemistry, and molecular biology underlying the production of and responses to these compounds. In particular, the methodical studies of Ando and his co-workers have been particularly valuable because their studies have focused on a single geometrid species, allowing a relatively complete story covering many aspects of pheromone production and use to be assembled.

In terms of the identification of new pheromones, one of the other most significant findings over the past decade has been the discovery of Type II pheromones in combination with Type I pheromones, and the numerous implications for the mechanisms involved in their biosynthesis and antennal reception. Although there are only a handful of documented examples of these mixed pheromones to date, it is certain that others will be discovered. The other area relating to Type II pheromone compounds that has really developed over the past decade has been the demonstration of the increasing structural diversity in these types of pheromones, such as the discoveries of epoxides of 1,3Z,6Z,9Z-tetraenes, trans-epoxides, and diepoxides. With the relatively small number of Type II pheromones identified to date, it is certain that further structural diversity is waiting to be found.

There are also numerous areas related to Type II pheromones about which we know little or nothing. For example, there is not yet any hard evidence that the unsaturated ketones are truly Type II pheromones, and we do not know the mechanisms by which the core structures of Type II pheromones are further modified, for example, by adding further unsaturations to the chain. In terms of biology, in many cases, particularly within the arctiids and noctuids, it has been demonstrated that the pheromone glands of females contain relatively large amounts of Type II pheromone compounds, but bioassays to work out the actual attractive blends either have not been done, or have failed to attract moths. Thus, much work remains to be done in understanding how Type II pheromone signals may act in concert with other types of signal. Overall, I fully anticipate that the next decades will provide many new discoveries and insights into this fascinating class of bioactive natural products.

References

Albert, von R., Bogenschütz, H. and Koning, E. (1984). Investigation on the use of pheromone-baited traps for monitoring the population dynamics of *Operophthera brumata* L. (Lepid., Geometridae). *Z. Angew. Entomol.*, **98**, 286–298.

Ando, T. (2003). www.tuat.ac.jp/~antetsu/review/e-List.pdf

Ando, T., Inomata, S. I. and Yamamoto, M. (2004). Lepidopteran sex pheromones. In *The Chemistry of Pheromones and Other Semiochemicals I. Vol. 239, Topics in Current Chemistry*, ed. S. Schulz. Berlin: Springer, pp. 51–96.

Ando, T., Kishi, H., Akashio, N., Qin, X. R., Saito, N., Abe, H. and Hashimoto, S. (1995). Sex attractants of geometrid and noctuid moths: chemical characterization and field test of monoepoxides of 6,9-dienes and related compounds. *J. Chem. Ecol.*, **21**, 299–311.

Ando, T., Matsuoka, K., Yamamoto, M., Muramatsu, M. and Naka, H. (2008). Identification of C21 Type II sex pheromone components and novel C20 and C22 trienyl biosynthetic precursors from a wasp moth, *Syntomoides imaon* (Arctiidae: Syntominae). In *Abstracts of the 25th Anniversary ISCE Meeting, International Society of Chemical Ecology*, State College PA, August 17–22, 2008. p. 224.

Ando, T., Ohsawa, H., Ueon, T., Kishi, H., Okamura, Y. and Hashimoto, S. (1993). Hydrocarbons with a homoconjugated polyene system and their monoepoxy derivatives: sex attractants of geometrid and noctuid moths distributed in Japan. *J. Chem. Ecol.*, **19**, 787–798.

Ando, T., Ohtani, K., Yamamoto, M., Miyamoto, T., Qin, X. R. and Witjaksono. (1997). Sex pheromone of Japanese giant looper, *Ascotis selenaria cretacea*: identification and field tests. *J. Chem. Ecol.*, **23**, 2413–2423.

Attygalle, A. B. (1998). Microchemical techniques. Chapter 7 in *Methods in Chemical*

Ecology. Vol. 1. Chemical Methods, ed. J.G. Millar and K.F. Haynes. New York: Chapman and Hall, pp. 207–293.

Attygalle, A.B., Svatoš, A., Wilcox, C. and Voerman, S. (1995). Gas-phase infrared spectroscopy for determination of double-bond configuration of some polyunsaturated pheromones and related compounds. *Anal. Chem.*, **67**, 558–567.

Becker, D., Cyjon, R., Cossé, A., Moore, I., Kimmel, T. and Wysoki, M. (1990). Identification and enantioselective synthesis of (Z,Z),6,9-*cis*-3S,4R-epoxynonadecadiene, the major sex pheromone component of *Boarmia selenaria*. *Tetrahedron Lett.*, **31**, 4923–4926.

Becker, D., Kimmel, T., Cyjon, R., Moore, I., Wysoki, M., Bestmann, H.J., Platz, H., Roth, K. and Vostrowsky, O. (1983). (3Z,6Z,9Z)-3,6,9-Nonadecatriene – a component of the sex pheromonal system of the giant looper, *Boarmia (Ascotis) selenaria* Schiffermüller (Lepidoptera: Geometridae). *Tetrahedron Lett.*, **24**, 5505–5508.

Bell, T.W. and Ciaccio, J.A. (1988). Alkylative epoxide rearrangement, application to stereoselective synthesis of chiral pheromone epoxides. *Tetrahedron Lett.*, **29**, 865–868.

Bell, T.W. and Ciaccio, J.A. (1993). Alkylative epoxide rearrangement. A stereospecific approach to chiral epoxide pheromones. *J. Org. Chem.*, **58**, 5153–5162.

Bell, T.W. and Meinwald, J. (1986). Pheromones of two arctiid moths (*Creatonotos transiens* and *C. gangis*): chiral components from both sexes and achiral female components. *J. Chem. Ecol.*, **12**, 385–409.

Bestmann, H.J., Brosche, T., Koschatzky, K.H., Michaelis, K., Platz, H., Roth, K., Süß, J., Vostrowsky, O. and Knauf, W. (1982). Pheromone-XLII. 1,3,6,9-Nonadecatetraen, das Sexualpheromon des Frostspanners *Operophtera brumata* (Geometridae). *Tetrahedron Lett.*, **23**, 4007–4010.

Bestmann, H.J., Janssen, E., Kern, F., Schäfer, D. and Vostrowsky, O. (1994). Pheromones. 97. The sex pheromone complex of the female arctiid moth *Tyria jacobaeae* (Lepidoptera, Arctiidae). *Z. Naturforsch. C*, **49**, 276–279.

Bestmann, H.J., Kern, F., Mineif, A., Platz, H. and Vostrowsky, O. (1992). Pheromones. 84. The sex pheromone complex of the arctiid moth *Arctia caja* (Lepidoptera: Arctiidae). *Z. Naturforsch. C*, **47**, 132–135.

Binda, M., Ferrario, P., Rossi, G., Tóth, M., Mori, K., Ninomiya, Y., Bengtsson, M., Rauscher, S. and Arn, H. (1990). Field tests to develop a sex pheromone formulation for *Hyphantria cunea* in Northern Italy. *Abstracts, Pheromones in Mediterranean Pest Management*, L'Organisation Internationale de Lutte Biologique/ Section Regionale Ouest Palearctique, Granada, Spain, Sept. 10–15, 1990. p. 43.

Bjostad, L.B., Wolf, W.A. and Roelofs, W.L. (1987). Pheromone biosynthesis in lepidopterans: desaturation and chain shortening. In *Pheromone Biochemistry*, ed. G.D. Prestwich and G.J. Blomquist. New York: Academic, pp. 77–120.

Blomquist, G.J., Nelson, D.R. and de Ronobales, M. (1987). Chemistry, biochemistry, and physiology of insect cuticular lipids. *Arch. Insect Biochem. Physiol.*, **6**, 227–265.

Blomquist, G.J., and Vogt, R. G. (eds.) 2003. Insect Pheromone Biochemistry and Molecular Biology. Elsevier Academic Press. Amsterdam. pp. 74.

Bogenschűtz, V.H., Knauf, W., Troger, E.J., Bestmann, H.J. and Vostrowsky, O. (1985). Pheromone 49: Freilandfange von Geometriden-Mannchen mit C19-Polyenen in Kiefernbestanden. *Z. Ang. Entomol.*, **100**, 349–354.

Brevet, J.-L. and Mori, K. (1992). Pheromone synthesis; CXXXIX. Enzymatic preparation of (2S,3R)-4-acetoxy-2,3-epoxybutan-1-ol and its conversion to the epoxy pheromones of the gypsy moth and the ruby tiger moth. *Synthesis*, **1992**, 1007–1012.

Brevet, J.-L. and Mori, K. (1993). Synthesis of both the enantiomers of (6*Z*,9*Z*)-*cis*-3,4-epoxy-6,9-heptadecadiene, the pheromone component of various geometrid moths. *Biosci. Biotechnol. Biochem.*, **57**, 1553–1556.

Buser, H. R., Guerin, P. M., Tóth, M., Szöcs, G., Schmid, A., Francke, W. and Arn, H. (1985). (*Z*,*Z*)-6,9-Nonadecadien-3-one and (*Z*,*Z*,*Z*)-3,6,9-nonadecatriene: identification and synthesis of sex pheromone components of *Peribatodes rhomboidaria*. *Tetrahedron Lett.*, **26**, 403–406.

Cabrera, A., Eiras, A., Gries, G., Gries, R., Urdaneta, N., Miras, B., Badji, C. and Jaffe, K. (2001). Sex pheromone of tomato fruit borer, *Neoleucinodes elegantalis. J. Chem. Ecol.*, **27**, 2097–2107.

Campbell, C. A. M., Tregidga, E. L., Hall, D. R., Ando, T. and Yamamoto, M. (2007). Components of the sex pheromone of the currant pug moth, *Eupithecia assimilata*, a re-emergent hop pest in UK. *Entomol. Exp. Appl.*, **122**, 265–269.

Cao, W. H., Charlton, R. E., Nechols, J. R. and Horak, M. J. (2003). Sex pheromone of the noctuid moth, *Tyta luctuosa* (Lepidoptera: Noctuidae), a candidate biological control agent of field bindweed. *Env. Entomol.*, **32**, 17–22.

Charlton R. E. and Roelofs, W. L. (1991). Biosynthesis of a volatile, methyl-branched hydrocarbon sex pheromone from leucine by arctiid moths (*Holomelina* spp.). *Arch. Insect Bioch. Physiol.*, **18**, 81–97.

Che, C. and Zhang, Z.-N. (2005). Concise total synthesis of (3*Z*,6*Z*,9*S*,10*R*)-9,10-epoxy-1,3,6-heneicosatriene, sex pheromone component of *Hyphantria cunea. Tetrahedron*, **61**, 2187–2193.

Chen, C. (1980). *Abstracts, International Congress of Entomology*, Kyoto, Japan, August 3–9, 1980.

Choi, M. Y., Lim, H., Park, K. C., Adlof, R., Wang, A. S., Zhang, A. and Jurenka, R. (2007). Identification and biosynthetic studies of the hydrocarbon sex pheromone in *Utetheisa ornatrix. J. Chem. Ecol.*, **33**, 1336–1345.

Chow, Y. S., Tsai, R. S., Gries, G., Gries, R. and Khaskin, G. (2001). Preliminary identification of the sex pheromone of *Orgyia postica* (Walker) (Lepidoptera: Lymantriidae). *Entomol. Sinica*, **8**, 13–20.

Clarke, C. A., Cronin, A., Francke, W., Phillip, P., Pickett, J. A., Wadhams, L. J. and Woodstock, C. M. (1996). Mating attempts between the scarlet tiger moth, *Callimorpha dominula* L., and the cinnabar moth, *Tyria jacobaeae* L. (Lepidoptera: Arctiidae), involve a common sex pheromone composition. *Experientia*, **52**, 636–638.

Coffelt, J. A., Vick, K. W., Sonnet, P. E. and Doolittle, R. E. (1979). Isolation, identification, and synthesis of a female sex pheromone of the navel orangeworm, *Amyelois transitella* (Lepidoptera: Pyralidae). *J. Chem. Ecol.*, **5**, 955–966.

Collins, C. W. and Potts, S. F. (1932). Attractants for the flying gypsy moth as an aid in locating new infestations. *USDA Technical Bull.*, **336**, 1–43.

Conner, W. E., Eisner, T., Vander Meer, R. K., Guerrero, A., Ghiringelli, D. and Meinwald, J. (1980). Sex attractant of an arctiid moth (*Utetheisa ornatrix*): a pulsed chemical signal. *Behav. Ecol. Sociobiol.*, **7**, 55–63.

Cossé, A. A., Cyjon, R., Moore, I., Wysoki, M. and Becker, D. (1992). Sex pheromone components of the giant looper, *Boarmia selenaria* Schiff. (Lepidoptera: Geometridae): identification, synthesis, electrophysiological evaluation and behavioral activity. *J. Chem. Ecol.*, **18**, 165–181.

Daterman, G. E., Peterson, L. J., Robbins, R. G., Sower, L. L., Daves, G. D. Jr. and Smith, R. G. (1976). Laboratory and field bioassay of the Douglas-fir tussock moth pheromone, (*Z*)-6-heneicosen-11-one. *Environ. Entomol.*, **5**, 1187–1190.

Davies, N. W., Meredith, G., Molesworth, P. P. and Smith, J. A. (2007). Use of the antioxidant butylated hydroxytoluene *in situ* for the synthesis of readily oxidized compounds. Application to the synthesis of the moth pheromone (Z,Z,Z)-3,6,9-nonadecatriene. *Aust. J. Chem.*, **60**, 848–849.

Déscoins, C. and Frérot, B. (1984). Taxonomic value of the chemical composition of the sex pheromone blends in arctiid moths and related families. *Abstracts of the International Congress of Entomology*, Hamburg. p. 467.

Déscoins, C., Lalanne-Cassou, B., Frérot, B., Malosse, C. and Renou, M. (1989). Comparative study of pheromonal secretions produced by arctiid and ctenuchid female moths (insects, Lepidoptera) from the neotropical area. *C. R. Acad. Sc. Paris Serie III*, **309**, 577–581.

Déscoins, C., Lalanne-Cassou, B., Malosse, C. and Milat, M.-L. (1986). Analysis of the sex pheromone produced by the virgin females of *Mocis latipes* (Guénée), Noctuidae, Catocalinae, from Guadeloupe (French Antilla). *C. R. Acad. Sc. Paris, Serie III*, **302**, 509–512.

Déscoins, C., Malosse, C., Renou, M., Lalanne-Cassou, B. and Le Duchat d'Aubigny, J. (1990). Chemical analysis of the pheromone blends produced by males and females of the neotropical moth, *Mocis megas* (Guenee) (Lepidoptera, Noctuidae, Catocalinae). *Experientia*, **46**, 536–539.

Ebata, T. and Mori, K. (1989). Synthesis of both the enantiomers of (Z)-*cis*-9,10-epoxy-6-heneicosene. *Agric. Biol. Chem.*, **53**, 801–804.

Einhorn, J., Boniface, B., Renou, M. and Milat, M.-L. (1984). Study on the sex pheromone of *Arctia villica* L. (Lepidoptera, Arctiidae). *C. R. Acad. Sci. Paris Serie III*, **298**, 573–576.

Einhorn, J., Lallemand, J.-Y., Zagatti, P., Gallois, M., Virelizier, H., Riom, J. and Menassieu, P. (1982). Isolation and identification of the sex pheromone blend of *Hyphantria cunea* (Drury) (Lepidoptera, Arctiidae). *C. R. Acad. Sci. Paris Serie II*, **294**, 41–44.

El-Sayed, A. M. (2008). The Pherobase: Database of Insect Pheromones and Semiochemicals. www.pherobase.com.

El-Sayed, A. M., Gibb, A. R. and Suckling, D. M. (2005a). Chemistry of the sex pheromone gland of the fall webworm, *Hyphantria cunea*, discovered in New Zealand. *N. Z. Plant Protection*, **58**, 31–36.

El-Sayed, A. M., Gibb, A. R., Suckling, D. M., Bunn, B., Fielder, S., Comeskey, D., Manning, L. A., Foster, S. P., Morris, B. D., Ando, T. and Mori, K. (2005b). Identification of sex pheromone components of the painted apple moth: A tussock moth with a thermally labile pheromone component. *J. Chem. Ecol.*, **31**, 621–646.

Francke, W., Brunnemann, U., Bergmann, J. and Plass, E. (1998). Semiochemistry at junctions: Volatile compounds from desert locusts, caddisflies, and geometrid moths. *Abstracts of the 2nd International Symposium on Insect Pheromones*, Wageningen, The Netherlands, 30 March–3 April 1998. pp. 71–73.

Francke, W., Gries, G., Gries, R., Haeussler, D., Moeller, K. and Plass, E. (1999). Patent application. Preparation of polyenes and polyene oxides as insect pheromones. Ger. Offen. (1999), 10 pp. CODEN: GWXXBX DE 19814330 A1 19991014 CAN 131:271802 AN 1999:667816.

Francke, W., Plass, E., Zimmermann, N., Tietgen, H., Tolasch, T., Franke, S., Subchev, M., Toshova, T., Pickett, J. A., Wadhams, L. J. and Woodcock, C. M. (2000). Major sex pheromone component of female herald moth *Scoliopteryx libatrix* is the novel branched alkene (6Z)-13-methylheneicosene. *J. Chem. Ecol.*, **26**, 1135–1149.

Frérot, B., Malosse, C., Chenon, R. D., Ducrot, P. H. and Cain, A. H. (1993). Identification of the sex pheromone components in *Pareuchaetes pseudoinsulata* Rego Barros (Lepidoptera, Arctiidae). *C. R. Acad. Sci. Paris, Serie III*, **317**, 1045–1050.

Frérot, B., Pougny, J. R., Milat, M.-L., Rollin, P. and Malosse, C. (1988b). Study of the pheromonal secretion of the tiger moth, *Cymbalophora pudica* (Esper) and enantioselectivity of its pheromonal perception. *C. R. Acad. Sci. Paris, Serie III*, **306**, 157–160.

Frérot, B., Renou, M., Malosse, C. and Déscoins. C. (1988a). Isolation and identification of pheromone compounds in female moths of *Tyria jacobaeae* (Lepidoptera, Arctiidae). Biological characterization of the absolute configuration of the main component. *Entomol. Exp. Appl.*, **46**, 281–289.

Fujii, T., Suzuki, M. G., Kawai, T., Tsuneizumi, K., Ohnishi, A., Kurihara, M., Matsumoto, S. and Ando, T. (2007). Determination of the pheromone-producing region that has epoxidation activity in the abdominal tip of the Japanese giant looper, *Ascotis selenaria cretacea* (Lepidoptera: Geometridae). *J. Insect Physiol.*, **53**, 312–318.

Gibb, A. R., Comeskey, D., Berndt, L., Brockerhoff, E. G., El-Sayed, A. M., Jactel, H. and Suckling, D. M. (2006). Identification of sex pheromone components of a New Zealand geometrid moth, the common forest looper *Pseudocoremia suavis*, reveals a possible species complex. *J. Chem. Ecol.*, **32**, 865–879.

Gibb, A. R., Pinese, B., Tenakanai, D., Kawi, A. P., Bunn, B., Ramankutty, P. and Suckling, D. M. (2007). (*Z*)-11-hexadecenal and (3*Z*,6*Z*,9*Z*)-tricosatriene: sex pheromone components of the red banded mango caterpillar *Deanolis sublimbalis*. *J. Chem. Ecol.*, **33**, 579–589.

Goller, S., Szöcs, G., Francke, W. and Schulz, S. (2007). Biosynthesis of (3Z,6Z,9Z)-octadecatriene: the main component of the pheromone blend of *Erannis bajaria* (Lepidoptera: Geometridae). *J. Chem. Ecol.*, **33**, 1505–1509.

Grant, G. G. (1977). Interspecific pheromone responses of tussock moths and some isolating mechanisms of eastern species. *Environ. Entomol.*, **6**, 739–742.

Grant, G. G., Millar, J. G., and Trudel, R. (2008). Pheromone identification of *Dioryctria abietivorella* (Lepidoptera: Pyrallidae) from an eastern North American population: geographic variation in pheromone response. *Can. Entomol.* **141**; 129–135.

Grant, G. G., Slessor, K. N., Liu, W. M. and Abou-Zaid, M. (2003). (Z,Z)-6,9-heneicosadien-11-one, labile sex pheromone of the whitemarked tussock moth, *Orgyia leucostigma*. *J. Chem. Ecol.*, **29**, 589–601.

Gries, G., Clearwater, J., Gries, R., Khaskin, G., King, S. and Schaefer, P. (1999). Synergistic sex pheromone components of white-spotted tussock moth, *Orgyia thyellina*. *J. Chem. Ecol.*, **25**, 1091–1104.

Gries, G., Gries, R., Underhill, E. W. and Humble, L. (1993). (6Z,9Z-3*R*,4*S*)-Epoxy-heptadecadiene: major sex pheromone component of the larch looper, *Semiothisa sexmaculata* (Packard) (Lepidoptera: Geometridae). *J. Chem. Ecol.*, **19**, 843–850.

Gries, R., Holden, D., Gries, G., Wimalaratne, P. D. C., Slessor, K. N. and Saunders, C. (1997a). 3Z-*cis*-6,7-*cis*-9,10-Diepoxy-heneicosene: novel class of lepidopteran pheromone. *Naturwissenschaften*, **84**, 219–221.

Gries, G., Slessor, K. N., Gries, R., Khaskin, G., Wimalaratne, P. D. C., Gray, T. G., Grant, G. G., Tracey, A. S. and Hulme, M. (1997b). (*Z*)6,(*E*)8-heneicosadien-11-one: synergistic sex pheromone component of Douglas-Fir tussock moth, *Orgyia pseudotsugata* (McDonnough) (Lepidoptera: Lymantridae). *J. Chem. Ecol.*, **23**, 19–34.

Gries, R., Khaskin, G., Clearwater, J., Hasman, D., Schaefer, P. W., Khaskin, E., Miroshnychenko, O., Hosking, G. and Gries, G. (2005). (Z,Z)-6,9-Heneicosadien-

11-one: major sex pheromone component of painted apple moth, *Teia anartoides*. *J. Chem. Ecol.*, **31**, 603–620.

Hai, T. V., Vang, L. V., Son, P. K., Inomata, S. and Ando, T. (2002). Sex attractants for moths of Vietnam: field attraction by synthetic lures baited with known lepidopteran pheromones. *J. Chem. Ecol.*, **28**, 1473–1481.

Hall, D. R., Beevor, P. S., Campon, D. G., Chamberlain, D. J., Cork, A., White, R., Almestre, A., Henneberry, T. J., Nandagopal, V., Wightman, J. A. and Rao, G. V. R. (1993). Identification and synthesis of new pheromones. *Bulletin de l'Organisation Internationale de Lutte Biologique/ Section Régionale Ouest Paléarctique* **16**, 1–9.

Hansson, B. S., Szöcs, G., Schmidt, F., Francke, W., Löfstedt, C. and Tóth, M. (1990). Electrophysiological and chemical analysis of the sex pheromone communication system of the mottled umber *Erannis defoliaria* (Lepidoptera: Geometridae). *J. Chem. Ecol.*, **16**, 1887–1897.

Haynes, K. F., Yeargan, K. V., Millar, J. G. and Chastain, B. B. (1996). Identification of the sex pheromone of *Tetanolita mynesalis* (Lepidoptera: Noctuidae), a prey species of a bolas spider, *Mastophora hutchinsoni*. *J. Chem. Ecol.*, **22**, 75–90.

Heath, R. R., Landolt, P. J., Leppla, N. C. and Dueben, B. D. (1988). Identification of a male-produced pheromone of *Anticarsia gemmatalis* (Hübner) (Lepidoptera: Noctuidae) attractive to conspecific males. *J. Chem. Ecol.*, **14**, 1121–1130.

Heath, R. R., Tumlinson, J. H, Leppla, N. C., McLaughlin, J. R., Dueben, B., Dundulis, E. and Guy, R. H. (1983). Identification of a sex pheromone produced by female velvetbean caterpillar moth. *J. Chem. Ecol.*, **9**, 645–656.

Hill, A. S., Kovalev, B. G., Nikolaeva, L. N. and Roelofs, W. L. (1982). Sex pheromone of the fall webworm moth, *Hyphantria cunea*. *J. Chem. Ecol.*, **8**, 383–396.

Hill, A. S. and Roelofs, W. L. (1981). Sex pheromone of the saltmarsh caterpillar moth, *Estigmene acrea*. *J. Chem. Ecol.*, **7**, 655–668.

Hoddle, M. S., Millar, J. G., Hoddle, C. D., Zou, Y., McElfresh, J. S. and Lesch, S. M. (2009). Field optimization of the sex pheromone of *Stenoma catenifer*: evaluation of lure types, trap height, male flight distances and number of traps needed per avocado orchard for detection. *J. Econ. Entomol.*, in press.

Hoddle, M. S., Millar, J. G., Hoddle, C. D., Zou, Y. and McElfresh, J. S. (2009). Synthesis and field evaluation of the sex pheromone of *Stenoma catenifer* (Lepidoptera: Elachistidae). *J. Econ. Entomol.*, **102**, 1460–1467.

Huang, W., Pulaski, S. P. and Meinwald, J. (1983). Synthesis of highly unsaturated insect pheromones: (Z,Z,Z)-1,3,6,9-heneicosatetraene and (Z,Z,Z)-1,3,6,9-nonadecatetraene. *J. Org. Chem.*, **48**, 2270–2274.

Jain, S. C., Dussourd, D. E., Conner, W. E., Eisner, T., Guerrero, A. and Meinwald, J. (1983). Polyene pheromone components from an arctiid moth (*Utetheisa ornatrix*): characterization and synthesis. *J. Org. Chem.*, **48**, 2266–2270.

Jurenka, R. A., Subchev, M., Abad, J. L., Choi, M. Y. and Fabrais, J. (2003). Sex pheromone biosynthetic pathway for disparlure in the gypsy moth, *Lymantria dispar*. *Proc. Natl. Acad. Sci. USA*, **100**, 809–814.

Jyothi, K. N., Prasuna, A. L. and Prasad, A. R. (2005). Evidence for presence of female produced pheromone components in male scent brush extract of castor semi-looper moth *Achaea janata* L. *Indian J. Exp. Biol.*, **43**, 335–341.

Kasang, G., Schneider, D. and Beroza, M. (1974). Biosynthesis of the sex pheromone disparlure by olefin-epoxide conversion. *Naturwissenschaften*, **61**, 130–131.

Kawai, T., Ohnishi, A., Suzuki, M. G., Fujii, T., Matsuoka, K., Kato, I., Matsumoto, S. and Ando, T. (2007a). Identification of a unique pheromonotropic neuropeptide including double FXPRL motifs from a geometrid species, *Ascotis selenaria cretacea*, which produces an epoxyalkenyl sex pheromone. *Insect Biochem. Mol. Biol.*, **37**, 330–337.

Kawai, T., Ohnishi, A., Suzuki, M. G., Fujii, T., Matsuoka, K., Kato, I., Matsumoto, S. and Ando, T. (2007b). Identification of a unique pheromonotropic neuropeptide including double FXPRL motifs from a geometrid species, *Ascotis selenaria cretacea*, which produces an epoxyalkenyl sex pheromone. [Erratum to document cited in CA147:068015]. *Insect Biochem. Mol. Biol.*, **37**, 1108.

Khrimian, A., Lance, D. R., Schwarz, M., Leonhardt, B. A. and Mastro, V. C. (2008). Sex pheromone of browntail moth, *Euproctis chrysorrhoea* (L.): Synthesis and field deployment. *J. Agric. Food Chem.*, **56**, 2452–2456.

Khrimian, A., Oliver, J. E., Hahn, R. C., Dees, N. H., White, J. and Mastro, V. C. (2004). Improved synthesis and deployment of (2S,3R)-2-(2Z,5Z-Octadienyl)-3-nonyloxirane, a pheromone of the pink gypsy moth, *Lymantria mathura*. *J. Agric. Food Chem.*, **52**, 2890–2895.

Knauf, W., Bestmann, H. J., and Vostrowsky, O. (1984). Responses of male winter moths (*Operophtera brumata*) to their sex attractant (3Z,6Z,9Z)-1,3,6,9-nonadecatetraene and to some structural analogues. *Entomol. Exp. Appl.*, **35**, 208–210.

Koenig, W. A. and Gehrcke, G. (1993). Gas chromatographic enantiomer separation with modified cyclodextrins: carboxylic acid esters and epoxides. *J. High Res. Chromatogr.*, **16**, 175–181.

Kovalev, B. G. and Nikolaeva, L. A. (1986). Attractiveness of some epoxy compounds for males of a tiger moth, *Phragmatobia fuliginosa* (Arctiidae), and of a geometrid moth, *Caustoloma* (*Therapis*) *flavicaria* (Geometriidae). *Zool. Zhurnal*, **56**, 802–804.

Kovalev, B. G., Vrkoc, Y., Fedoseev, N. Z., Weidenhoffer, Z. and Avdeeva, L. A. (1985). Isolation and identification of a component of the sex pheromone of *Orgyia gonostigma*. *Khim. Prir. Soedin.*, **102**, 277–278.

Krasnoff, S. B. and Roelofs, W. L. (1988). Sex pheromone released as an aerosol by the moth *Pyrrharctia isabella*. *Nature*, **333**, 263–264.

Krishnakumari, B., Prasuna, A. L., Jyothi, K. N., Valli, M. Y., Sighamony, S., Prasad, A. R. and Yadav, J. S. (1998). Behavioral and electrophysiological responses of *Achaea janata* Linn. (Lepidoptera: Noctuidae) males to synthetic female-produced sex pheromone components. *J. Entomol. Res.*, **22**, 197–202.

Kuenen, L. P. S., Mc Elflesh, J. S. and Millar, J. G. (2010). Identification of critical secondary components of the sex pheromone of the navel orange worm, *Amyellois transitella*. *J. Econ. Entomol.*, in press.

Lalanne-Cassou, B., Le Duchat d'Aubigny, J., Silvain, J.-F. and Déscoins, C. (1988). First trapping trials with the synthetic sex pheromone of *Mocis latipes* (Guenee) for trapping males in Guyana and Guadeloupe. *Colloques – Institut National de la Recherche Agronomique (Mediateurs Chim.: Comportement Syst. Lepidoptères, Appl. Agron.)*, **46**, 43–50.

Landolt, P. J. and Curtis, C. E. (1982). Interspecific sexual attraction between *Pyralis farinalis* L. and *Amyelois transitella* (Walker). *J. Kansas Ent. Soc.*, **55**, 248.

Landolt, P. J. and Heath, R. R. (1989). Lure composition, component ratio, and dose for trapping male *Mocis latipes* (Lepidoptera: Noctuidae) with synthetic sex pheromone. *J. Econ. Entomol.*, **82**, 307–309.

Landolt, P. J., Heath, R. R. and Leppla, N. C. (1986). (Z,Z,Z)-3,6,9-Eicosatriene and (Z,Z,Z)-3,6,9-heneicosatriene as sex pheromone components of a grass looper, *Mocis disseverans* (Lepidoptera: Noctuidae). *Environ. Entomol.*, **15**, 1272–1274.

Landolt, P.J., Tóth, M., Francke, W. and Mori, K. (1996). Attraction of moths (Lepidoptera: Arctiidae, Geometridae, Noctuidae) to enantiomers of several epoxydienes. *Fla. Entomol.*, **79**, 392–397.

Leal, W.S. (1998). Infrared and ultraviolet spectroscopy techniques. Chapter 6 in *Methods in Chemical Ecology, Vol. 1. Chemical Methods*, ed. J.G. Millar and K.F. Haynes. New York: Chapman and Hall, pp. 185–206.

Leal, W.S., Parra-Pedrazzoli, A.L., Kaissling, K.-E., Morgan, T.I., Zalom, F.G., Pesak, D.J., Dundulis, E.A., Burks, C.S. and Higbee, B.S. (2005). Unusual pheromone chemistry in the navel orangeworm: novel sex attractants and a behavioral antagonist. *Naturwissenschaften*, **92**, 139–146.

Lecheva, I. (1999). Possibilities of implementing synthetic sex pheromones against the geometrid *Erannis bajaria* (Lepidoptera: Geometridae). *Rasteniev`dni Nauki.*, **36**, 397–400.

Leonhardt, B.A., Mastro, V.C., Schwarz, M., Tang, J.D., Charlton, R.E., Pellegrini-Toole, A., Warthern, J.D. Jr., Schwalbe, C.P. and Cardé, R.T. (1991). Identification of sex pheromone of browntail moth, *Euproctis chrysorrhoea* (L.) (Lepidoptera: Lymantridae). *J. Chem. Ecol.*, **17**, 897–910.

Li, Z., Yao, E., Liu, T., Liu, Z., Wang, S., Zhu, H., Zhao, G. and Ren, Z. (1993). Structural elucidation of sex pheromone components of the geometrid *Semiothisa cinerearia* (Bremer et Gray) in China. *Chin. J. Chem.*, **11**, 251–256.

Lim, H., Park, K.C., Baker, T.C. and Greenfield, M.D. (2007). Perception of conspecific female pheromone stimulates female calling in an arctiid moth, *Utetheisa ornatrix*. *J. Chem. Ecol.*, **33**, 1257–1271.

Lizarraga, J.R. and Mori, K. (2001). Synthesis of (±)-leucomalure [(3Z,6R*,7S*,9R*,10S*)-*cis*-6,7-*cis*-9,10-diepoxy-3-henicosene], the major components of the female sex pheromone of the satin moth. *Nat. Prod. Lett.*, **15**, 89–92.

Maini, S. and Belifiori, D. (1996). *Noctua pronuba* and *Spilosoma luteum* males attraction to virgin females and synthetic sex pheromone of *Hyphantria cunea*. *Abstracts of the International Congress of Entomology, Firenze, 1996.* p. 190.

Marcelli, R., Montermini, A. and Maini, S. (1992). Preliminary trials with synthetic sex attractants of *Hyphantria cunea* Drury (Lepidoptera Arctiidae). *Inf. Fitopatol.*, **42**, 59–62.

Matsuoka, K., Tabunoki, H., Kawai, T., Ishikawa, S., Yamamoto, M., Sato, R. and Ando, T. (2006). Transport of a hydrophobic biosynthetic precursor by lipophorin in the hemolymph of a geometrid female moth which secretes an epoxyalkenyl sex pheromone. *Insect Biochem. Mol. Biol.*, **36**, 576–583.

Matsuoka, K., Yamamoto, M., Yamakawa, R., Muramatsu, M., Naka, H., Kondo, Y. and Ando, T. (2008). Identification of novel C20 and C22 trienoic acids from arctiid and geometrid female moths that produce polyenyl type II pheromone components. *J. Chem. Ecol.*, **34**, 1437–1445.

McDonough, L.M., Bailey, J.B., Hoffmann, M.P., Leonhardt, B.A., Brown, D.F., Smithhisler, C.L. and Olsen, K. (1986). *Sabulodes caberata* Guénée (Lepidoptera: Geometridae): components of its sex pheromone gland. *J. Chem. Ecol.*, **12**, 2107–2116.

McLaughlin, J.R. and Heath, R.R. (1989). Field trapping and observations of male velvetbean caterpillar moths and trapping of *Mocis* spp. (Lepidoptera: Noctuidae: Catacolinae) with calibrated formulations of sex pheromone. *Environ. Entomol.*, **18**, 933–938.

Mell, R. (1922). *Biologie und Systematik der Chinesichen Sphingiden.* Berlin: Friedländer.

Meyer, W. L. (1984). Sex pheromone chemistry and biology of some arctiid moths (Lepidoptera: Arctiidae): Enantiomeric differences in pheromone perception. *MS Thesis*, Cornell University, USA, 82 pp.

Millar, J. G. (2000). Polyene hydrocarbons and epoxides: a second major class of lepidopteran sex attractant pheromones. *Annu. Rev. Entomol.*, **45**, 575–604.

Millar, J. G., Giblin, M., Barton, D., Morrison, A. and Underhill, E. W. (1990c). Synthesis and field testing of enantiomers of 6Z,9Z-*cis*-3,4-epoxydienes as sex attractants for geometrid moths. Interactions of enantiomers and regioisomers. *J. Chem. Ecol.*, **16**, 2317–2339.

Millar, J. G., Giblin, M., Barton, D., Reynard, D. A., Neill, G. B. and Underhill, E. W. (1990d). Identification and field testing of female-produced sex pheromone components of the spring cankerworm, *Paleacrita vernata* Peck (Lepidoptera: Geometridae). *J. Chem. Ecol.*, **16**, 3393–3409.

Millar, J. G., Giblin, M., Barton, D. and Underhill, E. W. (1990a). 3Z,6Z,9Z-Trienes and unsaturated epoxides as sex attractants for geometrid moths. *J. Chem. Ecol.*, **16**, 2307–2316.

Millar, J. G., Giblin, M., Barton, D. and Underhill, E. W. (1990b). (3Z,6Z,9Z)-Nonadecatriene and enantiomers of (3Z,9Z)-*cis*-6,7-epoxy-nonadecadiene as sex attractants for two geometrid and one noctuid moth species. *J. Chem. Ecol.*, **16**, 2153–2166.

Millar, J. G., Giblin, M., Barton, D. and Underhill, E. W. (1991a). Chiral lepidopteran sex attractants: blends of optically active C20 and C21 diene epoxides as sex attractants for geometrid and noctuid moths (Lepidoptera). *Environ. Entomol.*, **20**, 450–457.

Millar, J. G., Giblin, M., Barton, D. and Underhill, E. W. (1991b). Synthesis and field screening of chiral monounsaturated epoxides as lepidopteran sex attractants and sex pheromone components. *J. Chem. Ecol.*, **17**, 911–929.

Millar, J. G., Giblin, M., Barton, D. and Underhill, E. W. (1992). Sex pheromone components of the geometrid moths *Lobophora nivigerata* and *Epirrhoe sperryi*. *J. Chem. Ecol.*, **18**, 1057–1068.

Millar, J. G., Giblin, M., Barton, D., Wong, J. W. and Underhill, E. W. (1991c). Sex attractants and sex pheromone components of noctuid moths *Euclidea cuspidea*, *Caenurgina distincta*, and geometrid moth *Eupithecia annulata*. *J. Chem. Ecol.*, **17**, 2095–2111.

Millar, J. G., Grant, G. G., McElfresh, J. S., Strong, W., Rudolph, C., Stein, J. D. and Moreira, J. A. (2005). (3Z,6Z,9Z,12Z,15Z)-pentacosapentaene, a key pheromone component of the fir coneworm moth, *Dioryctria abietivorella*. *J. Chem. Ecol.*, **31**, 1229–1234.

Millar, J. G., Hoddle, M., McElfresh, J. S., Zou, Y. and Hoddle, C. (2008). (9Z)-9,13-Tetradecadien-11-ynal, the sex pheromone of the avocado seed moth, *Stenoma catenifer. Tetrahedron Lett.*, **49**, 4820–4823.

Millar, J. G. and Underhill, E. W. (1986a). Short synthesis of 1,3Z,6Z,9Z-tetraene hydrocarbons. Lepidopteran sex attractants. *Can. J. Chem.*, **64**, 2427–2430.

Millar, J. G. and Underhill, E. W. (1986b). Synthesis of chiral bis-homoallylic epoxides. A new class of lepidopteran sex attractants. *J. Org. Chem.*, **51**, 4726–4728.

Millar, J. G., Underhill, E. W., Giblin, M. and Barton, D. (1987). Sex pheromone components of three species of *Semiothisa* (Geometridae), (Z,Z,Z)-3,6,9-heptadecatriene and two monoepoxydiene analogs. *J. Chem. Ecol.*, **13**, 1271–1283.

Miller, D. R., Millar, J. G., Grant, G. G., MacDonald, L. and DeBarr, G. L. (2009). (3Z, 6Z, 9Z, 12Z, 15Z)-Pentacosapentaene and (9Z, nE)-tetradecadienyl acetate: attractant lure blend for *Dioryctria ebeli* (Lepidoptera: Pyrallidae). *J. Entomol. Sci.*, in press.

Miyamoto, T., Yamamoto, M., Ono, A., Ohtani, K. and Ando, T. (1999). Substrate specificity of the epoxidation reaction in sex pheromone biosynthesis of the Japanese giant looper (Lepidoptera: Geometridae). *Insect Biochem. Mol. Biol.*, **29**, 63–69.

Mori, K. and Brevet, J.-L. (1991). Pheromone synthesis; CXXXIII. Synthesis of both the enantiomers of (3Z,6Z)-*cis*-6,7-epoxy-3,9-nonadecadiene, a pheromone component of *Erannis defoliaria*. *Synthesis*, **1991**, 1125–1129.

Mori, K. and Ebata, T. (1981). Synthesis of optically active pheromones with an epoxy ring, (+)-disparlure and the saltmarsh caterpillar moth pheromone [(Z,Z)-3,6-*cis*-9,10-epoxyheneicosadiene]. *Tetrahedron Lett.*, **22**, 4281–4282.

Mori, K. and Ebata, T. (1986). Synthesis of optically active pheromones with an epoxy ring, (+)-disparlure and both enantiomers of (3Z,6Z)-*cis*-9,10-epoxy-heneicosadiene. *Tetrahedron*, **42**, 3471–3478.

Mori, K. and Takeuchi, T. (1989). Synthesis of the enantiomers of (3Z,6Z)-*cis*-9,10-epoxy-heneicosatriene and (3Z,6Z)-*cis*-9,10-epoxy-eicosatriene, the new pheromone components of *Hyphantrea cunea*. *Liebigs Ann. Chem.*, **1989**, 453–457.

Muto, S. and Mori, K. (2003a). Synthesis of the four components of the female sex pheromone of the painted apple moth, *Teia anartoides*. *Biosci. Biotechnol. Biochem.*, **67**, 1559–1567.

Muto, S. and Mori, K. (2003b). Synthesis of all four stereoisomers of leucomalure, components of the female sex pheromone of the satin moth, *Leucoma salicis*. *Eur. J. Org. Chem.*, **2003**, 1300–1307.

Nakanishi, A. and Mori, K. (2005). New synthesis of the (3Z,6Z,9S,10R)-isomers of 9,10-epoxy-3,6-henicosadiene and 9,10-epoxy-1,3,6-henicosatriene, pheromone components of the female fall webworm moth, *Hyphantria cunea. Biosci. Biotechnol. Biochem.*, **69**, 1007–1013.

Odell, T. M., Xu, C.-H., Schaefer, P. W., Leonhardt, B. A., Yao, D.-F. and Wu, X.-D. (1992). Capture of gypsy moth, *Lymantria dispar* (L.), and *Lymantria mathura* (L.) males in traps baited with disparlure enantiomers and olefin precursor in the People's Republic of China. *J. Chem. Ecol.*, **18**, 2153–2159.

Ohmasa, Y., Wakamura, S., Kozai, S., Sugie, H., Horiike, M., Hirano, C. and Mori, S. (1991). Sex pheromone of the fruit-piercing moth, *Oraesia excavata* (Butler) (Lepidoptera: Noctuidae): isolation and identification. *Appl. Ent. Zool.*, **26**, 55–62.

Oliver, J. E., Dickens, J. C., Zlotina, M., Mastro, V. C. and Yurchenko, G. I. (1999). Sex attractant of the rosy Russian gypsy moth (*Lymantria mathura* Moore). *Z. Naturforsch. C*, **54**, 387–394.

Ostrauskas, H. (2004). Moths caught in pheromone traps for American white moth (*Hyphantria cunea* Dr.) (Arctiidae, Lepidoptera) in Lithuania during 2001. *Acta Zool. Lituanica*, **14**, 66–74.

Persoons, C. J., Vos, J. D., Yadav, J. S., Prasad, A. R., Sighomony, S., Jyothi, K. N. and Prasuna, A. L. (1993). Indo-Dutch cooperation on pheromones of Indian agricultural pest insects: sex pheromone components of *Diacrisia obliqua* (Arctiidae), *Achaea janata* (Noctuidae) and *Amsacta albistriga* (Arctiidae). *Bulletin de l'Organisation Internationale de Lutte Biologique/ Section Regionale Ouest Palearctique*, **16**, 136–140.

Pohnert, G. and Boland, W. (2000). Highly efficient one-pot double-Wittig approach to unsymmetrical (1Z,4Z,7Z)-homoconjugated trienes. *Eur. J. Org. Chem.*, **9**, 1821–1826.

Pu, G.-Q., Yamamoto, M., Takeuchi, Y., Yamazawa, H. and Ando, T. (1999). Resolution of epoxydienes by reversed phase chiral HPLC and its application to stereochemistry assignment of mulberry looper sex attractant. *J. Chem. Ecol.*, **25**, 1151–1162.

Qin, X. R., Ando, T., Yamamoto, M., Yamashita, M., Kusano, K. and Abe, H. (1997). Resolution of pheromonal epoxydienes by chiral HPLC, stereochemistry of separated enantiomers, and their field evaluation. *J. Chem. Ecol.*, **23**, 1403–1417.

Roelofs, W. and Bjostad, L. (1984). Biosynthesis of lepidopteran pheromones. *Bioorg. Chem.*, **12**, 279–298.

Roelofs, W. L., Hill, A. S., Linn, C. E., Meinwald, J., Jain, S. C., Herbert, H. J. and Smith, R. F. (1982). Sex pheromone of the winter moth, a geometrid with unusually low temperature precopulatory responses. *Science*, **217**, 657–659.

Rule, G. S. and Roelofs, W. L. (1989). Biosynthesis of sex pheromone components from linolenic acid in arctiid moths. *Arch. Insect Biochem. Physiol.*, **12**, 89–97.

Schal, C., Sevala, V. and Cardé, R. T. (1998). Novel and highly specific transport of a volatile sex pheromone by hemolymph lipophorin in moths. *Naturwissenschaften*, **85**, 339–342.

Schneider, D., Schulz, S., Kittmann, R. and Kanagaratnam, P. (1992). Pheromones and glandular structures of both sexes of the weed-defoliator moth *Pareuchaetes pseudoinsulata* Rego Barros (Lep., Arctiidae). *J. Appl. Entomol.*, **113**, 280–294.

Simmons, R. B., Conner, W. E. and Davidson, R. B. (1998). Identification of major components of the female pheromone glands of *Euchaetes egle* Drury (Arctiidae). *J. Lepid. Soc.*, **52**, 356–363.

Smith, R. G., Daterman, G. E. and Daves, G. D., Jr. (1975). Douglas-fir tussock moth: sex pheromone identification and synthesis. *Science*, **188**, 63–64.

Soulié, J., Boyer, T. and Lallemand, J.-Y. (1995a). Access to unsaturated chiral epoxides. Part II. Synthesis of a component of the sex pheromone of *Phragmatobia fuliginosa*. *Tetrahedron: Asymmetry*, **6**, 625–636.

Soulié, J., Péricaud, F. and Lallemand, J.-Y. (1995b). Access to unsaturated chiral epoxides. Part III. Synthesis of a component of the sex pheromone of *Boarmia selenaria*. *Tetrahedron: Asymmetry*, **6**, 1367–1374.

Soulié, J., Ta, C. and Lallemand, J.-Y. (1992). Access to unsaturated chiral epoxides. I. Bisallylic chiral epoxides. Application to the synthesis of lepidopteran pheromones. *Tetrahedron*, **48**, 443–452.

Steinbauer, M. J., Ostrand, F., Bellas, T. E., Nilsson, A., Andersson, F., Hedenstrom, E., Lacey, M. J. and Schiestl, F. P. (2004). Identification, synthesis and activity of sex pheromone gland components of the autumn gum moth (Lepidoptera: Geometridae), a defoliator of Eucalyptus. *Chemoecology*, **14**, 217–223.

Subchev, M. A., Ganev, J. A., Vostrowsky, O. and Bestmann, H. J. (1986). Screening and use of sex attractants in monitoring of geometrid moths in Bulgaria. *Z. Naturforsch.*, **41c**, 1082–1086.

Szöcs, G., Francke, W. and Tóth, M. (1998a). Winter "Love Story"… of geometrids. *Working Group Meeting, International Organization of Biological Control, Dachau, Germany.* Montfavet, France: IOBC Press, p. 15.

Szöcs, G., Otvos, I. and Sanders, C. (2001). *Erannis tiliaria* (Lepidoptera: Geometridae) males attracted to enantiomerically identical pheromone blend of *Erannis defoliaria*. *Can. Entomol.*, **133**, 297–299.

Szöcs, G., Plass, E., Francke, S., Francke, W., Zhu, J., Löfstedt, C., Sanders, C., Otvos, I., Subchev, M., Letcheva, I., Karpati, Z. and Tóth, M. (1998b). Homologous polyenes, or chiral epoxides: How are pheromones composed in winter geometrids (Lepidoptera)? *Abstracts, 2nd International Symposium on Insect Pheromones,* Wageningen, The Netherlands, 30 March–3April 1998; pp. 131–132.

Szöcs, G., Tóth, M., Bestmann, H. J. and Vostrowsky, O. (1984). A two-component sex attractant for males of the geometrid moth *Alsophila quadripunctata. Entomol. Exp. Appl.*, **36**, 287–291.

Szöcs, G., Tóth, M., Bestmann, H. J., Vostrowsky, O., Heath, R. R. and Tumlinson, J. H. (1987). Polyenic hydrocarbons as sex attractants for geometrids and amatids (Lepidoptera) found by field screening in Hungary. *Z. Naturforsch.*, **42c**, 165–168.

Szöcs, G., Tóth, M. and Francke, W. (2002). Sex attractants for lepidopterous species: pure enantiomers or racemates of epoxydienes. *Abstracts, International Society of Chemical Ecology 20th Annual Meeting,* Hamburg, August 3–7, 2002. p. 36.

Szöcs, G., Tóth, M., Francke, W., Francke, S. and Plass, E. (1996). Homologous polyenic hydrocarbons in the sex pheromones of winter geometrids (Lepidoptera). *Abstracts of the 13th Annual Meeting, International Society for Chemical Ecology,* Prague, 18–22 August, 1996; p. 178.

Szöcs, G., Tóth, M., Francke, W., Schmidt, F., Philipp, P., König, W. A., Mori, K., Hansson, B. S. and Löfstedt, C. (1993). Species discrimination in five species of winter-flying geometrids (Lepidoptera) based on chirality of semiochemicals and flight season. *J. Chem. Ecol.*, **19**, 2721–2735.

Szöcs, G., Tóth, M., Karpati, Z., Zhu, J., Löfstedt, C., Plass, E. and Francke, W. (2004). Identification of polyenic hydrocarbons from the northern winter moth, *Operophtera fagata*, and development of a species specific lure for pheromone traps. *Chemoecology*, **14**, 53–58.

Szöcs, G., Tóth, M. and Mori, K. (2005). Absolute configuration of the major sex pheromone component of the satin moth, *Leucoma salicis*, verified by field trapping test in Hungary. *Chemoecology*, **15**, 127–128.

Tamaki, Y., Sugie, H., Ando, Y., Yamashita, A., Ooya, S., Kamiwada, H., Suzuki, H. and Fukumoto, T. (1996). Insect attractants for *Aedia leucomelas*. Jap. *Kokai Tokkyo Koho*, **96**, 295–602.

Tan, Z. H., Gries, R., Gries, G., Lin, G. Q., Pu, G. Q., Slessor, K. N. and Li, J. (1996). Sex pheromone components of mulberry looper, *Heremophila atrilineata* Butler (Lepidoptera: Geometridae). *J. Chem. Ecol.*, **22**, 2263–2271.

Toshova, T., Subchev, M., Plass, E. and Francke, W. (2003). *Scoliopteryx libatrix* (L.) (Lep., Noctuidae) male reaction to the synthetic main sex pheromone component and its isomers – wind tunnel and field investigations. *J. Appl. Entomol.*, **127**, 195–199.

Tóth, M., Buser, H. R., Guerin, P. M., Arn, H., Schmidt, F., Francke, W. and Szöcs, G. (1992). *Abraxas grossulariata* L. (Lepidoptera: Geometridae): identification of (3Z,6Z,9Z)-3,6,9-heptadecatriene and (6Z,9Z)-6,9-cis-3,4-epoxyheptadecadiene in the female sex pheromone. *J. Chem. Ecol.*, **18**, 13–25.

Tóth, M., Buser, H. R., Pena, A., Arn, H., Mori, K., Takeuchi, T., Nikolaeva, L. N. and Kovalev, B. G. (1989). Identification of (3Z,6Z)-1,3,6–9,10-epoxyheneicosatriene and (3Z,6Z)-1,3,6–9,10-epoxyeicosatriene in the sex pheromone of *Hyphantria cunea. Tetrahedron Lett.*, **30**, 3405–3408.

Tóth, M., Szöcs, G., Francke, W., Guerin, P. M., Arn, H. and Schmid, A. (1987). Field activity of sex pheromone components of *Peribatodes rhomboidaria. Entomol. Exp. Appl.*, **44**,199–204.

Tóth, M., Szöcs, G., Francke, W., Schmidt, F., Philipp, P., Löfstedt, C, Hansson, B. S. and Farag, A. I. (1994). Pheromonal production of and response to optically active epoxydienes in some geometrid moths (Lepidoptera: Geometridae). *Z. Naturforsch.*, **49c**, 516–521.

Tóth, M., Szöcs, G., Löfstedt, C., Hansson, B. S., Schmidt, F. and Francke, W. (1991). Epoxyheptadecadienes identified as sex pheromone components of *Tephrina arenacearia* Hbn. (Lepidoptera: Geometridae). *Z. Naturforsch.*, **46c**, 257–263.

Underhill, E. W., Millar, J. G., Ring, R. A., Wong, J. W., Barton, D. and Giblin, M. (1987). Use of a sex attractant and an inhibitor for monitoring winter moth and Bruce spanworm populations. *J. Chem. Ecol.*, **13**, 1319–1330.

Underhill, E. W., Palaniswamy, P., Abrams, S. R., Bailey, B. K., Steck, W. F. and Chisholm, M. D. (1983). Triunsaturated hydrocarbons, sex pheromone components of *Caenurgina erechtea. J. Chem. Ecol.*, **9**, 1413–1423.

Van den Dool, H. and Kratz, P. D. (1963). A generalization of the retention index system including linear temperature programmed gas-liquid partition chromatography. *J. Chromatogr.*, **11**, 463–471.

Wakamura, S., Arakaki, N., Ono, H. and Yasui, H. (2001a). Identification of novel sex pheromone components from a tussock moth, *Euproctis pulverea. Entomol. Exp. Appl.*, **100**, 109–117.

Wakamura, S., Arakaki, N., Yamamoto, M., Hiradate, S., Yasui, H., Kinjo, K., Yasuda, T., Yamazawa, H. and Ando, T. (2005). Sex pheromone and related compounds in the Ishigaki and Okinawa strains of the tussock moth *Orgyia postica* (Walker) (Lepidoptera: Lymantriidae). *Biosci. Biotechnol. Biochem.*, **69**, 957–965.

Wakamura, S., Arakaki, N., Yamamoto, M., Hiradate, S., Yasui, H., Yasuda, T. and Ando, T. (2001b). Posticlure: a novel trans-epoxide as a sex pheromone component of the tussock moth, *Orgyia postica* (Walker). *Tetrahedron Lett.*, **42**, 687–689.

Wakamura, S., Arakaki, N., Yamazawa, H., Nakajima, N., Yamamoto, M. and Ando, T. (2002). Identification of epoxyhenicosadiene and novel diepoxy derivatives as sex pheromone components of the clear-winged tussock moth *Perina nuda. J. Chem. Ecol.*, **28**, 449–467.

Wang, S. and Zhang, A. (2007). Facile and efficient syntheses of (3Z,6Z,9Z)-3,6,9-nonadecatriene and homologues: pheromone and attractant components of Lepidoptera. *J. Agric. Food Chem.*, **55**, 6929–6932.

Watanabe, H., Tabunoki, H., Miura, N., Sato, R. and Ando, T. (2007). Analysis of odorant-binding proteins in antennae of a geometrid species, *Ascotis selenaria cretacea*, which produces lepidopteran Type II sex pheromone components. *Invert. Neuroscience*, **7**, 109–118.

Wei, W., Miyamoto, T., Endo, M., Murukawa, T., Pu, G. Q. and Ando, T. (2003). Polyunsaturated hydrocarbons in the hemolymph: biosynthetic precursors of epoxy pheromones of geometrid and arctiid moths. *Insect Biochem. Mol. Biol.*, **33**, 397–405.

Wei, W., Yamamoto, M., Asato, T., Fujii, T., Pu, G. Q. and Ando, T. (2004). Selectivity and neuroendocrine regulation of the precursor uptake by pheromone glands from hemolymph in geometrid female moths, which secrete epoxyalkenyl sex pheromones. *Insect Biochem. Mol. Biol.*, **34**, 1215–1224.

Wimalaratne, P. D. C. and Slessor, K. N. (2004). Chiral synthesis of (Z)-3-*cis*-6,7-*cis*-9,

10-diepoxyhenicosenes, sex pheromone components of the satin moth, *Leucoma salicis. J. Chem. Ecol.*, **30**, 1225–1244.

Witjaksono, Ohtani, K., Yamamoto, M., Miyamoto, T. and Ando, T. (1999). Responses of Japanese giant looper male moth to synthetic sex pheromone and related compounds. *J. Chem. Ecol.*, **25**, 1633–1642.

Witzgall, P., Lindblom, T., Bengtsson, M. and Tóth, M. (2004). The Pherolist. www-pherolist.slu.se/pherolist.php

Wong, J. W., Palaniswamy, P., Underhill, E. W., Steck, W. F. and Chisholm, M. D. (1984a). Novel sex pheromone components from the fall cankerworm moth, *Alsophila pometaria. J. Chem. Ecol.*, **10**, 463–473.

Wong, J. W., Palaniswamy, P., Underhill, E. W., Steck, W. F. and Chisholm, M. D. (1984b). Sex pheromone components of fall cankerworm moth, *Alsophila pometaria.* Synthesis and field trapping. *J. Chem. Ecol.*, **10**, 1579–1596.

Wong, J. W., Underhill, E. W., MacKenzie, S. L. and Chisholm, M. D. (1985). Sex attractants for geometrid and noctuid moths. Field trapping and electroantennographic responses to triene hydrocarbons and monoepoxydiene derivatives. *J. Chem. Ecol.*, **11**, 727–756.

Wunderer, H., Hansen, K., Bell, T. W., Schneider, D. and Meinwald, J. (1986). Sex pheromones of two Asian moths (*Creatonotos transiens, C. gangis*; Lepidoptera – Arctiidae): behavior, morphology, chemistry and electrophysiology. *Exp. Biol.*, **46**, 11–27.

Yadav, J. S., Valli, M. Y. and Prasad, A. R. (1998). Total synthesis of enantiomers of (3Z,6Z)-*cis*-9,10-epoxy 1,3,6-heneicosatriene – the pheromonal component of *Diacrisia obliqua. Tetrahedron*, **54**, 7551–7562.

Yadav, J. S., Valli, M. Y. and Prasad, A. R. (2001). Isolation, identification, synthesis, and bioefficacy of female *Diacrisia obliqua* (Arctiidae) sex pheromone blend. An Indian agricultural pest. *Pure Appl. Chem.*, **73**, 1157–1162.

Yamakawa, R., Do, N. D., Adachi, Y., Kinjo, M. and Ando, T. (2009). (6Z, 9Z, 12Z,)-6, 9, 12-octadecatriene and (3Z, 6Z, 9Z, 12Z)-3, 6, 9, 12-icosatetraene, the novel sex pheromones produced by emerald moths. *Tetrahedron Lett.* **50**, 4738–4740.

Yamamoto, M., Do, N. D., Komoya, T., Adachi, Y., Kinjo, M. and Ando, T. (2007). Abstract P-112, 4th Asia-Pacific Conference on Chemical Ecology, Tsukuba, Japan, Sept. 10–14, 2007.

Yamamoto, M., Kiso, M., Yamazawa, H., Takeuchi, J. and Ando, T. (2000). Identification of chiral sex pheromone secreted by giant geometrid moth, *Biston robustum* Butler. *J. Chem. Ecol.*, **26**, 2579–2590.

Yamamoto, M., Yamazawa, H., Nakajima, N. and Ando, T. (1999). A convenient preparation of optically active diepoxyhenicosene (leucomalure), lymantrid sex pheromone, by chiral HPLC. *Eur. J. Org. Chem.*, **7**, 1503–1506.

Yamamoto, M., Yamakawa, R., Oga, T., Takei, Y, Kinjo, M. and Ando, T. (2008). Synthesis and chemical characterization of hydrocarbons with a 6,9,11, 3,6,9,11, or 1,3,6,9-polyene system, pheromone candidates in lepidoptera. *J. Chem. Ecol.*, **34**, 1057–1064.

Yamazawa, H., Nakajima, N., Wakamura, S., Arakaki, N., Yamamoto, M. and Ando, T. (2001). Synthesis and characterization of diepoxyalkenes derived from (3Z,6Z,9Z)-trienes: lymantriid sex pheromones and their candidates. *J. Chem. Ecol.*, **27**, 2153–2167.

Yamazawa, H., Yamamoto, M., Karasawa, K. I., Pu, G.-Q. and Ando, T. (2003).

Characterization of geometrid sex pheromones by electrospray ionization time-of-flight mass spectrometry. *J. Mass Spectrom.*, **38**, 328–332.

Yao, E., Li, Z., Luo, Z., Shang, Z., Yin, K. and Hong, B. (1991). Report on structural elucidation of sex pheromone components of a tea pest (*Ectropis obliqua* Prout). *Prog. Nat. Sci.*, **1**, 566–569.

Yasui, H., Wakamura, S., Arakaki, N., Irei, H., Kiyuna, C., Ono, H., Yamazawa, H. and Ando, T. (2005). Identification of a sex pheromone component of the geometrid moth *Milionia basalis pryeri*. *J. Chem. Ecol.*, **31**, 647–656.

Yin, K. S., Hong, B. B., Shang, Z. Z., Yao, E. Y. and Li, Z. M. (1994). Sex pheromone of *Ectropis obliqua* Prout. *Prog. Nat. Sci.*, **4**, 732–740.

Zhang, Q. H. and Schlyter, F. (1996). High recaptures and long sampling range of pheromone traps for fall web worm moth *Hyphantria cunea* (Lepidoptera: Arctiidae) males. *J. Chem. Ecol.*, **22**, 1783–1796.

Zhang, Q. H., Schlyter, F., Chu, D., Ma, X. Y. and Ninomiya, Y. (1998). Diurnal and seasonal flight activity of males and population dynamics of fall webworm moth, *Hyphantria cunea* (Drury) (Lep., Arctiidae) monitored by pheromone traps. *J. Appl. Entomol.*, **122**, 523–532.

Zhang, Z. B., Wang, Z. M., Wang, Y. X., Liu, H. Q., Lei, G. X. and Shi, M. (1999). A facile synthetic method for chiral 1,2-epoxides and the total synthesis of chiral pheromone epoxides. *Tetrahedron: Asymmetry*, **10**, 837–840.

Zhang, Z. B., Wang, Z. M., Wang, Y. X., Liu, H. Q., Lei, G. X. and Shi, M. (2000). A simple synthetic method for chiral 1,2-epoxides and the total synthesis of a chiral pheromone epoxide. *J. Chem. Soc. Perkin 1*, **2000**, 53–57.

Zhu, J. W., Löfstedt, C., Philipp, P., Francke, W., Tammaru, T. and Haukioja, E. (1995). A sex pheromone component novel to the Geometridae identified from *Epirrita autumnata*. *Entomol. Exp. Appl.*, **75**, 159–164.

19

Volatile hydrocarbon pheromones from beetles

Robert J. Bartelt

The members of the Coleoptera, the largest insect order, are as diverse in their pheromone chemistry as they are in their appearance and biology. As in most other insect groups, long-range beetle pheromones tend to be oxygenated compounds, but some species use hydrocarbons. The hydrocarbon pheromones discussed here are distinguished from typical, cuticular hydrocarbons in that they are sex-specific, are small enough to be volatile (e.g., <25 carbons), and, in almost all cases, have been shown to elicit long-range behavioral responses. Most that have so far been encountered are male-produced and are attractive to both sexes, but there are exceptions. These hydrocarbons have proven to be chemically and biologically interesting, and some have become important in practical pest management. Biochemically, they include both polyketides and terpenoids. There is an extensive literature involving the hydrocarbon pheromones of sap beetles, and an effort is made here to give comprehensive coverage of this research. A smaller number of articles have been published about volatile hydrocarbon pheromones in other beetle families, and this information is covered in the latter portion of the chapter.

Sap beetles

Members of two sap beetle genera, *Carpophilus* and *Colopterus* (Coleoptera, Nitidulidae), have been found to use volatile hydrocarbons as long-range pheromones. These beetles are small in size (<5 mm) and cryptic in habits, but they are frequently abundant, feeding mostly on fruits or other plant materials that are ripening, decomposing, or fermenting. Several species are cosmopolitan and infest a variety of agricultural products, both before and after harvest. Some affected crops are figs, dates, stone fruits, and corn. Both adults and larvae cause damage by direct feeding (Hinton, 1945; Lindgren and Vincent, 1953). The beetles also introduce harmful microorganisms, such as those that produce mycotoxins (reviews by Dowd, 1991, 1995; Wicklow, 1991) and cause fruit breakdown (Kable, 1969). Some species transmit tree diseases such as oak wilt (Dorsey and Leach, 1956). On the other hand, nitidulids are valuable pollinators of tropical fruits such as sugar apples (*Annona* spp.) (Nagel *et al.*, 1989). Reference works describing these beetles include

Hinton (1945) and Connell (1956, 1991). Williams *et al.* (1983) compiled a useful bibliography for *Carpophilus*.

Pheromones have now been studied in ten *Carpophilus* species. Four of these belong to the nitidulid complex attacking fruits in southern California but also occur widely throughout the world; these are *Ca. hemipterus* (L.), *Ca. mutilatus* Erichson, *Ca. freemani* Dobson, and *Ca. obsoletus* Erichson. *Ca. lugubris* Murray is a particularly common sap beetle in corn in the Midwestern United States; *Ca. antiquus* Melsheimer, *Ca. brachypterus* Say, and *Ca. sayi* Parsons also occur in this region but are less common. *Ca. dimidiatus* (F.) is a major member of the sap beetle complex affecting corn in the southeastern United States (where *Ca. freemani* and *Ca. mutilatus* are abundant as well). Finally, *Ca. davidsoni* Dobson is an Australian species that is currently a major pest of peaches, apricots, and other stone fruits. The pheromone of *Colopterus truncatus* Randall, a North American vector of oak wilt disease, has also been investigated.

Nitidulid pheromone research has included laboratory and field behavior, analytical and synthetic chemistry, physiological and biochemical studies, and practical uses in agriculture. Below, laboratory research is summarized first, followed by field studies.

First evidence of pheromones and wind tunnel bioassay

The first evidence for long-range pheromones in nitidulids was for *Ca. hemipterus* and was obtained in a laboratory wind tunnel (Bartelt *et al.*, 1990a). Flying beetles of both sexes were clearly more attracted to containers with males feeding on artificial diet than to containers with females on diet or to diet alone. This result was corroborated by tests with volatiles collected from feeding males and females onto porous-polymer filters.

Wind-tunnel tests on collected volatiles also provided the initial pheromone information for five other species: *Ca. lugubris* (Bartelt *et al.*, 1991), *Ca. freemani* (Bartelt *et al.*, 1990b), *Ca. obsoletus* (Petroski *et al.*, 1994a), *Ca. mutilatus* (Bartelt *et al.*, 1993a), and *Ca. davidsoni* (Bartelt and James, 1994). The general procedure was to stock a warm, well-lit wind tunnel with at least several hundred mixed-sex beetles. The beetles initially rested in corners but began to fly about and become responsive to volatiles after several hours without food, and dozens would be in the air at any instant. Two odor sources to be compared (usually treated pieces of filter paper) were hung side by side in the upwind end of the wind tunnel, and the numbers of beetles alighting on these during a set period of time (typically 2–5 min) were recorded. Beetles encountering an active odor plume would approach the source with a characteristic upwind, casting flight and alight, often remaining on the source until the test was over. As many as 30–50 tests could be conducted with one group of beetles during a bioassay session of several hours, allowing a break of 1–5 min between tests. While the assay procedure did not permit detailed observation of individual behavior, it did allow rapid evaluation of samples, and the en masse approach of beetles to an odor source was similar to field responses subsequently seen around pheromone traps.

Pheromone identification

The volatile collections from feeding beetles and derived chromatographic fractions were analyzed by gas chromatography (GC). The key objective was to associate the pheromone activity of samples with individual GC peaks, and consistently encountered sex-specific compounds were regarded as possible pheromone components. Volatile collections, chromatographic fractions, and purified compounds were evaluated in the wind tunnel to verify the association between particular GC peaks and pheromone activity whenever possible. However, only the male/female GC comparisons could be used for *Ca. antiquus*, *Ca. brachypterus*, *Ca. sayi*, *Ca. dimidiatus*, and *Co. truncatus* because of limited beetle numbers, and verification of activity had to await field testing of synthetic compounds.

Chromatographic pheromone purification usually involved two steps (Bartelt *et al.*, 1990a). The first was column chromatography or high-performance liquid chromatography (HPLC) on silica, which separated the volatile collections into fractions by polarity. The second step was HPLC on a silica column coated with silver nitrate, which separated compounds by numbers of double bonds and substitution patterns. For every species, the least polar (hexane) fraction from the silica column contained at least one male-specific compound (by GC), suggesting each had a hydrocarbon pheromone. Long retention by silver-nitrate HPLC indicated that these had multiple double bonds, and this chromatographic step nicely separated the male-specific compounds from more saturated, inert impurities (e.g., Bartelt *et al.*, 1990b, 1992b; Bartelt and James, 1994; Bartelt and Weisleder, 1996).

Monitoring the purification of the first pheromone (*Ca. hemipterus*) by wind-tunnel bioassay was complicated by a synergistic relationship with food odors. It was difficult to demonstrate the activity of the pheromone once it became chromatographically separated from the food scent that was also present in the volatile collections (Bartelt *et al.*, 1990a). Consequently, food scent, or a synthetic version of this, was routinely added to each bioassay treatment, and the food scent itself became the control when pheromone activity was being evaluated. A variety of esters, alcohols, and other compounds were found to synergize the pheromone of *Ca. hemipterus* (Dowd and Bartelt, 1991). Synergistic effects were seen with all the species.

Identification of the pheromone components always relied on mass spectrometry (MS) and sometimes on proton nuclear magnetic resonance (NMR) (Bartelt *et al.*, 1990a,b, 1993a,b) and ultraviolet (UV) spectroscopy (Bartelt *et al.*, 1990a) and on MS of saturated derivatives (prepared by catalytic hydrogenation) (Bartelt *et al.*, 1990a,b, 1991, 1992b). The approach was to use the available (sometimes limited) spectral and chromatographic data to propose likely structures and then to synthesize these alternatives until one was found to match the natural compound by all criteria. For the initial species, a large amount of spectral and chemical data was required to solve the unusual and somewhat complicated pheromone structures. However, as the library of model compounds grew and certain biosynthetic generalizations became evident, it became easier to "guess" structures and synthesize the proper compounds from only the GC retention times and mass spectra.

Summary of male-specific hydrocarbons

Twenty-three male-specific conjugated triene and tetraene hydrocarbons were identified from the ten *Carpophilus* and one *Colopterus* species (Table 19.1). These include relatively abundant components with clear activity in at least one species and also minor components of uncertain biological importance. Species often share compounds, but the overall combinations and ratios are usually distinctive. All compounds have their alkyl branches on alternate carbons, and all double bonds have the *E* configuration. Each of these hydrocarbons was previously unknown. Small amounts of *Z* isomers were found in natural samples (Bartelt *et al.*, 1990a, 1992b), but these were believed to be artifacts because the "freshest" samples always had smaller amounts and because pure synthetic all-*E* tetraenes isomerized slightly during storage to produce *Z* isomers.

In seven of the species studied, the main pheromone components were tetraenes. The major component and one minor component in *Ca. hemipterus* were identified as tetraenes **5** and **7**, respectively (Bartelt *et al.*, 1990a). Bartelt *et al.* (1992b) identified nine additional tetraenes and one triene. Williams *et al.* (1995) discovered that *Ca. brachypterus* males produce five of the tetraenes also present in *Ca. hemipterus* (**5–8** and **18**), but in slightly different proportions. The only male-specific compound encountered from *Ca. obsoletus* was tetraene **7**, one of the minor components from *Ca. hemipterus* and *Ca. brachypterus* (Petroski *et al.*, 1994a). The major component of the *Ca. lugubris* pheromone was identified as tetraene **8** (Bartelt *et al.*, 1991), which again was a minor component in *Ca. hemipterus*. Subsequently, *Ca. lugubris* was found to emit a minor amount of tetraene **7** (Bartelt *et al.*, 1993b) and even smaller amounts of tetraenes **21** and **22** (Bartelt *et al.*, 2004). *Ca. sayi*, which is difficult to distinguish from *Ca. lugubris* by appearance, also has the same four tetraenes (**7**, **8**, **21** and **22**) and in similar proportions, but *Ca. sayi* has trienes **2** and **3**, in addition. The 17-carbon tetraene **9** is the only known pheromone component of *Ca. antiquus* (Bartelt *et al.*, 1993b), and it is also the major component of *Ca. dimidiatus*, which also has a minor amount of the 18-carbon tetraene **23** (Bartelt *et al.*, 1995b).

In the other four species, the main components were trienes. Eight male-specific hydrocarbons were identified from *Ca. freemani*, the most abundant of which were triene **2** and tetraene **8** (Bartelt *et al.*, 1990b). The two most abundant components of *Ca. davidsoni* were likewise **2** and **8**, but the ratio was different from that in *Ca. freemani*. Altogether, 15 trienes and tetraenes were identified from *Ca. davidsoni*, at levels spanning nearly five orders of magnitude (Bartelt and James, 1994; Bartelt and Weisleder, 1996). The major and minor pheromone components of *Ca. mutilatus* were found to be trienes **4** and **3**, respectively (Bartelt *et al.*, 1993a). Both occur in very different ratios in *Ca. freemani* and *Ca. davidsoni*. Male *Ca. mutilatus* also emit tetradecanal (Bartelt *et al.*, 1993a; Bartelt and Weisleder, 1996), which was not encountered in any of the other nitidulids. Two trienes (**1** and **10**) and two tetraenes (**5** and **18**) were identified from male *Co. truncatus* (Cossé and Bartelt, 2000). Major triene **1** was not found in any other species, but the three minor components had been encountered previously. GC with electroantennographic detection (GC-EAD) was used to aid pheromone analysis for *Co. truncatus*.

Table 19.1 *Structures and relative abundances of male-specific hydrocarbons in 11 nitidulid species.*[1]

Structure	Species										
	hm	br	ob	lg	sa	dv	fr	mt	an	dm	tr
Components with demonstrated activity for at least one species											
1	-	-	-	-	-	-	-	-	-	-	**100**
2	3	-	-	-	2	**100**	**100**	-	-	-	-
3	-	-	-	-	4	7	2	**10**	-	-	-
4	-	-	-	-	-	1	<1	**100**	-	-	-
5	**100**	**100**	-	-	-	<1	-	-	-	-	3
6	8	11	-	-	-	<1	-	-	-	-	-
7	13	6	**100**	14	18	9	<1	-	-	-	-
8	2	3	-	**100**	**100**	31	4	-	-	-	-
9	-	-	-	-	-	<1	-	-	**100**	**100**	-
Minor components of uncertain biological importance											
10	-	-	-	-	-	<1	-	-	-	-	2
11	-	-	-	-	-	<1	<1	-	-	-	-
12	-	-	-	-	-	<1	1	-	-	-	-
13	-	-	-	-	-	<1	-	-	-	-	-
14	3	-	-	-	-	-	-	-	-	-	-
15	<1	-	-	-	-	-	-	-	-	-	-
16	<1	-	-	-	-	-	-	-	-	-	-
17	-	-	-	-	-	<1	-	-	-	-	-
18	6	4	-	-	-	-	-	-	-	-	<1
19	1	-	-	-	-	-	-	-	-	-	-
20	1	-	-	-	-	-	-	-	-	-	-
21	1	-	-	<1	2	<1	-	-	-	-	-
22	-	-	-	2	4	<1	<1	-	-	-	-
23	-	-	-	-	-	-	-	-	-	7	-

[1]For each species, components normally used in field baits shown in bold type. Abbreviations: hm = *Ca. hemipterus*, br = *Ca. brachypterus*, ob = *Ca. obsoletus*, lg = *Ca. lugubris*, sa = *Ca. sayi*, dv = *Ca. davidsoni*, fr = *Ca. freemani*, mt = *Ca. mutilatus*, an = *Ca. antiquus*, dm = *Ca. dimidiatus*, tr = *Co. truncatus*.

Synthesis and spectral properties

A general method for synthesizing the 23 compounds in Table 19.1 is illustrated in Figure 19.1 (Bartelt *et al.*, 1990c, 1993a,b). The starting material was always an aldehyde. One or more Wittig–Horner condensations each extended the chain by two carbons, introduced an *E* double bond and a methyl or ethyl branch, and provided a functional group (an ester) at which subsequent elaboration would be directed. This ester was reduced to an alcohol with lithium aluminum hydride and the resulting alcohol was oxidized to an aldehyde with

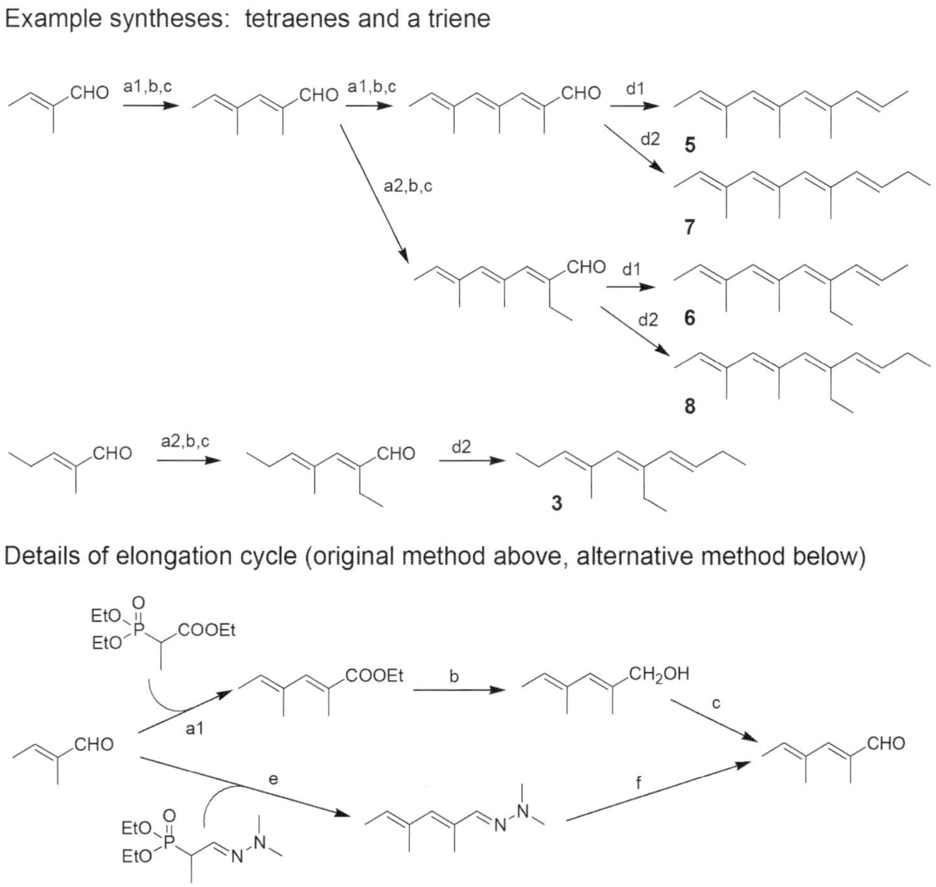

Example syntheses: tetraenes and a triene

Details of elongation cycle (original method above, alternative method below)

Figure 19.1 Synthetic scheme for example pheromone components. Abbreviations: a1 and a2 = Wittig–Horner condensations with triethyl 2-phosphonopropionate and triethyl 2-phosphonobutyrate, respectively; b = reduction of ester with lithium aluminum hydride; c = partial oxidation of alcohol with manganese dioxide to aldehyde; d1 and d2 = Wittig condensations with ethyltriphenylphosphonium bromide and propyltriphenylphosphonium bromide, respectively; e = condensation with dimethylhydrazone phosphonate reagent; f = hydrolysis under acidic conditions. Compound numbers are as in Table 19.1.

manganese dioxide to prepare for the next chain-extending cycle. The final reaction was a Wittig condensation that created the final, di-substituted double bond and attached a terminal alkyl group of proper size. Improved, alternative procedures for the final Wittig reaction with greater *E* selectivity were subsequently reported (Petroski, 1997; Petroski and Bartelt, 2007).

Recently, tetraene **8** was prepared using an alternative chain-elongation scheme that involves dimethylhydrazone phosphonate reagents (Figure 19.1) instead of the usual carboxylic ester phosphonates (Petroski and Bartelt, 2007). With this method the chain-terminating functional group is always in the oxidation state of an aldehyde and does not require the usual reduction and partial oxidation reactions. Instead, only a hydrolysis step is required to release the aldehyde function after chain elongation (Figure 19.1).

Each synthetic all-*E* hydrocarbon has been characterized with respect to GC retention (on a non-polar DB-1 capillary column), mass spectrum, and proton NMR spectrum. The chromatographic and spectral data are located in the same references as the identifications and syntheses. Interpretation of mass spectra of these compounds has been discussed (Bartelt *et al.*, 1992b). Proton NMR spectra are diagnostic for the compounds and were invaluable in confirming the structures of the synthetic hydrocarbons. Nuclear Overhauser NMR experiments allowed the configurations at the tri-substituted double bonds to be firmly established (Bartelt *et al.*, 1990a, 1991, 1992b). Carbon-13 spectra were obtained for compounds **2** (Petroski *et al.*, 1994b), **4**, **7**, and **8** (Bartelt and Weisleder, 1996), and **5–8** (Bartelt *et al.*, 1990c).

Laboratory activity of synthetic pheromones

The activity of synthetic hydrocarbons was established in the wind tunnel for six of the eleven species (references given in Table 19.2). The most abundant components in each species were invariably attractive. However, many of the less abundant male-specific hydrocarbons were not active, either alone or in combination with other attractants and regardless of dose. Thus, of the 12 male-specific hydrocarbons in *Ca. hemipterus*, only tetraenes **5**, **6**, **7**, and **8** (Table 19.1) showed activity (Bartelt *et al.*, 1992b). Tetraenes **5** and **7** were synergistic with each other and essentially accounted for the attractiveness of the whole male-derived hydrocarbon mixture (Bartelt *et al.*, 1990a). Similarly, two of the eight identified compounds from *Ca. freemani*, triene **2** and tetraene **8**, were synergistic and fully accounted for the activity of the male hydrocarbons (Bartelt *et al.*, 1990b).

The *Z* isomers were generally not active in the wind-tunnel bioassays (Bartelt *et al.*, 1990a, 1992b), but large amounts of the synthetic (2*E*,4*E*,6*Z*) isomer of triene **2**, which was never found in natural samples, did partially inhibit the response of *Ca. freemani* to its major pheromone component (all-*E* **2**) in the wind tunnel (Petroski and Weisleder, 1997). Petroski and Weisleder (1999) showed that certain unnatural, cyclopropyl or methoxy analogs of triene **2** could greatly reduce laboratory responses of *Ca. freemani* to its pheromone.

Table 19.2 *Summary of literature demonstrating activity of nitidulid pheromones versus appropriate controls, and ancillary research issues.*

Pheromone	Bioassay	Field test location	Other issues	Reference
Ca. hemipterus	wt	-	chem	Bartelt *et al.*, 1990a
	wt	-	chem	Bartelt *et al.*, 1992b
	f	California	cr, tp, seas	Bartelt *et al.*, 1992a
	f	Australia	cr, seas	James *et al.*, 1993
	f	Ohio	cr, tp	Williams *et al.*, 1993
	f	Israel	cr, tp	Blumberg *et al.*, 1993
	f	California	cr, tp	Bartelt *et al.*, 1994a
	f	Australia	cr	James *et al.*, 1994
	f	California	cr, seas	Bartelt *et al.*, 1994b
	f	Australia	cr	Bartelt and James, 1994
	f	California	mult, cr	Bartelt *et al.*, 1995a
	f	Ohio	cr	Williams *et al.*, 1995
	f	South Carolina	cr, seas	Bartelt *et al.*, 1995b
Ca. brachypterus	f	Ohio	chem, cr	Williams *et al.*, 1995
Ca. obsoletus	wt, f	California	chem, seas	Petroski *et al.*, 1994a
	f	California	cr, seas	Bartelt *et al.*, 1994b
	f	California	cr	Bartelt *et al.*, 1995a
	f	Brazil	cr, wh	Abreu, 1997
Ca. lugubris	wt, f	Illinois	chem	Bartelt *et al.*, 1991
	f	Ohio	tp	Lin *et al.*, 1992
	f	Ohio	cr, tp	Williams *et al.*, 1993
	f	Illinois	cr	Bartelt *et al.*, 1993b
	f	Ohio	cr	Williams *et al.*, 1995
Ca. sayi	f	Minnesota	chem, cr, tp	Bartelt *et al.*, 2004
Ca. davidsoni	wt, f	Australia	chem, cr	Bartelt and James, 1994
	f	Australia	cr, seas	James *et al.*, 1994
Ca. freemani	wt, f	California	chem, tp	Bartelt *et al.*, 1990b
	f	Ohio	cr, tp	Williams *et al.*, 1993
	f	Illinois	cr	Bartelt *et al.*, 1993b
	f	California	cr, seas	Bartelt *et al.*, 1994b
	f	California	mult, cr	Bartelt *et al.*, 1995a
	f	Ohio	cr	Williams *et al.*, 1995
	f	South Carolina	cr, seas	Bartelt *et al.*, 1995b
Ca. mutilatus	wt, f	California	chem, seas	Bartelt *et al.*, 1993a
	f	Australia	cr, seas	James *et al.*, 1993
	f	Israel	cr, tp	Blumberg *et al.*, 1993
	f	California	tp	Bartelt *et al.*, 1994a

Table 19.2 (*cont.*)

Pheromone	Bioassay	Field test location	Other issues	Reference
	f	Australia	cr	James *et al.*, 1994
	f	California	cr, seas	Bartelt *et al.*, 1994b
	f	Australia	cr	Bartelt and James, 1994
	f	California	mult, cr	Bartelt *et al.*, 1995a
	f	South Carolina	cr, seas	Bartelt *et al.*, 1995b
Ca. antiquus	f	Illinois	chem, cr	Bartelt *et al.*, 1993b
	f	Ohio	cr	Williams *et al.*, 1995
Ca. dimidiatus	f	South Carolina	chem, cr, seas	Bartelt *et al.*, 1995b
	f	Brazil	cr, wh	Abreu, 1997
Co. truncatus	f	Illinois	chem, cr	Cossé and Bartelt, 2000
	f	Illinois, Minnesota	cr, tp, seas	Kyhl *et al.*, 2002

Bioassay: wt = wind tunnel, f = field traps. Other issues: chem = pheromone chemistry, mult = multi-species lures, cr = cross attraction, seas = seasonal patterns, tp = trapping parameters such as height, type, pheromone dose, pheromone formulation/aging, or host-related synergist, wh = warehouse environment.

Factors affecting pheromone emission

Pheromone emission from *Carpophilus* males usually began 1–4 days after placement of adults onto diet medium (Bartelt and James, 1994; Bartelt *et al.*, 1993b; Nardi *et al.*, 1996). Separated males produced pheromone at far higher rates than males in groups (Petroski *et al.*, 1994a; Bartelt and James, 1994; Bartelt *et al.*, 1993b, 1995b). For example, mean emission from single *Ca. dimidiatus* males in flasks with artificial diet was 0.6 µg per day, but the amount was only 0.001 µg per male per day from groups of 50 males (Bartelt *et al.*, 1995b). In *Ca. davidsoni*, pheromone composition as well as pheromone amount varied with beetle numbers. The ratio of tetraenes to trienes in the pheromone from groups of males was only about 20% as high as from individual males (Bartelt and James, 1994). Finally, mating can influence pheromone production; emission from individual *Ca. freemani* males decreased by about 90% within one day of introducing females (Nardi *et al.*, 1996). Typical pheromone production rates for isolated *Carpophilus* males on artificial diet were in the range of 0.1 to 3 µg per day (Petroski *et al.*, 1994a; Bartelt and James, 1994; Bartelt *et al.*, 1995b; and unpublished data).

Site of pheromone production

The site of pheromone production in male *Ca. freemani* was found to be the posterior portion of the abdomen (Dowd and Bartelt, 1993). Microscopic examination revealed a male-specific tissue within the abdomen that was composed of very large round cells (100 µm

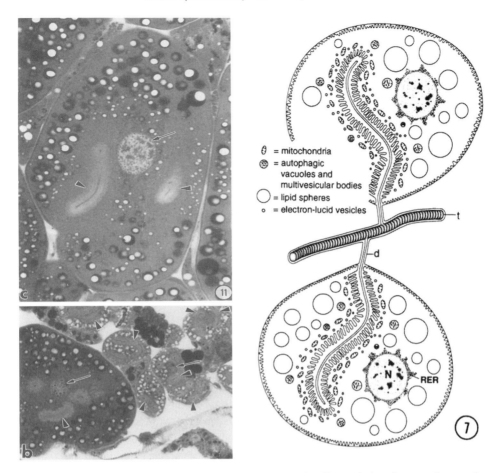

= mitochondria

= autophagic vacuoles and multivesicular bodies

= lipid spheres

o = electron-lucid vesicles

Figure 19.2 Pheromone secreting cells from *Ca. freemani* males. Transmission electron micrographs are shown at left. The upper micrograph is from a male that was emitting pheromone on the day the sample was prepared; cell diameter is about 100 μm. The lower micrograph is from a male that ceased emitting pheromone after mating and shows cells in various stages of shrinkage. Both pictures are at the same scale. Arrows indicate nuclei, and wedges point to portions of the secretory "end apparatus". Diagrammatic view of the pheromone-secreting cells is shown at right. Abbreviations: N = nucleus, RER = rough endoplasmic reticulum, t = tracheal branch, and d = ductule, which is the outlet from the end apparatus that transits each cell. Drawing by J. B. Nardi; micrographs and drawing reprinted with permission from Nardi *et al.*, 1996.

diameter). These cells were connected directly to the tracheal system (Dowd and Bartelt, 1993). Electron microscopy revealed cell structures and organelles typical for secretory cells and demonstrated that the secretory "end apparatus" of each cell was connected by a fine duct to a tracheal branch (Nardi *et al.*, 1996). Thus, pheromone emission is apparently through the tracheal system. The secretory cells were about ten times larger in diameter in beetles emitting pheromone than in those that were not. These features are shown in Figure 19.2.

Pheromone biosynthesis and composition

The pheromone hydrocarbon structures suggested a polyketide biosynthesis involving acetate (Ac), propionate (Pr), and butyrate (Bu) acyl units (Bartelt *et al.*, 1990b, 1991, 1992b) (Figure 19.3). This suggestion was verified experimentally with 12-carbon triene **2** (Table 19.1) from *Ca. freemani* (Petroski *et al.*, 1994b) and later with the tetraenes **7** and **8** of *Ca. davidsoni* and triene **4** of *Ca. mutilatus* (Bartelt and Weisleder, 1996). The final acyl unit loses its carboxyl carbon during biosynthesis (Bartelt and Weisleder, 1996). The bio-synthetic experiments exploited the tolerance of these beetles to large amounts of organic acids in their diets. Acids labeled with deuterium or carbon-13 became incorporated into the pheromones of the beetles, and locations of the heavy isotope atoms were then deter-mined by NMR and MS. The biochemical reactions were proposed to be as in fatty acid biosynthesis, except that the double bonds are not reduced (Petroski *et al.*, 1994b). The actual biochemical mechanisms remain unstudied.

Generalizing, the following biosynthetic "rules" apply to all 23 structures in Table 19.1:

(1) The trienes are composed of four acyl biosynthetic units, while the tetraenes have five.
(2) The first biosynthetic unit may be either Ac, Pr, or Bu.
(3) The second unit is always Pr.
(4) The third, fourth (and if a tetraene, the fifth) units may be Pr or Bu.
(5) The carboxyl carbon of the final acyl unit is removed.

Another biosynthetic issue is the origin of complex hydrocarbon blends, such as the 11 different tetraenes of *Ca. hemipterus* (Figure 19.3). While there could be separate biosyn-thetic systems for each of the *Ca. hemipterus* tetraenes, a more parsimonious explanation, which fits the data, is that a single biosynthetic system exists with imperfect selectivity for acyl units (Bartelt *et al.*, 1992b). If the most abundant tetraene, **5**, represents the "normal" product, then tetraenes **6**, **7**, **14**, and **18** represent instances of one acyl substitution (or bio-synthetic "mistake"). There are six possible ways in which two of these substitutions could exist in one compound, and these are represented by **8**, **15**, **16**, **19**, **20**, and **21**. Occurrence of two substitutions would be rarer than just one, and the observed abundances reflected this expectation. Tetraenes with three or four substitutions would be even rarer, and these were not detected. Substitutions were never observed for the second acyl unit, which was unfailingly propionate. Related arguments can be made for the patterns of hydrocarbons in *Ca. freemani* and *Ca. davidsoni* (Bartelt *et al.*, 1990b; Bartelt and Weisleder, 1996).

The functional pheromone of *Ca. hemipterus* appears to be a subset of the male-specific hydrocarbons (**5–8** in Figure 19.3, Bartelt *et al.*, 1992b). The compounds of *Ca. hemipterus* that were active had structural features in common; as drawn in Figure 19.3, the left-hand portions of the active compounds were identical. Petroski and Vaz (1995) used computer-based molecular modeling to correlate molecular shape with biological activity.

If some of the male-specific hydrocarbons are merely biosynthetic artifacts, then lack of behavioral activity in some cases would not be surprising. However, the inactive structures would represent a reservoir of male-specific compounds that could become pheromone

Aspects of pheromone biosynthesis

Constituent acyl units and final locations of their carbons for tetraene 8

Proposed steps for chain elaboration with acyl units

Biosynthetic and bioassay summary for *Ca. hemipterus* tetraenes

1st	2nd	3rd	4th	5th	Structure	Relative abundance	Bioassay activity
		Acyl unit				Relative	Bioassay
Main pattern of acyl units (tetraene **5**)							
Ac	Pr	Pr	Pr	Pr	**5**	100	Yes
One acyl substitution, relative to **5**							
Pr*	Pr	Pr	Pr	Pr	**8**	6	No
Ac	Pr	Bu*	Pr	Pr	**14**	3	No
Ac	Pr	Pr	Bu*	Pr	**6**	8	Yes
Ac	Pr	Pr	Pr	Bu*	**7**	13	Yes
Two acyl substitutions, relative to **5**							
Pr*	Pr	Bu*	Pr	Pr	**19**	1	No
Pr*	Pr	Pr	Bu*	Pr	**20**	1	No
Pr*	Pr	Pr	Pr	Bu*	**21**	1	No
Ac	Pr	Bu*	Bu*	Pr	**15**	0.2	No
Ac	Pr	Bu*	Pr	Bu*	**16**	0.5	No
Ac	Pr	Pr	Bu*	Bu*	**8**	2	Yes

Figure 19.3 Biosynthesis of pheromone components from acyl precursors (upper) and conclusions regarding *Ca. hemipterus* tetraenes (lower). Abbreviations: Ac = acetate, Pr = propionate, Bu = butyrate; [X] is an unknown acyl carrier such as coenzyme A; deviations from the main acyl incorporation pattern indicated by (*).

components over evolutionary time once the beetles began to sense them. Natural selection operating on both the frequency of particular acyl substitutions during biosynthesis and on the antennal sensitivity to various structures could eventually shift pheromone composition. Such shifts must have occurred during speciation in the genus. It is striking that the major components of all *Carpophilus* species in Table 19.1 occur in other species as minor male-specific compounds. The genus could be very useful for studying the evolution of pheromone systems.

Field activity of pheromones

Attractiveness of the synthetic pheromones in the field was very clear for all species (literature listed in Table 19.2). As a practical matter for field testing, only the more abundant male-specific compounds were formulated in synthetic blends unless there was bioassay evidence that additional, minor ones should also be included. Males and females were caught in similar numbers. Six example data sets are shown in Figure 19.4. There was always a remarkable synergism with food odors, whether the pheromone worked reasonably well by itself (e.g., with *Ca. dimidiatus*, *Ca. freemani*, and *Ca. antiquus*) or whether the response to the pheromone alone was almost imperceptible (e.g., with *Ca. hemipterus*, *Ca. mutilatus*, and *Ca. lugubris*). In the example for *Ca. hemipterus*, the combination attracted over 75 times more beetles than either the pheromone or food alone. The numbers of beetles captured in traps with pheromone plus food could be dramatic. The upper portion of Figure 19.4 shows a trap in a date garden in California that caught over 100,000 *Ca. mutilatus* in just three days (Bartelt *et al.*, 1994a).

A model for pheromone function under natural conditions is that a male beetle would begin to release its pheromone within hours or days of arriving at a site suitable for feeding and colonization. His pheromone, together with volatiles from the host site, would be highly attractive to other beetles flying in that area, and newly arriving males would soon begin to emit their own pheromone. But beetles respond to the general location of a pheromone source rather than only to a precise spot (James *et al.*, 1996b; and unpublished data); thus adjacent, uninfested food sources would become occupied and the infestation would spread. Pheromonal mechanisms would probably retard or prevent further aggregation wherever densities become too large. As noted above, crowding generally reduces pheromone emission rate, and it is known that pheromone emission can be suppressed by mating (Nardi *et al.*, 1996) and that changes in blend composition can occur due to crowding (Bartelt and James, 1994).

Interactions among species in the field

Multiple nitidulid species were present in most trapping locations, and much information was acquired about the species specificity of the pheromones (references in Table 19.2). A summary of cross-attraction results is given in Table 19.3. Species showed a general preference for their own blends, but significant cross-attraction often occurred when pheromones

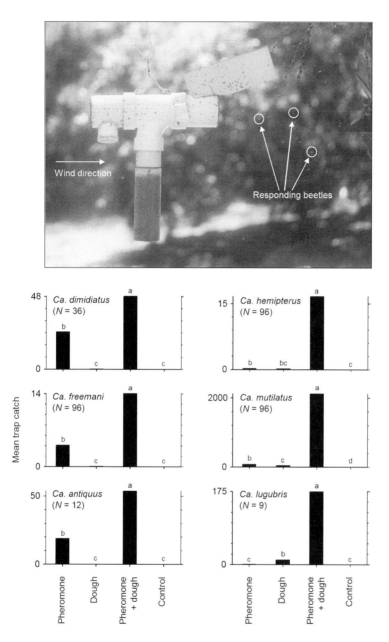

Figure 19.4 Field activity of *Carpophilus* pheromones. Above: Wind-directed funnel trap (Dowd *et al.*, 1992) in a California date garden. The fin orients the trap so that the opening (just below the fin) is accessible to beetles approaching the trap from down wind (dozens in flight at right). The pheromone and food baits are in the trap section at the left; screens prevent the insects from contacting the baits; attracted beetles collect in the bottle. Trap contains about $100,000$ *Ca. mutilatus*, captured over a period of three days. Photograph by Richard S. Vetter. Below: Synergistic responses of six *Carpophilus* species to their pheromones plus fermenting bread dough in the field. In each panel, means accompanied by the same letter were not significantly different in statistical analysis; 'N' is the number of data points (trap catches) for each treatment. References: *Ca. dimidiatus* (Bartelt *et al.*, 1995b); *Ca. hemipterus*, *Ca. freemani*, and *Ca. mutilatus* (Bartelt *et al.*, 1994b); *Ca. antiquus* (Bartelt *et al.*, 1993b); and *Ca. lugubris* (Bartelt *et al.*, 1991).

Table 19.3 Cross-attraction in the field to pheromones of eleven species.[1]

Responding species	Pheromone tested							
	hm / br	ob	lg / sa	dv	fr	mt	an / dm	tr
Species for which pheromones are known								
Ca. hemipterus (hm)	+++	+	+	++	±	–	–	nt
Ca. brachypterus (br)	+++	+	±	nt	–	–	–	–[b]
Ca. obsoletus (ob)	++	+++	++	nt	–	–	nt	nt
Ca. lugubris (lg)	+	++	+++	nt	+	–	–	–[b]
Ca. sayi (sa)	+	+	+++	nt	+	–	–	nt
Ca. davidsoni (dv)	+	+	+	+++	+	+	nt	nt
Ca. freemani (fr)	±	±	+	nt	+++	+	–	nt
Ca. mutilatus (mt)	+	+	+	+	+	+++	–	nt
Ca. antiquus (an)	+	+	++	nt	+	–	+++	+
Ca. dimidiatus (dm)	±	±	±	nt	±	–	+++	nt
Co. truncatus (tr)	+++	++	++	nt	±	–	–	+++
Species for which pheromones are not known								
Ca. humeralis	++	+	+	nt	±	±	+	nt
Ca. corticinis	–	+	++	nt	+	–	–	nt
Ca. marginatus	+	+	+	nt	+	–	–	nt
Ca. marginellus	+	+	++	++	++	–	±	nt
Ca. gaveni	±	nt	nt	+	nt	+	nt	nt
Stelidota geminata	–	nt	–	nt	–	nt	nt	nt
Epuraea (Haptoncus) luteolus	–	–	–	nt	–	–	–	nt
Glischrochilus quadrisignatus	–	nt	–	nt	–	nt	nt	–

[1] Relative degree of attraction: +++ > ++ > + > ± > –, where "+++" = response to species' own pheromone plus dough and "–" = response to dough control; "nt" = not tested. [b]Attraction increased significantly when triene **1** was omitted from blend (only **5** and **10** present).

had components in common. For example, the pheromones of *Ca. hemipterus* and *Ca. lugubris* contain minor amounts of the only known pheromone component of *Ca. obsoletus* (tetraene **7**), and *Ca. obsoletus* was readily attracted to these synthetic pheromones as well as to its own (Bartelt *et al.*, 1994b). Cross-attraction was seen between genera: *Co. truncatus* responded well to the pheromones of *Ca. hemipterus*, *Ca. obsoletus*, and *Ca. lugubris*, which emit tetraene **5** or homolog **6**. *Colopterus* and *Carpophilus* are closely related, belonging to the same subfamily (Carpophilinae).

The pheromones of *Ca. hemipterus* and *Ca. brachypterus* differ only slightly in proportions of components, and both species responded about equally to both synthetic mixtures (Williams *et al.*, 1995). Pheromones in two other pairs of species, *Ca. antiquus* plus *Ca. dimidiatus* and *Ca. lugubris* plus *Ca. sayi*, are also quite similar in composition. Even with nearly identical pheromones, cross-attraction would not necessarily occur in nature if the species have temporal, geographic, or ecological separation. For example, *Ca. brachypterus* flies earlier in spring than *Ca. hemipterus* in Illinois (unpublished), and *Ca. antiquus* has a more northern range than *Ca. dimidiatus* (Bartelt *et al.*, 1993b, 1995b). *Ca. sayi* is a specialist on hardwood tree wounds while *Ca. lugubris* is a wide-ranging generalist on overripe fruits and fermenting vegetable matter (Connell, 1956), thus separation can occur by habitat even if geographic ranges and seasonal timing overlap.

Some of the observed cross-attraction may have been an artifact of unnaturally high pheromone releases from septa; for example, the response threshold of *Ca. mutilatus* to the pheromone of *Ca. hemipterus* was at least ten times higher than to its own (Bartelt *et al.*, 1994a). Furthermore, the ratios of components emitted from septa shifted somewhat over time (Bartelt and James, 1994); therefore, the preference for a species' own blend may not have been as clear as possible, which would make cross-attraction appear relatively strong.

One case of cross-attraction seems to have a kairomonal basis. *Ca. antiquus* responded very clearly to the *Ca. lugubris* pheromone, yet these species do not have any components in common. In sampled corn ears, *Ca. antiquus* was seldom found except when *Ca. lugubris* was also present (Bartelt *et al.*, 1993b). Thus, *Ca. antiquus* appears to use the *Ca. lugubris* pheromone as a cue for locating a host site. There is also some evidence for allomonal effects. Many species are attracted to tetraene **5**, which is present at a low level from *Co. truncatus* (Cossé and Bartelt, 2000), but the major component of *Co. truncatus*, triene **1**, inhibited responses to **5** by other species in a field study (Cossé and Bartelt, 2000). Triene **1** by itself was slightly attractive to *Co. truncatus*, but it did not enhance, and sometimes even reduced, attraction to the key minor component, **5** (Cossé and Bartelt, 2000; Kyhl *et al.*, 2002).

Field data may provide clues about unstudied pheromones. *Ca. corticinis* Erichson, *Ca. marginellus* Motschulsky, and *Ca. sayi* Parsons were particularly attracted to the pheromone of *Ca. lugubris* and to other pheromones containing tetraene **7** (Bartelt *et al.*, 1995b; James *et al.*, 1995; Williams *et al.*, 1995). *Ca. humeralis* (F.) was generally attracted to pheromones that contained tetraenes (Bartelt *et al.*, 1994b; Blumberg *et al.*, 1993), while *Ca. gaveni* Dobson responded best to pheromones with trienes (James *et al.*, 1995). Efforts to demonstrate pheromones in the laboratory from *Ca. marginellus* and *Ca. humeralis* have been unsuccessful. However, in a systematic wind-tunnel investigation on *Ca. humeralis* (Zilkowski and Bartelt, 1999), tetraene **5** was the most attractive of 18 compounds tested and had a response threshold of just 10 pg, comparable to the best pheromone responses in other species. It is anticipated that *Ca. humeralis* would emit **5** if the proper laboratory environmental conditions were finally discovered.

Sap beetles of other genera were also captured during trapping studies. However, *Stelidota geminata* (Say), *Epuraea* (*Haptoncus*) *luteolus* (Erichson), and *Glischrochilus* spp., which belong to subfamilies other than Carpophilinae, were captured only because of the food coattractants and did not respond significantly to any of the *Carpophilus* pheromones (Bartelt *et al.*, 1994b; Kyhl *et al.*, 2002; Williams *et al.*, 1993).

Field trapping parameters

A number of factors affected the performance of synthetic pheromones in the field, such as trap type, trap placement, kind of food-related odors, and pheromone formulation. The traps used in most studies were of a wind-directed, funnel design (Figure 19.4) and were constructed from PVC plumbing pipe, screen wire, and other common materials (Dowd *et al.*, 1992). The modular design allows any trap section to be modified, if necessary, without having to replace the entire trap. Other successful trap designs included Mason jars with plastic funnels attached at the mouth (known as "NIT" traps, Williams *et al.*, 1993), Japanese beetle traps (Williams *et al.*, 1993), and various bucket, funnel, and other traps of commercial or laboratory design (Dowd, 2005; Hossain *et al.*, 2007; James *et al.*, 1996a).

Effects of trap height varied among species. For *Ca. hemipterus* and *Ca. mutilatus*, wind-directed "Dowd" traps 3 m above the ground captured 3–4 times more beetles than those 0.3 m above the ground, but *Ca. humeralis*, responding to the pheromones of *Ca. hemipterus* and *Ca. mutilatus*, were caught 16 times more frequently at 0.3 m than at 3 m (Bartelt *et al.*, 1994a). In Ohio, *Ca. lugubris* were trapped more effectively in "NIT" traps at 1 m above the ground than at ground level (Williams *et al.*, 1993).

Many kinds of food-related materials were used successfully in the field as pheromone synergists. These included fermenting fig juice (Bartelt *et al.*, 1990b, 1992a), fermenting fruit juice in a gel formulation (James *et al.*, 1998), rotting grapefruit (Blumberg *et al.*, 1993), empirical blends of synthetic compounds (Bartelt *et al.*, 1992a), blends of synthetic compounds that are typical of yeast fermentations (Lin *et al.*, 1992) or that were specifically formulated to mimic overripe peaches and peach juice (Bartelt and Hossain, 2006), and fermenting bread dough (see Williams *et al.*, 1993). Bread dough has been the "standard" source of food odors in the research studies because it is easy to use, effective, and fairly reproducible. Except where indicated above, it was used in all of the cited field studies.

Rubber septa were used as the pheromone dispensers. A practical problem was that components of multi-component pheromones sometimes varied greatly in volatility. Therefore, the amounts of the less volatile components were increased so that the septa would release blends similar to the natural proportions listed in Table 19.1 (e.g., Bartelt and James, 1994; Williams *et al.*, 1995). The synthetic pheromones used in field tests were typically 70–80% pure. The major impurities were *Z* isomers of the pheromone components (Bartelt *et al.*, 1990c), which were not inhibitory at such low levels.

Effects of pheromone dose on trap catch have been studied. The usual amount of pheromone per septum was 500 μg. In California, all doses of *Ca. hemipterus* pheromone from 15 to 15 000 μg per septum were significantly active in the field, and for *Ca. mutilatus*, all doses from 50 to 15 000 μg were active (Bartelt *et al.*, 1994a). Attractiveness increased with pheromone dose throughout the range for both species. Similar trends were noted for these species in Australia when 500 - and 5000 -μg doses were compared (James *et al.*, 1994). With *Carpophilus* beetles, high pheromone doses never became repellent.

Septa were normally used in the field for two weeks and then replaced. Activity decreased during this time, inferred from trapping experiments in which septa of different ages were present simultaneously. In an all-season study in southern California, trap catches of *Ca. freemani*, which has a relatively volatile pheromone, were only 9–13% as high during the second week as they were during the first week (Bartelt *et al.*, 1994b). Catches decreased more slowly when less volatile pheromones were used. Catches of *Ca. hemipterus* declined to 54–66% during the second week, and catches of *Ca. mutilatus* to 36–39% (Bartelt *et al.*, 1994b). Other studies provided similar values (Bartelt *et al.*, 1992a, 1994a).

The existence of cross-attraction suggested that it might be possible to combine several pheromones into one lure, which would simplify sap beetle management, and this idea was explored in field tests. In California, all two-way and three-way combinations of the pheromones for *Ca. hemipterus*, *Ca. mutilatus*, and *Ca. freemani* were as effective as the separate pheromones for all species except that there was a slight decrease in the response of *Ca. freemani* (Bartelt *et al.*, 1995a). Three-way pheromone combinations were also used successfully in Australia for *Ca. davidsoni*, *Ca. mutilatus*, and *Ca. hemipterus* (James *et al.*, 1996a, 2000c).

Practical uses

There has been considerable research on the use of the pheromones as agricultural tools. Practical potential increased when the pheromones became commercially available in 2002 from Great Lakes IPM, Vestaburg, Michigan. Population monitoring has been an important application. *Carpophilus* species and *Co. truncatus* can be difficult to detect because of their small size and cryptic nature, but the pheromones have provided a rapid, simple, and sensitive tool for doing this. Flight periods of many species have been documented with pheromone traps during long-term studies in California, South Carolina, and Minnesota in the United States and in Australia (references noted in Table 19.2). James *et al.* (2000a, b) surveyed horticultural regions throughout Australia with *Carpophilus* pheromones to learn about species composition and seasonality and whether insecticides were achieving adequate control. Trap catches, together with additional data for temperature and rainfall, were concluded to be useful for predicting economic damage to Australian stone fruits by

Ca. davidsoni (James *et al.*, 1997). Abreu (1997) demonstrated in Brazil that *Ca. dimidiatus* and *Ca. obsoletus* could be detected in cocoa warehouses with pheromone traps. Monitoring traps have also been used to survey beetles in different habitats for infestation by natural enemies such as nematodes (Dowd *et al.*, 1995); these efforts resulted in the discovery of a new nematode species (Poinar and Dowd, 1997). Dowd (2000) used the pheromone of *Ca. lugubris* to compare infestations of these beetles in Bt and non-Bt sweet corn in Illinois. Dispersal of oak wilt fungus by *Co. truncatus* and *Ca. sayi* has been investigated with the aid of pheromone traps (Ambourn *et al.*, 2005).

The pheromones could be used in conjunction with various microorganisms to manipulate or manage beetle populations. An autoinoculating device has been invented that causes feral beetles to come into contact with various microorganisms or materials, of the experimenter's choosing (Vega *et al.*, 1995). For example, *Ca. lugubris* in field corn can be induced to pick up a formulation of *Bacillus subtilis* Ehrenberg (Cohn), a harmless species of soil bacteria. The beetles then disperse into the crop and colonize ears, typically those that have been previously damaged, introducing *B. subtilis* as they go. These sites would normally be subject to infection by *Aspergillus flavus*, which produces aflatoxin, but when *B. subtilis* is present, *A. flavus* does not become established. Thus, an abundant but economically minor field-corn pest could be employed to protect the crop from a highly dangerous fungal metabolite, with minimal input from the grower. In other crops, where beetle damage is of greater economic significance, direct control could be exerted by inoculating attracted beetles with insecticidal nematodes (Dowd *et al.*, 1995) or a fungus such as *Beauveria bassiana* (Balsamo) Vuillemin (Vega *et al.*, 1995; Dowd and Vega, 2003). The advantage of such inoculation, rather than immediate killing with an insecticide, is the potential for horizontal transmission of the pathogen to other beetles in the immediate area, magnifying the effect of the treatment.

The pheromones can also be used to enhance a beneficial activity of the beetles. Nitidulids are primary pollinators of custard apple (*Annona*) flowers, but poor pollination has often been a problem (Gazit *et al.*, 1982; Nagel *et al.*, 1989). A new concept was to place pheromone lures in *Annona* trees to increase beetle density and enhance pollination, and Peña *et al.* (1999) found that pollination and fruit set were significantly improved when the pheromones were employed.

Protection of Australian stone fruit orchards

The most intensive effort to use pheromones in insect management has been in Australia, where an attract-and-kill strategy has been developed for protecting stone fruit crops. Historically, the Oriental fruit moth, *Grapholita molesta* Busck, was the major stone fruit pest, and heavy insecticide applications kept both this moth and, coincidentally, the *Carpophilus* beetles at acceptable levels. However, the widespread adoption of pheromone-based mating disruption for *G. molesta* control released the *Carpophilus* beetles from insecticide pressure, and these beetles, *Ca. davidsoni* in particular, became the dominant stone fruit pests (James *et al.*, 1994). Late applications of broad-spectrum insecticides often gave unsatisfactory beetle control and could potentially leave residues on harvested fruit. Therefore, new management methods were needed.

The massive captures of *Ca. mutilatus* in California date gardens (Bartelt *et al.*, 1994a, b), the new knowledge about the *Ca. davidsoni* pheromone, and the fact that both sexes of beetles are attracted (rather than just males as with moths) suggested that crop protection might be possible through pheromone use. Over the course of a decade, research was conducted toward this goal and a number of different trap configurations and placements were tried (James *et al.*, 1996b, 2001; Hossain and Williams, 2005; Hossain *et al.*, 2005, 2006). Two key conclusions were reached: First, control efforts should begin well before fruit ripening; the concept was to substantially reduce the beetle population before the crop became attractive and susceptible to beetle damage. Second, attract-and-kill stations, which were baited with pheromone and cases of insecticide-treated, overripe peaches, could have a protective influence extending over hundreds of meters. Hossain *et al.* (2006) found that with three stations per hectare, peach crops were protected as effectively as with insecticide. However, the method was regarded as impractical by growers because it would be messy and cumbersome to maintain the masses of overripe peaches at the stations. Therefore, a synthetic version of overripe-peach odor was developed that was equivalent in attractiveness and was able to replace actual fruit in the bait stations (Bartelt and Hossain, 2006; unpublished information). The synthetic blend was typical of fermentation volatiles, with ethanol as the main ingredient, along with small amounts of acetaldehyde, ethyl acetate, 2-methyl-1-propanol, 2-methyl-1-butanol, and 3-methyl-1-butanol. Development of the pheromone-based fruit-protection technology continues.

Flea beetles

It has been discovered that the crucifer flea beetle, *Phyllotreta cruciferae* Goeze, a significant pest of oilseed *Brassica* and other cruciferous crops in North America and Europe, uses a hydrocarbon pheromone. The first published evidence for a pheromone in *P. cruciferae* was that canola plants infested by unsexed adults were more attractive to both males and females than damaged plants only, in both laboratory and field bioassays (Peng and Weiss, 1992). Subsequently, Peng *et al.* (1999) determined with field bioassays that the males were the attractive sex, fitting the pattern of a male-produced aggregation pheromone.

Collection of volatiles from males feeding on pieces of cabbage revealed a blend of six male-specific sesquiterpenes, including hydrocarbons **24–28** and ketone **29**, shown in Figure 19.5 (Bartelt *et al.*, 2001). Compound **24** was the most abundant. One or more of these were thought likely to account for the pheromonal activity reported earlier. Hydrocarbon **28** [(+)-γ-cadinene] is known from citronella oil (Herout and Sýkora, 1958), but structures **24–27** and **29** were new. The opposite enantiomers of hydrocarbons **24** and **25** were previously isolated from the trees Nordmann fir (*Abies nordmanniana*) and silver fir (*Abies alba*) (Bartelt *et al.*, 2001; Khan and Pentegova, 1988; Khan *et al.*, 1989), and the opposite enantiomer of **27**, known as *ar*-himachalene, is found in Himalayan cedar (Pandey and Dev, 1968). Neither compound **26** nor its enantiomer were known from nature, but

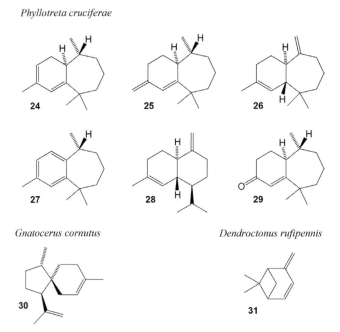

Phyllotreta cruciferae

24   25   26

27   28   29

Gnatocerus cornutus   *Dendroctonus rufipennis*

30   31

Figure 19.5 Sex-specific terpenoids emitted by three beetle species.

the antipode of the beetle compound had previously been prepared by modifying a plant-derived compound (Joseph and Dev, 1968). Ketone **29** was novel. The same six compounds, plus two structurally related alcohols, were also isolated from males of three other flea beetle species, *Aphthona flava* Guillebeau, *A. czwalinae* (Weise), and *A. cyparissiae* (Koch) (Bartelt *et al.*, 2001). In addition, hydrocarbons **24** and **25** were detected from the eggplant flea beetle, *Epitrix fuscula* Crotch (Zilkowski *et al.*, 2006).

Racemic **24**, **25**, **27**, and **29** were synthesized from simple precursors (Bartelt *et al.*, 2003). Ketone **29** was the initial synthetic target and was the key intermediate leading to the other compounds. The synthesis supported the previously reported structures and relative stereochemistry (Bartelt *et al.*, 2001) and provided material for initial field testing.

Assignment of correct, absolute configurations for **24**, **25**, **27**, and **29** from beetles was unexpectedly complicated. The initial conclusion (Bartelt *et al.*, 2001) was based on chemical conversion of **24**, **25**, and **29** to **27**, without affecting the asymmetric center on the 7-membered ring and then comparison of **27** to the literature with respect to optical rotation. Knowing the absolute configuration of **27** would also establish the configurations of the other three. The individual enantiomers of **24**, **25**, **27**, and **29** were synthesized by Muto *et al.* (2004), using citronellal of known configuration as the chiral starting material. This work supported the basic structures and relative stereochemistry, but disconcertingly, the absolute configurations determined by Muto *et al.* (2004) were concluded to be exactly the reverse of those proposed originally. This contradiction was resolved by Mori (2005).

The solvent used by Pandey and Dev (1968) to measure the optical rotation of **27** was chloroform, whereas that used by Bartelt *et al.* (2001) was hexane (because several of the beetle-derived compounds encountered in the study deteriorated in chloroform). Mori (2005) discovered that the optical rotation of **27** in hexane is similar in magnitude, but opposite in sign, to that in chloroform, which accounted for the previous discrepancy. Eventually, a method was found to separate the enantiomers of **27** by chiral GC, and the GC results corroborated Mori's conclusions. The structural assignments of **24**, **25**, **27**, and **29** determined by Muto *et al.* (2004) are shown in Figure 19.5 and are considered definitive.

A blend of racemic **24**, **25**, **27**, and **29**, along with (+)-**28** obtained from citronella oil, was prepared and formulated so that the emitted compounds would be in the proportions found from *P. cruciferae*. The blend was field tested in Canada by Soroka *et al.* (2005). (Compound **26** was not available in sufficient amounts for inclusion in the study.) They found that the mixture was attractive to both sexes, supporting the original idea that the compounds serve as a pheromone. Furthermore, the blend was synergized by allyl isothiocyanate, a mustard oil from the hosts that was previously found to be attractive to *P. cruciferae* (see Soroka *et al.*, 2005, for discussion). This biological pattern was like that for the nitidulids, but with completely different chemicals.

Tóth *et al.* (2005) performed additional field tests. They found that the synthetic blend used in North America was also effective in Hungary and that synergism by allyl isothiocyanate was again very strong. Electrophysiological (GC–EAD) experiments indicated that of the male-specific compounds, only **24**, **25**, and **28** were sensed by the beetle antennae and only **24** gave a strong response. Thus, a series of component-subtraction experiments was deployed in the field to explore the importance of the various compounds, starting with the blend used in Canada. During this project the pure enantiomers prepared by Muto *et al.* (2004) became available and added another dimension to testing. Conclusions were that the enantiomer of **24** found in the beetles was the key attractant; that compounds **25**, **27**, **28**, and **29** had little, if any, effect on trap catch; and that the enantiomers not present in the beetle were not inhibitory, meaning that the more easily synthesized racemic compounds could serve as effective attractants. Tóth *et al.* (2005) also reported attraction of other *Phyllotreta* species: *P. vittula* (Redtenbacher), *P. nodicornis* (Marsham), *P. procera* (Redtenbacher), *P. ochripes* (Curtis), *P. nigripes* (F.), and *P. nemorum* (L.). There was a general preference for blends containing **24**, but results were not always signficant, probably due to low beetle numbers. Much opportunity exists for additional pheromone research with this group.

Other Coleoptera

There are other examples of volatile hydrocarbon pheromones in beetles that have been researched less extensively. One of these is from the broad-horned flour beetle, *Gnatocerus cornutus* (F.) (Coleoptera: Tenebrionidae), a common stored-product pest. Tebayashi *et al.* (1998) found that males produce at least one compound that was attractive to both sexes in laboratory bioassays and at sub-nanogram levels. The compound was identified as the sesquiterpene (+)-acoradiene. The beetle-derived enantiomer was subsequently synthesized

(Tashiro *et al.*, 2004), and the originally reported stereochemistry was amended. The final structure (**30**) is shown in Figure 19.5. The compound was not previously known from nature, but the opposite enantiomer occurs in the plant *Juniperus rigida*.

A second example is the spruce beetle, *Dendroctonus rufipennis* (Kirby) (Coleoptera: Scolytidae), a major cause of mortality in mature spruce stands. Gries *et al.* (1992) found that the terpene verbenene (**31** in Figure 19.5), was emitted from the beetles, predominantly from females, and concluded that it is a pheromone component. The oxygenated compounds seudenol, frontalin, and 1-methyl-cyclohex-2-en-1-ol, were previously identified as pheromone components in this species. Attraction to verbenene alone was demonstrated in field traps, and it enhanced captures to the other pheromone components. The absolute configuration of verbenene has not been investigated.

Finally, a number of relatively small, straight-chain hydrocarbons, often with double bonds, have been reported from beetles in various families. Females of the rove beetle, *Aleochara curtula* (Goeze) (Coleoptera, Staphylinidae) possess four alkenes, (Z)-7- and (Z)-9-heneicosene and (Z)-7- and (Z)-9-tricosene, which were lacking from males (Peschke and Metzler, 1987). These elicited mating attempts from males when exposed to models treated with the alkenes. Long-range attraction was not assessed, but these alkenes are volatile enough to be sensed from a distance.

In the tenebrionid beetle *Parastizopus transgariepinus*, males emitted 1-tridecene as a sex attractant while performing a peculiar "head standing" behavior, and only females responded to it in laboratory bioassays (Geiselhardt *et al.*, 2008). The compound was determined to be released from the aedeagal gland. Curiously, 1-tridecene was also detected in the defensive secretion from the pygidial gland of both sexes, along with other compounds such as quinones and terpenes.

Teasing out the biological functions of such hydrocarbons can clearly be subtle and complicated. Encountering compounds that serve more than one function is not unprecedented, and Blum (1996) discussed many examples. It is likely that additional research will discover many examples of hydrocarbons serving as volatile pheromones in beetles, whether this is their only function or just one of several functions.

References

Abreu, J. M., de. (1997). Attraction of *Carpophilus dimidiatus* and *C. obsoletus* (Coleoptera: Nitidulidae) to aggregation pheromones in stored cocoa. *Agrotrópica*, **9**, 41–48.

Ambourn, A. K., Juzwik, J. and Moon, R. D. (2005). Seasonal dispersal of the oak wilt fungus by *Colopterus truncatus* and *Carpophilus sayi* in Minnesota. *Plant Dis.*, **89**, 1067–1076.

Bartelt, R. J., Carlson, D. G, Vetter, R. S. and Baker, T. C. (1993a). Male-produced aggregation pheromone of *Carpophilus mutilatus* (Coleoptera: Nitidulidae). *J. Chem. Ecol.*, **19**, 107–118.

Bartelt, R. J., Cossé, A. A., Zilkowski, B. W., Weisleder, D. and Momany, F. A. (2001). Male-specific sesquiterpenes from *Phyllotreta* and *Aphthona* flea beetles. *J. Chem. Ecol.*, **27**, 2397–2423.

Bartelt, R.J., Dowd, P.F. and Plattner, R.D. (1991). Aggregation pheromone of *Carpophilus lugubris*: New pest management tools for the nitidulid beetles. In *Naturally Occurring Pest Bioregulators*, ed. P.A. Hedin. American Chemical Society Symposium Series 449. Washington, DC: American Chemical Society, pp. 27–40.

Bartelt, R.J., Dowd, P.F., Plattner, R.D. and Weisleder, D. (1990a). Aggregation pheromone of driedfruit beetle, *Carpophilus hemipterus*: windtunnel bioassay and identification of two novel tetraene hydrocarbons. *J. Chem. Ecol.* **16**, 1015–1039.

Bartelt, R.J., Dowd, P.F., Shorey, H.H. and Weisleder, D. (1990b). Aggregation pheromone of *Carpophilus freemani* (Coleoptera: Nitidulidae): a blend of conjugated triene and tetraene hydrocarbons. *Chemoecology*, **1**, 105–113.

Bartelt, R.J., Dowd, P.F., Vetter, R.S., Shorey, H.H. and Baker, T.C. (1992a). Responses of *Carpophilus hemipterus* (Coleoptera: Nitidulidae) and other sap beetles to the pheromone of *C. hemipterus* and host-related coattractants in California field tests. *Environ. Entomol.*, **21**, 1143–1153.

Bartelt, R.J. and Hossain, M.S. (2006). Development of synthetic food-related attractant for *Carpophilus davidsoni* and its effectiveness in the stone fruit orchards in southern Australia. *J. Chem. Ecol.*, **32**, 2145–2162.

Bartelt, R.J. and James, D.G. (1994). Aggregation pheromone of Australian sap beetle, *Carpophilus davidsoni* (Coleoptera: Nitidulidae). *J. Chem. Ecol.*, **20**, 3207–3219.

Bartelt, R.J., Kyhl, J.F., Ambourn, A.K., Juzwik, J. and Seybold, S.J. (2004). Male-produced aggregation pheromone of *Carpophilus sayi*, a nitidulid vector of oak wild disease, and pheromonal comparison with *Carpophilus lugubris*. *Agric. For. Entomol.*, **6**, 39–46.

Bartelt, R.J., Seaton, K.L. and Dowd, P.F. (1993b). Aggregation pheromone of *Carpophilus antiquus* (Coleoptera: Nitidulidae) and kairomonal use of *C. lugubris* pheromone by *C. antiquus*. *J. Chem. Ecol.*, **19**, 2203–2216.

Bartelt, R.J., Vetter, R.S., Carlson, D.G. and Baker, T.C. (1994a). Influence of pheromone dose, trap height, and septum age on effectiveness of pheromones for *Carpophilus mutilatus* and *C. hemipterus* (Coleoptera: Nitidulidae) in a California date garden. *J. Econ. Entomol.*, **87**, 667–675.

Bartelt, R.J., Vetter, R.S., Carlson, D.G. and Baker, T.C. (1994b). Responses to aggregation pheromones for five *Carpophilus species* (Coleoptera: Nitidulidae) in a California date garden. *Environ. Entomol.*, **23**, 1534–1543.

Bartelt, R.J., Vetter, R.S., Carlson, D.G., Petroski, R.J. and Baker, T.C. (1995a). Pheromone combination lures for *Carpophilus* (Coleoptera: Nitidulidae) species. *J. Econ. Entomol.*, **88**, 864–869.

Bartelt, R.J., Weaver, D.K. and Arbogast, R. (1995b). Aggregation pheromone of *Carpophilus dimidiatus* (F.) (Coleoptera: Nitidulidae) and responses to *Carpophilus* pheromones in South Carolina. *J. Chem. Ecol.*, **21**, 1763–1779.

Bartelt, R.J. and Weisleder, D. (1996). Polyketide origin of pheromones of *Carpophilus davidsoni* and *C. mutilatus* (Coleoptera: Nitidulidae). *Bioorg. Med. Chem.*, **4**, 429–438.

Bartelt, R.J., Weisleder, D., Dowd, P.F. and Plattner, R.D. (1992b). Male-specific tetraene and triene hydrocarbons of *Carpophilus hemipterus*: structure and pheromonal activity. *J. Chem. Ecol.*, **18**, 379–402.

Bartelt, R.J., Weisleder, D. and Plattner, R.D. (1990c). Synthesis of nitidulid beetle pheromones: alkyl-branched tetraene hydrocarbons. *J. Agric. Food Chem.*, **38**, 2192–2196.

Bartelt, R.J., Weisleder, D. and Momany, F.A. (2003). Total synthesis of himachalene sesquiterpenes of *Aphthona* and *Phyllotreta* flea beetles. *Synthesis*, **2003**, 117–123.

Blum, M. S. (1996). Semiochemical parsimony in the Arthropoda. *Annu. Rev. Entomol.*, **41**, 353–374.

Blumberg, D., Goldenberg, S., Bartelt, R. J. and Williams, R. N. (1993). Responses to synthetic aggregation pheromones, host-related volatiles, and their combinations by *Carpophilus* spp. (Coleoptera: Nitidulidae) in laboratory and field tests. *Environ. Entomol.*, **22**, 837–842.

Connell, W. A. (1956). *Nitidulidae of Delaware*. Bulletin Number 318 (Technical). Newark, Delaware: University of Delaware Agricultural Experiment Station. 67 pp.

Connell, W. A. (1991). Sap beetles (Nitidulidae, Coleoptera). In *Insect and Mite Pests in Food*, Vol. 1, ed. J. R. Gorham. Agricultural Handbook Number 65, United States Department of Agriculture and United States Department of Health and Human Services. Washington, DC: United States Government Printing Office, pp. 151–174.

Cossé, A. A. and Bartelt, R. J. (2000). Male-produced aggregation pheromone of *Colopterus truncatus*: structure, electrophysiological, and behavioral activity. *J. Chem. Ecol.*, **26**, 1735–1748.

Dorsey, C. K. and Leach, J. G. (1956). The bionomics of certain insects associated with oak wilt with particular reference to the Nitidulidae. *J. Econ. Entomol.*, **49**, 219–230.

Dowd, P. F. (1991). Nitidulids as vectors of mycotoxin-producing fungi. In *Aflatoxin in Corn: New Perspectives*, ed. O. Shotwell and C. R. Hurburgh, Jr. Research Bulletin 599, Iowa Agriculture and Home Economics Experiment Station. Ames, Iowa: Iowa State University, pp. 335–342.

Dowd, P. F. (1995). Sap beetles and mycotoxins in maize. *Food Addit. Contam.*, **12**, 497–508.

Dowd, P. F. (2000). Dusky sap beetles (Coleoptera: Nitidulidae) and other kernel damaging insects in Bt and non-Bt sweet corn in Illinois. *J. Econ. Entomol.*, **93**, 1714–1720.

Dowd, P. F. (2005). Suitability of commercially available insect traps and pheromones for monitoring dusky sap beetles (Coleoptera: Nitidulidae) and related insects in Bt sweet corn. *J. Econ. Entomol.*, **98**, 856–861.

Dowd, P. F. and Bartelt, R. J. (1991). Host-derived volatiles as attractants and pheromone synergists for driedfruit beetle, *Carpophilus hemipterus*. *J. Chem. Ecol.*, **17**, 285–308.

Dowd, P. F. and Bartelt, R. J. (1993). Aggregation pheromone glands of *Carpophilus freemani* (Coleoptera: Nitidulidae) and gland distribution among other sap beetles. *Ann. Entomol. Soc. Am.*, **86**, 464–469.

Dowd, P. F., Bartelt, R. J. and Wicklow, D. T. (1992). Novel insect trap useful in capturing sap beetles (Coleoptera: Nitidulidae) and other flying insects. *J. Econ. Entomol.*, **85**, 772–778.

Dowd, P. F., Moore, D. E., Vega, F. E., McGuire, M. R., Bartelt, R. J., Nelson, T. C. and Miller, D. A. (1995). Occurrence of a mermithid nematode parasite of *Carpophilus lugubris* (Coleoptera: Nitidulidae) in central Illinois. *Environ. Entomol.*, **24**, 1245–1251.

Dowd, P. F. and Vega, F. E. (2003). Autodissemination of *Beauveria bassiana* by sap beetles (Coleoptera: Nitidulidae) to overwintering sites. *Biocontrol Sci. Technol.*, **13**, 65–75.

Gazit, S., Galon, I. and Podoler, H. (1982). The role of nitidulid beetles in natural pollination of annona in Israel. *J. Am. Hortic. Soc.*, **107**, 849–852.

Geiselhardt, S., Ockenfels, P. and Peschke, K. (2008). 1-Tridecene: male-produced sex pheromone of the tenebrionid beetle *Parastizopus transgariepinus*.

Naturwissenschaften, **95**, 247–251.

Gries, G., Borden, J. H., Gries, R., Lafontaine, J. P., Dixon, E. A., Wieser, H. and Whitehead, A. T. (1992). 4-methyl-6,6-dimethylbicyclo[3.1.1]hept-2-ene (vebenene): new aggregation pheromone of the scolytid beetle *Dendroctonus rufipennis*. *Naturwissenschaften*, **79**, 367–368.

Herout, V., and Sýkora, V. (1958). The chemistry of cadinenes and cadinols. *Tetrahedron*, **4**, 246–255.

Hinton, H. E. (1945). *A Monograph of the Beetles Associated with Stored Products*. Norwich, UK: Jarrold and Sons. 443pp.

Hossain, M. S. and Williams, D. G. (2005). Potential for reduced-dose of synthetic aggregation pheromone of *Carpophilus* spp. for an attract and kill strategy in stone fruit orchards. *Plant Prot. Q.*, **98**, 126–128.

Hossain, M. S., Williams, D. G., Hossain, M. A. B. M. and Norng, S. (2007). Comparison of trap designs for use with aggregation pheromone and synthetic co-attractant in a user-friendly attract and kill system to control *Carpophilus* spp. (Coleoptera: Nitidulidae). *Aust. J. Entomol.*, **46**, 244–250.

Hossain, M. S., Williams, D. G., Mansfield, C., Bartelt, R. J., Callinan, L. and Il'ichev, A. L. (2006). An attract-and-kill system to control *Carpophilus* spp. in Australian stone fruit orchards. *Entomol. Exp. Appl.*, **118**, 11–19.

Hossain, M. S., Williams, D. G. and Milmer, A. D. (2005). *Carpophilus* spp. (Coleoptera: Nitidulidae) responses to aggregation pheromone plus decomposed fruit as co-attractant in stone fruit orchards in southern Australia. *Gen. Appl. Entomol.*, **34**, 33–42.

James, D. G., Bartelt, R. J. and Faulder, R. J. (1994). Attraction of *Carpophilus* spp. (Coleoptera: Nitidulidae) to synthetic aggregation pheromones and host-related coattractants in Australian stone fruit orchards: Beetle phenology and pheromone dose studies. *J. Chem. Ecol.*, **20**, 2805–2819.

James, D. G., Bartelt, R. J., Faulder, R. J. and Taylor, A. (1993). Attraction of Australian *Carpophilus* spp. (Coleoptera: Nitidulidae) to synthetic pheromones and fermenting bread dough. *J. Aust. Entomol. Soc.*, **32**, 339–345.

James, D. G., Bartelt, R. J. and Moore, C. J. (1996a). Trap design effect on capture of *Carpophilus* spp. (Coleoptera: Nitidulidae) using synthetic aggregation pheromones and a coattractant. *J. Econ. Entomol.*, **89**, 648–653.

James, D. G., Bartelt, R. J. and Moore, C. J. (1996b). Mass-trapping of *Carpophilus* spp. (Coleoptera: Nitidulidae) in stone fruit orchards using synthetic aggregation pheromones and a coattractant: development of a strategy for population suppression. *J. Chem. Ecol.*, **22**, 1541–1556.

James, D. G., Faulder, R. J. and Bartelt, R. J. (1995). Fauna and seasonal abundance of *Carpophilus* spp. (Coleoptera: Nitidulidae) in four stone fruit growing regions of southeastern Australia as determined by pheromone-trapping. *J. Aust. Entomol. Soc.*, **34**, 327–333.

James, D. G., Faulder, R. J., Vogele, B., Bartelt, R. J. and Moore, C. J. (1997). Phenology of *Carpophilus* spp. (Coleoptera: Nitidulidae) in stone fruit orchards as determined by pheromone trapping: Implications for prediction of crop damage. *Aust. J. Entomol.*, **36**, 165–173.

James, D. G., Faulder, R. J., Vogele, B., and Moore, C. J. (2000a). Pheromone-trapping of *Carpophilus* spp. (Coleoptera: Nitidulidae): fauna, abundance and seasonality in some Australian horticultural regions. *Plant Prot. Q.*, **15**, 57–61.

James, D. G., Faulder, R. J., Vogele, B., and Moore, C. J. (2000b). Pheromone-trapping of *Carpophilus* spp. (Coleoptera: Nitidulidae) in stone fruit orchards near Gosford,

New South Wales: fauna, seasonality and effect of insecticides. *Aust. J. Entomol.*, **39**, 310–315.

James, D. G., Moore, C. J., Faulder, R. J. and Vogele, B. (1998). An improved coattractant for pheromone trapping of *Carpophilus* spp. (Coleoptera: Nitidulidae). *Aust. J. Entomol.*, **39**, 83–85.

James, D. G., Vogele, B., Faulder, R. J., Bartelt, R. J. and Moore, C. J. (2001). Pheromone-mediated mass trapping and population diversion as strategies for suppressing *Carpophilus* spp. (Coleoptera: Nitidulidae) in Australian stone fruit orchards. *Agric. For. Entomol.*, **3**, 41–47.

James, D. G., Vogele, B., Faulder, R. J. and Moore, C. J. (2000c). Efficacy of multispecies pheromone lures for *Carpophilus davidsoni* Dobson and *Carpophilus mutilatus* Erichson (Coleoptera: Nitidulidae). *Aust. J. Entomol.*, **39**, 83–85.

Joseph, T. C. and Dev, S. (1968). Studies in sesquiterpenes – XXXII: structure of himachalene monohydrochloride and the preparation of trans-himachalenes. *Tetrahedron*, **24**, 3853–3859.

Kable, P. F. (1969). Brown rot of stone fruits on the Murrumbidgee irrigation areas: 1. Aetiology of the disease in canning peaches. *Aust. J. Agric. Res.*, **20**, 301–316.

Khan, V. A. and Pentegova, V. A. (1988). Volatile components of the oleoresin of *Abies alba*. *Chem. Nat. Comp.*, **1988**, 162–166.

Khan, V. A., Tkachev, A. V. and Pentegova, V. A. (1989). New sesquiterpenoids from the oleoresin of *Abies alba*. *Chem. Nat. Comp.*, **1989**, 606–611.

Kyhl, J. F., Bartelt, R. J., Cossé, A. A., Juzwik, J. and Seybold, S. J. (2002). Semiochemical-mediated flight responses of sap beetle vectors of oak wilt, *Ceratocystis fagacearum*. *J. Chem. Ecol.*, **28**, 1527–1547.

Lin, H., Phelan, P. L. and Bartelt, R. J. (1992). Synergism between synthetic food odors and the aggregation pheromone for attracting *Carpophilus lugubris* in the field (Coleoptera: Nitidulidae). *Environ. Entomol.*, **21**, 156–159.

Lindgren, D. L. and Vincent, L. E. (1953). Nitidulid beetles infesting California dates. *Hilgardia*, **22**, 97–118.

Mori, K. (2005). Synthesis of (R)-ar-turmerone and its conversion to (R)-ar-himachalene, a pheromone component of the flea beetle: (R)-ar-himachalene is dextrorotatory in hexane, while levorotatory in chloroform. *Tetrahedron: Asymmetr.*, **16**, 685–692.

Muto, S., Bando, M. and Mori, K. (2004). Synthesis and stereochemistry of the four himachalene-type sesquiterpenes isolated from the flea beetle (*Aphthona flava*) as pheromone candidates. *Eur. J. Org. Chem.*, **2004**, 1946–1952.

Nagel, J., Peña, J. E. and Habeck, D. (1989). Insect pollination of atemoya in Florida. *Fla. Entomol.*, **72**, 207–211.

Nardi, J. B., Dowd, P. F. and Bartelt, R. J. (1996). Fine structure of cells specialized for secretion of aggregation pheromone in a nitidulid beetle *Carpophilus freemani* (Coleoptera: Nitidulidae). *Tissue Cell*, **28**, 43–52.

Pandey, R. C. and Dev, S. (1968). Studies in sesquiterpenes – XXX: synthesis of ar-himachalene and himachalanes. *Tetrahedron*, **24**, 3829–3839.

Peña, J. E., Castiñeiras, A., Bartelt, R. and Duncan, R. (1999). Effect of pheromone bait stations for sap beetles (Coleoptera: Nitidulidae) on *Annona* spp. fruit set. *Fla. Entomol.*, **82**, 475–480.

Peng, C., Bartelt, R. J. and Weiss, M. J. (1999). Male crucifer flea beetles produce an aggregation pheromone. *Physiol. Entomol.*, **24**, 98–99.

Peng, C. and Weiss, M. J. (1992). Evidence of an aggregation pheromone in the flea beetle, *Phyllotreta cruciferae* (Goeze) (Coleoptera: Chrysomelidae). *J. Chem. Ecol.*, **18**, 875–884.

Peschke, K. and Metzler, M. (1987). Cuticular hydrocarbons and female sex pheromones of the rove beetle, *Aleochara curtula* (Goeze) (Coleoptera: Staphylinidae). *Insect Biochem.*, **17**, 167–178.

Petroski, R. J. (1997). Improved preparation of *Carpophilus freemani* aggregation pheromone. *Synthetic Commun.*, **27**, 3279–3289.

Petroski, R. J. and Bartelt, R. J. (2007). Direct aldehyde homologation utilized to construct a conjugated-tetraene hydrocarbon insect pheromone. *J. Agric. Food Chem.*, **55**, 2282–2287.

Petroski, R. J., Bartelt, R. J. and Vetter, R. S. (1994a). Male-produced aggregation pheromone of *Carpophilus obsoletus* (Coleoptera: Nitidulidae). *J. Chem. Ecol.*, **20**, 1483–1493.

Petroski, R. J., Bartelt, R. J. and Weisleder, D. (1994b). Biosynthesis of (2E,4E,6E)-5-ethyl-3-methyl-2,4,6-nonatriene: The aggregation pheromone of *Carpophilus freemani* (Coleoptera: Nitidulidae). *Insect Biochem. Mol. Biol.*, **24**, 69–78.

Petroski, R. J. and Vaz, R. (1995). Insect aggregation pheromone response synergized by "host-type" volatiles: molecular modeling evidence for close proximity binding of pheromone and coattractant in *Carpophilus hemipterus* (L.) (Coleoptera: Nitidulidae). In *Computer-Aided Molecular Design*, ed. C. H. Reynolds, M. K. Holloway and H. K. Cox. American Chemical Society Symposium Series 589. Washington, DC: American Chemical Society, pp. 197–210.

Petroski, R. J. and Weisleder, D. (1997). Inhibition of *Carpophilus freemani* Dobson (Coleoptera: Nitidulidae) aggregation pheromone response by a Z-double-bond pheromone analog. *J. Agric. Food Chem.*, **45**, 943–945.

Petroski, R. J. and Weisleder, D. (1999). Reduction of *Carpophilus freemani* Dobson (Coleoptera: Nitidulidae) aggregation pheromone response by synthetic analogues. *J. Agric. Food Chem.*, **47**, 1189–1195.

Poinar, G. O. and Dowd P. F. (1997). *Psammomermis nitidulensis* n. sp. (Nematoda: Mermithidae) from sap beetles (Coleoptera: Nitidulidae) with biological observations and a key to the species of *Psammomermis*. *Fund. Appl. Nematol.*, **20**, 207–211.

Soroka, J. J., Bartelt, R. J., Zilkowski, B. W. and Cossé, A. A. (2005). Responses of flea beetle *Phyllotreta cruciferae* to synthetic aggregation pheromone components and host plant volatiles in field trials. *J. Chem. Ecol.*, **31**, 1829–1843.

Tashiro, T., Kurosawa, S. and Mori, K. (2004). Revision of the structure of the major aggregation pheromone of the broad-horned flourbeetle (*Gnatocerus cornutus*) to (1S,4R,5R)-α-acoradiene by its synthesis. *Biosci. Biotechnol. Biochem.*, **68**, 663–670.

Tebayashi, S., Hirai, N., Suzuki, T., Matsuyama, S., Nakakita, H, Nemoto, T. and Nakanishi, H. (1998). Identification of (+)-acoradiene as an aggregation pheromone for *Gnatocerus cornutus* (F.). *J. Stored Prod. Res.*, **34**, 99–106.

Tóth, M., Csonka, É., Bartelt, R. J., Cossé, A. A., Zilkowski, B. W., Muto, S.-E. and Mori, K. (2005). Pheromonal activity of compounds identified from male *Phyllotreta cruciferae*: field tests of racemic mixtures, pure enantiomers, and combinations with allyl isothiocyanate. *J. Chem. Ecol.*, **31**, 2705–2720.

Vega, F. E., Dowd, P. F. and Bartelt, R. J. (1995). Dissemination of microbial agents using an autoinoculating device and several insect species as vectors. *Biol. Control*, **5**, 545–552.

Wicklow, D. T. (1991). Epidemiology of *Aspergillus flavus* in corn. In *Aflatoxin in Corn: New Perspectives*, ed. O. L. Shotwell and C. R. Hurburgh, Jr. Research Bulletin 599, Iowa Agriculture and Home Economics Experiment Station. Ames, Iowa: Iowa State University, pp. 315–328.

Williams, R. N., Ellis, M. S. and Bartelt, R. J. (1995). Efficacy of *Carpophilus* aggregation
  pheromones on nine species in northeastern Ohio and identification of the
  pheromone of *C. brachypterus*. *Entomol. Exp. Appl.*, **77**, 141–147.

Williams, R. N., Fickle, D. S., Bartelt, R. J. and Dowd, P. F. (1993). Responses by adult
  Nitidulidae (Coleoptera) to synthetic aggregation pheromones, a coattractant, and
  effects of trap design and placement. *Eur. J. Entomol.*, **90**, 287–294.

Williams, R. N., Fickle, D. S., Kehat, M., Blumberg, D. and Klein, M. G. (1983).
  Bibliography of the genus Carpophilus Stephens (Coleoptera: Nitidulidae). Research
  Circular 278, Ohio Agricultural Research and Development Center. Wooster,
  Ohio: Ohio State University. 95 pp.

Zilkowski, B. W. and Bartelt, R. J. (1999). Cross-attraction of *Carpophilus humeralis*
  to pheromone components of other *Carpophilus* species. *J. Chem. Ecol.*, **25**,
  1759–1770.

Zilkowski, B. W., Bartelt, R. J., Cossé, A. A. and Petroski, R. J. (2006). Male-produced
  aggregation pheromone compounds from the eggplant flea beetle (*Epitrix
  fuscula*): identification, synthesis, and field bioassays. *J. Chem. Ecol.*, **32**,
  2543–2558.

20

Future directions in hydrocarbon research

Abraham Hefetz, Claude Wicker-Thomas
and Anne-Geneviève Bagnères

The research on insect hydrocarbons as semiochemicals has taken a huge leap forward in the past three decades. A crude citation search using the terms "hydrocarbon*" and "pheromon*" (Thompson ISI) dating from 1965 to 1982, when the first large review of the chemical ecology of hydrocarbons was published (Howard and Blomquist, 1982), revealed 10 citations. A comparable search from 1983 to 2005, the publishing year of the second review by the same authors (Howard and Blomquist, 2005), and using the same terms, revealed 483 citations. Despite the explosion of knowledge regarding all facets of hydrocarbon biology, we are only now beginning to understand their complex role in communication. Here we attempt to indicate some areas of future hydrocarbon research that are expected to enhance our understanding of these structurally simple but biologically important molecules.

The ubiquitous occurrence of hydrocarbons and their multifaceted function in insects present the biggest challenge

Hydrocarbons provide an excellent model system for studying the evolution of semiochemicals by co-option, but also present an immense challenge to research. The ubiquitous occurrence of hydrocarbons in the epicuticle and tissues that possess cuticular intima has facilitated their isolation and structure elucidation, but is also one of the major obstacles in deciphering their role in communication. Other confounding factors include the fact that in most cases they occur as highly complex mixtures and, at least for cuticular hydrocarbons, multifaceted functions have been assigned to them. For example, apart from their function in providing a water impermeable layer (Hadley, 1994; Rourke and Gibbs, 1999; Chapter 6, this book), cuticular hydrocarbons have been assigned a role as communicative agents (Howard, 1993), specifically in nestmate recognition (reviewed in Hefetz, 2007; Chapter 11, this book), as determinants of reproductive skew in social insects (reviewed in Monnin, 2006; Chapter 13, this book), and as sex pheromones (reviewed in Howard and Blomquist, 1982; Howard and Blomquist, 2005; Chapters 4, 15, 17, and 18, this book). In all of these cases, the pheromones are part of the cuticular hydrocarbon complex present on the epicuticle, which presents the challenge of isolating and identifying the active components. In some cases, a subset of the complex mixture (one or several components) has been implicated (Monnin *et al.*, 1998;

Peeters *et al.*, 1999; Heinze *et al.*, 2002; De Biseau *et al.*, 2004; Endler *et al.*, 2004), while in other cases, the entire complex has proven to be important (Liebig *et al.*, 2000; Cuvillier-Hot *et al.*, 2002; Dietemann *et al.*, 2005; Hartmann *et al.*, 2005; Lommelen *et al.*, 2006). At least for nestmate recognition in *Linepithema humile* and *Aphaenogaster cockerelli* it was shown that it is not the specific composition that is important, but rather that it is sufficient that the interacting individuals constitute two different classes of hydrocarbons in order to elicit the proper response (Greene and Gordon, 2007). Chemosensory sensilla capable of deciphering complex mixtures have been described in the ant *Camponotus japonicus* (Ozaki *et al.*, 2005; Chapter 10, this book).

However, at least with respect to social insects, the inference of whether the functional unit is a subset of the entire hydrocarbon profile is mostly based on correlative evidence alone. There are several major handicaps in investigating such a structure–function relationship. So far, we have acquired only a limited ability to use purified hydrocarbons from biological sources for lack of efficient and reliable separation methods for such complex mixtures. Many of the branched hydrocarbons coelute in gas chromatography even under optimum conditions for separating components. Assembling hydrocarbon mixtures from synthetic compounds is also nearly impossible because of the lack of the necessary synthetic branched hydrocarbons, in particular dimethylalkanes. Clearly, there is a need for the chemical synthesis of additional methyl-branched hydrocarbons for use in behavior studies, and the first review on the chemical synthesis of long-chain hydrocarbons (Millar, Chapter 8, this book) should facilitate this. Moreover, branched hydrocarbons may have one or more chiral centers (Chapter 18, this book). To date there has been virtually no information regarding which enantiomers are present in the biological samples, or whether chirality is important at all. Another obstacle lies in the proper performance of the bioassay in providing causative evidence for the role of cuticular hydrocarbons as communicative agents. Traditional application of solvent extraction may provide erroneous results since insects are very sensitive to changes in the cuticular lipid layer. The recently published method of dry coating of hydrocarbons has paved the way to further studies (Torres *et al.*, 2007).

In most of the studies, hydrocarbon analysis has been accomplished by GC/MS for identification and gas chromatography for quantification (see Chapter 2, this book). In recent years, analyses have increasingly used SPME as the preferred mode of extraction. Both the extraction and analytical procedures tend to exclude the very-high-molecular-weight hydrocarbons that are often present on the cuticle. Since, in many cases, communication is achieved via contact pheromones, these high-molecular weight hydrocarbons should not be ignored. The advent of a better capillary column makes such analyses possible today. For example, in the ant *Formica truncorum* GC/MS-analysis of cuticular hydrocarbons using regular capillary columns and a moderate temperature program revealed the highest molecular weight compound detected to be hentriacontane (n-C_{31}: mw 436); by using more resistant columns and high-temperature programs, however, compounds as high as trimethylpentatetracontane (triMe-C_{45}: mw 674) were detected (Akino, 2006).

The multifaceted function of cuticular hydrocarbons necessitates further theoretical as well as empirical clarification. For example, their use as both fertility signals and nest-mate recognition pheromones may seem contradictory, since the first function requires within-nest idiosyncrasy, i.e., for discriminating fertile from sterile individuals, while the second function requires within-colony odor uniformity. The response threshold hypothesis (Le Conte and Hefetz, 2008) attempts to resolve this apparent conflict in function, as well as to provide a suitable framework for future experiments to test specific parts of the hypothesis.

Distribution of hydrocarbons, regulation and transport

In many species, hydrocarbons not only occur on the cuticular surface, but also accumulate in specific exocrine glands. However, to date in all known cases, the glands sequester rather than synthesize de novo these hydrocarbons (Soroker *et al.*, 1994, for ants; Katzav-Gozansky *et al.*, 1997, for bees; Schal *et al.*, 1998, for moths).

In ants, the postpharyngeal gland (PPG) is known for stocking hydrocarbons that are occasionally exchanged with those in the epicuticle (Bagnères and Morgan, 1991; Meskali *et al.*, 1995; Soroker and Hefetz, 2000; see also Chapter 5, this book). Although the profiles of the hydrocarbons emanating from the PPG are largely congruent with those of the epicuticle, there are occasionally some differences, in particular with the relative amounts of straight-chain alkanes. This may reflect differences in metabolism between the two organs, (i.e., as selective degradation in one but not in the other), as well as selective sequestration of specific hydrocarbons. The honeybee Dufour gland is endowed with hydrocarbons, which it sequesters from the hemolymph (Katzav-Gozansky *et al.*, 1997). Dufour glands of ants are renowned for producing volatile hydrocarbons ranging from octane (n-C_8) to nonadecane (n-C_{19}) (Blum and Hermann, 1978). It is not known in these cases whether the compounds are synthesized in the gland or sequestered from the hemolymph. This is particularly interesting in ants since two distinct glands accumulate hydrocarbons, but of different chain lengths. For example, in *Cataglyphis niger* the major PPG hydrocarbons are 3-methylheptacosane and 3-methylnonacosane, whereas those of the Dufour gland are n-tridecane (n-C_{13}) and n-pentadecane (n-C_{15}) (Hefetz and Lenoir, 1992). Assuming that both components are synthesized in the fat body and transported through the hemolymph to the appropriate gland bound to a lipophorin, it would be interesting to determine whether there are different lipophorins for low- and high-molecular-weight hydrocarbons and, if not, whether the glands possess a specific sequestration system that determines the hydrocarbon content of the gland. Little is known about insect lipophorin isoforms and binding specificity (Sevala *et al.*, 2000; Fan *et al.*, 2004), and virtually nothing about the glandular uptake mechanisms.

Even more intriguing are the inter-caste variations in cuticular hydrocarbon profiles in social insects (Bonavita-Cougourdan *et al.*, 1993; Haverty *et al.*, 1996; Greene and Gordon, 2003; Chapter 12, this book). Since these differences are quantitative rather than qualitative, regulation must be at the transport and deposition levels. Apart from the fact that hydrocarbon deposition is likely to occur through the pore canals, we know little of the process. Although inter-castes sometime avoid physical contact, in species where physical contact does take place a between-caste passive transfer of the hydrocarbons must certainly occur. This necessitates an effective clearing system, of which we know nothing. Differential analysis of body parts in ants also revealed an uneven hydrocarbon distribution throughout the body surface (Bonavita-Cougourdan, 1988). Such partitioning poses many questions. How does the insect overcome the possible diffusion of particular hydrocarbons to the "wrong body section"? What regulates such selective deposition and, most importantly, does this have a functional meaning or is it a random process? All of the above questions can be addressed by careful time-dependent studies using either stable or radioactive isotopes.

Perception and integration of complex hydrocarbon mixtures

Despite the wealth of behavioral evidence on the role on hydrocarbons as behavior-modifying chemicals, little is known either of their perception by the peripheral nervous system or of how the information they convey is integrated by the central nervous system. There are two obstacles to obtaining such information. Conventional EAG recordings are very difficult because of the highly lipophilic nature of hydrocarbons, rendering their dissolution in recording electrolytes impossible. Moreover, most of the behaviorally-active hydrocarbons are long chain and, therefore, non-volatile, again making traditional EAG recording difficult. The use of coupled GC/EAD attempts to overcome these problems, but unfortunately only a few studies have used this technique for hydrocarbons (D'Ettorre *et al.*, 2004). The recent identification of various amphipathic proteins from chemosensilla lymph may provide a solution. For example, in the ant *Camponotus japonicus* the cuticular hydrocarbon antennal chemosensory sensilla possess a chemosensory protein (CSP) that can act as a carrier in a recording electrode to present hydrocarbons to the odorant receptors (Chapter 10, this book). The wider use of such amphipathic proteins will surely enhance the badly needed evidence of how the biologically-active hydrocarbons are perceived by insects.

Another problem, in particular in social insects, is the complexity of the biologically-active hydrocarbon mixtures. For example, hydrocarbon mixtures implicated in ant nestmate recognition often comprise dozens of compounds, and it is clear that the mixture is important for discriminating nestmates from non-nestmates. This has been a real stumbling block because measuring so many components seemed to be technically impossible. The pioneering study with *Camponotus japonicus* has now provided the first clue to how it works. There is a specific nestmate recognition sensillum that comprises many receptors that are presumably tuned to the specific nestmate recognition cues. Moreover, they are activated by any non-nestmate conspecific mixture, but not by that of nestmates (Ozaki *et al.*, 2005).

The deciphering of pheromone blends in the antennal lobes and other parts of the brain is starting to unfold (Hansson and Anton, 2000). While the first level of information processing of complex hydrocarbon mixtures seems to be at the peripheral multi-receptor level

(Ozaki *et al.*, 2005), further processing would appear to be achieved at the antennal lobe (Riffell *et al.*, 2009). It is important to note that many insects respond to very subtle differences in hydrocarbon blends, in particular when signals such as fertility signals or nestmate-recognition signals in social insects are involved. Both neural-ensemble recording and neural imaging may be recruited to answer important questions, such as whether all hydrocarbons present in a mixture are neuronally active, or does the information converge or segregate at the antennal lobe level?

Gene and hormonal regulation of hydrocarbons

The emerging importance of hydrocarbons as pheromones will surely draw the attention of researchers to their mode of regulation. Although we know the site of the biosynthesis of hydrocarbons and some of the key processes, knowledge of the enzymes involved in hydrocarbon biosynthesis is far from complete (Chapters 3 and 4, this book). Characterization of the enzymes involved is needed, in order to unravel both the hormonal control of hydrocarbon biosynthesis, and the genes involved in their expression. The positions of the methyl-branches in methyl-branched alkanes can be very important for communication, yet virtually nothing is known about the enzymology of how specificity occurs. Focusing on particular key enzymes such as the desaturases and elongases, with their substrate specificity, not only will enable the examination of whether they are under hormonal control, but will also unravel their genomics. By comparing them with various gene banks available for insects it may be possible to select candidate genes for such key enzymes. Identification of the appropriate genes will provide a tool for in-depth analysis of hydrocarbon biosynthesis, in particular where their composition has far-reaching consequences, such as cases where hydrocarbons are implied as fertility signals in social insects or as sex-pheromones in Diptera. Since, in most cases, the queen caste is the fertile individual and workers are sterile, hydrocarbon-associated gene expression may be linked to developmental and caste-specific gene expression. Moreover, since in many species queenless workers develop ovaries as well as biosynthesize the hydrocarbon fertility signal, the on and off switch of gene expression can be readily investigated.

A large step has been taken through the development of two powerful techniques: microarray analysis and RNA interference (RNAi). Microarray analysis provides information on the expression of a large panel of genes (including biosynthesis and regulatory genes) in related organisms, permitting comparative studies of changes in pheromone composition with those of gene expression. Furthermore, it may permit the disclosure of networks of genes that are important for pheromone regulation in various animal models. The second technique, RNAi, has been well developed in *Drosophila melanogaster*, where all the genes can be readily silenced. Through its use it is possible to fully repress a gene of interest in selected tissues, while retaining its expression at normal levels in other tissues. This is very useful in the study of pleiotropic genes which can, for instance, be specifically knocked-out in oenocytes, the site of hydrocarbon production. The RNAi knock-down technique has now been extended to other species. Extensive use of the *Drosophila* system with all its available mutants may pave the way to understanding hydrocarbon genetics in many other insect species.

Conclusion

New developments in analytical chemistry in the coming years will undoubtedly elicit new separation techniques that will allow easier isolation of methyl-branched hydrocarbons, irrespective of the branching position or array of unsaturated hydrocarbons often found in complex mixtures. Since chiral columns for separation of long-chain hydrocarbons do not yet exist, other separation techniques such as capillary electrophoresis or LC-GC-MS with the possibility of trapping compounds or classes of compounds could soon become available. Such separation and recovery methods will certainly allow better behavioral experimentation with isolated natural compounds/mixtures, an essential step to enable the acquisition of causative evidence to support the current correlative data.

One fascinating aspect of hydrocarbon evolution as semiochemicals lies in the documented chemical mimicry systems between parasite and host (Chapter 14, this book). Some systems operate by camouflage and passive transport, others by loss of parasite specific compounds, some use de novo synthesis of the same host mixture or part of the mixture, and others are as yet unrevealed. To date we are mostly limited to the description of the host–parasite chemical coevolution (or arms race), but we hope in the near future to be able to associate biosynthetic pathways as well as gene expression with such evolutionary processes.

With the increasing use of molecular techniques that permit the characterization of a huge number of genes in a very short time, the pheromone biosynthesis and regulation pathways will undoubtedly be deciphered within the next one or two decades – for *D. melanogaster*, at least. What will be the impact on science? Research on other organisms will help address important questions: how do pheromones evolve? What is their role in population isolation leading to speciation? How can pest insects be controlled without harming the environment? These and related questions pertaining to the role of hydrocarbons as pheromones will undoubtedly be a major concern in the current century, and can be answered only through comparative studies of the wealth of taxa that use hydrocarbons as communicative agents.

References

Akino, T. (2006). Cuticular hydrocarbons of *Formica truncorum* (Hymenoptera: Formicidae): description of new very long chained hydrocarbon components. *Appl. Entomol. Zool.*, **41**, 667–677.

Bagnères, A.-G. and Morgan, E. D. (1991). The postpharyngeal glands and the cuticle of Formicidae contain the same characteristic hydrocarbons. *Experientia*, **47**, 106–111.

Blum, M. S. and Hermann, H. R. (1978). Venoms and venom apparatuses of the Formicidae: Myrmeciinae, Ponerinae, Dorylinae, Pseudomyrmecinae, Myrmicinae and Formicinae. In *Arthropod Venoms*, ed. S. Bettini. Berlin: Springer, pp. 801–869.

Bonavita-Cougourdan, A. (1988). Interindividual variability and idiosyncrasy in social behaviours in the ant *Camponotus vagus* Scop. *Ethology*, **77**, 58–66.

Bonavita-Cougourdan, A., Clément, J.-L. and Lange, C. (1993). Functional subcaste discrimination (foragers and brood-tenders) in the ant *Camponotus vagus* Scop.: Polymorphism of cuticular hydrocarbon patterns. *J. Chem. Ecol.*, **19**, 1461–1477.

Cuvillier-Hot, V., Gadagkar, R., Peeters, C. and Cobb, M. (2002). Regulation of reproduction in a queenless ant: aggression, pheromones and reduction in conflict. *Proc. R. Soc. Lond., Ser. B: Biol. Sci.*, **269**, 1295–1300.

D'Ettorre, P., Heinze, E., Schulz, C., Francke, W. and Ayasse, M. (2004). Does she smell like a queen? Chemoreception of a cuticular hydrocarbon signal in the ant *Pachycondyla inversa*. *J. Exp. Biol.*, **207**, 1085–1091.

De Biseau, J. C., Passera, L., Daloze, D. and Aron, S. (2004). Ovarian activity correlates with extreme changes in cuticular hydrocarbon profile in the highly polygynous ant, *Linepithema humile*. *J. Insect Physiol.*, **50**, 585–593.

Dietemann, V., Liebig, J., Hölldobler, B. and Peeters, C. (2005). Changes in the cuticular hydrocarbons of incipient reproductives correlate with triggering of worker policing in the bulldog ant *Myrmecia gulosa*. *Behav. Ecol. Sociobiol.*, **58**, 486–496.

Endler, A., Liebig, J., Schmitt, T., Parker, J. E., Jones, G. R., Schreier, P. and Hölldobler, B. (2004). Surface hydrocarbons of queen eggs regulate worker reproduction in a social insect. *Proc. Natl. Acad. Sci. USA*, **101**, 2945–2950.

Fan, Y. L., Schal, C., Vargo, E. L. and Bagnères, A.-G. (2004). Characterization of termite lipophorin and its involvement in hydrocarbon transport. *J. Insect Physiol.*, **50**, 609–620.

Greene, M. J. and Gordon, D. M. (2003). Cuticular hydrocarbons inform task decisions. *Nature*, **423**, 32.

Greene, M. J. and Gordon, D. M. (2007). Structural complexity of chemical recognition cues affects the perception of group membership in the ants *Linephithema humile* and *Aphaenogaster cockerelli*. *J. Exp. Biol.*, **210**, 897–905.

Hadley, N. F. (1994). *Water relations of terrestrial arthropods*. San Diego, CA: Academic Press.

Hansson, B. S. and Anton, S. (2000). Function and morphology of the antennal lobe: New developments. *Annu. Rev. Entomol.*, **45**, 203–231.

Hartmann, A., D'Ettorre, P., Jones, G. R. and Heinze, J. (2005). Fertility signaling – the proximate mechanism of worker policing in a clonal ant. *Naturwissenschaften*, **92**, 282–286.

Haverty, M. I., Grace, J. K., Nelson, L. J. and Yamamoto, R. T. (1996). Intercaste, intercolony, and temporal variation in cuticular hydrocarbons of *Coptotermes formosanus* Shiraki (Isoptera: Rhinotermitidae). *J. Chem. Ecol.*, **22**, 1813–1834.

Hefetz, A. (2007). The evolution of hydrocarbon pheromone parsimony in ants (Hymenoptera: Formicidae) – interplay of colony odor uniformity and odor idiosyncrasy. A review. *Myrmecological News*, **10**, 59–68.

Hefetz, A. and Lenoir, A. (1992). Dufour's gland composition in the desert ant *Cataglyphis*: species specificity and population differences. *Z. Naturforsch.*, **47**, 285–289.

Heinze, J., Stengl, B. and Sledge, M. F. (2002). Worker rank, reproductive status and cuticular hydrocarbon signature in the ant *Pachycondyla cf. inversa*. *Behav. Ecol. Sociobiol.*, **52**, 59–65.

Howard, R. W. (1993). Cuticular hydrocarbons and chemical communication. In *Insect lipids: chemistry, biochemistry and biology*, ed. D. W. Stanley-Samuelson and D. R. Nelson. Lincoln and London: University of Nebraska Press, pp. 179–226.

Howard, R. W. and Blomquist, G. J. (1982). Chemical ecology and biochemistry of insect hydrocarbons. *Annu. Rev. Entomol.*, **27**, 149–172.

Howard, R. W. and Blomquist, G. J. (2005). Ecological, behavioral, and biochemical aspects of insect hydrocarbons. *Annu. Rev. Entomol.*, **50**, 371–393.

Katzav-Gozansky, T., Soroker, V. and Hefetz, A. (1997). The biosynthesis of Dufour's gland constituents in queens of the honeybee (*Apis mellifera*). *Invertebr. Neurosci.*, **3**, 239–243.

Le Conte, Y. and Hefetz, A. (2008). Primer pheromones in social hymenoptera. *Annu. Rev. Entomol.*, **53**, 523–542.

Liebig, J., Peeters, C., Oldham, N. J., Markstadter, C. and Hölldobler, B. (2000). Are variations in cuticular hydrocarbons of queens and workers a reliable signal of fertility in the ant *Harpegnathos saltator*? *Proc. Natl. Acad. Sci. USA*, **97**, 4124–4131.

Lommelen, E., Johnson, C. A., Drijfhout, F. P., Billen, J., Wenseleers, T. and Gobin, B. (2006). Cuticular hydrocarbons provide reliable cues of fertility in the ant *Gnamptogenys striatula*. *J. Chem. Ecol.*, **32**, 2023–2034.

Meskali, M., Bonavita-Cougourdan, A., Provost, E., Bagnères, A.-G., Dusticier, G. and Clément, J.-L. (1995). Mechanism underlying cuticular hydrocarbon homogeneity in the ant *Camponotus vagus* (Scop.) (Hymenopetra: Formicidae): role of postpharyngeal glands. *J. Chem. Ecol.*, **21**, 1127–1148.

Monnin, T. (2006). Chemical recognition of reproductive status in social insects. *Ann. Zool. Fenn.*, **43**, 515–530.

Monnin, T., Malosse, C. and Peeters, C. (1998). Solid-phase microextraction and cuticular hydrocarbon differences related to reproductive activity in queenless ant *Dinoponera quadriceps*. *J. Chem. Ecol.*, **24**, 473–490.

Ozaki, M., Wada-Katsumata, A., Fujikawa, K., Iwasaki, M., Yokohari, F., Satoji, Y., Nisimura, T. and Yamaoka, R. (2005). Ant nestmate and non-nestmate discrimination by a chemosensory sensillum. *Science*, **309**, 311–314.

Peeters, C., Monnin, T. and Malosse, C. (1999). Cuticular hydrocarbons correlated with reproductive status in a queenless ant. *Proc. R. Soc. Lond., Ser. B: Biol. Sci.*, **266**, 1323–1327.

Riffell, J. A., Lei, H., Christensen, T. A. and Hildebrand, J. G. (2009). Characterization and coding of behaviorally significant odor mixtures. *Curr. Biol.*, **19**, 335–340.

Rourke, B. C. and Gibbs, A. G. (1999). Effects of lipid phase transitions on cuticular permeability: Model membrane and in situ studies. *J. Exp. Biol.*, **202**, 3255–3262.

Schal, C., Sevala, V. and Cardé, R. T. (1998). Novel and highly specific transport of a volatile sex pheromone by hemolymph lipophorin in moths. *Naturwissenschaften*, **85**, 339–342.

Sevala, V. L., Bagnères, A. G., Kuenzli, M., Blomquist, G. J. and Schal, C. (2000). Cuticular hydrocarbons of the dampwood termite, *Zootermopsis nevadensis:* Caste differences and role of lipophorin in transport of hydrocarbons and hydrocarbon metabolites. *J. Chem. Ecol.*, **26**, 765–789.

Soroker, V. and Hefetz, A. (2000). Hydrocarbon site of synthesis and circulation in the desert ant *Cataglyphis niger*. *J. Insect Physiol.*, **46**, 1097–1102.

Soroker, V., Vienne, C., Hefetz, A. and Nowbahari, E. (1994). The postpharyngeal gland as a 'gestalt' organ for nestmate recognition in the ant *Cataglyphis niger*. *Naturwissenschaften*, **81**, 510–513.

Torres, C. W., Brandt, M. and Tsutsui, N. D. (2007). The role of cuticular hydrocarbons as chemical cues for nestmate recognition in the invasive Argentine ant (*Linepithema humile*). *Insectes Soc.*, **54**, 363–373.

Index